AF697820

WITHDRAWN

Biology, Ecology and Management of Aquatic Plants

Developments in Hydrobiology 147

Series editor
H. J. Dumont

Biology, Ecology and Management of Aquatic Plants

Proceedings of the 10th International Symposium on Aquatic Weeds, European Weed Research Society

Edited by

Joseph Caffrey, Philip R.F. Barrett, Maria Teresa Ferreira, Ilidio S. Moreira, Kevin J. Murphy & Philip Max Wade

Reprinted from Hydrobiologia, volume 415 (1999)

Kluwer Academic Publishers
Dordrecht / Boston / London

Library of Congress Cataloging-in-Publication Data

A C.I.P. Catalogue record for this book is available from the Library of Congress.

ISBN 0-7923-6182-2

Published by Kluwer Academic Publishers,
P.O. Box 17, 3300 AA Dordrecht, The Netherlands

Sold and distributed in North, Central and South America
by Kluwer Academic Publishers,
101 Philip Drive, Norwell, MA 02061, U.S.A.

In all other countries, sold and distributed
by Kluwer Academic Publishers Group,
P.O. Box 322, 3300 AH Dordrecht, The Netherlands

SB
614
.I58
1999

Printed on acid-free paper

All Rights reserved
© 1999 Kluwer Academic Publishers

No part of the material protected by this copyright notice may be reproduced
or utilized in any form or by any means, electronic or mechanical,
including photocopying, recording or by any information storage and
retrieval system, without written permission from the copyright owner.

Printed in the Netherlands

Hydrobiologia **415:** v–viii, 1999.
J. M. Caffrey, P. R. F. Barrett, M. T. Ferreira, I.S. Moreira, K. J. Murphy & P. M. Wade (eds),
Biology, Ecology and Management of Aquatic Plants.

Contents

Ecology

Invasive aquatic and riparian plants

Management

Integrated Plant Control

Physical and Biomanipulative Control

Chemical Control

Hydrobiologia **415:** ix–x, 1999.
J. M. Caffrey, P. R. F. Barrett, M. T. Ferreira, I.S. Moreira, K. J. Murphy & P. M. Wade (eds), Biology, Ecology and Management of Aquatic Plants.

Symposium Sponsors

Fundação para a Ciência e Tecnologia

Instituto da Água

Instituto de Hidráulica, Engenharia Rural e Ambiente

Instituto de Conservação da Natureza

Direcção-Geral de Protecção das Culturas

AgrEvo

Caixa Geral de Depósitos

Special Edition Referees

Agami, M.
Ali, M.
Anderson, L.
Bailey, J.
Barrett, P.
Best, E.
Bini, M.
Boone, P.
Boylen, C.
Caffrey, J.
Carrapico, F.
Child, L.
Cilliers, C.
Clayton, J.
Coops, H.
Dawson, H.
Dominy, F.
Dutarte, A.
Eaton, J.
Ferreira, T.
Fox, A.
Froud-Williams, R.
Gopal, B.
Grosse, W.
Haller, W.
Hanbury, R.
Hardwick, L.K.
Haslam, S.
Haury, J.
Hellsten, S.
Holland, D.
Hollander den, N.
Hollingsworth, P.
Hootsmans, M.
Hroudova, Z.
Husak, S.
Janaeur, G.
Jones, I.
Kelly-Quinn, M.
Kvet, J.
Lambi, C.
Maberly, S.
Madsen, J.
Mitchell, D.
Moreira, I.
Muller, S.
Murphy, K.
Newman, J.
Nilsson, N.
Ozimek, T.
Peltre, M.
Pieterse, A.
Pot, R.
Prach, K.
Reynolds, J.
Ridge, I.
Sabbitini, R.
Schwarz, A.
Spencer, W.
Stratford, K.
Sutton, D.
Thomaz, S.
Tiley, G.
Triest, L.
van Donk, E.
van Vierssen, W.
Villizzi, I.
Wade, M.
Whitton, B.
Willby, N.

Hydrobiologia **415:** 1–5, 1999.
J. M. Caffrey, P. R. F. Barrett, M. T. Ferreira, I.S. Moreira, K. J. Murphy & P. M. Wade (eds),
Biology, Ecology and Management of Aquatic Plants.
© 1999 *Kluwer Academic Publishers. Printed in the Netherlands.*

Propagule bank and plant emergence of macrophytes in artificial channels of a temperate irrigation area in Argentina

L. W. Acosta*, M. R. Sabbatini, O. A. Fernández & M. A. Burgos
CERZOS and Departamento de Agronomía, Universidad Nacional del Sur, (8000) Bahía Blanca, Argentina

Key words: submerged plants, macrophytes, propagule bank, irrigation and drainage channels, south America, aquatic weeds

Abstract

The irrigation district of the lower valley of Río Colorado (62° 37′ W, 39° 23′ S) was examined during 1995 and 1996 to evaluate propagule bank abundance and regeneration in five irrigation and five drainage channel sites. Six 15 cm deep cores from each site were collected in late winter and sectioned into two subsamples. One was sieved to identify taxa and propagule density and the other was placed under controlled laboratory conditions to record emergence over 90 days. Mean propagule density of *Potamogeton pectinatus* (260 tubers. m^{-2}, 173 rhizome nodes. m^{-2} and 246 seeds. m^{-2}) and *Zannichellia palustris* (968 seeds. m^{-2}) was higher in irrigation than in drainage channels. By contrast, higher values in drainage channels were recorded for *Ruppia maritima* (4352 seeds. m^{-2}) and *Chara contraria* (35 633 oospores. m^{-2}). Plant emergence in *P. pectinatus* was independent of channel type. Higher and lower values were registered in irrigation channels for *Z. palustris* and the other two species, respectively. Plant emergence in relation to propagule density for all the species ranged from 24% to 56%, and no differences were found between irrigation and drainage channels. Results are discussed in relation to the established vegetation and the different management and environmental conditions of both types of channels.

Introduction

The irrigation district of the lower valley of Río Colorado (VIRC, Figure 1) is located in the southern part of the province of Buenos Aires, Argentina, where approximately 90 000 ha belonging to 1500 producers are currently under various intensities of irrigation. The land is dedicated to the cultivation of vegetables, cereals, fruits and production of fodder. Water comes from the Colorado River and is distributed by a network of 5441 km and 3738 km of irrigation and drainage channels, respectively. The landscape is almost flat with a slight incline (0.002%), and as a result water drainage is very slow and thus easily blocked by the growth of aquatic weeds. Mean annual temperature is 15 °C, with a maximum mean value in January of 21.8 °C and a minimum mean value of 7.5 °C in July.

Potamogeton pectinatus L. and *Chara contraria* A Braun ex Kütz. are the main submerged macrophytes in the irrigation district of VIRC, representing a serious problem for proper water management. *Zannichellia palustris* L. and *Ruppia maritima* L. are also commonly present; however, their abundance is lower and at present do not constitute an important aquatic weed problem (Fernández et al., 1987; Sabbatini et al., 1998).

Propagule bank data can provide information about species composition, relative abundance and plant distribution between and within wetlands (Welling et al., 1988). Germination and growth is an important step in the reproduction cycle of the macrophytes, which may be affected by numerous environmental factors such as hydration, aeration, temperature and light (Barrat-Segretain, 1996). These environmental conditions are different between drainage and irrigation channels of VIRC; for example, water conductivity is significant lower in irrigation (1.1–1.9 mS cm^{-1}) than in drainage channels (5.8–6.6 mS cm^{-1}), irrigation have faster flowing than drainage channels and the irrigation network is dried out for three months during the winter for management purposes (cleaning,

* Author for correspondence

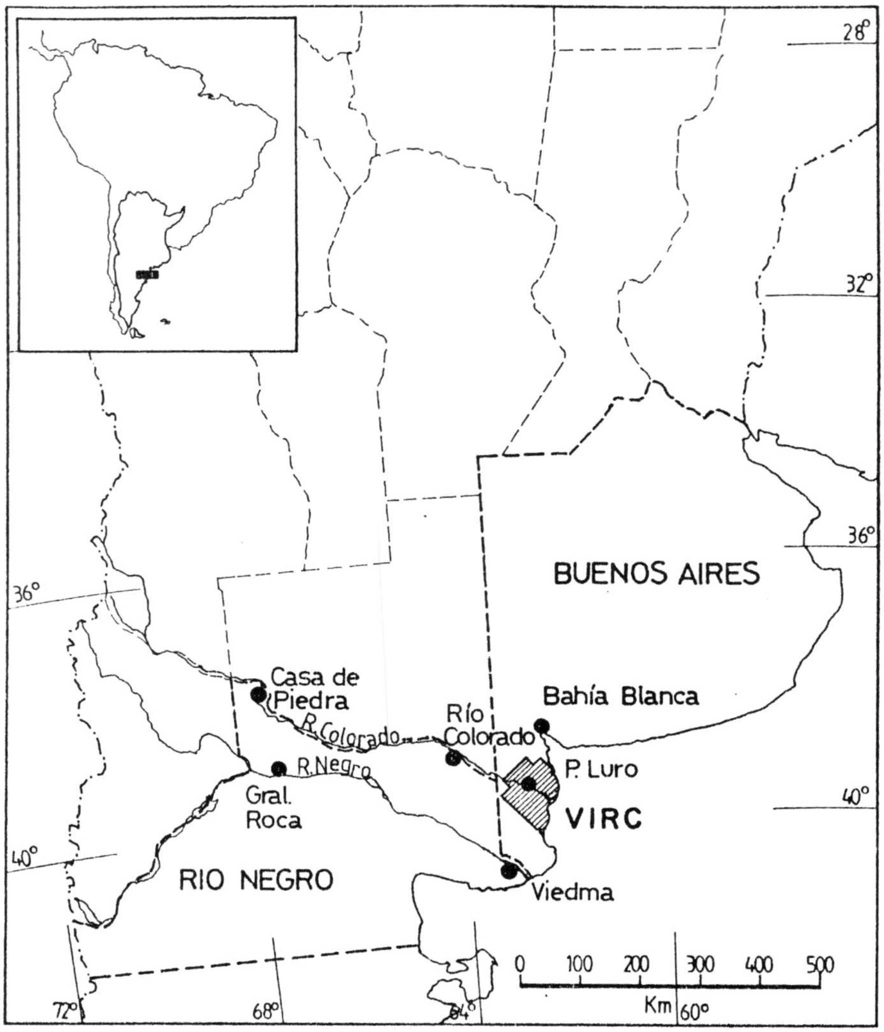

Figure 1. Irrigation district of VIRC in Argentina.

remodelling). In contrast, water is always present in the drainage system (Fernández et al., 1998).

Understanding sediment propagule abundance and emergence of submerged aquatic plants would help in decisions about management procedures. The present study was performed to evaluate propagule bank abundance and regeneration of submerged macrophytes in both irrigation and drainage channels of VIRC.

Materials and methods

Six sediment samples (177 cm^2 $\times$ 15 cm deep) were collected in late winter during 1995 and 1996 from each of five irrigation and five drainage channel sites of VIRC. In the laboratory, each sample was vertically sectioned into two subsamples. One subsample was placed in pots (201 cm^2 $\times$ 3 cm deep) and submerged in an aerated aquarium (2500 cm^2 $\times$ 34 cm deep) filled with water. A 16:8 h light-dark (65.8 μE cm^{-2}. s^{-1}) photoperiod and a temperature of 20 $\pm$ 5 °C were maintained throughout 90 days, and emergence checked periodically. Seedlings were removed as soon as they could be identified. The other subsample was sieved (150 μ–0.8 cm) to identify taxa and estimate propagule bank density. Propagule taxa identification was based on morphological characteristics described by Cáceres (1978), Van Vierssen

Table 1. Propagule abundance and plant emergence in sediment samples collected in irrigation channel (IC) and drainage channel (DC) sites. Ppec.tub: tubers of *P. pectinatus*, Ppec.rhi: rhizome nodes of *P. pectinatus*, Ppec.seed: seeds of *P. pectinatus*, Rmar.seed: seeds of *R. maritima*, Zpal.seed: seeds of *Z. palustris*, Ccon.oosp: oospores of *C. contraria*. Scheffé test ($n = 30$: six samples × 5 sites), *: $p < 0.05$, **: $p < 0.01$, ns: non-significant differences

Species code	Propagules (m^{-2})			Plant emergence (m^{-2})			Emergence (%)		
	IC	DC		IC	DC		IC	DC	
Ppec.tub	260	142	**	143	67	ns	54	50	ns
Ppec.rhi	173	125	ns	97	66	ns	56	53	ns
Ppec.seed	246	172	*	59	52	ns	24	34	ns
Rmar.seed	200	4352	**	53	1151	**	43	56	ns
Zpal.seed	968	595	**	273	179	**	33	47	ns
Ccon.oosp	6726	35633	**	1606	10299	**	24	28	ns

(1982) and Kantrud (1990, 1991). Data was analysed using Nested-ANOVA, test of Scheffé and regression analysis.

In both seasons, the existing vegetation was determined in the target sites twice, in mid (December–January) and late season (March–April). Within each section, 10 random grapnel hauls were taken to sample submerged macrophytes species, using an eight-pronged plant grapnel on a 5-m cord which was thrown into the water channels at right angles to the bank and dragged back along the bottom (Murphy et al., 1981). The abundance of each species in each subsample was rated on a scale of 1 = scarce, 2 = common and 3 = abundant.

Weed control procedures were not made during the experiment in drainage channel sites. Irrigation channel sites were dredged during the winter irrigation recess, using the same procedure (excavators) applied annually in most of the irrigation channels of VIRC. In these channels, sediment samples were taken one month after dredging.

Results

Table 1 shows the propagule density and plant emergence in the sampled sites. Except for rhizome nodes of *P. pectinatus*, species propagule abundance was different between irrigation and drainage channels. Tuber and seed bank density of *P. pectinatus* were 46% and 26% higher in irrigation than in drainage channels, respectively, and *Z. palustris* had 39% more seeds in irrigation channels. In contrast, seeds of *R. maritima* and oospores of *C. contraria* were 81% and 95% higher in drainage than irrigation channels, respectively. *C. contraria* was the most abundant taxon occurring in 78% of irrigation and 87% of drainage channel total propagule bank.

Plant emergence was always a proportion of the total bank, ranging from a minimum of 24% for the oospores of *C. contraria* to a maximum of 56% for the rhizomes of *P. pectinatus* and seeds of *R. maritima* (Table 1). No differences were observed between the two types of channels. Seedling density originating from the propagule bank of *R. maritima* and *C. contraria* was 95% and 84% higher in drainage than in irrigation channels, respectively. For *Z. palustris*, seedling density was 31% higher in irrigation channels, and for *P. pectinatus*, it did not differ between channels being independent of the type of propagule considered (seed, tuber and rhizome).

Standing plant abundance showed that *P. pectinatus* and *C. contraria* were common species at the sampled sites, no difference was found between irrigation and drainage channels. *Z. palustris* was scarce in both kinds of channels, whereas *R. maritima* was absent in irrigation channels and scarce in drainage channels. Other macrophytes species that occur in VIRC in small amounts are *Potamogeton pusillus, Myriophyllum aquaticum, Azolla filiculoides* and filamentous algae (Fernández et al., 1998; Sabbatini et al., 1998), but they were not present at the target sites.

Regression analysis showed dependent relationships between propagule bank density × plant emergence in *P. pectinatus* tubers, *R. maritima* seeds and

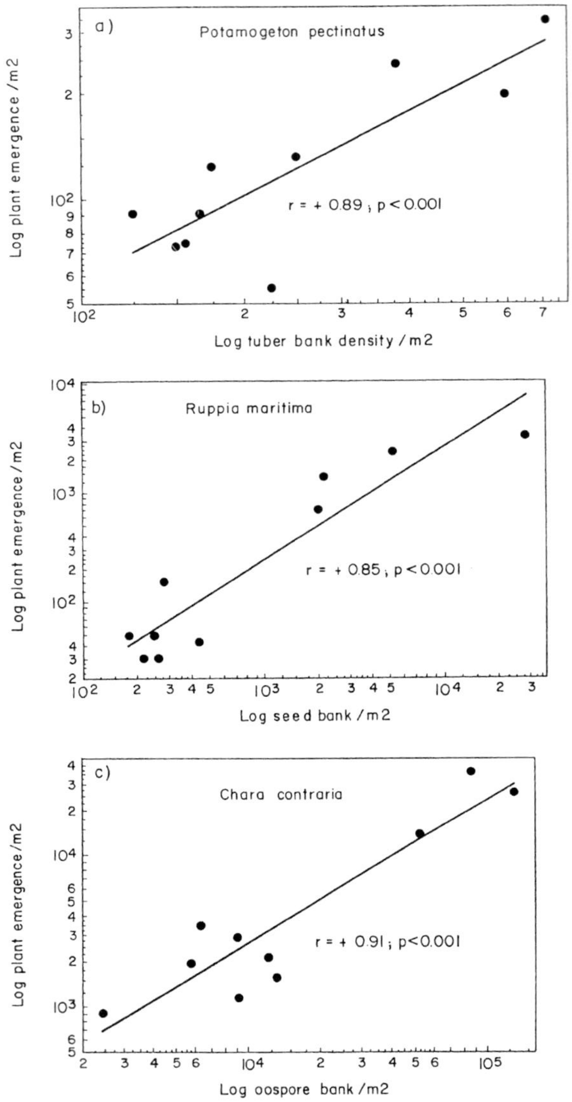

Figure 2. Relationships between propagule bank richness and plant emergence in sediment samples collected from five irrigation and five drainage channels sites of VIRC (a) *P. pectinatus* tubers, (b) *R. maritima* seeds and (c) oospores of *C. contraria*.

C. contraria oospores (Figure 2), but not in the other propagules.

Discussion

The total propagule bank observed in the two channel systems varied from 8500 to 41 000 propagules per m^2. The macroalga *C. contraria* dominated the bank accounting for 82% of all propagules found. Numerous sporelings of the macroalga also emerged from the sediment in both channel types (Table 1). *Chara* is known to be invasive and able to survive in disturbed habitats (Wade, 1990; Westcott et al., 1997). *C. contraria* occurs in VIRC, frequently forming extensive monospecific stands in the channel network (Sabbatini et al., 1986).

In the channels of VIRC, the annuals *R. maritima* and *Z. palustris* are able to maintain a rich seed bank. This is in agreement with other studies of annual macrophyte species, which tend to accumulate a large persistent seed bank, population survival being dependent on its successful germination (Bonis et al., 1995; Westcott et al., 1997).

In drainage channels of VIRC, the perennial *P. pectinatus*, regardless of its dominance in the existing vegetation, has a seed bank significantly lower in density than that of the other species. Similarly, Westcott et al. (1997) found that perennial macrophytes, for example *Potamogeton* spp., although predominant in the shoreline marshes in Lake Ontario, were rare as seeds and seedlings. Also, Wiegleb et al. (1991) reported that although a high number of seeds is usually produced in *Potamogeton* species, germination fails in the most cases and sexual reproduction does not contribute much to the maintenance of populations. Kautsky (1990) pointed out that the main regenerative strategy for *P. pectinatus* is its tuber bank, virtually no seed bank existing. In our study, 38% of the propagule bank of *P. pectinatus* was represented by seeds, and 24% of plant emergence by seedlings, showing that in the channel system of VIRC, the seed bank of *P. pectinatus* probably contribute to maintain the population. The density of the tuber bank of *P. pectinatus* in VIRC is lower than the values registered for other water bodies, for example in the Netherlands and Camargue, France (377 to 3975 tubers m^{-2}: Van Wijk, 1988) and in the Askö area, Sweden (52–1092 tubers m^{-2}: Kautsky, 1987).

Kautsky (1990), Scott et al. (1993) and Westcott et al. (1997) found no correlation between propagule bank and standing vegetation for aquatic macrophytes. It is interesting to point out that *R. maritima*, while absent in the standing vegetation of the irrigation channel target sites, was abundant in the bank (Table 1). *R. maritima*, usually found in highly saline drainage channels of VIRC (Sabbatini et al., 1998), is a well-known brackish water species (Costa & Seeliger, 1989). The significant correlation obtained between propagule bank abundance and plant emergence in *P. pectinatus* tubers, *R. maritima* seeds and *C. contraria* oospores (Figure 2), would mean that propagule inputs to the bank provoke changes in plant emergence density. Propagule inputs could change every season with different factors, as for example water temperature or weed control procedures, affecting plant emergence on those species.

This study found differences between the propagule bank of the two studied types of channels, showing that the banks of *P. pectinatus* and *Z. palustris* were more abundant in irrigation channels, while the banks of *R. maritima* and *C. contraria* were more abundant in drainage channels. Sabbatini et al. (1998) found differences between the standing vegetation of irrigation and drainage channels of VIRC, pointing out that its occurrence and abundance are closely related to environmental conditions, the primary factors being water conductivity, winter drought, water depth, water movement and management regime. Owing to these differences in management and environmental conditions, irrigation channels are more disturbed habitats than drainage channels. For example, most of the irrigation channels receive a deep dredge with excavators during the winter, whereas most of the drainage channels are only dredged every 3–4 years. Likewise, irrigation channels have faster water movement, lower water conductivity and are dry for about three months during winter. These differences in the conditions of both types of channels no doubt affect the bank species composition and emergence.

Acknowledgements

This work was part-supported by a Government of Argentina CONICET postgraduate scholarship to L. W. Acosta. Helpful comments and suggestions of anonymous reviewers significantly improved the paper.

References

Barrat-Segretain, M.H., 1996. Strategies of reproduction, dispersion, and competition in river plants: a review. Vegetatio 123: 13–37.

Bonis, A., J. Lepart & P. Grillas, 1995. Seed bank dynamic and coexistence of annual macrophytes in a temporary and variable habitat. Oikos 74: 81–92.

Cáceres, E. J., 1978. Contribución al conocimiento de carófitos del centro de Argentina Acad. Nac. Cienc. 2: 316–372.

Costa, C. S. B. & V. Seeliger, 1989. Vertical distribution and resource allocation of *Ruppia maritima* L. in a southern Brazilian estuary. Aquat. Bot. 33: 123–129.

Fernández, O. A., J.H. Irigoyen, M.R. Sabbatini & R.E. Brevedan, 1987. Aquatic plant mgmt in drainage channels of southern Argentina. J. Aquat. Plant Manage. 25: 65–67.

Fernández, O. A., K. J. Murphy, A. López Cazorla, M. R. Sabbatini, M. A. Lazzari, J. C. J. Domaniewski & J. H. Irigoyen, 1998. Interrelationships of fish and channel environmental conditions with aquatic macrophytes in an Argentine irrigation system. Hydrobiologia 380: 18–25.

Kantrud, H. A., 1990. Sago pondweed (*Potamogeton pectinatus* L.): a literature review. United States Department of the Interior, Fish & Wildlife Service. Resource publication 176: 91 pp.

Kantrud, H. A., 1991. Wigeongrass (*Ruppia maritima* L.): a literature review. United States Department of the Interior, Fish & Wildlife Serv. Res. 10: 58 pp.

Kaustky, L., 1987. Life cycles of three populations of *Potamogeton pectinatus* L. at different degrees of wave exposure in Askö area, northern Baltic proper. Aquat. Bot. 27: 177–186.

Kaustky, L., 1990. Seed and tuber banks of aquatic macrophytes in Askö area, northern Baltic proper. Holarct. Ecol. 13: 143–148.

Murphy, K. J., R. G. Hanbury & J. Eaton, 1981. The ecological effects 2-methylthio triazine herbicides used for aquatic weed control in navigable canals: 1. Effects on aquatic flora and water chemistry. Arch. Hydrobiol. 91: 231–294.

Sabbatini, M. R., J. H. Irigoyen & O. A. Fernández, 1986. Phenology and biomass dynamic of *Chara contraria* A. Braun ex Kütz. in drainage channels of a temperate irrigation area in Argentina. 7th International Symposium on Aquatic Weeds. Loughborough, Leicestershire, England. 15–19 September 1986: 285–289.

Sabbatini, M. R., K. J. Murphy & J. H. Irigoyen, 1998. Vegetation-environment relationships in irrigation channel systems of southern Argentina. Aquat. Bot. 60: 119–133.

Scott, D. W., R. J. M. Dwayne & A. K. Paul, 1993. Relationships of marsh seed banks to vegetation patterns along environmental gradients. Freshwat. Biol. 29: 361–370.

Van Vierssen, W., 1982. The ecology of communities dominated by *Zannichellia* taxa in western Europe I. Characterisation and autoecology of the *Zannichellia* taxa. Aquat. Bot. 12: 103–155.

Van Wijk, R. J., 1988. Ecological studies on *Potamogeton pectinatus* L. I. General characteristics. biomass production and life cycles under field conditions. Aquat. Bot. 31: 211–258.

Wade, P. M., 1990. The colonisation of disturbed freshwater habitats by *Characeae*. Folia Geobot. Phytotax., Praha 25: 275–278.

Welling, C. H., R. L. Pederson & A. G. Van der Valk, 1988. Recruitment from the seed bank and the development of zonation of emergent vegetation during a drawdown in a prairie wetland. J. Ecol. 76: 483–496.

Westcott, K., T. H. Whillans & M. G. Fox, 1997. Viability and abundance of seeds of submerged macrophytes in the sediment of disturbed and reference shoreline marshes in Lake Ontario. Can. J. Bot. 75: 451–456.

Wiegleb, G., H. Brux & W. Herr, 1991. Human impact on the ecological performance of *Potamogeton* species in north-western Germany. Vegetatio 97: 161-172.

Hydrobiologia **415:** 7–11, 1999.
J. M. Caffrey, P. R. F. Barrett, M. T. Ferreira, I.S. Moreira, K. J. Murphy & P. M. Wade (eds),
Biology, Ecology and Management of Aquatic Plants.
© 1999 *Kluwer Academic Publishers. Printed in the Netherlands.*

Modelling *Potamogeton pectinatus*: for better or for worse

Michiel J. M. Hootsmans
Department of Environmental Science & Water Resources, IHE-Delft, P.O. Box 3015, 2601 DA Delft, The Netherlands
Present address: Amsterdam Water Supply, Location Leiduin, Vogelenzangseweg 21, 2114 BA Vogelenzang, The Netherlands
Tel: [+31] - (0) - 23 523 35 38; Fax: [+31] - (0) 23 528 14 60. E-mail: m.hootsman@gw.amsterdam.nl

Key words: macrophyte growth, simulation model, weed management, *Potamogeton pectinatus*

Abstract

The SAGA plant growth model is introduced in connection with integrated weed management. SAGA describes the seasonal development of *Potamogeton pectinatus* L. The model is mainly driven by light climate characteristics (water layer extinction, periphyton shading) and their effects on plant photosynthesis and tuber production. It could be a tool for evaluating several weed control methods (mowing, shading, grazing) as it was calibrated and validated quite well for a vegetation growing in Lake Veluwe (The Netherlands). The SAGA model was used to provide management advice for a *P. pectinatus* vegetation growing profusively in two Argentine irrigation schemes. The model's re-calibration to local conditions was hampered by some unexplained variation in seasonal biomass cycle data. Nevertheless, the results obtained were considered quite acceptable for comparative scenario analysis. The simulated vegetation dynamics suggest that the mowing regime currently applied in the channels is not an optimal solution. Neither is an artifically increased turbidity, created by the introduction of benthivorous common carp (*Cyprinus carpio* L.). Their combination, however, may lead to quite satisfactory results, allowing a repetition of the mowing operation to be postponed by two years. Direct grazing of the vegetation by grass carp (*Ctenopharyngodon idella* Val.) also proved a potential management tool with substantial effect.

Introduction

The Sagopondweed Growth Analysis model (SAGA) is a plant growth simulation tool describing growth and development of *Potamogeton pectinatus* L. in detail (Hootsmans, 1994). This cosmopolitan species may be regarded as a desired macrophyte as well as a notorious weed. SAGA was developed as part of a larger study into the functioning of a shallow eutrophicated lake in The Netherlands (Van Vierssen et al., 1994) and could be used to provide management advice for lake vegetation restoration. In the current paper, the model is introduced, and some of its potential for application in irrigation channel weed management is demonstrated in a case study from Argentina.

Model outline and calibration

The SAGA model is mainly driven by light climate characteristics (irradiance, water layer light extinction, periphyton shading) and their effects on plant photosynthesis and tuber production (see Figure 1). Ageing of the vegetation and its effects on photosynthesis have also been incorporated. The model could be calibrated and validated quite well for a *P. pectinatus* vegetation growing in Lake Veluwe (The Netherlands; see Figure 2, and Hootsmans, 1994). The SAGA model was subsequently used to provide management advise for a *P. pectinatus* vegetation growing in two Argentine irrigation schemes, as part of a European Union cooperative research project (Hootsmans, 1996). Weed development in these systems is often quite profuse, requiring intensive (and costly) channel management (see e.g. Fernandez et al., 1987).

Extensive data sets on plant biomass development were collected in several irrigation and drain-

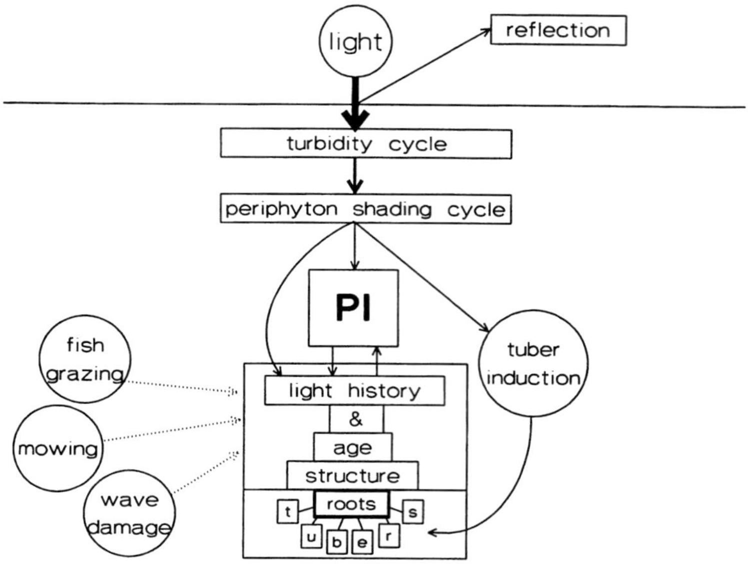

Figure 1. Overview of the main interactions and components present in SAGA.

age channels, together with detailed information on seasonal changes in light extinction (water turbidity), periphyton shading and plant photosynthesis, as well as the effects of common carp (*Cyprinus carpio* L.) and grass carp (*Ctenopharyngodon idella* Val.); see Hootsmans (1996), Hootsmans et al. (1996) and Sidorkewicj et al. (1996).

Subsequently, several model parameters were set to the values observed for the Argentine population. These included light response curve data, tuberbank characteristics, the observed starting date of above-ground biomass growth and regional light climate data (irradiance and day-length, turbidity, periphyton shading regime, channel depth).

The ensuing model response followed the actual biomass development of the Argentinean population quite well, although there obviously remains quite some unexplained year-to-year variation (see Figure 3). It has to be noted that the model could easily accomodate the five-fold difference between the peak biomass levels reached by the Dutch and the Argentine populations of *P. pectinatus*, without any fundamental change in its structure. As this result was considered quite acceptable, SAGA was used for further modelling studies of management options. These involved mowing of above-ground biomass, increasing water turbidity (which in reality can be achieved by introducing common carp), feeding on weeds by grass carp and some combinations of these options.

Mowing was implemented as a user-defined moment at which a certain fraction of the above-ground biomass is removed, starting from the uppermost layers of the vegetation. The mowing option is mathematically identical to the storm-dependent wave damage routine in the SAGA version described by Hootsmans (1994). Grass carp grazing is simulated as a more complicated type of mowing. Every day, a fish-density dependent potential fraction of above-ground biomass is removed. The actual removal fraction is dependent of fish activity, which is implemented as a multiplier varying between 0–1. In freshwater with sufficient oxygen, grass carp feeding activity appears to be largely governed by water temperature. Intake is limited noticeably below 20 °C (Opuszynski, 1972; Van Zon, 1973, 1977, 1989; Rottman, 1977), reaching zero uptake below 15 °C. In the irrigation systems studied, mean water temperatures are 20 °C or above during the major part of the growing season (December–March; Hootsmans, 1996). In the model, fish activity increases from 0 to 1 between 1 and 30 November, decreasing again to zero between 1 and 30 April. Above 20 °C, grass carp daily consumption (as plant freshweight) amounts to about their body freshweight (Opuszynski, 1972; Van Zon, 1973; Rottman, 1977). Finally, shading due to increased turbidity caused by

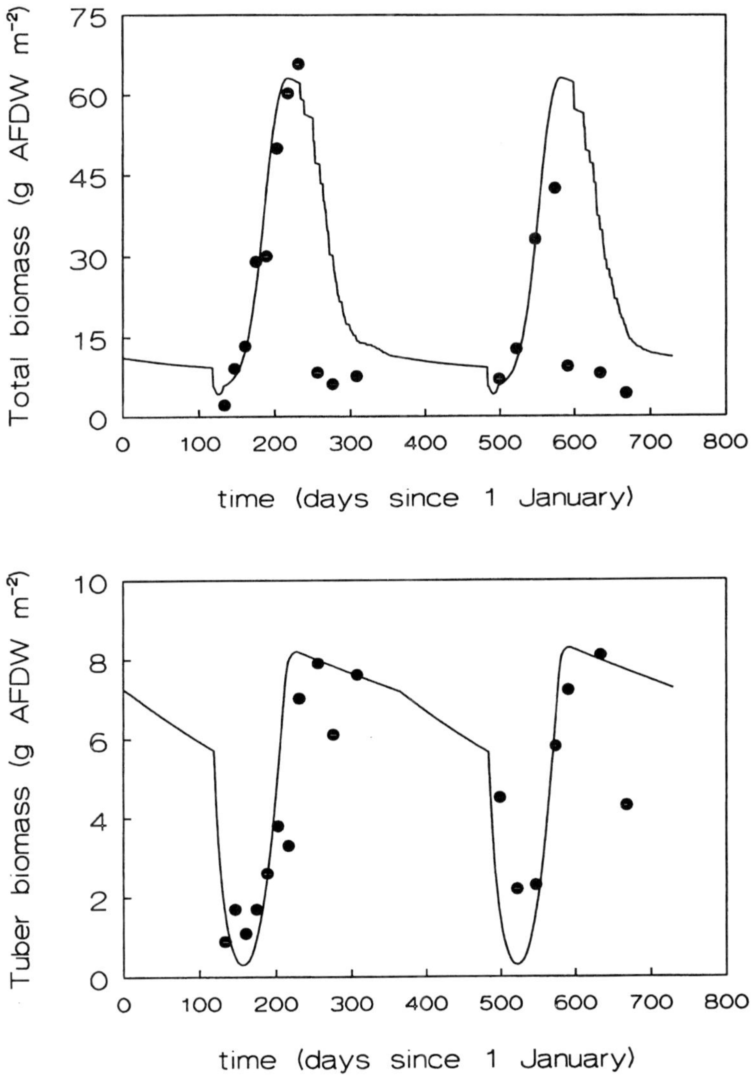

Figure 2. Total biomass (ash-free dryweight, AFDW, above) and tuber biomass (below) of *Potamogeton pectinatus* in Lake Veluwe, The Netherlands, observed in 1986 and 1987 (dots) and predicted by the SAGA model (continuous line).

common carp benthic feeding is implemented through the extinction coefficient which may change according to a user-defined seasonal pattern.

Results and discussion

In channel management, it is mostly the above-ground biomass that is causing problems. Thus, in Figure 4, the results of the various simulations for a 1 m deep drainage channel are presented in terms of the percentage of peak above-ground biomass that is removed due to various management options. The simulated vegetation dynamics show that the currently applied, costly mowing regime in the channels will not lead to sustained control: as in reality in the study area, to be effective it has to be repeated annually ('MOW' in Figure 4; note that this option implies mowing in the first year only). Obviously, in serious cases in which immediate results are required, mowing can be regarded as the only option, besides measures like the application of herbicides.

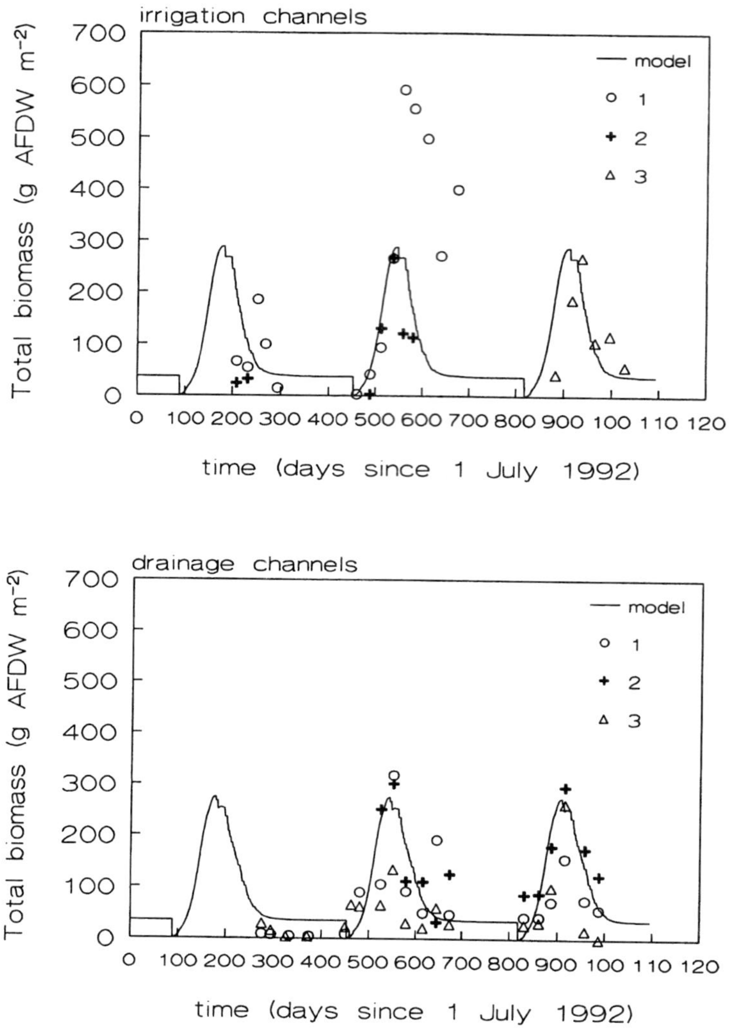

Figure 3. Simulated and observed biomass cycles of *P. pectinatus* in several irrigation channels (1,2 and 3, above) and drainage channels (1,2 and 3, below) in Argentina. The start of the growing season in 1992 is on July 1. Note the difference in vertical scale between Figures 2 and 3.

Increased turbidity, possibly obtained by the introduction of common carp ('SHADE'), at first glance appears not very effective. This is likely due to the high irradiance levels in the area. Still, its (sustained) effect of 25% removal is more noticeable than that in the second and third year of the 'MOW' strategy.

Direct grazing of the vegetation by grass carp ('FISH') also proved to be a promising management tool. The stocking density of 300 kg ha^{-1} as used here is high but appears not excessive for limiting the development of a fast growing vegetation (see also Van Zon, 1989; Spencer, 1994).

A combination of 'SHADE' and 'MOW' ('SHADMOW') yields quite interesting results. The outcome is not simply predicted from the results of the separate treatments. Mowing once, combined with the introduction of common carp that keep turbidity sufficiently high, seems very effective. It allows a repetition of the mowing operation to be postponed by at least two years, depending on the local requirements regarding unhampered water flow. The SAGA model also predicts that in relatively shallow channels (0.5 m) the light climate conditions for the vegetation are such that mowing, fish grazing or shading have little if any sustained effect. Keeping the mean depth of main

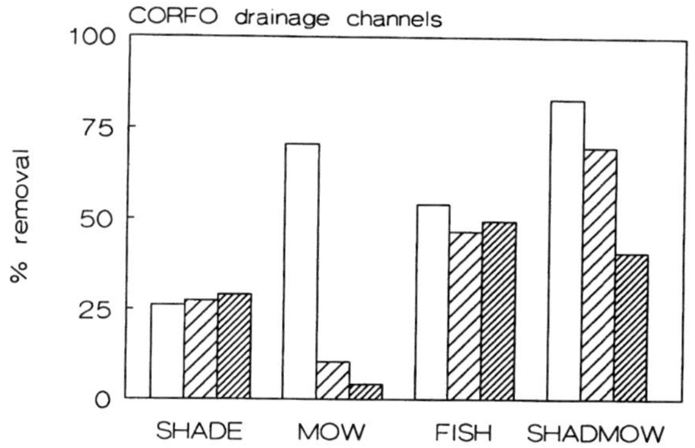

Figure 4. The percentage of the above-ground peak biomass removed by four different management options, relative to the 'no control' option, in a 1 m deep drainage channel in Argentina. Results from model simulations of three consecutive years (different hatching). SHADE = application of a turbidity increase from the currently observed 2 m^{-1} to a value of 4 m^{-1} due to the introduction of a sufficiently dense population of common carp; MOW = removing 70% of the aboveground biomass by mowing once in late spring, in the *first* season only; FISH = effect of the presence of a population of grass carp (300 kg ha^{-1}); SHADMOW= combination of SHADE and MOW.

channels at least at 1 m may be an important boundary condition for successful weed management.

Obviously, a model is not reality, it can only provide an image of the level of our understanding of the concerted effect of the various factors involved in weed development. A logical follow-up would be a large scale application of some of the options discussed, but this is not easily achieved: new, promising management techniques may fail and who will then cover the losses of the farmers involved? In this context, international donors, supporting research for development, could also provide funding for the field testing of above mentioned ideas.

Acknowledgements

The author gratefully acknowledges the work and assistance provided by all the partners involved in the research project 'Biological management of irrigation channel weed problems in semi-arid agriculture' (European Union STD 3 programme contract no. TS3*-CT92-0125). Prof. Dr Patrick Denny, Prof. Dr Wim Van Vierssen and Dr Jan Vermaat critically read the manuscript.

References

Fernandez, O. A., J. H. Irigoyen, M. R. Sabbatini & R. E. Brevedan, 1987. Aquatic plant management in drainage canals of southern Argentina. J. Aquat. Plant Mgmt 25: 65–67.

Hootsmans, M. J. M., 1994. A growth analysis model for *Potamogeton pectinatus*. Geobotany 21: 250–286.

Hootsmans, M. J. M., 1996. Biological management of irrigation channel weed problems in irrigated semi-arid agriculture. Final Report EU STD-3 project TS3*-CT92-0125: 91 pp. + 2 annexes (a summary can be obtained through Internet: http://www.ihe.nl/dew/research/arg.htm).

Hootsmans, M. J. M., A. A. Drovandi, N. S. Perez & F. Wiegman, 1996. Photosynthetic plasticity in *Potamogeton pectinatus* L. from Argentina: strategies to survive adverse light conditions. Hydrobiologia 340: 1–5.

Opuszynski, K., 1972. Use of phytophagous fish to control aquatic plants. Aquaculture 1: 61–74.

Rottman, R., 1977. Management of weedy lakes and ponds with grass carp. Fisheries 2: 8–14.

Sidorkewicj, N. S., A. C. L. Cazorla & O. A. Fernandez, 1996. The interaction between *Cyprinus carpio* L. and *Potamogeton pectinatus* L. under aquarium conditions. Hydrobiologia 340: 271–275.

Spencer, D. F., 1994. Estimating the impact of triploid grass carp on sago pondweed in the Byrnes Canal – Implications for biological control in Northern California irrigation systems. Ecol. Mod. 72: 187–204.

Van Vierssen, W., M. J. M. Hootsmans & J. E. Vermaat, 1994. Lake Veluwe, a macrophyte-dominated system under eutrophication stress. Geobotany 21.

Van Zon, J. C. J., 1973. Studies on the biological control of aquatic weeds in The Netherlands. Proc. 3rd Int. Symp. Biol. Contr. Weeds: 31–38.

Van Zon, J. C. J., 1977. Grass carp (*Ctenopharyngodon idella* Val.) in Europe. Aquat. Bot. 3: 143–155.

Van Zon, J. C. J., 1989. Status of the use of grass carp (*Ctenopharyngodon idella* Val.). Proc. 5th Int. Symp. Biol. Contr. Weeds: 249–260.

Hydrobiologia **415**: 13–19, 1999.
J. M. Caffrey, P. R. F. Barrett, M. T. Ferreira, I.S. Moreira, K. J. Murphy & P. M. Wade (eds), Biology, Ecology and Management of Aquatic Plants.
© 1999 *Kluwer Academic Publishers. Printed in the Netherlands.*

Effects of *Cyprinus carpio* on *Potamogeton pectinatus* in experimental culture: the incidence of the periphyton

N. S. Sidorkewicj*, A. C. López Cazorla, O. A. Fernández, G. C. Möckel & M. A. Burgos
CERZOS and Departamento de Agronomía, Universidad Nacional del Sur, 8000 Bahía Blanca, Argentina E-mail: nsidorke@criba.edu.ar

Key words: common carp, sago pondweed, epiphyton, light attenuation

Abstract

The effect of one-year-old common carp (*Cyprinus carpio* L.) on plants and seedlings of *Potamogeton pectinatus* L., and on periphyton development was studied in 100 l glass aquaria. Two 30-day experiments were conducted using a muddy sediment or a pebbly bottom. In both cases, three treatments based on different fish size (two fish/aquarium) were assayed. The control had no fish. In both experiments, chlorophyll content of the periphyton increased from the beginning to the end of the period, for the three fish treatments. Light attenuation by periphyton was high, with final values 12–30% higher than in the control. Periphytic communities acclimated to reduced light conditions when the bottom was muddy. It can be concluded that, in muddy conditions, small carp affected plant growth by shade stress, which is achieved by a combination of increase in turbidity and the developing of a leaf periphytic cover. Under pebbly conditions, plant damage was caused by collisions and the growth of epiphytic algae. Medium and large fish consumed plants in both experiments. Seedlings were affected by herbivory in all cases.

Introduction

Common carp, *Cyprinus carpio* L., is a benthivorous fish that has received considerable attention as a result of impacts on submerged macrophytes following its introduction to new habitats outside its native range (King & Hunt, 1967; McCrimmon, 1968; Fletcher et al., 1985; Brumley, 1991). The adverse effects of this fish upon aquatic vegetation have been historically attributed to physical disturbance, causing plant uprooting (Crivelli, 1983) and an increase in abiogenic turbidity (Sidorkewicj et al., 1996, 1998, 1999a, b; Fernández et al., 1998). More recently, bottom-up and top-down effects induced by carp have also been taken into account as mechanisms indirectly affecting the biomass of submerged macrophytes through changes in phytoplankton densities (Roberts et al., 1995; Sidorkewicj et al., 1999a, b). Less known are the effects that carp may have on vegetation through alteration in epiphytic periphyton.

Recent studies have demonstrated the importance of periphyton in terms of productivity (Lowe, 1996), as components of aquatic food webs (Otto, 1982) and as agents involved in the decline of submerged macrophytes through shading (Sand-Jensen, 1977; Phillips et al., 1978). To study the direct effects of different sizes of juvenile common carp on plants and seedlings of *Potamogeton pectinatus* L. (sago pondweed), and the indirect effects that operate through the development of epiphytic cover, two laboratory experiments with contrasting sediment conditions were performed. Acclimation of periphytic communities to decreasing light availability, caused by increasing turbidity, was also evaluated.

Materials and methods

Experimental setup

Two consecutive 30-day experiments were conducted using twenty 100-l glass aquaria (0.40 m × 0.60 m × 0.50 m height). In the first experiment (Experi-

* Author for correspondence

ment 1), muddy sediment (76% sand, 14% clay, 10% loam) was collected from a drainage channel in Argentina and distributed in the tanks in a 5 cm-thick bed. In the second experiment (Experiment 2), a 5 cm-thick layer of washed fine gravel was instead used. In both cases, the tanks were filled with tap water. In each experiment, four plants and two seedlings of *P. pectinatus*, individually contained in 200 ml pots, were introduced into each aquarium. For the epiphytic periphyton study, eight transparent plastic strips (2 × 8 cm) were suspended vertically in the water column of each aquarium 5 cm from surface. The vertical position allowed an equal colonization on both sides of the strip (Soto Perez, 1994). Nine months old common carp (age 0^+) were weighed and added to 15 aquaria in size-matched pairs, using three size classes as treatments (n = 5). Mean fresh weights (± s.d.) were: 6.6 ± 0.9 g (T_1), 14.0 ± 0.7 g (T_2) and 21.0 ± 1.3 g (T_3) for Experiment 1, and 15.2 ± 2.7 g (T_1), 27.7 ± 3.1 g (T_2) and 42.5 ± 0.7 g (T_3) for Experiment 2. Values for fish total length were 79 ± 5 mm (T_1), 101 ± 2 mm (T_2) and 115 ± 5 mm (T_3) in Experiment 1, and 101 ± 7 mm (T_1), 124 ± 4 mm (T_2) and 142 ± 7 mm (T_3) in Experiment 2. In each case, five aquaria with no fish were used as control. Light supply was provided by fluorescent lighting (40 W; 566 μE m^{-2} s^{-1} at water surface), with a 12 h photoperiod. Water aeration was provided by pumps. No fish food or fertiliser was added.

Sampling and analysis

At the beginning and at the end of the experiments, determinations were made of aboveground dry weight (PDW; 70 °C, 48 h) and total foliage length (TL) of plants. TL measurements were performed by adaptation of a line intersect method (Tennant, 1975), commonly used to measure root length (Sidorkewicj, 1998). Seedlings were measured in height (H). Weekly, two plastic strips were removed randomly from each aquarium; one was used for biomass and light attenuance measurements and the other was used to determine chlorophyll content. Light transmittance (PAR; LI-COR 192SB and METEX M-4650) was measured by submerging the intact strips in tap water under full sunlight. Attenuance (A) was calculated as proportional transmittance reduction relative to transmittance through a clean strip. For biomass measurements, the material attached to both sides of the strip was scraped off with a razor blade and filtered through pre-ashed and pre-weighed Whatman GF/C filters (1.2 μm). Determinations were then made of dry weight (DW; 105 °C, 24 h), ash weight (AW; 520 °C, 2 h) and ash-free dry weight (AFDW = DW − AW). For chlorophyll analysis, the material removed from the second strip was filtered through non pre-ashed GF/C filters. Chlorophyll *a* (Chl. *a*) and *b* (Chl. *b*) contents were determined following Wintermans & De Mots (1965). As all periphyton measurements corresponded to colonization on both sides of the strips, calculations were made to refer the data to one side. Weekly, dissolved oxygen (WTW OXI 96), pH, water temperature and conductivity (WTW LF 96-A) were measured. Changes in organic and inorganic tripton, phytoplankton density, dissolved nutrients, turbidity and underwater light attenuation were also recorded (Sidorkewicj et al., 1999a).

Statistics

When residuals were not normally distributed or variances were not homogeneous, data were transformed using log (x) or log (x + 1). Differences in aboveground dry weight of plants (PDW_f - PDW_i), plant total length (TL_f - TL_i) and seedling height (H_f - H_i), based on the mean value per aquarium, were analysed by means of a one-way ANOVA ($p \leq 0.05$). The comparison among consecutive mean values was carried out by means of an *a priori*, one-tailed t test (LSD; $p \leq 0.05$). As was indicated above, periphyton measurements were performed weekly in independent and randomly removed plastic strips, allowing the use of a split-plot factorial design in time for DW, AW, AFDW, Chl. *a*, Chl. *b* and A (n = 5; $p \leq 0.05$) (Steel & Torrie, 1980). The significance of the differences between mean values was determined by SNK test ($p \leq 0.05$). Periphyton growth parameters were estimated by applying to the AFDW data the exponential growth model: $B_t = B_0 . \exp^{r(t-t)_0}$, where B_0 and B_t are biomass (AFDW; mg cm^{-2}) at the beginning of the experiment and at time t (days), respectively, and r is the unrestricted growth rate (d^{-1}). Between treatment values of r were compared by analysis of covariance (ANCOVA, $p \leq 0.05$; Steel & Torrie, 1980). Rectangular hyperbolae were fitted to the attenuance-AFDW periphyton datasets: $A = (a \times B) \times [(c + B)]^{-1}$, where A is the attenuance, B is the biomass (AFDW; mg cm^{-2}), and a and c are the parameters (slope and asymptote, respectively) of the rectangular hyperbola. The fitted curves were compared between treatments by means of an F statistic ($p \leq 0.05$), described by Vermaat & Hootsmans (1994a).

Results

The average instantaneous growth rate of the fish used in the study (Bagenal & Tesch, 1978) was −0.30% in Experiment 1 and −0.36% in Experiment 2. Physico-chemical parameters did not exhibit differences between treatments (one-way ANOVA; $p > 0.10$). Mean (± s.d.) oxygen concentration, water temperature, conductivity and pH were 6.5 ± 1.7 mg l^{-1}, 23.2 ± 1.2 °C, 1.1 ± 0.2 mS cm^{-1} and 7.9 ± 0.2, respectively in Experiment 1, and 9.3 ± 2.2 mg l^{-1}, 17.3 ± 1.3 °C, 0.4 ± 0.0 mS cm^{-1} and 8.6 ± 0.3, respectively in Experiment 2. In Experiment 1, snails (*Littoridina parchappii*), which were present in the sediment, remained during all the experimental period in control tanks, but disappeared from fish aquaria within two days.

Vegetation

Initial PDW and TL (mean ± s.d.) were 43.9 ± 22.2 mg and 3.7 ± 1.1 m, respectively, in Experiment 1 and 35.3 ± 11.3 mg and 3.0 ± 0.7 m, respectively, in Experiment 2. Seedlings measured 28.3 ± 6.3 cm in Experiment 1 and 17.6 ± 3.5 cm in Experiment 2. Effects of fish on vegetation for both experiments are shown in Table 1. No uprooting was noticed. The lack of significant differences between control-T_1-T_2 plants in Experiment 2 may be explained by a high within-treatment variation. Herbivory was evidenced in pebbly conditions by the appearance of green faecal pellets at the bottom of all fish aquaria.

Periphyton

Experiment 1

DW, AW and AFDW increased from the beginning to the end of the experiment (Figure 1 A–C). The three variables showed a significant interaction between treatments and time. No significant differences in estimated growth rates were detected; however, higher values were observed in the fish treatments (0.011–0.012 d^{-1}) than in the control (0.005 d^{-1}). Chl. *a* and Chl. *b* concentration increased from the beginning to the end of the experiment in the fish treatments (Figure 1 D, E). The interaction between treatments and time was significant for both type of chlorophyll. The chlorophyll content in the organic matter (μg Chl. $(a + b)$ (mg AFDW)$^{-1}$), averaged over the whole period (± s.d.), was greater in the fish treatments (T_1: 12.9 ± 3.6; T_2: 16.7 ± 4.0; T_3: 12.9 ± 4.5) than in the control (9.1 ± 4.1). Light attenuance increased through the experiment (Figure 1F), with final values 20–21% in T_1 and T_2, and 30% in T_3, higher than the control. The treatment × time interaction was not significant. Rectangular hyperbolae fitted to the attenuance-AFDW datasets were significant only for the fish treatments, with high values of R^2: 0.74 for T_1, 0.92 for T_2, and 0.84 for T_3. Curves were significantly different, ascending faster in direct relation to the fish size (Figure 2).

Experiment 2

DW, AW and AFDW showed growing values as fish size increased (Figure 1G–I). Only in AFDW, a significant treatment × time interaction effect was detected. Instantaneous growth rates did not differ statistically between treatments; however, values tended to be higher according to fish size (T_1: 0.003; T_2: 0.006; T_3: 0.012 d^{-1}). The three fish treatments showed values higher than the control for both types of chlorophyll (Figure 1J, K). In Chl. *a*, a significant treatment × time interaction was detected. The chlorophyll content in the organic matter was greater than in the previous experiment, with higher values in the fish treatments (T_1: 28.7 ± 10.7; T_2: 24.1 ± 5.5; T_3: 27.5 ± 8.7 μg Chl. $(a + b)$ [mg AFDW]$^{-1}$) than in the control (17.9 ± 4.0 μg Chl. $(a + b)$ [mg AFDW]$^{-1}$). Attenuance was of the same order of magnitude than in Experiment 1 (Figure 1L). Final values were 12% in T_1 and T_2, and 22% in T_3, higher than control attenuance. The interaction between treatments and time was non significant. Attenuance-AFDW data showed high dispersion, so the adjustment of hyperbolic curves was not possible.

Discussion

Vegetation was severely affected by all fish treatments in both experiments. In Experiment 1, the main negative effect caused by small carp was probably shade stress, due to the typical growth response of *P. pectinatus* plants to reduced light conditions (Van Wijk et al., 1988). This would have resulted from the development of a dense periphytic cover and the high values of water turbidity recorded (up to 94 NTU; Sidorkewicj et al., 1999a). In Experiment 2, although some plant growth occurred, fish appeared to cause plant damage by combined collisions, as indicated by fallen leaves, and growth of epiphytic algae, which resulted in a light attenuance increase of 12%. Medium and large fish

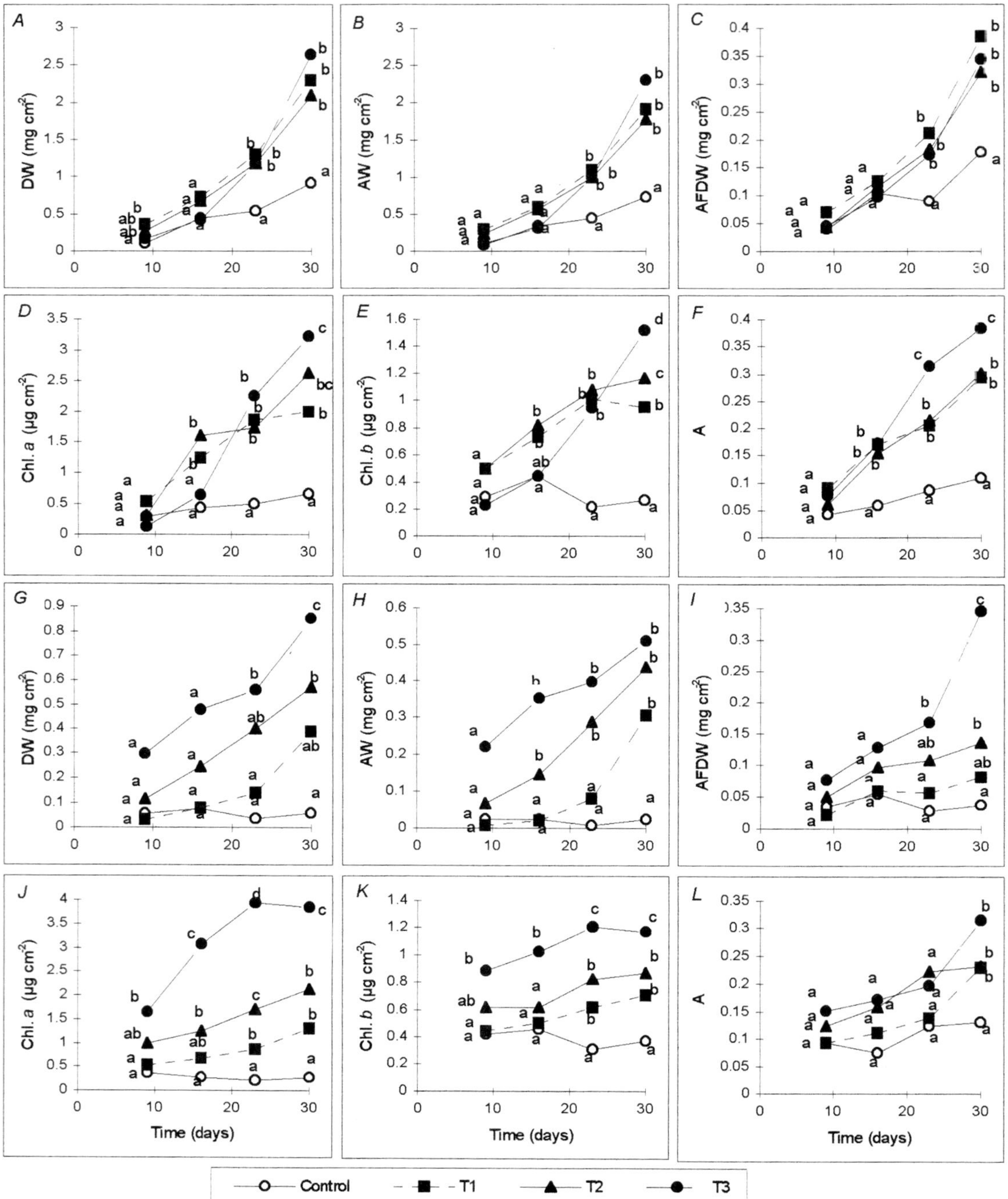

Figure 1. Mean values of periphyton biomass (dry weight: DW; ash weight: AW; ash-free dry weight: AFDW), chlorophyll *a* and *b* concentration (Chl. *a*, Chl. *b*) and light attenuance (A) in Experiment 1 (A–F) and Experiment 2 (G–L). T_1: smaller-size fish treatment; T_2: medium-size fish; T_3: largest-size fish. Different letters (a-d) indicate significant differences between treatments, at each sampling date, with SNK test ($p \leq 0.05$).

Table 1. Effects of control and fish (T_1: smaller-size; T_2: medium-size; T_3: larger-size) treatments on vegetation in Experiment 1 and Experiment 2. Plus and minus signs represent parameters increase and decrease, respectively, throughout 30 days. Different **letters (a–c)** indicate significant differences between treatments with LSD test ($p \leq 0.05$)

		Control	T_1	T_2	T_3
Experiment 1:	Plant length:	+ 116% **a**	–11% **b**	–95% **c**	–95% **c**
		(healthy; luxuriant growth; uniformly branched)	(upper canopy; basal chlorosis and necrosis)	(several plants disappeared including the root system; remaining plants: bare and browsed shoots)	
Plant dry weight:		+ 234% **a**	–5% **b**	–57% **c**	–57% **c**
Seedling height:		+ 47% **a**	–18% **b**	—	
		(healthy; uniformly branched)	(browsed)	(disappeared)	
Experiment 2:	Plant length:	+ 247% **a**	+ 160% **a**	+ 170% **a**	–13% **b**
		(healthy)	(not browsed; leaves fallen at the bottom of the aquaria)	(browsed; few fallen leaves)	(strongly browsed; absence of fallen leaves)
Plant dry weight:		+ 346% **a**	+ 166% **a**	+ 203% **a**	–76% **b**
Seedling height:		+ 44% **a**	–29% **b**	–25% **b**	—
		(healthy)	(strongly browsed)	(strongly browsed)	(disappeared)

affected plant growth mainly by herbivory in both experiments. Seedlings were affected by herbivory in all cases, indicating that carp prefer soft vegetal tissues.

In Experiment 1, the predominant fraction in the periphytic strata was from inorganic origin, as was expected from the high values of inorganic tripton concentration in the three treatments (up to 178 mg l^{-1}; Sidorkewicj et al., 1999a). A stimulus to the development of periphytic algae by fish was also observed, in spite of the high water turbidity recorded (maximum values: 94, 254 and 329 NTU in T_1, T_2 and T_3, respectively; Sidorkewicj et al., 1999a). Contrarily, Robertson et al. (1997) reported for two billabongs in Australia that carp had a negative effect on the development of the autotrophic components of the periphyton, under similar water turbidities to those recorded here. In the present study, the increased development of epiphytic algae in fish aquaria seems to be related to a top-down effect through the consumption by the fish of *L. parchappii* (a molluscan species which selectively feeds on periphyton; Cazzaniga, 1981), and maybe other benthic organisms. Probably, an additional bottom-up effect through the enhancement of dissolved nutrient concentration by fish excretion and sediment bioturbation (Sidorkewicj et al., 1999a) was achieved. In Experiment 2, on the contrary, only a bottom-up effect through fish excretion would be responsible for the observed increase in the density of attached algae. The higher chlorophyll content in the organic matter, by comparison with the other experiment, indicates a more suitable environment in terms of light availability. In both experiments, the increase in the final chlorophyll concentration with fish size, was probably due to a higher excretion as a result of increased consumption of vegetal matter (Table 1). In addition, in Experiment 1, a higher excretion could have occurred as a consequence of an increase in the consumption of zooplankton (Sidorkewicj et al., 1999a) and zoobenthos. The body size of fish may determine not only the quantities but also the proportions of food items consumed (Lammens & Hoogenboezem, 1991), since what they are able to eat is closely related to the diameter of the chewing cavity (9% SL; Sibbing, 1988, 1991). In both experiments, a significant increase in phytoplankton biomass was also observed in all fish treatments (Sidorkewicj et al., 1999a), and this phenomenon seems not to have had negative effects on periphyton development through competition for nutrients or allelopathy.

The periphytic growth observed in the present study was exponential, contrary to the logistic fit reported elsewhere (see Rodriguez, 1987). According to the cited author, if the underlying process is logistic-like, using an exponential model will produce underes-

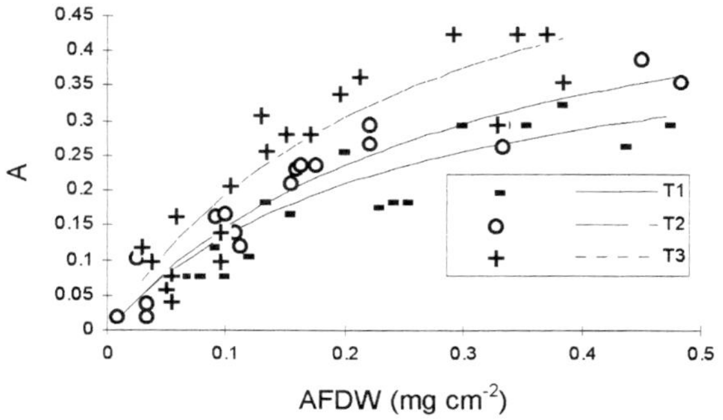

Figure 2. Attenuance-biomass datasets and fitted rectangular hyperbolae obtained in Experiment 1. Treatment abbreviations as in Figure 1.

timation of the specific growth rate. However, the error would be small if only the early phase, when growth is almost exponential, is used in the estimation. The lack of differences between growth rates agree with Vermaat & Hootsmans (1994b) who reported absence of differences between growth rates corresponding to different levels of irradiation. In the present study, differences in underwater light availability were caused by increasing turbidity values as fish size increased (Sidorkewicj et al., 1999a).

Epiphyte shading has been reported by Sand-Jensen (1977) and Phillips et al. (1978) as a main cause in the decline of submerged vegetation. The light attenuation recorded in the present study was high in both experiments (12–30%), in spite of the differences in biomass amounts reached in both cases. Differences between the attenuance-biomass curves in Experiment 1 indicated an efficient acclimation of the periphytic communities to different light conditions, in relation with the increasing turbidity levels of the three treatments. This means that, for equivalent biomass, the periphytic communities cause higher light attenuance as underwater light extinction increase, which is relevant for submerged macrophytes growing in turbid waters, typical of carp habitats. Similar results were found by Vermaat & Hootsmans (1994b) in different levels of irradiance.

The results obtained can help to acknowledge the ecological consequences of the recent introduction of *C. carpio* in the irrigation system of the Valle Inferior del Río Colorado (VIRC, Argentina), where negative effects of this fish upon submerged vegetation have been found (Fernández et al., 1998; Sidorkewicj, 1998; Sidorkewicj et al., 1998, 1999b). In this system, Drovandi (1993) and Soto Perez (1994) found that periphyton is an agent involved in the light attenuation experienced by submerged macrophytes, but they did not establish a correlation between carp populations and epiphytic development. In the present study, a high percentage of plant destruction was due to carp herbivory. However, previous studies performed in the VIRC (Sidorkewicj, 1998; Sidorkewicj et al., 1998, 1999b) revealed that, although a certain plant consumption by fish existed, other processes such as increases in abiogenic and biogenic turbidity were involved in plant damage. In such situations, the increased light attenuation caused by periphyton enhancement could be crucial, adding a new stress factor to vegetation.

Acknowledgements

The authors thank the Commission of the European Communities for financial support (STD3 Programme Contract No TS3*-CT92-0125), and CONICET (Argentina) for a scholarship given to the first author. Suggestions of M.R. Sabbatini, N. Winzer and R. Camina are also appreciated.

References

Bagenal, T. B. & F. W. Tesch, 1978. Age and growth. In Bagenal, T. (ed.), Methods for Assessment of Fish Production in Fresh Waters. IBP Handbook No 3, 3rd edn. Blackwell Scientific Publications, Oxford: 101–136.

Brumley, A. R., 1991. Cyprinids of Australasia. In Winfield, I. J. & J. S. Nelson (eds), Cyprinid Fishes, Systematics, Biology and Exploitation. Fish and Fisheries Series 3. Chapman & Hall, London: 264–283.

Cazzaniga, N. J., 1981. Caracterización química y faunística de canales de drenaje del Valle Inferior del Río Colorado. Ecosur 8: 25–46.

Crivelli, A. J., 1983. The destruction of aquatic vegetation by carp. A comparison between Southern France and the United States. Hydrobiologia 106: 37–41.

Drovandi, A. A., 1993. Aquatic weed problems in two irrigation schemes: effect of light climate change on the growth and development of *Potamogeton pectinatus* L. in the south of Argentina. M.Sc. thesis. IHE, Delft, The Netherlands: 71 pp.

Fernández, O. A., K. J. Murphy, A. López Cazorla, M. R. Sabbatini, M.A. Lazzari, J.C.J. Domaniewski & J. H. Irigoyen, 1998. Interrelationships of fish and channel environmental conditions with aquatic macrophytes in an Argentine irrigation system. Hydrobiologia 380: 15–25.

Fletcher, A. R., A. K. Morison & D. J. Hume, 1985. Effects of carp, *Cyprinus carpio* L., on communities of aquatic vegetation and turbidity of waterbodies in the Lower Goulburn River Basin. Aust. J. mar. Freshwat. Res. 36: 311–327.

King, D. R. & G. S. Hunt, 1967. Effect of carp on vegetation in a Lake Erie Marsh. J. Wildlife Mgmt 31: 181–188.

Lammens, E. H. R. R. & W. Hoogenboezem, 1991. Diets and feeding behaviour. In Winfield, I. J. & J. S. Nelson (eds), Cyprinid Fishes, Systematics, Biology and Exploitation. Fish and Fisheries Series 3. Chapman & Hall, London: 353–376.

Lowe, R. L., 1996. Periphyton patterns in lakes. In Stevenson, R. J., M. L. Bothwell & R. L. Lowe (eds), Algal Ecology, Aquatic Ecology Series. Academic Press, San Diego, California: 57–76.

McCrimmon, H. R., 1968. Carp in Canada. Bull. Fish. Res. Bd Can. 165: 93 pp.

Otto, C., 1982. Adaptations to benthic freshwater herbivory. In Wetzel, R. G. (ed.), Periphyton of Freshwater Ecosystems. Dr W. Junk Publishers, The Hague: 199–205.

Phillips, G. L., D. F. Eminson & B. Moss, 1978. A mechanism to account for macrophyte decline in progressively eutrophicated waters. Aquat. Bot. 4: 103–126.

Roberts, J., A. Chick, L. Oswald & P. Thompson, 1995. Effect of carp, *Cyprinus carpio* L., an exotic benthivorous fish, on aquatic plants and water quality in experimental ponds. Aust. J. mar. Freshwat. Res. 46: 1171–1180.

Robertson, A. I., M. R. Healey & A. J. King, 1997. Experimental manipulations of the biomass of introduced carp (*Cyprinus carpio*) in billabongs. II. Impacts on benthic properties and processes. Mar. Freshwat. Res. 48: 445–454.

Rodriguez, M. A., 1987. Estimating periphyton growth parameters using simple methods. Limnol. Oceanogr. 32: 458–464.

Sand-Jensen, K., 1977. Effects of epiphytes on eelgrass photosynthesis. Aquat. Bot. 3: 55–63.

Sibbing, F. A., 1988. Specializations and limitations in the utilization of food resources by the carp, *Cyprinus carpio*: a study of oral food processing. Envir. Biol. Fishes 22: 161–178

Sibbing, F. A., 1991. Food capture and oral processing. In Winfield, I. J. & J. S. Nelson (eds), Cyprinid Fishes, Systematics, Biology and Exploitation. Fish and Fisheries Series 3. Chapman & Hall, London: 377–412.

Sidorkewicj, N. S., 1998. Efecto de *Cyprinus carpio* L. (Pisces: Cyprinidae) sobre la vegetación sumergida y la calidad del ambiente acuático. Tesis Doctoral. Univ. Nac. del Sur, Argentina: 268 pp.

Sidorkewicj, N. S., A. C. López Cazorla & O. A. Fernández, 1996. The interaction between *Cyprinus carpio* L. and *Potamogeton pectinatus* L. under aquarium conditions. Hydrobiologia 340: 271–275.

Sidorkewicj, N. S., H. Freije & A. C. López Cazorla, 1999a. Effects of young common carp (*Cyprinus carpio* L.) on seston dynamics in experimental culture. Ecol., Envir. Conserv. 5 (in press).

Sidorkewicj, N. S., M. R. Sabbatini, A. C. López Cazorla & O. A. Fernández, 1999b. Efecto de la carpa común sobre la vegetación sumergida y la calidad del agua. Actas XIII Congreso Latinoamericano de Malezas (in press).

Sidorkewicj, N. S., A. C. López Cazorla, K. J. Murphy, M. R. Sabbatini, O. A. Fernández & J. C. J. Domaniewski, 1998. Interaction of common carp with aquatic weeds in Argentine drainage chanels. J. Aquat. Plant Mgmt 36: 5–10.

Soto Perez, N., 1994. Periphyton and seston dynamics and their implications for light climate and photosynthesis by *Potamogeton pectinatus* L. in an argentinan irrigation scheme. M.Sc. thesis. IHE, Delft, The Netherlands: 57 pp.

Steel, R. G. D. & J. H. Torrie, 1980. Principles and Procedures of Statistics: a Biometrical Approach, 2nd edn. McGraw-Hill, U.S.A.: 633 pp.

Tennant, D., 1975. A test of a modified line intersect method of estimating root length. J. Ecol. 63: 995–1001.

Van Wijk, R. J., E. M. J. Van Goor & J. A. C. Verkley, 1988. Ecological studies on *Potamogeton pectinatus* L. II. Autoecological characteristics, with emphasis on salt tolerance, intraspecific variation and isoenzyme patterns. Aquat. Bot. 32: 239–260.

Vermaat, J. E. & M. J. M. Hootsmans, 1994a. Intraspecific variation in *Potamogeton pectinatus* L.: a controlled laboratory experiment. In Van Vierssen, W., M. Hootsmans & J. Vermaat (eds), Lake Veluwe, a Macrophyte-dominated System under Eutrophication Stress. Kluwer Academic Publishers, Dordrecht: 26–39.

Vermaat, J. E. & M. J. M. Hootsmans, 1994b. Periphyton dynamics in a temperature-light-gradient. In Van Vierssen, W., M. Hootsmans & J. Vermaat (eds), Lake Veluwe, a Macrophyte-dominated System under Eutrophication Stress. Kluwer Academic Publishers, Dordrecht: 193–212.

Wintermans, J. F. G. M. & A. De Mots, 1965. Spectrophotometric characteristics of chlorophylls a and b and their phaeophytins in ethanol. Biochim. Biophys. Acta 109: 448–453.

Hydrobiologia **415:** 21–27, 1999.
J. M. Caffrey, P. R. F. Barrett, M. T. Ferreira, I.S. Moreira, K. J. Murphy & P. M. Wade (eds),
Biology, Ecology and Management of Aquatic Plants.
© 1999 Kluwer Academic Publishers. Printed in the Netherlands.

Survival strategy of *Pistia stratiotes* L. in the Djoudj National Park in Senegal

Nick G. den Hollander[1,2], Ingrid W. Schenk[1], Sara Diouf[3], Martin J. Kropff[2] & Arnold H. Pieterse[1]
[1]*Royal Tropical Institute, Department of Agriculture and Enterprise Development, Mauritskade 63, 1092 Amsterdam, The Netherlands*
[2]*Wageningen Agricultural University, Department of Theoretical Production Ecology, Bornsesteeg 47, 6708 PD Wageningen, The Netherlands*
[3]*Ministère de l'Environnement, Direction des Parcs Nationaux Senegal, Dakar-Fann, Senegal*

Key words: Pistia stratiotes, seed formation, flowering, habitat factors

Abstract

The water level in the Senegal River is now regulated due to the construction of two high dams (1986 and 1988). In consequence, large parts of the floodplains have permanently dried up and the brackish water in the lower part of the river has become fresh. The Djoudj National Park, situated approximately 60 km upstream from Saint Louis in Senegal in the former floodplain, is artificially flooded when inlets from the river are opened during the rainy season. This change in ecological conditions has led to an annually recurrent, explosive development of the floating aquatic weed *Pistia stratiotes* L. in the Park during the first half of the dry season. A rapid increase in the salt content of the water in the Park, in the course of the dry season (due to the high salt content of the soils and evaporation), eventually causes massive die back of *Pistia*. Rapid seed production by *Pistia* permits it to survive these unfavourable conditions, i.e. high salinity coupled with drought during the second half of the dry season. This applies in particular to an area where the water is fertilised by excreta from a pelican colony. Flowering of *Pistia* appears to be markedly enhanced when vegetative reproduction is inhibited by crowding.

Introduction

Following the construction of two high dams on the Senegal River (the Diama Dam near the mouth of the river, 25 km north of the city of Saint Louis in Senegal, operational since 1988, and the Manantali Dam in Mali, operational since 1986), marked ecological changes have occurred in the region. The river's water level is regulated and, as a consequence, large parts of the floodplains have dried up permanently. On the other hand, an approximately 30-km-long and 6-km-wide stretch of the river near the Diama dam has become submerged throughout the year, due to dyke construction. In addition, water in the lower part of the river has become fresh as tidal salt water inflow is prevented.

These environmental changes have also affected the Djoudj National Park which, from an ecological point of view, is a valuable wetland area. This park (surface area: 16 000 ha) is located in the former floodplain, about 60 km north east of Saint Louis (Le Parc National des Oiseaux du Djoudj, 1995). It was created in 1971 as a sanctuary for the 366 different species of African and European birds using the area as a breeding and wintering site. The Park has been cut off from the river by embankments constructed in connection with the Diama dam. To imitate ecological conditions which prevailed before the construction of the Diama dam, the Park is flooded via inlets which are opened during the wet season and closed at the beginning of the dry season, in October. In the course of the dry season the water level in the Park drops, due to evaporation. Eventually, most waterbodies in the Park dry out completely until the rains start in June and the inlets are opened again. Due to the salty soils there is an increase in salinity of the water after flooding.

Salinity further increases during the dry season, due to evaporation.

As a consequence of these ecological changes, conditions in the Park have become favourable for an explosive development of *Pistia stratiotes* L. (water lettuce), a free-floating aquatic weed which occurs throughout tropical and sub-tropical regions (Holm et al., 1970; Cook et al., 1974). This applies in particular to an area in the Djoudj River, the Park's main waterway, where water is fertilised by excreta from a pelican colony. In this stretch of the river, dense *Pistia* mats are formed during the first months of the dry season and seriously disturb the ecosystem.

Pistia reproduces vegetatively by forming stoloniferous off-shoots and generatively by producing seeds. In areas where conditions for *Pistia* growth are unfavourable during certain parts of the year (in the Djoudj Park, increasing salinity and drought in the course of the dry season), sexual reproduction is essential for its survival (Harley, 1990). The seeds will sink to the bottom sediment, permitting survival in the regenerative stage in the many areas of the Park which dry out completely. Germination occurs after post-rainfall flooding.

In this study the annual advance and retreat of *Pistia* in the Djoudj River has been monitored. Special attention was paid to the relationship between its development and seed formation in relation to increasing salinity of the water during the dry season. Since preliminary observations seemed to indicate that *Pistia* plants in areas with high plant density flowered more profusely, tests were undertaken to assess whether crowded conditions accelerate flowering in *Pistia*.

Methods

Research was conducted in the first six months of the dry season (from September 1997 to March 1998). The main study area was the Djoudj River, which, when the inlets are opened, forms the main connection between the Senegal River and the salty 'Grand Lac' (see Figure 1). During previous seasons, high densities of *Pistia* had been found in this area.

Conductivity of the water (μS/cm), as an indicator of salinity, was measured by means of an ATI Orion 130 portable conductivity meter at 18 different sites in the Djoudj River.

Four sites, Khoyoje, Colony, Tiquel and Petit Djoudj, were selected for more frequent measurements. The density of the *Pistia* vegetation at these sites was recorded by estimating the percentage of water surface covered by *Pistia* plants. The four sites were selected on the basis of preliminary measurements, indicating marked ecological differences. Petit Djoudj was the most eastern site in the Djoudj River where *Pistia* plants occurred. At the beginning of the dry season conductivity at Khoyoje was relatively low. This also applied to the site near the pelican colony. However, it may be expected that the water near this site is nutrient rich due to pelican excrement. Conductivity at Tiquel and Petit Djoudj was relatively high.

From the last week of January until the second week of March 10 adult *Pistia* plants were collected weekly at 4 different sites in the Djoudj park surrounding the pelican colony. Plant size, number of leaves, juvenile plants, flowers and seed capsules were determined.

Seeds were collected between January and March 1998. Part of these seeds were placed in four 1-L bottles (50 seeds per bottle) and kept at the Biological Station in the Djoudj Park. Two bottles were filled with water from a site near the Djoudj River with a relatively high conductivity (7000 μS/cm), where adult plants had already died. Two other bottles were filled with tap water (conductivity 100 μS/cm). One bottle with water from the site near the Djoudj River and one with tap water were placed in the dark whereas the two remaining bottles were placed in the light. The remaining seeds were kept dry. On September 2 1998 they were placed in 12 petri dishes (5 cm) filled with tap water in the laboratory of the Department of Theoretical Production Ecology of the Wageningen Agricultural University in the Netherlands. Each petri dish contained 50 seeds. Subsequently, these petri dishes were stored in a germination chamber at a temperature of 30 °C (12 hours in the light and 12 hours in the dark). Germination rates were assessed after two weeks and again after four weeks.

To study the effect of crowding, which occurs through vegetative reproduction, 7 large (2 m×2 m) and 7 small (0.50 m×0.50 m) plots were fenced using bamboo poles, wire and mats of dried *Typha* plants. These 14 plots, located alongside the river bank, where water depth varied between 0.80 m and 1.20 m, were established according to a random block design. A large plot was always adjacent to a small plot, giving seven 'plot pairs'. The distance between the various 'plot pairs' differed from 3 to 6 metres. All plant material was removed from the plots. It was not possible for floating *Pistia* seedlings or any other floating plants

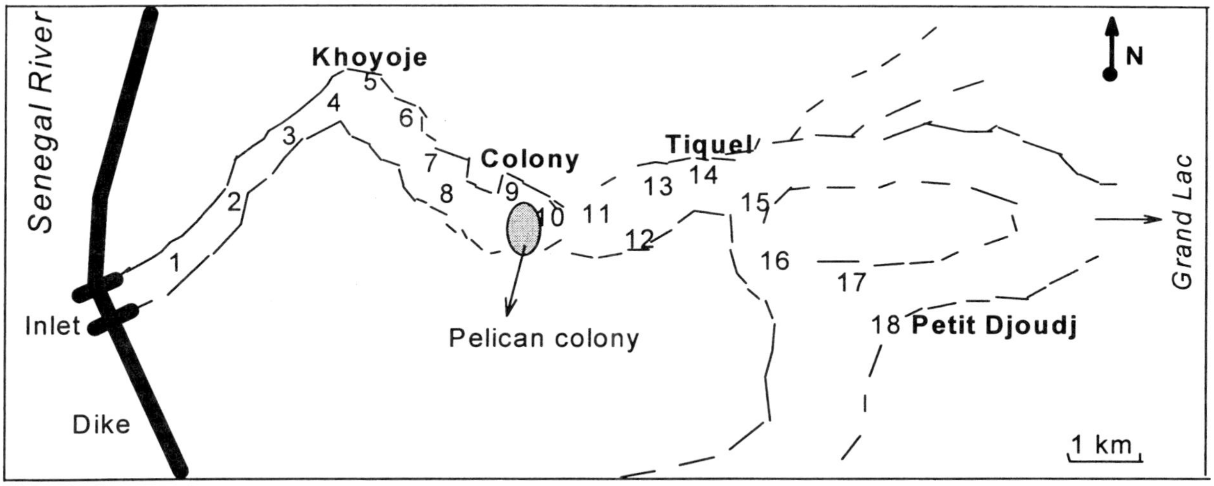

Figure 1. Selected water sampling sites in the Djoudj River, the main waterway in the Djoudj National Park.

to enter the plots. Seedlings rising from the bottom were removed.

One *Pistia* plant was placed in each plot (the mother plant, plant number 1). These plants were all of the same size, approximately the same age and did not bear flowers. Stolons (off-shoots producing daughter plants) had been removed. For 63 days the number of plants and flowers in all plots was determined weekly and each plant was numbered in order of succession. Consequently, it was possible to observe the development of individual plants. The diameter, number of leaves, number of stolons, and number of flowers (and fruits) for each plant were determined. A daughter plant was counted and numbered as soon as fifty per cent of its contour was outside the contour of the mother plant. When more than 50% of a leaf had decayed, it was no longer included in the count.

On day 49, 75% of the plants were removed from the large plots. The remaining 25% were left in a 1 m × 1 m area within the 2 m × 2 m plots. Two plants were placed in the open space in each plot to assess whether this would initiate a change from generative (flowering) to vegetative production (formation of stolons). These plants were selected from the oldest ten plants (numbered one to ten) to guarantee their full development. For all plots log-logistic curves were fitted to observe data using the GENSTAT statistical package (Dayne & Lane, 1987)

Results

Conductivity measurements showed a gradient from low values near the inlet in the western part of the Park to high levels towards 'Grand Lac' in the eastern part (Figure 2). Particularly at the more eastern sites, conductivity increased markedly in the course of the study period (up to 11 000 μS/cm in early February).

Data from the four selected sites are presented in Figure 3. In Khoyoje conductivity remained rather low during the study period and there were hardly any *Pistia* plants. A very large development of *Pistia* was observed near the pelican colony, where conductivity values were higher than at Khoyoje. At Tiquel, where conductivity values had increased rapidly, there was first an early increase in *Pistia*, but subsequently the plants started to die. Conductivity values also increased rapidly at Petit Djoudj to relatively high values but here hardly any *Pistia* plants were observed.

Pistia plants near the pelican colony started to flower before September 22 and the first fruits were found on November 2. The first seeds, trapped in the root system of the plants, were found here in January. No flowers, fruits or seeds were found at Khoyoje, Tiquel and Petit Djoudj. Plants at these sites were much smaller than plants found near the pelican colony.

Plants sampled between January and March 1998 around the pelican colony at four different sites markedly suffered from the increase in the conductivity level. Plant size, number of leaves per plant, number of flowers, and seeds per plant decreased

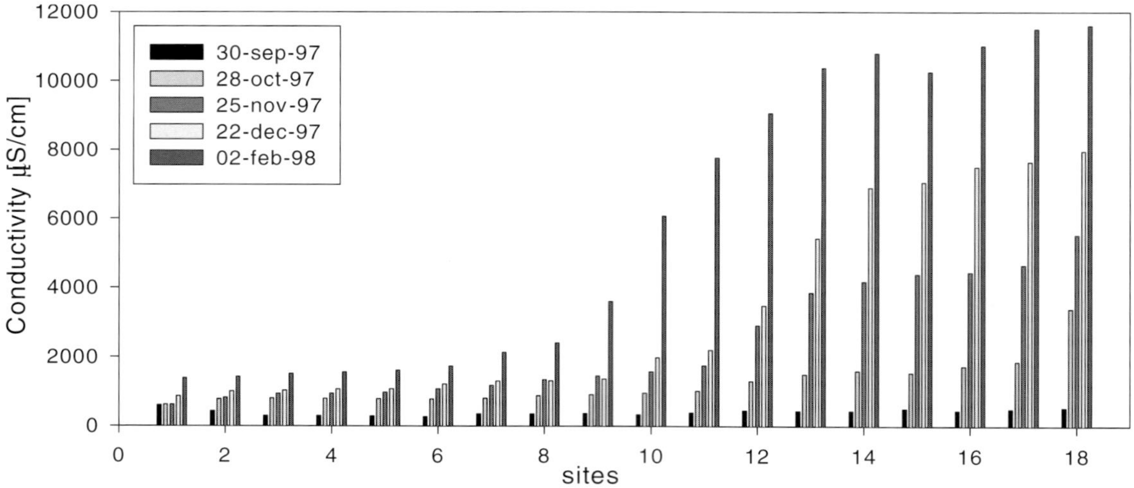

Figure 2. Conductivity of the water at the 18 sites in the Djoudj River at different dates.

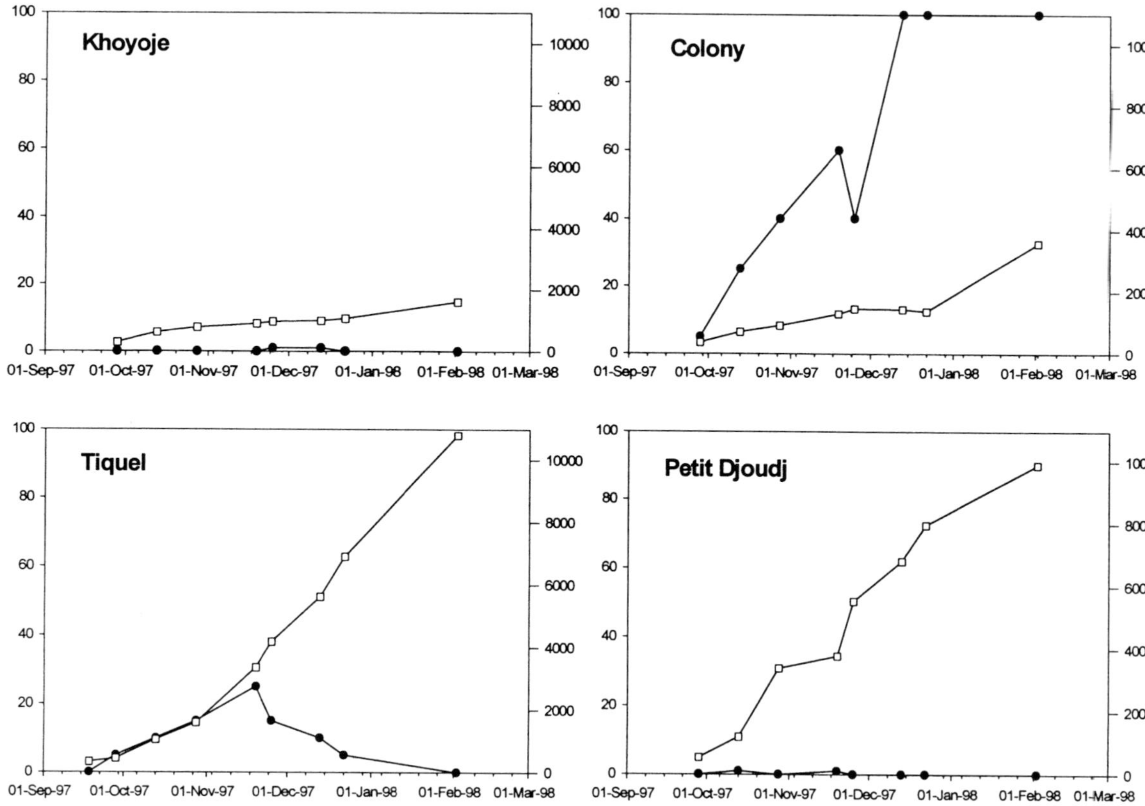

Figure 3. Conductivity of the water [μS/cm] ○ at four selected sites in the Djoudj River, and the percentage of water surface near these sites covered by *Pistia* ● at different dates.

at most of the sampling sites. The diameter of the largest plant, found in January, was 27 cm, whereas in November plants were observed with a diameter of up to 45 cm. Seed production continued until the plants died but salt stress did not seem to induce additional flower production. A slight increase in ripe seed capsules and capsules already burst open was observed.

Pistia seeds placed in a bottle containing water from the site near the Djoudj River (where adult plants had already died due to high salinity) germinated neither in the light nor in the dark. Seeds placed in a bottle with tap water in the light started germinating after approximately one week. After two weeks approximately 25% of the seeds had germinated. Seeds in the bottle with tap water, kept in the dark, did not germinate. After a few weeks the bottle with water from the site near the Djoudj River, which was kept in the light, toppled over causing water stirring. Subsequently, after about one week, a small number of seeds started to germinate in this bottle as well. The seedlings remained alive for over 6 weeks but did not increase in size.

Seeds which had been kept dry in the laboratory for more than 5 months showed a germination percentage of approximately 60% after two weeks exposure to moist conditions and approximately 90% after four weeks.

Results of the experiments in the fenced plots are presented in Figure 4. The small plots were entirely covered by plants around day 20. Although some flowers had been observed before day 20, the plants started flowering profusely from this day onwards and the number of juveniles per plant decreased until about one juvenile per adult plant. The number of plants further increased to almost 20 plants on day 50. The large plots were entirely covered by *Pistia* plants around day 40. Subsequently, the number of plants increased exponentially up to at least day 60 when the experiment was terminated. The number of juvenile plants per adult plant decreased as the number of plants per plot increased. Profuse flowering in the large plots occurred later than in the small plots, approximately 50 days after the commencement of the experiment.

Plants which had been placed in the open space following the removal of 75% of the plants continued producing flowers until day 60 when the experiment was terminated, but also started to produce slightly more off-shoots.

Discussion

Pistia, which has a relatively low salt tolerance (Haller, 1974), started to die at Tiquel when conductivity values exceeded the plants' tolerance level (about 3500–4000 μS/cm). Plants at Tiquel were much smaller than those found near the pelican colony, probably due to a lower level of nutrients in the water and/or salinity stress. As plants at Tiquel did not produce seeds (they had not flowered) before dying, it may be assumed that *Pistia* plants found here at the beginning of the dry season did not germinate from a seed bank at this site. Seeds could have been brought here via the waterflow, when the inlets from the river were opened in the wet season, or by animals (pelicans use *Pistia* as nesting material).

Plants near the pelican colony flowered and formed seeds before the second half of the dry season, when conductivity values at this site also reached a level which brought about a negative effect on their growth. Eventually, at the end of the dry season, these *Pistia* plants died (den Hollander, 1996, pers. obs.). Guiral (1994) measured high P (1.05 mol/l) and N (3.55 mol/l) values near the pelican colony, whereas close to Khoyoje this author measured 0.32 mol/l P and approximately zero mol/l N. These preliminary results suggest that a high nutrient content of the water near the pelican colony, and the relatively low conductivity values during the first half of the dry season, make it possible for *Pistia* plants to win the race between seed production and the adult plant death due to high salinity levels. As at the beginning of the dry season many *Pistia* seedlings were found at this site, *Pistia* seeds having apparently risen up from the bottom sediment.

At Khoyoje nutrients may constitute a limiting factor for *Pistia* development. At Tiquel and Petit Djoudj the high salinity level was probably responsible for the decline in the *Pistia* population. As at Petit Djoudj very few plants were observed, perhaps also a low level of nutrients (this site is located at a greater distance from the pelican colony) prevents early *Pistia* development. However, an absence of seeds may also be responsible for the low number of *Pistia* plants at Petit Djoudj, as well as at Khoyoje.

Based on the effect of stirring on germination in the *in vitro* experiment it is conceivable that lack of oxygen in the water from the site near the Djoudj River, at the end of the dry season, prevents the newly produced seeds from germinating after sinking to the bottom. Datta & Biswas (1970) also indicated that a low oxygen concentration may inhibit *Pistia* seed germination.

It is evident that light is an important factor in connection with the germination of *Pistia* seeds as there was no germination in the bottles kept in the dark. Pieterse et al. (1981) also reported that *Pistia* seeds do not germinate in the dark and Harley (1990) observed

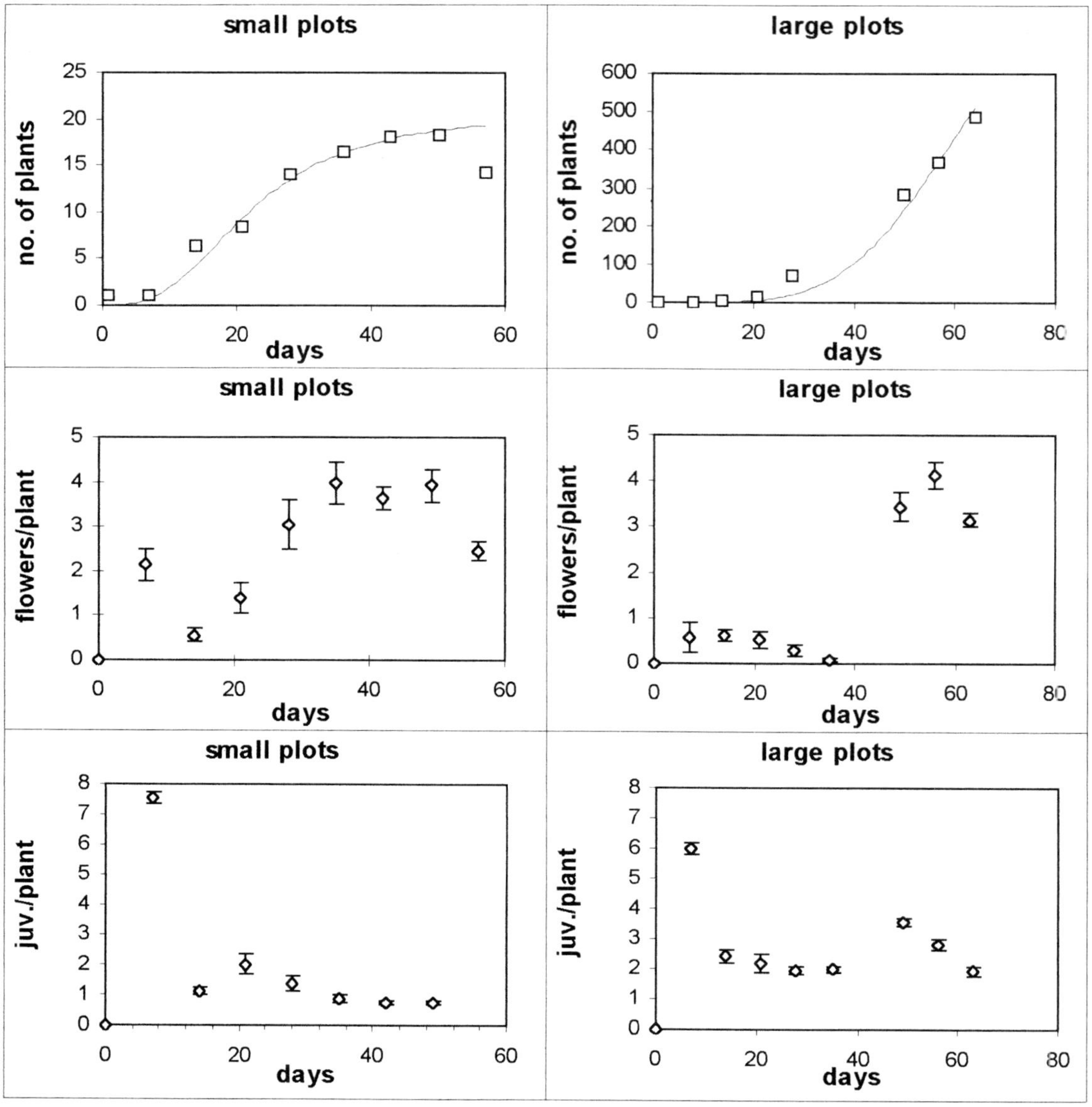

Figure 4. Growth curves, number of flowers per plant and number of juvenile plants per plant in the small and in the large plots. Recording started on September 29, 1997.

that germination took longer and that the germination percentage was lower in the dark compared to the light. The *in vitro* experiments have also demonstrated that *Pistia* seeds are able to withstand a dry period of over 5 months without losing their viability, which implies that drying out of the waterbodies in the Djoudj Park is not a limiting factor for their survival.

We consider that ecological conditions affecting *Pistia* survival and reproduction in the Djoudj Park are somewhat unusual for this species. Although the species is able to multiply vegetatively via the formation of daughter plants (Pieterse, 1985), seed formation seems the only way for *Pistia* to survive under these specific circumstances of the Djoudj Park.

In the experimental plots, flowering started when they were entirely overgrown. Moreover, the number of juvenile plants per adult plant decreased when flowering started. This indicates that a shift from vegetative reproduction to generative reproduction is induced by crowding. As plants transferred to an open space continued to flower, at least until day 60 when the experiment was terminated, but also produced slightly more off-shoots, it seems that this shift is reversible.

It seems likely that the high nutrient content of the water near the pelican colony leads to a combination of two factors which result in massive seed production, a rapid growth and crowding.

It still remains to be clarified how crowding brings about an effect on flowering. It could be connected with changes in the hormone balance. In this context it should be noted that Pieterse (1982, 1985) reported induction of flowering in *Pistia* by certain plant growth regulators, including gibberellic acid, salicylic acid and the chelating agent EDDHA (ethylene-di-*o*-hydroxyphenylacetic acid). Further work is required to investigate this issue.

Acknowledgements

This research was financed within the framework of the EU (DG XII INCO/DC) project no. IC18-CT96-0080 and the Rumphius Foundation. Thanks are due to Marthe Mbengue, Charles Diémé, Indega Bindia and Mandiaye Sall for their assistance. Professor Dr Georg Janauer of the University of Vienna in Austria and Mr. Seppo Hellsten of VTT in Oulu, Finland, partners within the framework of the EU project no. IC18-CT96-0080, are acknowledged for their advice.

References

Cook, C. D. K., B. J. Gut, E. M. Rix, J. Schneller & M. Seitz, 1974. Water Plants of the World. A Manual for the Identification of the Genera of Freshwater Macrophytes. Dr W. Junk Publishers, The Hague.

Datta, S. C. & K. K. Biswas, 1970. Germination pattern and seedling morphology of *Pistia stratiotes*. L. Phyton 27: 157–161.

Dayne, R. W & P. W. Lane (eds), 1987. Genstat 5 Release 3 Reference Manual, Clarendon Press, Oxford, England.

Guiral, D., 1994. Situation, étude et contrôle des végétations aquatiques dans le parc national du Djoudj. ORSTOM Centre de Montpellier, France.

Haller, W. T., 1974. Effects of salinity on growth of several aquatic macrophytes. Ecology 55: 891–894.

Harley, K. L. S., 1990. Production of viable seeds by water lettuce, *Pistia stratiotes* L., in Australia. Aquat. Bot. 36: 277–279.

Holm, L. G., L. W. Weldon & R. D. Blackburn, 1970. Aquatic weeds. Pest Articles & News Summaries 16: 576–89.

Le Parc National des Oiseaux du Djoudj, 1995. Ziegler Beckmann GmbH., Cologne.

Pieterse, A. H., 1982. A review of chemically induced flowering in *Lemna gibba* G3 and *Pistia stratiotes* L. Aquat. Bot. 13: 21–28.

Pieterse, A. H., 1985. *Pistia stratiotes*. In: Halevy, A.H. (ed.), IV Handbook of Flowering. CRC Press, Boca Raton: 94–96.

Pieterse, A. H., L. de Lange & L. Verhagen, 1981. A study on certain aspects of seed germination and growth of *Pistia stratiotes* L. Acta Bot. Neerl. 30: 47–57.

Hydrobiologia **415:** 29–33, 1999.
J. M. Caffrey, P. R. F. Barrett, M. T. Ferreira, I.S. Moreira, K. J. Murphy & P. M. Wade (eds), Biology, Ecology and Management of Aquatic Plants.
© 1999 *Kluwer Academic Publishers. Printed in the Netherlands.*

Comparative ecology of *Monochoria korsakowii* and *M. vaginalis*

Hidenobu Kunii[1] & Kazuhiro Okibe[2]
[1]*Research Center for Coastal Lagoon Environments, Shimane University, Matsue 690-8504, Japan*
[2]*Graduate School of Science, Shimane University, Matsue 690-8504, Japan*

Key words: aquatic weed, conservation, life strategy, rice field, transpiration, vulnerable species

Abstract

Two native annual species, *Monochoria korsakowii* Regel et Maack and *M. vaginalis* (Burm. fil.) Kunth, occur in Japan. Although they show similar habitat preferences, these plants are now in contrasting situations in Japan; while *M. vaginalis* is a noxious weed, *M. korsakowii*, a once common weed of rice fields, has diminished its distributional range and is now listed as a vulnerable species. For the purpose of conservation and management of *M. korsakowii*, the life history traits of these species were investigated with particular emphasis on the seasonal growth cycles, patterns of dry matter distribution and photosynthetic and transpiration rates. We found significant differences in phenology and seasonal growth between *M. korsakowii* and *M. vaginalis*. *M. korsakowii* germinated about a month earlier and flowered one and a half months later than *M. vaginalis*. While shoot to root dry weight ratio in *M. korsakowii* was larger than 1.0, that in *M. vaginalis* was smaller than 1.0 during growth. Differences were also found in leaf life span, numbers of leaves and flowers per individual plant and transpiration rate. These findings support the belief that the decline of *M. korsakowii* relates, to some extent, to the recent changes in agricultural practices in Japan.

Introduction

Monochoria is a small genus of Pontederiaceae, comprising eight species, usually found at the edges of ponds, rivers and irrigation ditches or rice fields where there is standing water for at least past of the year. In Japan, there occur two native annual species, *Monochoria korsakowii* Regel et Maack and *M. vaginalis* (Burm. fil.) Kunth (Kadono, 1994), and rice fields are considered to be the main habitat for these two species (Wang et al., 1996a). Although they show similar habitat preferences, these plants are now in contrasting situations in Japan. Whilst *M. vaginalis* is a noxious weed, *M. korsakowii*, a once common weed of rice fields, has rapidly diminished its distributional range and is now listed as a vulnerable species. In contrast to *M. vaginalis*, however, little attention has been paid to the ecology of *M. korsakowii* except for some recent studies (Wang et al., 1996a, b, c, 1997).

Substantial changes have occurred in Japanese rice field irrigation practices over the last 30 years (Hasegawa & Tabuchi, 1995). In the new-style system, rice fields are well-drained and occasionally dry up even during the cultivation period. It is known that new agricultural practices have caused a considerable decrease in freshwater fish, amphibians, birds and insects (e.g. Saitoh et al., 1988; Fasola & Ruiz, 1996; Fujioka & Lane, 1997). Marked changes have occurred also in the periods of planting and harvesting of rice in Japan, and Shimizu (1998) reported the impact of changing agricultural practices on the diversity of aquatic weeds. For the purpose of conservation and management of *M. korsakowii*, information about the differences in the phenological and growth traits between these two species is quite important. In this context, we studied the differences in phenology and seasonal growth between *M. korsakowii* and *M. vaginalis* growing under the same environmental conditions.

Materials and methods

Seeds of both species were collected in a previous year from the plants growing in the campus pond, Shimane Prefecture, Japan. For the phenological examination

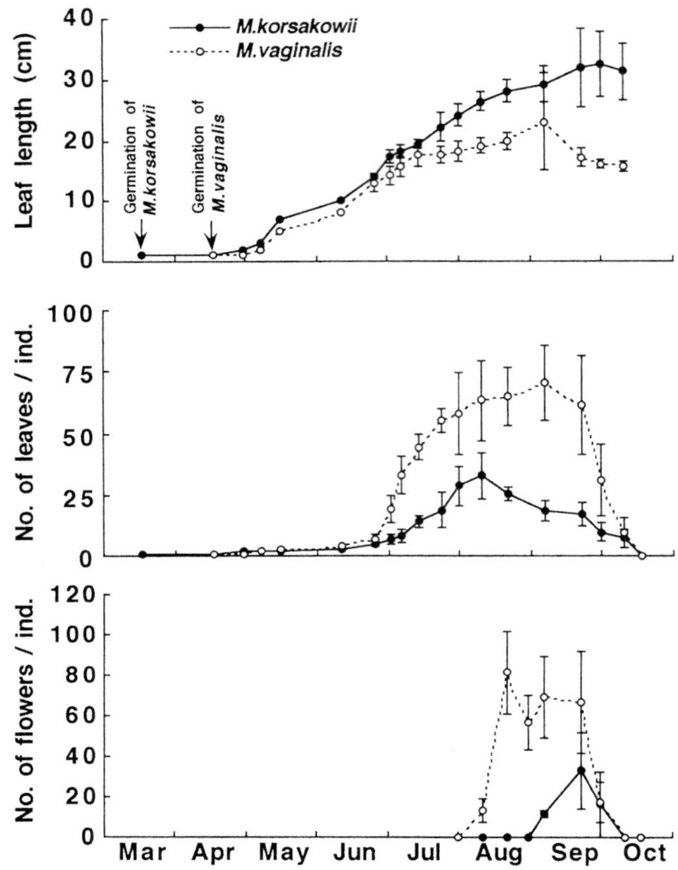

Figure 1. Seasonal changes of leaf length and the numbers of leaves and inflorescences of *M. korsakowii* and *M. vaginalis* grown under outdoor conditions. Each value is shown as the mean of three individuals and vertical bars represent S.D. of the mean.

and determination of seasonal growth, seeds were sown in large containers in early March 1997 and their germination was observed at weekly intervals under outdoor conditions. After they had developed leaves longer than 5 cm, each juvenile plant was transplanted into a plastic pot (25.5 cm in diameter and 24.5 cm in depth), in which the water level was maintained at 1 cm above the soil. For each species, 18 plastic pots were prepared, of which three were used for the successive phenological examination and the remaining 15 for the determination of seasonal growth. For the phenological examination, all of the leaves were marked and all inflorescences and fruits were checked at weekly intervals from early March to October 1997. For the determination of seasonal growth, triplicate samples were harvested at monthly intervals and the samples were separated into each organ type (roots, petioles, dead leaf-blades, living leaf-blades and flowers/fruits) and then weighed after air drying (80 °C for 48 h). Specific leaf areas (SLA) and total lamina area per individual plant were also obtained each month.

Besides the phenological observation mentioned above, photosynthetic and transpiration rates of both species were measured outdoors using a portable CO_2-H_2O gas analyzer (SPB-4, Shimadzu, Japan) in early October 1998. Leaves of each plant were chosen at random and then measured. During the measurements, photon flux density and leaf temperature were almost constant between 1,420 and 1,630 μmol m^{-2} s^{-1} and 33.7 and 35.1 °C, respectively.

Results

Significant differences were found in phenology and seasonal growth traits between *M. korsakowii* and *M. vaginalis* (Figure 1). Germination of *M. korsakowii*

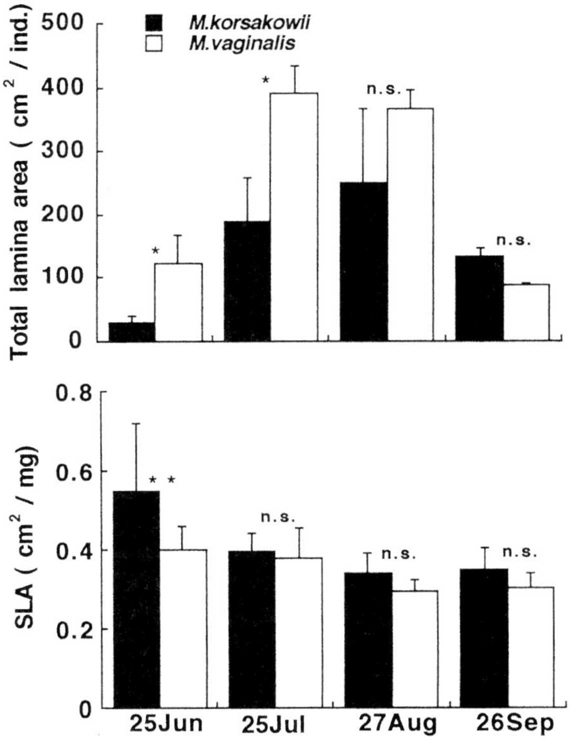

Figure 2. Seasonal changes of total lamina area and SLA in *M. korsakowii* and *M. vaginalis* grown in pots in outdoor conditions. Each value is shown as the mean of three individuals and vertical bars represent S.D. of the mean. Significant differences between *M. korsakowii* and *M.vaginalis* are indicated by asterisks (*t*-test, *, $p < 0.05$; **, $p < 0.01$).

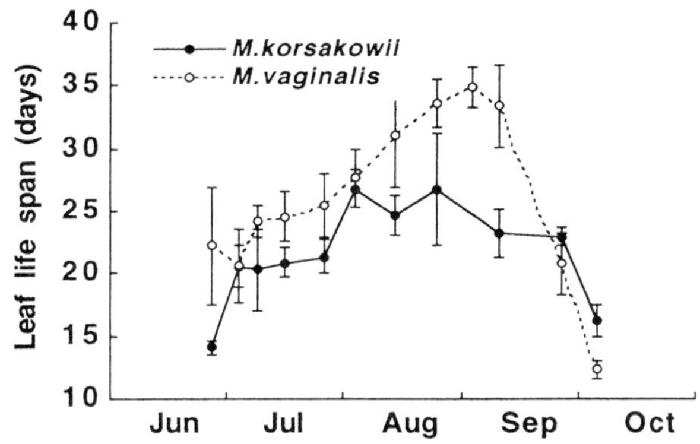

Figure 3. Seasonal changes in leaf life span in *M. korsakowii* and *M. vaginalis* grown in pots in outdoor conditions. Each value is shown as the mean of three individuals and vertical bars represent S.D. of the mean.

Table 1. Comparison of photosynthetic and transpiration rates between *M. korsakowii* and *M. vaginalis*. See text for experimental conditions

	M. korsakowii ($n = 9$)	*M. vaginalis* ($n = 9$)	*p*
Photosynthesis (μ molCO_2m^{-2}s^{-1})	10.78 ± 2.55	9.20 ± 1.11	n.s.
Transpiration (mmolH_2Om^{-2}s^{-1})	5.90 ± 0.18	5.38 ± 0.44	0.002

occurred in mid-March, a month earlier than that of *M. vaginalis*. In contrast, flowering of *M. korsakowii* began in mid-September, about one and a half months later than that of *M. vaginalis*. Since both plants died off in mid-October, the growth period was longer in *M. korsakowii* than in *M. vaginalis*.

During growth, *M. korsakowii* was always taller than *M. vaginalis* and produced a smaller number of large leaves than *M. vaginalis*. Compared to *M. vaginalis*, total lamina area per plant was low in *M. korsakowii* in both June and July but no difference was found in either August or September (Figure 2). There were no significant differences in SLA except for that obtained in June. Annual mean leaf life span of *M. korsakowii* (21.6 days) was significantly shorter ($p<0.05$) than that of *M. vaginalis* (25.9 days) (Figure 3).

At their maximum dry weight, plants of *M. korsakowii* and *M. vaginalis* allocated ca. 40% and ca. 60% of their dry mass to roots, and thus the shoot-to-root dry weight ratio was 1.4 and 0.6, respectively (Figure 4). In addition, it was found that maximum reproductive allocation was much higher in *M. vaginalis* (6.6% on average) than in *M. korsakowii* (1.1% on average). Average number of seeds per plant was estimated to be ca. 8,200 in *M. korsakowii* and ca. 29,700 in *M. vaginalis*.

While there was a significant difference in transpiration rate between *M. korsakowii* and *M. vaginalis*, no significant differences were found in photosynthetic rate under the same environmental conditions (Table 1).

Discussion

In many regions elsewhere in Japan, rice fields are irrigated and farmers plant rice seedlings in May and June, and they harvest rice around September. Because of the selective breeding of rice, the harvest time is becoming earlier. In terms of life-cycle strategy, the precocious *M. vaginalis*, which can produce seeds before the harvesting of rice, is considered to fit in well

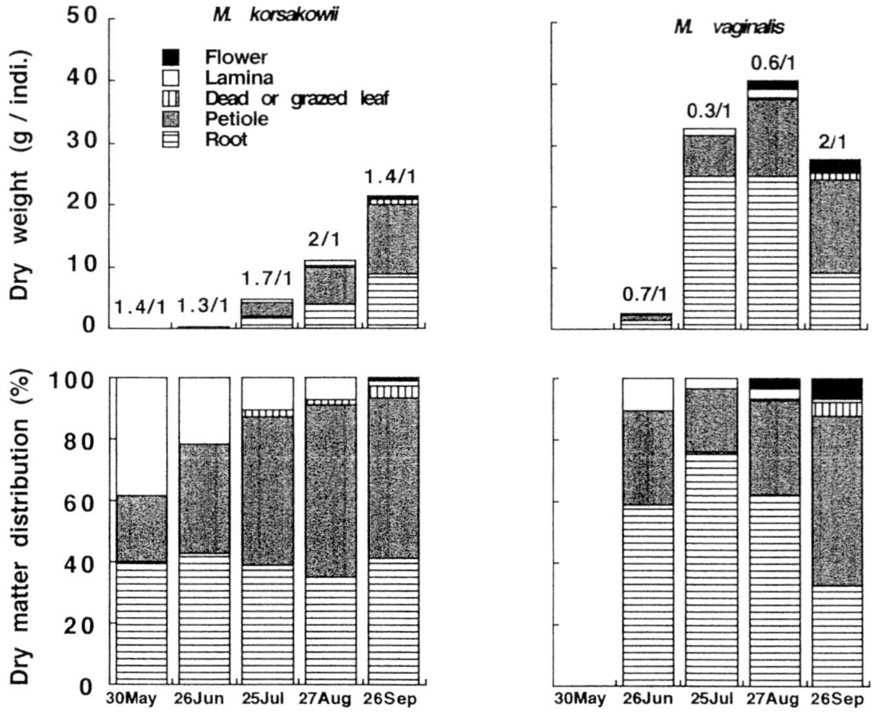

Figure 4. Seasonal changes in dry weight and dry matter distribution of each organ for *M. korsakowii* and *M. vaginalis* grown in pots in outdoor conditions. Each value is shown as the mean of three individuals and vertical bars represent S.D. of the mean. Numerical figures above each column show shoot-to-root ratio.

with the recent agricultural cycle, and therefore, the plant is liable to become noxious. In contrast, *M. korsakowii* blooms later than *M. vaginalis* and it may be cut down before producing seeds.

Rice fields in Japan were traditionally supplied with water and drained via shallow earth ditches. In the last 30 years, however, more than 80% of rice fields have been converted to a new irrigation system in which water is typically supplied through underground pipes via taps and is drained into deep, concrete-sided ditches (Fujioka & Lane, 1997). In the old-style system, the water level in rice fields is close to that of ditches and thus the soils are apt to be water-logged or moist all the year round. In the new-style system, however, rice fields are well-drained and water occasionally dries up even during the cultivation of rice. Under such water conditions, *M. korsakowii*, which shows poorer root development and a higher transpiration rate than *M. vaginalis*, seems to be less tolerant of water stress than *M. vaginalis*.

The present results suggest that the interaction between changed agricultural practices and subtle differences in the phenology and patterns of resource allocation of two closely related species appear to have resulted in the near extinction of one and the promotion to noxious weed status of the other. Further attention should be paid in the future to the responses of these two plants to herbicides, water depth, cutting, etc., which are not considered in the present study.

Acknowledgements

We thank F. Koike, Y. Sugimura, S. Harada, K. Tsuji and I. Yamamoto for their help during the course of this study.

References

Fasola, M. & X. Ruiz, 1996. The value of rice fields as substitutes for natural wetlands for waterbirds in the Mediterranean region. Colonial Waterbirds19: 122–128.

Fujioka, M. & S. J. Lane, 1997. The impact of changing irrigation practices in rice fields on frog populations of the Kanto Plain, central Japan. Ecol. Res. 12: 101–108.

Hasegawa, S. & T. Tabuchi, 1995. Well-facilitated paddy fields in Japan. In T. Tabuchi & S. Hasegawa (eds), Paddy Fields of the

World. Japanese Society of Irrigation, Drainage and Reclamation Engineering, Tokyo: 103–124.

Kadono, Y., 1994. Aquatic plants in Japan. Bun-ichi Sogo Shuppan, Tokyo, 179 pp.

Saitoh, K., O. Katano & K. Koizumi, 1988. Movement and spawning of several freshwater fishes in temporary waters around paddy fields. Jap. J. Ecol. 38: 35–47. (In Japanese with English synopsis.)

Shimizu, N., 1998. What is the plant species diversity in paddy field ecosystem? In National Institute of Agro-Environmental Sciences (ed.), Biological Diversity in Paddy Field Ecosystem. Youkendou, Tokyo: 82–126. (In Japanese.)

Wang, G., T. Kusanagi & K. Itoh, 1996a. Isozyme variation in *Monochoria korsakowii* Regel et Maack and *M. vaginalis* (Burm.f.) Kunth (Pontederiaceae). Weed Research (Japan) 41: 255–263. (In Japanese with English summary.)

Wang, G., T. Kusanagi & K. Itoh, 1996b. Photoperiodic response of flowering in *Monochoria korsakowii* Regel et Maack and *M. vaginalis* (Burm.f.) Kunth. Weed Res. (Japan) 41: 241–246. (In Japanese with English summary.)

Wang, G., T. Kusanagi & K. Itoh, 1996c. Environmental factors relating to dormancy breaking, germination and emergence of seeds in *Monochoria korsakowii* Regel et Maack and *M. vaginalis* (Burm.f.) Kunth. Weed Res. (Japan) 41: 247–254. (In Japanese with English summary.)

Wang, G., T. Kusanagi & K. Itoh, 1997. Effects of time of sowing, shading and compound fertilizer application on growth and seed production of *Monochoria korsakowii* Regel et Maack and *M. vaginalis* (Burm.f.) Kunth. J. Weed Sci. Technol. 42: 135–143. (In Japanese with English summary.)

Hydrobiologia **415:** 35–40, 1999.
J. M. Caffrey, P. R. F. Barrett, M. T. Ferreira, I.S. Moreira, K. J. Murphy & P. M. Wade (eds), Biology, Ecology and Management of Aquatic Plants.
© 1999 *Kluwer Academic Publishers. Printed in the Netherlands.*

Competition between three submerged macrophytes, *Elodea canadensis* Michx, *Elodea nuttallii* (Planch.) St John and *Lagarosiphon major* (Ridl.) Moss

C. S. James*, J. W. Eaton & K. Hardwick
School of Biological Sciences, Jones Building, University of Liverpool, P.O. Box 149, Liverpool L69 3BX, U.K.

Key words: competition, pH, *Elodea canadensis*, *Elodea nuttallii*, *Lagarosiphon major*

Abstract

During photosynthesis, submersed aquatic plant stands create ‘envelopes’ of raised pH and dissolved O_2 concentrations and depleted dissolved CO_2 concentrations. These may extend beyond the plant stands, suppressing the photosynthetic performance and increasing the photorespiratory stress of adjacent species. In competition between three aquatic macrophytes, *Elodea canadensis*, *Elodea nuttallii* and *Lagarosiphon major*, we hypothesised that a competitive advantage is achieved by those species that have the best stress generation/toleration mechanisms. pH and free CO_2 conditions were measured under laboratory conditions at two plant densities and compared with measurements of photosynthesis and respiration at various pH and free CO_2 concentrations. *L. major* raised pH faster and photosynthesised at a greater rate under the experimental conditions and may gain competitive advantage at high pH. No clear differences in photosynthetic behaviour were observed between the *Elodea* spp. and it is concluded that a wider range of factors than those studied here are probably involved in competition between the three species.

Introduction

In the last 100 years, increasing world-wide travel and trade has resulted in the introduction of many aquatic plant species to countries beyond their native range, where their subsequent spread has sometimes been extremely rapid. For example, *Elodea canadensis* was introduced into the British Isles in the nineteenth century and became a widely distributed and troublesome alien species in many situations (Simpson, 1984). The introduction of *Elodea nuttallii* in 1966 has resulted in its displacement of *E. canadensis* from many waters where the latter had previously become well established (Simpson, 1990). It was observed that displacement often occurred over a relatively short time period of one to two years (Lund, 1979; Simpson, 1984). More recently, a species native to South Africa, *Lagarosiphon major* has been reported to be actively displacing *E. nuttallii* and appears to be competitively superior to *Elodea* spp. in at least some habitats.

* Author for correspondence

The three species are morphologically similar, all having long, flexuous leafy stems initially rooted in the substrate, but often becoming detached later in the life cycle. Reproduction in the British Isles is only vegetative as all known populations are female (Simpson, 1986). In view of the similarities in morphology and reproduction of the three species, what is driving these displacements?

Extensive research on the dynamics of terrestrial vegetation has shown that a species ultimate competitive success in a habitat may depend upon its ability to capture and utilise limiting resources. In the aquatic environment, the availabilities of dissolved inorganic carbon, light and macronutrients are critical for the photosynthesis and growth of submerged macrophytes. Despite the high concentrations of organic and inorganic substances often found in lowland water bodies, supply may nevertheless limit plant performance. This is due to the slow diffusion rates of substances through the laminar-flow boundary layers which surround submerged surfaces, especially where mass water flow is slow or absent (Losee & Wetzel,

1988). Supply of inorganic nutrients and dissolved inorganic carbon may be restricted by the slow replenishment of depleted resources from the surrounding water body. Conversely, released substances such as dissolved oxygen (DO) can accumulate within the boundary layer. Dissolved inorganic carbon (DIC) utilisation and the production of oxygen result in a shift towards high pH and DO supersaturation in waters adjacent to the active plant surfaces. Field measurements show increases in pH and DO above atmospheric equilibrium and associated decreases in free CO_2 within plant stands (e.g. Frodge & Pauley, 1990; Jones et al., 1996). It is hypothesised that these conditions of raised pH and DO, and lowered DIC may, especially under conditions of low mass water flow, extend beyond the plant stand itself and, hence, could interfere with the growth and consequently the competitive ability of adjacent plants.

It is suggested that a competitive advantage will be achieved by those species that have the best stress generation/ toleration mechanisms, where stress is defined as raised pH and DO, and lowered free CO_2. The aim of this work was to investigate the generation of pH and CO_2 stress under laboratory conditions and to compare the physiological tolerances of the species to increasing pH, in relation to the previously described species displacements.

Methods

Stress generation under controlled laboratory conditions

For determination of stress generation under laboratory conditions, measurements were made in monocultures of each species grown at two densities for four weeks. Forty, 10 cm apical lengths of shoot of *E. canadensis*, *E. nuttallii* and *L. major* were selected and planted singly in 250 ml plastic cups filled with canal sediment. More shoots than required for the experiment were cultured to allow for sub-sampling at the start of the experiment and for death of some replicates. The cups were placed in 10 l green plastic buckets filled with tap water previously aerated for 24 h. These were then nursery cultured at 15 °C under an irradiance of 80 μmol m^{-2} s^{-1} PAR on a 16:8, light:dark cycle for two weeks prior to the start of measurements to allow establishment of the shoots. Following the nursery culturing, shoots were transferred to clean buckets at the starting densities of two plants (low density) and four plants (high density), with four replicates per treatment. Initial dry weights (g) were approximated by sub-sampling (n = 5) cultured material. As starting lengths of each species were standardised to 10 cm lengths, the mean calculated dry weights (g) of shoots of *L. major* (0.213 g) following nursery culturing were greater than either *E. canadensis* (0.074 g) or *E. nuttallii* (0.112 g). Differences in mean weights of the starting material between the *Elodea* spp. were largely due to the faster growth rate of *E. nuttallii* during nursery culturing. Species were then grown for four weeks under the light and temperature conditions as described for nursery culturing. Nutrient supplements were added weekly at 100 mg m^{-3} P (as potassium dihydrogen orthophosphate) and 1 g m^{-3} N (as ammonium nitrate). pH measurements were made 2 cm below the water surface in the centre of each bucket, every 2–4 d initially, and for the final week once. Measurements were taken centrally to avoid areas shaded by the sides of the buckets, and at 2 cm in depth to avoid effects on pH due to exchange with the atmosphere at the water surface. This is also close to the position of maximum biomass accumulation during the study. Carbon fractions (free CO_2, bicarbonate and carbonate) were calculated from weekly measurements of alkalinity, pH, temperature and conductivity, following the method of Mackereth et al. (1978).

The photosynthetic response to increasing pH

For physiological comparisons between species, experimental plant material was cultured under controlled environmental conditions in three 50 l glass aquaria at 15 °C and a light intensity of approximately 80 μmol m^{-2} s^{-1} PAR on a 16:8, light:dark cycle. pH values of between 8.5 and 9.5 were recorded during culturing. Photosynthetic rates were measured at approximately 1 pH unit intervals between pH 6.0 and 9.7, to cover the range in which the plants are found in the field, using samples of leaves removed from each species (n = 19–22). For consistency, leaves were taken 3 cm below the apical tip. Oxygen evolution or uptake was measured using a Clark-type oxygen electrode (Hansatech, Kings Lynn, U.K.) at 15°C, and an irradiance of 290 μmol m^{-2} s^{-1} PAR (shown in preliminary work to be just beyond the saturating level for all three species) for evolution, with darkness used to determine uptake. Determinations were made in Forsberg solution (Forsberg, 1965) with the omission of buffering agents, as these have been found to inhibit uptake of bicarbonate at high pH (Prins et al., 1982).

The pH was measured at the beginning and end of each period of oxygen recording to provide the pH range (mean change of 0.17 pH units) over which readings were taken. From pH, total CO_2, temperature and conductivity, free CO_2 concentrations in the Forsberg medium were calculated.

Results

When cultured in the laboratory, all three species progressively created conditions of high pH in the surrounding medium (Figure 1), mostly rising to maxima around pH 9.7, but in the case of high density *E. nuttallii* and *L. major*, to pH 10.1. Initial density-dependent differences in the rates of pH increase were observed between high and low densities with highest planting densities resulting in the fastest increase in pH. However, no significant differences in pH were observed between either species or high and low density treatments by day 14. Total CO_2, free CO_2 and bicarbonate decreased and carbonate concentrations increased over the experimental period for all treatments. For all treatments within 7 d free CO_2 concentrations within the medium were less than 1 μmol m^{-3}.

The three species all photosynthesised over a wide pH range, but as expected, net rates decreased with rising pH. Smoothed lines were fitted to the data using locally weighted regression analysis (Figure 2). Photosynthetic compensation points for all three species were at pH 9.7 ($\pm$ 0.1 pH unit). A visual comparison of 99% confidence limits for *E. nuttallii* and *L. major* revealed little overlap within the range pH 6.0–8.5 (Figure 3), with the latter species photosynthesising at a consistently higher rate. Above pH 8.5, photosynthetic rates of *L. major* were higher than those of the *E. nuttallii.*, although overlap of confidence intervals suggests that these rates were not significantly different. The confidence intervals for *E. canadensis* (not shown here) closely overlapped those of *E. nuttallii* over the entire pH range measured. Figure 4 shows the comparable photosynthetic performance of the three species expressed in relation to log free CO_2. Regression analysis revealed significant differences ($P < 0.001$) in the elevation of response lines between *L. major* and *Elodea* spp. Significant differences ($P = 0.05$) were also found between the slopes of the response lines for *E. nuttallii* and *L. major*. No significant differences in the slopes of the response lines were observed between the two *Elodea* spp.

Increases in pH were not found to have any effect upon respiration rates, as measured by oxygen uptake in the darkness.

Discussion

Results show that all three species created similarly stressful conditions of high pH and low CO_2 to extents which agree with previous observations in mature field populations (e.g. Frodge et al., 1990; Jones et al., 1996). Maximum pH values found for both low- and high- density treatments are within the range in which the physiological studies indicate that the species are not able to achieve net positive photosynthesis. This corresponds with growth data (not presented here) in which all three species were found to have very low growth rates over the experimental duration. Respiration rates were not found to be affected by pH, as expected, since the major factor influencing respiration rates is DO, which is not significantly affected by pH.

The minimum free CO_2 concentrations found in laboratory studies here were close to the lower end of the free CO_2 compensation point range (0.7–25 μmol l^{-1}) reported for various submerged macrophytes (Bowes & Salvucci, 1989). Consequently bicarbonate was likely to be the main source of carbon. When the pH reached pH 9.7, net positive photosynthesis ceased, but some activity presumably continued, utilising bicarbonate and releasing OH^-. This could explain why higher values, in excess of pH 10, were observed in some treatments.

L. major maintained markedly higher photosynthetic rates than either of the two *Elodea* spp. However, as free CO_2 depletion became more intense, differences between the species decreased, although there was evidence that *L. major* was still able to maintain slight positive net photosynthesis at extreme CO_2 depletion

Maximum photosynthetic rates were low compared with measurements made by previous authors using the same technique with *E. nuttallii* rates almost half those of previous findings (e.g. Jones et al., in press). This is possibly a symptom of physiological stress and a consequent down-regulation in the photosynthetic rate. A corresponding increase in respiration rates was not observed.

There is some indication that the competitive success of *L. major* may be a consequence of greater toleration to pH stress. Overall, photosynthetic rates of

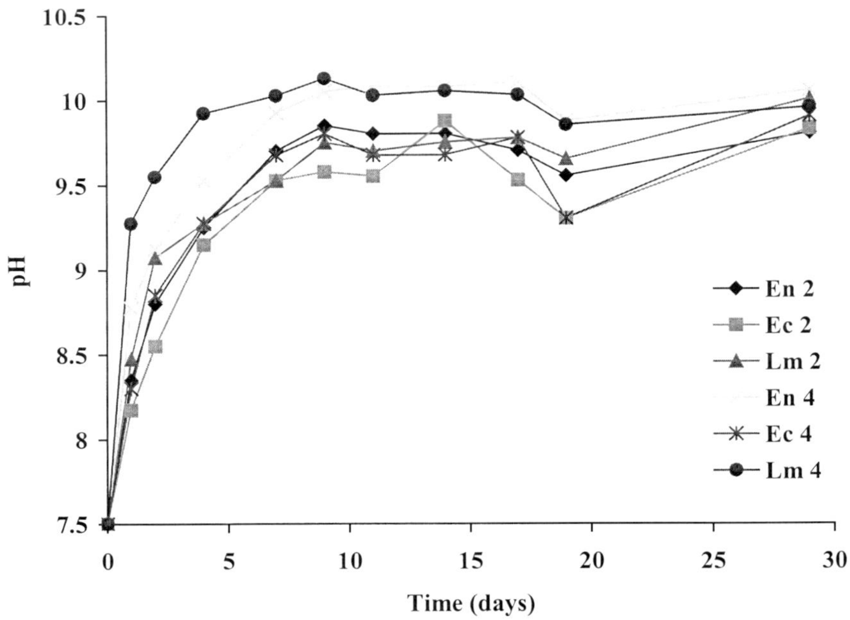

Figure 1. Change in pH of cultures at high (four plants per culture) and low (two plants per culture) planting densities of *E. canadensis* (EC), *E. nuttallii* (EN) and *L. major* (LM).

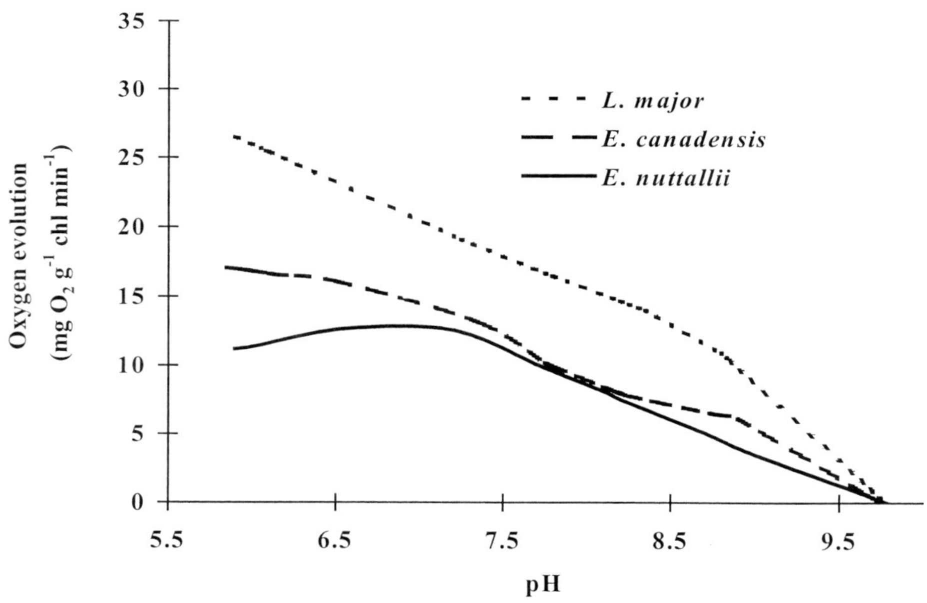

Figure 2. Least weighted regression lines for photosynthetic rates in response to increasing pH of the incubation medium.

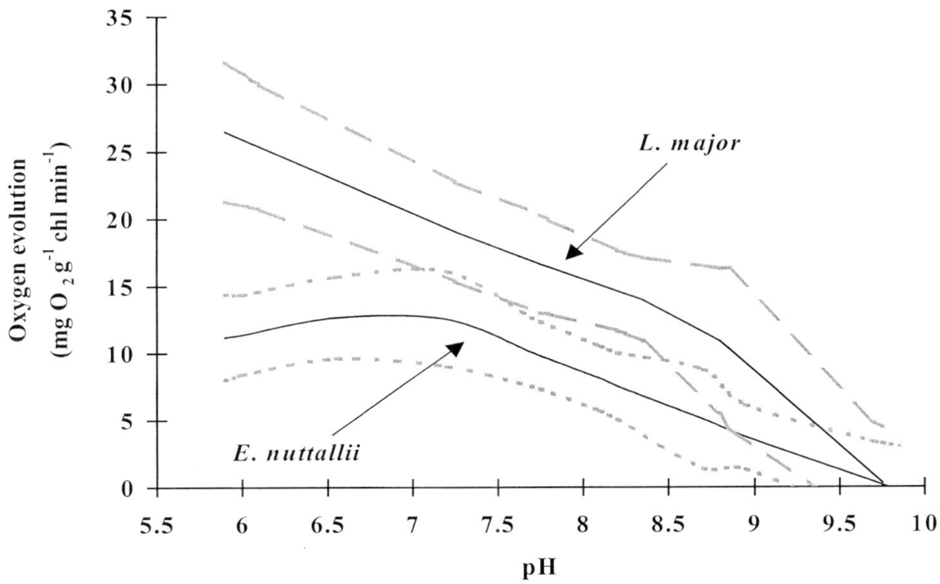

Figure 3. Photosynthetic rates of *E. nuttallii* and *L. major* in response to increasing pH of the incubation medium showing 99% critical intervals for least weighted regression lines.

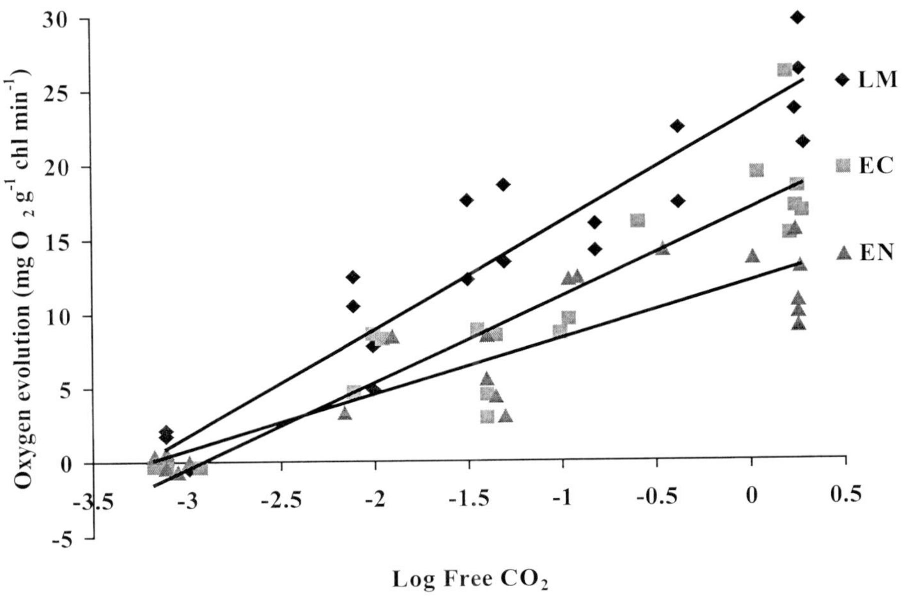

Figure 4. Relationship between log free CO_2 concentrations in the incubation medium and net photosynthesis of *E. canadensis* ($R^2 = 0.8306$), *E. nuttallii* ($R^2 = 0.7809$) and *L. major* ($R^2 = 0.9082$).

L. major were found to be greater than in *Elodea* spp. Thus, *L. major* may be successful in out-competing *Elodea* spp. as a result of its ability to photosynthesise and consequently grow, particularly under very stressful conditions of high pH and low free CO_2, perhaps through more efficient bicarbonate utilisation than the other species.

These studies do not explain, however, the apparent competitive success of *E. nuttallii* in displacing *E. canadensis*. The conditions created and tolerated by these two species were largely similar. It is likely that the competitive ability will depend upon a combination of factors including morphological and seasonal characteristics not studied here. Simpson (1990) concluded that *E. nuttallii* formed taller growing canopies than *E. canadensis* which then suppressed the latter by shading. Kunii (1984) emphasised the potential for early seasonal growth of *E. nuttallii* to confer a competitive advantage on that species over species with slower starting characteristics and therefore later accessing of space and of resources such as light and nutrients. It is possible that such features operate in combination with the physiological difference reported here to drive the species displacements observed under field conditions.

Acknowledgements

This work was funded by a Natural Environmental Research Council postgraduate studentship to C. James, for which we are grateful.

References

Bowes, G. & M. E. Salvucci, 1989. Plasticity in the photosynthetic carbon metabolism of submersed aquatic macrophytes. Aquat. Bot. 34: 233–266.

Forsberg, C., 1965. Nutritional studies of *Chara* in axenic cultures. Physiol. Pl. 18: 275–290.

Frodge, J. D., G. L. Thomas & G. B. Pauley, 1990. The effects of canopy formation by floating and submergent aquatic macrophytes on the water quality of two Pacific Northwest lakes. Aquat. Bot. 38: 231–248.

Jones, J. I., J. W. Eaton & K. Hardwick, (in press). The effect of changing environmental variables in the surrounding water on the physiology of *Elodea nuttallii*. Aquat. Bot.

Jones, J. I., J. W. Eaton & K. Hardwick, 1996. Diurnal carbon restrictions on the photosynthesis of dense stands of *Elodea nuttallii* (Planch.) St. John. Hydrobiologia 340: 11–16.

Kunii, H., 1984. Seasonal growth and profile structure development of *Elodea nuttallii* (Planch.) St. John in a pond Ojaka-ike, Japan. Aquat. Bot. 18: 239–247.

Losee, R. F. & R. G. Wetzel, 1988. Water movement within submersed littoral vegetation. Verh. int. Ver. theor. angew. Limnol. 23: 62–66.

Lund, J. W. G., 1979. The mystery of *Elodea* Michx in Great Britain. Watsonia 12: 338.

Mackereth, F. J. H., J. Heron & J. F. Talling, 1978. Water Analysis: Some Revised Methods for Limnologists. Freshwater Biological Association, Scientific Publication No. 36, Ambleside: 120 pp.

Prins, H. B. A., J. F. H. Snel, P. E. Zanstra & R. S. Helder, 1982. The mechanism of bicarbonate assimilation by the polar leaves of *Potamogeton* and *Elodea*. CO_2 concentrations at the leaf surface. Pl. Cell Envir. 5: 207–214.

Simpson, D. A., 1984. A short history of the introduction and spread of *Elodea* Michx in the British Isles. Watsonia 15: 1–9.

Simpson, D. A., 1986. Taxonomy of *Elodea* in the British Isles. Watsonia 16: 1–9.

Simpson, D. A., 1990. Displacement of *Elodea canadensis* Michx by *Elodea nuttallii* (Planch.) H. St. John in the British Isles. Watsonia 18: 173–177.

Hydrobiologia **415:** 41–45, 1999.
J. M. Caffrey, P. R. F. Barrett, M. T. Ferreira, I.S. Moreira, K. J. Murphy & P. M. Wade (eds), Biology, Ecology and Management of Aquatic Plants.
© 1999 *Kluwer Academic Publishers. Printed in the Netherlands.*

Metals in sediment and water of three reed (*Phragmites australis* (Cav.) Trin. ex Stend.) stands

J. Santos Oliveira[1,*], João Almeida Femandes[2], Carla Alves[3], João Morais[1] & Paula Urbano[3]
[1] *GDEH/FCT/UNL, Campus da FCT, Quinta da Torre, 2825 Monte de Caparica, Portugal*
[2] *Universidade de Évora, Largo dos Colegiais 2, 7001 Évora Codex, Portugal*
[3] *U.L.H.T, Campo Grande 376, 1700 Lisboa, Portugal*

Key words: Phragmites, eutrophication, wetlands, nutrient

Abstract

In the frame of the research program EUREED (Effects of interaction between eutrophication and major environmental factors on the ecosystem stability of reed (*Phragmites australis*) vegetation in European land-water ecotones), three reed stands, established near Alcácer do Sal, Lagoa de Albufeira and Montargil, were studied. The stands were monitored monthly between December 1993 and December 1994 for productivity, soil and water chemistry. The relationship between the concentration of some metals in soil and water and its influence on the stability of the reed vegetation was studied. Environmental factors, such as salinity, negative redox potential on the root system and high Cu, Na and Zn concentrations in soil did not affect the normal development of reed vegetation and its capability to recover after mechanical damage.

Introduction

In Portugal, the study of the ecology of wetland macrophytes is limited and mainly focused in those that are reported as aquatic weeds. In halophytic environments, the submersion rate and salinity regulate the typical distribution of the most important species in non-disturbed environments. These factors determine that some species of the genera *Scirpus, Juncus, Potamogeton* and *Phragmites* are observed in almost monotypic populations. In negatively disturbed environments, it is possible to observe the invasion by species such as *Paspalum vaginatum* or some other exotic weeds (Ferreira, 1994a, b).

In freshwater environments, the most important factors are the climate, the depth and duration of submersion and the nutrient availability. The occurrence of disturbance factors also determines the distribution of species. Example of these influences is the colonisation of disturbed environments by *Typha* sp. instead of the less resilient *Phragmites australis* (Ferreira, 1994 a, b). The stability of reed communities on land-water ecotones throughout Europe was the aim of the EUREED project. For this project, three stands in Portugal have been studied to determine the influence of nutrient availability in the soil and soil solution on the growth and stability of reed communities. Three well characterised stations have been selected, representing a eutrophic brackish tidally influenced stand (Alcácer do Sal), a freshwater eutrophic stand (Lagoa de Albufeira), and a freshwater mesotrophic stand (Montargil). The vegetation growth, together with soil and water chemistry have been monitored during a life cycle period (Dec. 93–Dec. 94).

The main hypothesis of the EUREED project states that the eutrophic status of the reed stand could be one of the main reasons for the die-back processes that have been observed in northern and central Europe. Such processes had never been recorded in Portugal, even in eutrophic stands.

* Author for correspondence

Description of sites studied

The sampling station of Alcácer do Sal is located directly near a small tributary of the Sado river (38° 22′ 11″ N, 8° 29′ 06″ W), with clay soils. In the

Table 1. Sampling dates in each sampling site from December 1993 to December 1994

Sampling site	Sampling date										
	Dec 93	Jan 94	Feb 94	Mar 94	Apr 94	May 94	Jun 94	Jul 94	Aug 94	Sep 94	Dec 94
Alcácer do Sal	19	11	15	17	13	30	13	10	30	—	20
Lagoa de Albufeira	—	—	—	—	20	—	2 and 15	11	—	1	21
Montargil	20	12	16	21	14	—	1 and 19	11	31	—	21

nearby nontidal alluvial plain, rice is intensively cultivated. The reed (*Phragmites australis*) is the dominant macrophyte species in the area. Other wetland species identified were *Juncus effusus* and *Typha angustifolia*. Pesticides and fertilisers used in the agricultural fields may affect the reed community. The climate is slightly atlantic (mediterranic influenced) with moderate winters and hot summers. The tide cycle is the main factor affecting flow and water level variation.

The sampling station of Montargil is located directly near a small tributary of the Montargil reservoir (39°10′ 01″ N, 8°10′ 01″ W), with sandy-podzolic soils. The lower flooded area of the drainage basin where the sampling station is located is dominantly colonised by reed. The upper region of the basin is used for low income agriculture and intensive forestry. The reed is expanding on the permanently flooded soils and is the dominant macrophyte species in the area. Regular cut-offs are made by the nearby farmers in order to control its rapid expansion. Other wetland species identified were *Juncus effusus* and *Typha angustifolia*. The reed community shows a high regeneration capacity after seasonal cuts. The climate is slightly continental with fresh winters and hot summers. Precipitation dominates flow and level variations.

The sampling station of Lagoa de Albufeira is located on a managed waterfowl conservation area, created by the damming of the upstream area of a coastal lagoon, about 20 km far from Setúbal (38° 31′ 42″ N, 9° 8′ 22″ W) where, formerly, rice used to be cultivated intensively. Since the dike construction in 1991, a permanently flooded freshwater table of several hectares was created and quickly colonised by reed. The main land occupation is forestry and urban development. The climate is slightly atlantic with moderate winters and summers. This stand was monitored between April and December of 1994. There is no significant surface flow and the main factors controlling water level are precipitation and the reservoir water management.

Material and methods

The research program was designed on field analysis and sampling, to characterise the biological, physical and chemical evolution of the stand. Sampling dates are presented in Table 1. The following field procedures were performed, from December 1993 to December 1994: collection of 40 cm depth and 10 cm diameter cores of soil, collection of superficial and interstitial water samples, collection of soil gas samples, collection of litter, counting, measuring and weighing of plants in a 1 square m quadrate. The quadrate dimensions were chosen after statistical analysis of the minimal representative area, following the criteria of plant density and average plant height. The cut-off quadrates where monitored for regrowth intensity in the following sampling dates.

There were made field measurements of pH, conductivity, dissolved oxygen, redox potential and temperature in water, using Horiba Water Checker UIO and Orion Portable Meter model 290A. There were collected water samples for laboratory determination of N–NH_4^+, N–NO_2^-, N–NO_3^-, NKjeldahl, P–$P0_4^{3,-}$, P-Total, $S0_4^{2,-}$, total alkalinity, chloride, Ca, Mg, Na, K, Mn, Fe, Cu, Ni and Zn. Sampling, sample preservation and analytical methods were performed according to standard procedures (APHA, AWWA, WPCF, 1985).

There were made field measurements of redox potential in 40 cm soil profile and temperature using an

Orion Portable Meter model 290A. There were collected 40 cm depth and 10 cm diameter cores of soil (two per quadrate) for laboratory determination of moisture, organic matter, N–NH_4^+, N–NO_3^-, N-Kjeldahl, P–PO_4^{3-}, P-Total, pH (1:5 in water), pH (1:5 in KCl 1N), Ca, Mg, Na, K, Mn, Fe, Cu, Ni and Zn. The soil cores were sealed with Parafilm 'M' laboratory film and kept and transported to laboratory at low temperature (4–8 °C). At the laboratory, one of the cores was divided in 5 cm length pieces. Each piece was sealed and kept at low temperature until analytical procedure was performed. The other core was invoiced to the Agronomy Institute, where some pedological characteristics of soil (Cation Exchange Capacity, Basic Cation Saturation Percentage and Texture) were determined.

Results and discussion

Organic matter and nutrients in sediment

The observed distributions of organic matter and nutrients in the sediment were previously described (Fernandes et al., 1995; Oliveira et al., 1996a, b; Oliveira & Fernandes, 1998) and can be summarised in Figure 1.

The organic matter content in the sediment seems to be dependent on the flow characteristics and on the contribution of the surrounding vegetation. The distribution along the 40 cm sediment profile is depth dependent, being the higher values found in the surface layers. There were not found strong time dependent variations. However, all the sites showed their minimum on the summer, showing the influence of temperature on the organic matter decaying processes. The patterns of depth and time variations of the N-Kjeldahl content in sediment were similar to those observed for the organic matter. In what concerns the total phosphorus content in the sediment, there were not observed depth dependent variations. The time and depth distributions of total phosphorus in the sediment do not seem to be dependent on the organic matter accumulation/decaying processes.

In what concerns the time and depth distributions for each of the soluble nitrogen compounds analysed (ammonia and nitrate), there were not observed depth dependent variations for the ammonia, and nitrate seems to be present in higher concentrations in the layer between 20 cm and 25 cm. The annual average values obtained for the ammonia were not significantly different in the three sites. In all sites, nitrate was always present in concentrations higher than 15 mmol kg^{-1} and presented its absolute minimum concentrations in late spring, and relative maximum concentrations in middle summer.

Soluble phosphates presented similar time distributions in all sites. Between May and July it was never detected in concentrations above the detection limit (1 mmol kg^{-1}). This point to the existence of a severe phosphate limitation in the season of more intense plant growth. This phosphate limitation seems to affect the nitrate absorption by the reed vegetation, leading to the accumulation of nitrate in the sediment, verified in all sites in July. All sites presented the same average phosphate concentration of 15 mmol kg^{-1} in a 40 cm sediment profile.

Metals in sediment and water

The obtained results (average values of the upper 20 cm of the soil core) are summarised in Figure 2. No significant differences were found in the analytical results of the 5 cm layers of the core even when the redox potential was strongly affected by the tide in Alcácer.

Considering the results obtained at Alcácer, the most important observations are:

In soil, Mg (225 mmol kg^{-1}), Na (177 mmol kg^{-1}), K (229 mmol kg^{-1}) and Fe (473 mmol kg^{-1}) were the most abundant metals found, in terms of annual average concentration. The less abundant were Cu and Ni. Cu, Zn and Ni were detected in higher concentrations than they are normally present in soils.

The geological nature of the Sado basin including a rich copper and zinc pyrites area and the mining facilities there located can explain the observed values for Cu (1.1 mmol kg^{-1}), Ni (0.2 mmol kg^{-1}) and Zn (7.2 mmol kg^{-1}). In water, Ca (20 mmol l^{-1}), Mg (5.0 mmol l^{-1}) and Na (11 mmol l^{-1}) were the dominant metals found. K (1.6 mmol l^{-1}) and Fe (1.4 mmol l^{-1}) were also present in important concentrations. Ni was never detected in concentrations above the detection limit (0.1 μmol l^{-1}).

Considering the results obtained at Montargil, the most important observations are:

In soil, Na (194 mmol kg^{-1}), K (139 mmol kg^{-1}), and Fe (241 mmol kg^{-1}) were the most abundant metals found, Mg (36 mmol kg^{-1}) was also detected in significant concentrations. Ni was never detected in concentrations above the detection limit (1 mol kg^{-1}). In water, Na (1.0 mmol l^{-1}) was the most abundant metal found. Ca, Mg, K and Fe were also present in

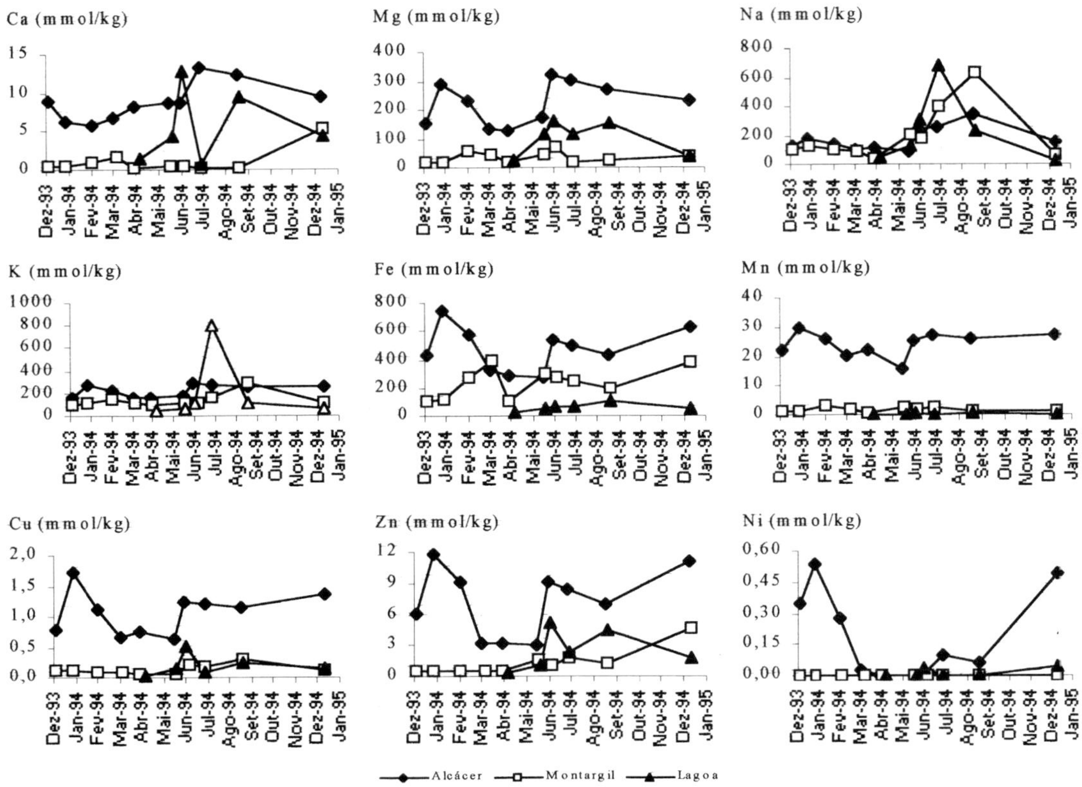

Figure 1. Time distribution for organic matter, N-Kjeldahl, total phosphorus, soluble phosphates, ammonia and nitrates in the sediment (dry weight). Average values for a 40 cm sediment core.

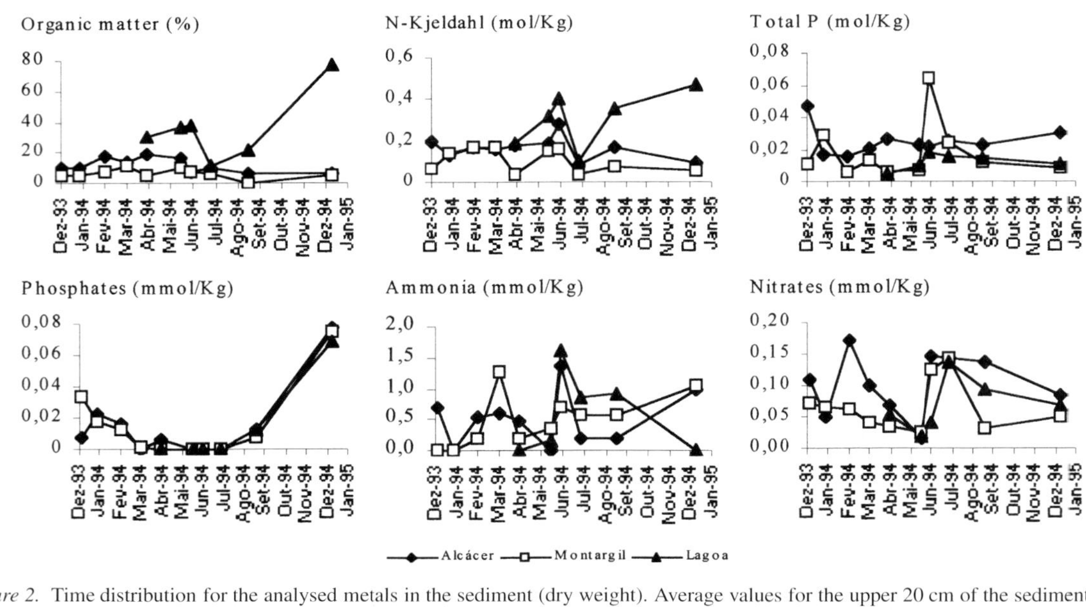

Figure 2. Time distribution for the analysed metals in the sediment (dry weight). Average values for the upper 20 cm of the sediment core.

concentrations between 0.1 and 0.2 mmol l^{-1}. Ni and Cu were never detected in concentrations above the detection limit (0.1 μmol l^{-1}).

Considering the results obtained at Lagoa de Albufeira, the most important observations are:

In soil, Mg (103 mmol kg^{-1}), Na (238 mmol kg^{-1}) and K (203 mmol kg^{-1}, were the most abundant metals found. Fe (60 mmol kg^{-1}) was also detected in significant concentrations. In water, Na (2.3 mmol l^{-1} was the dominant metal found. Ca (0.2 mmol l^{-1}), Mg (0.8 mmol l^{-1}) and K (0.1 mmol l^{-1}) were also present. Ni and Mn were never detected in concentrations above the detection limit (0.1 μmol l^{-1}).

Conclusions

In contrast to the results reported by Ostendorp (1989) and Den Hartog et al. (1989) for many reed stands in Europe, no signs of die-back are reported in Portugal. On the contrary, reed behaves generally as a weed plant, as reported by Havens et al. (1997) in the United States, demanding intensive control by cut-offs. The stand of Albufeira was chosen exactly because of the explosive colonisation by reed in a within 3 year period after the damming of the area. The trophic status of the studied stands is stable as follow-up observations of the sites demonstrated. Soluble phosphates are clearly the limiting factor for reed development in all stands during the growth season.

The variations of the stands biological characteristics (growth pattern and plant density), related to environmental factors, showed a similar pattern of response, pointing to a stronger dependence on light and temperature monthly variations than to other particular factors. The resilience of the stands after mechanical damages, demonstrated by the strong recovery observed, can be explained by the important nutrient reserves of the root system. Alcácer is the stand with the higher metal content in soil and water and the one where the reed stand seems to be stronger. The variation of metals concentration in soil seems to be dependent on the biological cycle of *Phragmites*. In the three stands, almost all the metals presented their lowest concentrations in soil between March and May corresponding to the period of more intensive plant growth. Ca, Mg, Na, K and Mn reached their maximum concentrations between June and August, when the plant growth was slowing. Fe, Cu, Zn and Ni reached their maximum concentrations in January, that is the month of highest total metal concentration in soil and minimum vegetative activity.

Natural reed stands can, from the obtained results, be considered as important regulators of the metal concentrations in water and sediments, pointing to the possible use of this stands for the mitigation of heavy metal diffuse pollution of lotic and lentic environments (at least during the growth period), as well as artificial stands, as the case of Albufeira demonstrates. This conclusion is of particular importance for the Portuguese situation, where the growth period follows the rainy season.

Acknowledgements

European Union Environment Program (1990–1994) sponsored the EUREED project.

References

APHA, AWWA, WPCF, 1985. Standard Methods for the Examination of Water and Wastewater 16th edn. Arnold E. Greenberg, R. Rhodes Trussel, Lenore S. Clesceri & Mary Ann H. Franson (eds). APHA, Washington DC, U.S.A.

Den Hartog, C., J. Kvet & H. Sukopp, 1989. Reed. A common species in decline. Aquat. Bot. 35: 1–4.

Fernandes, J. P., J. F. Oliveira, C. Alves, A. Urbano & J. Morais, 1995. Nutrient balances in tidal and freshwater reed stands in relation to the trophic status of the stand. Oral presentation Symposium Eutrophication: causes, consequences and remediation. Instituto de Zoologia Dr. Augusto Nobre, Porto, Portugal.

Ferreira, M. T., 1 994a. Aquatic and marginal vegetation of the River Divor and its relation to land use. Verh. int. Ver. Limnol. 25: 2309–2315.

Ferreira, M. T., 1994b. Macrófitas lóticos do Alto Alentejo: Tipologia e interferência dos ecossistemas agr'airios envolventes. Recursos Hídricos. 15: 46–58.

Havens, K. J., W. I. Priest & H. Berquist, 1997. Investigation and long-term monitoring of *Phragmites australis* within Virginia's constructed wetland sites. Envir. Mnt. 21: 599–605.

Oliveira, J. F., J. P. Fernandes, C. Alves, A. Urbano & J. Morals, 1996a. Uso de zonas húimidas povoadas com *Phragmites* sp. para tratamento terciário de águas residuais – caso de estudo da Lagoa de Albufeira. In APRH, ABES (eds), 30° Congresso da Água / VII SILUBESA, Comunicações 3. APRH, Lisboa: 387–397.

Oliveira, J. F., J. P. Fernandes, C. Alves, A. Urbano & J. Morais, 1996b. Heavy metals and nutrient balances in a eutrophic tidally influenced reed (*Phragmites australis*) stand – the case study of Alcácer do Sal. In Ferrante, A. J. & C.A. Brebbia (eds), Coastal Environment–Environmental problems in Coastal Regions. Computational Mechanics Publications, Southampton: 19 1–200.

Oliveira, J. F. & A. Fernandes, 1998. Portugal. In Vymazal, J.H., Brix, P.F., Cooper, M.B., Green, R. Harberl (eds), Constructed Wetlands for Wastewater Treatment in Europe. Backhuys Publishers, Leiden, The Netherlands.: 227–240.

Ostendorp, W., 1989. 'Die-back' of reeds in Europe – A critical review of literature. Aquat. Bot. 35: 5–26.

Hydrobiologia **415:** 47–53, 1999.
J. M. Caffrey, P. R. F. Barrett, M. T. Ferreira, I.S. Moreira, K. J. Murphy & P. M. Wade (eds),
Biology, Ecology and Management of Aquatic Plants.
© 1999 Kluwer Academic Publishers. Printed in the Netherlands.

Heavy metal content of common reed (*Phragmites australis*/Cav./Trin. ex Steudel) and its periphyton in Hungarian shallow standing waters

G. Lakatos[1,*], M. Kiss[1] & I. Mészáros[2]
[1]*Department of Applied Ecology , Kossuth University, P.O. Box 22, H-4010 Debrecen, Hungary*
[2]*Department of Botany, Kossuth University, P.O. Box 14, H-4010 Debrecen, Hungary*

Key words: common reed, periphyton, heavy metal content

Abstract

In order to reveal the actual effects of common reed (*Phragmites australis* (Cav) Trin. ex Steudel) and its periphyton complex on the water quality, studies on the heavy metal contents of common reed and its periphyton are important. This paper discusses the results of studies on heavy metal contents of the host plant-periphyton complex carried out in Lake Balaton, Lake Velence, Lake Fertő and Kis-Balaton Reservoir. A second goal was to study the effects of reed stands in different condition (healthy, transitional, die-back) on the structure of its periphyton and their role in water quality indication. Simultaneously with the reed and periphyton sampling, water samples were taken for investigation. We determined the wet mass of periphyton and after drying at 105 °C; the dry mass was measured and ash contents of reed and its periphyton samples were also determined. The concentration of 27 (Na, Mg, Ca, K, Zn, Cu, Cd, Cr, Pb, Mo, etc.) cations was determined from the reed, periphyton and water samples by ICP-AES method. The heavy metal content of reed and periphyton was especially high at places impacted by high anthropogenic loadings. On the basis of our results, the biofilter role of periphyton-reed complex can be confirmed. Since the main reasons of reed decline have not been revealed, future research has to focus on involving reed-periphyton studies also, which may provide good basis to find the most appropriate ways to protect and restore the reed communities of shallow standing waters.

Introduction

The significance of the littoral zone and the aquatic plant-periphyton community in the structure and function of shallow waters has already been recognized (Wetzel, 1964), but only some papers have given detailed information on them especially in Central Europe (Dokulil, 1975; Sommer, 1977; Padisák, 1982; Lakatos, 1989; Lakatos & Biró, 1991; Lakatos et al., 1998).

Common reed (*Phragmites australis* (Cav.) Trin. ex Steudel) is a frequent species in different regions of Europe (Haslam, 1978). In Hungary, extensive reed stands occur in the shorelines of shallow waters. The most important plant communities in the littoral zone of shallow lakes and reservoirs are the reed communities (*Scirpo-Phragmitetum*) which play multiple roles in the management of water quality.

The function of reed stands in water purification is not restricted to the removal of plant nutrients, but includes the elimination of organic pollutants, heavy metals and pathogenic microorganisms. The influence of the littoral zone on water quality improvement is largely increased by the existence of periphytic organisms (periphyton) on the submerged parts of reed plants. In order to reveal the effects of this complex on the water quality, studies of heavy metal contents of common reed and its periphyton are required.

Increasing loading of nutrients in shallow lakes, leading to excessive eutrophication have affected reed stands negatively and resulted in accelerated die-back of reed in many parts of central Europe. Although the phenomenon of reed decline was first observed in Hungary more than 25 years ago, serious damage has been reported only since the 1980s (Kovács et al., 1978; Dinka, 1986; Lakatos, 1989) mostly from large lakes (Lake Balaton, Lake Fertő and Lake Velence).

* Author for correspondence

Much work has been undertaken on the causes of this phenomenon (Ostendorp, 1989; Dinka et al., 1995; Armstrong et al., 1996).

In this paper, we discuss the results of studies on heavy metal contents of the host plant-periphyton complex in Lake Balaton, Lake Velence, Lake Fertő and Kis-Balaton Reservoir (Kis-Balaton Water Protection System). We aimed to reveal the effects of reed stands in different conditions (healthy, transitional state, die-back) on the structure of periphyton and its role in water quality indication.

Materials and methods

Lake Balaton (surface area c. 600 km^2) is one of the largest lakes in Central Europe. The littoral area is relatively narrow, but, due to the long shoreline, this area (c. 100 km^2) it is still significant and about 25% of it is covered by common reed (*Phragmites australis* (Cav.) Trin. ex Steudel) (Simonian et al., 1995). In the littoral zone of Lake Balaton the reed-periphyton complex has been studied in four transects in the bays at Paloznak, Bozsa, Szigliget and Keszthely for seven years (1989–1996). The reed stands in Paloznak and Bozsa bays can be considered healthy and undisturbed stands of the lake. The reed stands in the bays at Keszthely and Szigliget are much richer in nutrients and heavy metals than in the other areas of the lake. The reed-periphyton samples were taken near the open water in each transect (Lakatos & Bíró, 1991).

Lake Velence (Velencei-tó) is the second largest sodic lake in Hungary. The surface area of the lake is about 26 km^2 and previously the reed stands covered 10 km^2 of the water surface, but during the rehabilitation treatment of the 1980s, one third of the reed stands were dredged out. In the case of Lake Velence, 14 sampling points were selected at the open water side of reed stands in regions with different water quality (Lakatos, 1978). Unfortunately, reed decline can also be observed in the shoreline of this lake.

Lake Neusiedl (Fertő) is the second largest eutrophic shallow lake in Central Europe since its total area reaches 300 km^2, of which about 200 km^2 is covered by the *Phragmites* belt (Dokulil et al., 1980). The larger part of the lake belongs to Austria, while Hungary has the smaller section. During the study of this lake, reed-periphyton samples were taken at 30 sampling plots, both in the Hungarian and Austrian parts of the lake (Lakatos, 1989). Studies in the frame of EUREED Project were started in 1994. In the Hungarian part of the lake, 10 sampling sites located in areas of different water quality and of reed stands with different conditions were chosen. The outer area of the water can be characterized by healthy reed stands, the transitional water area by reed stands in transitional state, while the inner lakes and the reed belt can be characterized by die-back reed stands (Dinka et al., 1995).

The Kis-Balaton Water Protection System (KBWPS) was constructed in the lower section of the River Zala for the reduction of the external nutrient load to Lake Balaton. The KBWPS consists of two main parts, the first one having an area of 18 km^2 and average depth of 1.14 m (Pomogyi, 1993; Korponai et al., 1997). In the open water regions, reed-periphyton samples and substrate samples were taken from 22 sites.

Samples of reed stem tissue and periphyton were taken together with water samples. Underwater part of reed stem was cut and covered with a plastic tube. The lower part of the tube was closed while the stem was taken out of the water. In the laboratory, from a given area of this stem, periphyton was scraped for different analyses. After drying it at 105 °C, the dry mass was determined and the ash contents of periphyton and reed samples were estimated also. The concentration of 27 (Ag, Al, Au, Ar, B, Ba, Ca, Cd, Cr, Cu, Fe, K, La, Li, Mg, Mn, Mo, Na, Ni, P, Pb, Rb, S, Sr, Ti, V, Zn) elements was determined from the periphyton, reed and water samples by Inductively Coupled Plasma Atomic Emission Spectrometric (ICP-AES) method (Spectroflame ICP atomic emission spectrometer, Spectro GmbH). From these measurable elements, we had chosen some of the most important heavy metals (Cd, Cr, Cu, Fe, Mn, Mo, Pb, Zn). Correlations between the heavy metal contents of periphyton and its host plant and the concentration factors (Cf) were also calculated.

Chemical relationships of the four water bodies were investigated by multivariate statistical analysis. On the basis of mean heavy metal concentrations in the periphyton and the submerged reed stems (substrate), respectively, water bodies were classified by cluster analysis on percent similarity coefficients. Dendrograms were made by using the weighted centroid (median) clustering method (Kovach, 1993).

Results

On the basis of the ash content of periphyton, samples from Lake Velence and Lake Fertő (Lake Neusiedl)

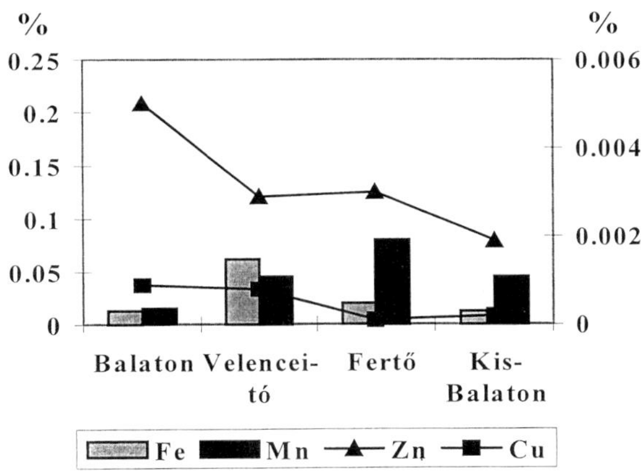

Figure 1. Heavy metal accumulation of green reed (*Phragmites australis*) in the littoral zone.

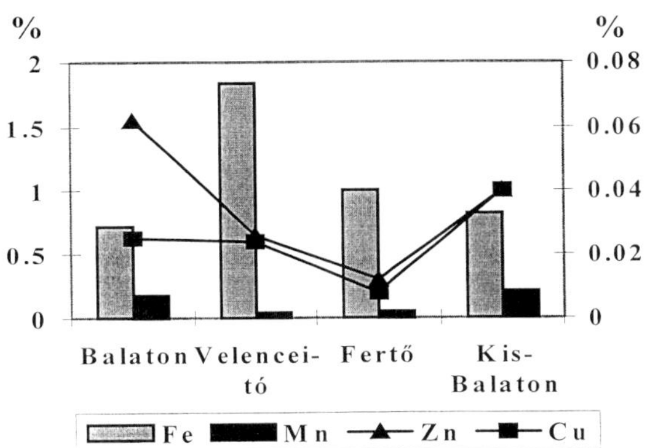

Figure 2. Heavy metal accumulation of reed-periphyton in the littoral zone.

are grouped into the same category and can be described by the dominance of inorganic components (Lakatos, 1983). Samples collected in Lake Balaton and KBWPS (Kis-Balaton) belong to the organic-inorganic category. The ash content of green reed samples are 5–15 times smaller than that of the old reed-periphyton.

The lowest iron concentration was measured in the periphyton and green reed samples from Lake Balaton (Figures 1 and 2), and the highest iron concentration samples from Lake Velence. In contrast to this, the manganese content of the samples of periphyton collected in Lake Balaton and Kis-Balaton was about threefold higher than in the samples of the other lakes. Similarly to the manganese, the highest concentrations of zinc and copper were detected in samples of Lake Balaton and Kis-Balaton which indicated twofold accumulation in comparison with the green reed samples. These values are six and eight times higher than in the periphyton samples collected in Lake Fertő (Figure 2).

The heavy metal content of the periphyton and green reed has reached higher values near to the bank than to the open-water side of reed stands in Lake Balaton. In case of iron and zinc contents, positive correlation can be found between the green reed and its periphyton (Figure 3).

Similarly to the reed-epiphyton studies of 1979 in Lake Balaton, values for concentration factors (Cf) have been calculated (Figure 4). Decreases can be observed in the sodium, iron and zinc contents in case of the reed-periphyton, while the concentration factor of manganese turned out to be larger. In the periphyton, 10 ppm concentration was measured for lead, which remains under the limit of traceability even if the water sample is subjected to a treatment of tenfold concentration. The reed-periphyton plays an important role in the adsorption and retention of cations, especially of heavy metals. The precipitated calcium carbonate and the various iron compounds are good adsorbents, thus, the periphyton functions as trapping for heavy metals and plant nutrients. Its contribution to the material circulation cannot entirely be regarded as passive process, since the living organisms of the periphyton, through their biochemical and enzymological processes, plays an essential role in it. They also provide basis for fish nutrition in the food network.

In case of KBWPS, among the cations detected in the inorganic fractions of the reed-periphyton, calcium was found in the highest content. The differences in the calcium content depended on the rate of the so called incrustation and excrustation. Iron, as a microelement was observed in the second or third highest concentrations in the periphyton; it exceeded the sodium and potassium contents significantly. The calcium, iron and magnesium play important roles in the cycle of dissolved ortho-phosphate (phosphorus), for example in the processes of precipitation and absorption.

To reveal the possible reasons of reed decline, healthy and diseased reed stems were analyzed. In case of some elements, significant differences were found between the healthy and diseased reed stem. The results of the study on element content of reed did not help us much in recognizing the direct causes of the decline of reed community, but lower content of molybdenum and sulphur of the diseased reed stems are warning (Kovács et. al., 1989; Lakatos, 1993).

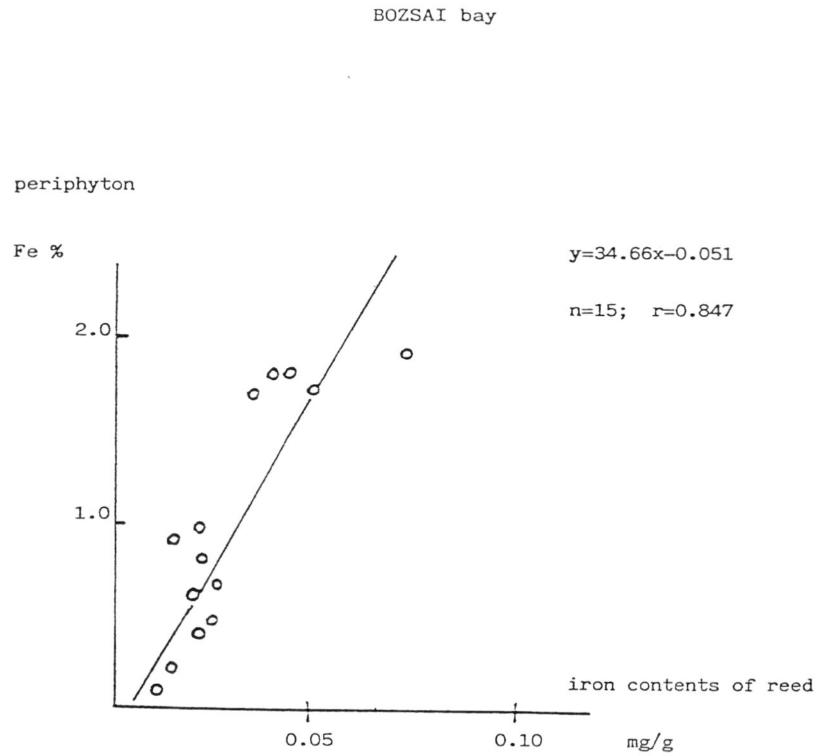

Figure 3. Correlation between the iron content of green reed and its periphyton (Lake Balaton).

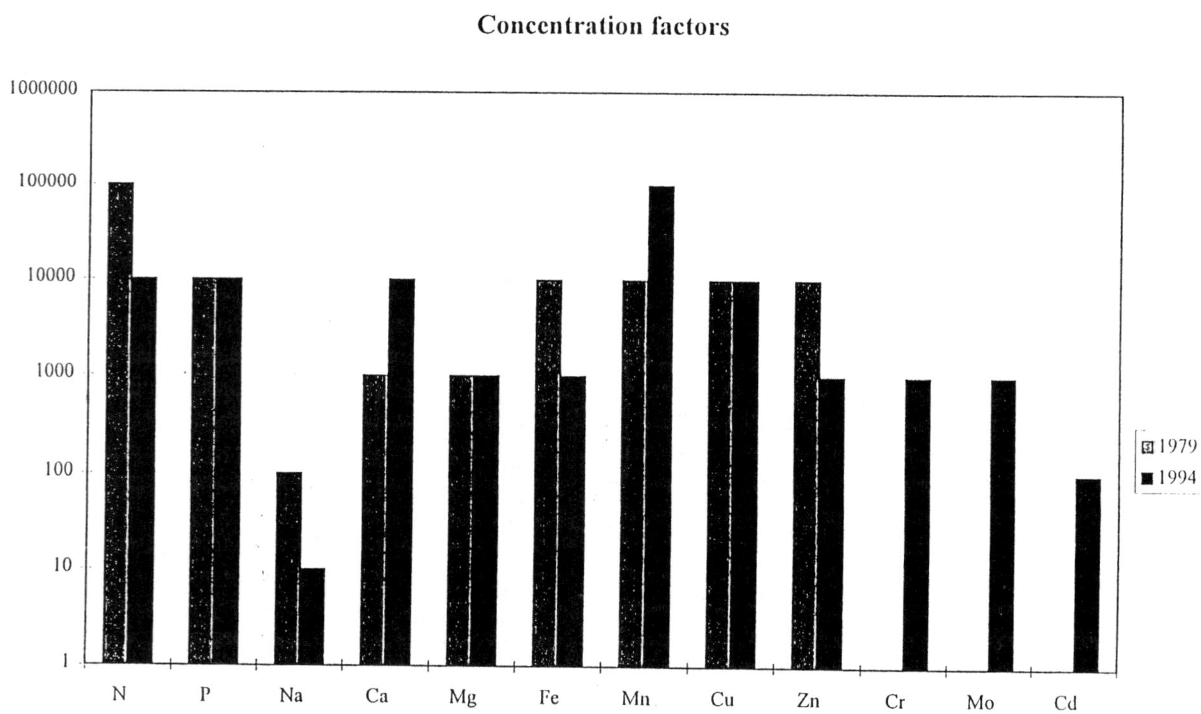

Figure 4. Concentration factors for the different elements in periphyton in 1979 and 1994 in Lake Balaton.

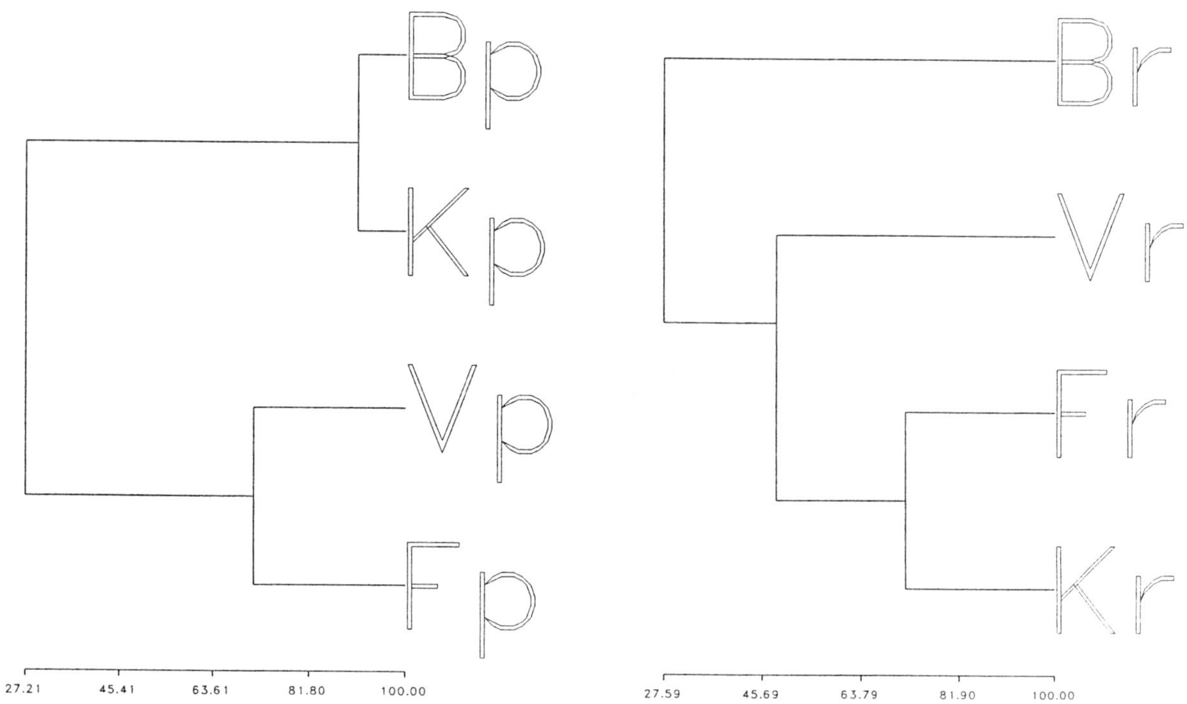

Figure 5. Relationship of Lake Balaton (B), Kis-Balaton (K), Fertö (F) and Lake Velence (V) on the basis of heavy metal composition of the periphyton (*p*) and submerged reed stems (*r*). Weighted centroid (median) clustering on percent similarity coefficients.

Results of the multivariate approach to the chemical relationships of the four water bodies are given in Figure 5. Cluster analysis revealed high degree of similarity between Lake Balaton and Kis-Balaton periphyton and somewhat lower between Lake Velence and Lake Neusiedl. However, the two groups seem to represent two different types of periphyton composition, the similarity being only c. 27% between the two.

Contrastingly, the dendrogram based on the heavy metal composition of submerged reed stems (periphyton substrate) seems to reflect a markedly different pattern, where reeds from Kis-Balaton and Lake Neusiedl were the most similar forming one loose cluster with reeds from Lake Velence, whilst the most distinct was the stand in Lake Balaton.

Discussion

Referring to the phenomenon of biogenic decalcination, among the elements of the reed-periphyton, concentration of the calcium proved to be the highest in all cases. Comparing the element concentration in the reeds near the shore, higher calcium concentration was found in the open-water side in Lake Balaton. The order of macro-ions according to their amount is the following: Ca>Mg>Al>K>Na. The iron content of the periphyton is also significant and it was detected in the second largest concentration (1.0%, in case of Keszthely Bay). The order of heavy metals examined in the periphyton was: Fe>Mn>Zn>Cr>Cu>Mo>Pb>Cd. The large chromium and molybdenum content of the periphyton, measurable with the application of ICP analysis, is quite uncommon and remarkable, although the common reed can withstand extreme environmental conditions, including the presence of toxic heavy metals, such as Zn and Cd (Dushenkov et al., 1995).

In Lake Neusiedl, the ash content is higher than in the other investigated lakes and do not decrease below 50% (Szczepanska, 1970; Kowalczewski, 1975). Although the water of the Lake Neusiedl is characterized by the dominance of Na and Mg, the reed-periphyton contained Ca in high concentration, which exceeds 10%. Mg reached relatively high values only in the samples taken from the southern part of lake.

Differences in the heavy metal content of periphyton appeared not only among the shallow water bodies, but among the different water areas of a given lake also. The heavy metal content of the periphyton was especially high in water areas effected by large anthropogenic load or located near to sewage effluent. In case of Lake Velence, the heavy metal concentration in the periphyton is high in the brown water area of the lake, which is still in natural condition, and in the green water area, where considerable planktonic eutrophication takes place due to the loading of nutrients and pollutants (Lakatos, 1978).

The concentration of Mn, Cr and Pb was considerably higher in the western part of Lake Neusiedl, which is more intensively used for recreation. These cations and their concentration factors indicate the impacts of the external loading and express good evidence for the filtering, accumulating and eliminating role of the reed-periphyton complex. For the periphyton, the analysis showed positive correlation between the individual heavy metals and the magnesium which underlines the importance of the adsorption processes (Allen, 1971; Mickle & Wetzel, 1978).

The high calcium, iron and magnesium content of the periphyton and the favourable circumstances of oxygen supply ensures the proper functioning of reed-periphyton as biofilters and chemical traps, through which they effect the water quality.

Conclusion

The most important plant communities in the littoral zone of the lakes studied are the reed communities. The effectivity of reed belts in the improvement of water quality is enhanced by the periphyton. The reed-periphyton complex plays an important role in the adsorption and retention of cations, especially of heavy metals.

Differences in the heavy metal content of periphyton and its host plant were found among the shallow water bodies. Comparing the heavy metal content of the four large shallow waters in Hungary, the highest iron content can be detected in the samples collected in Kis-Balaton, and the concentration of manganese is also higher than in Lake Velence and Lake Neusiedl, while the concentrations of zinc and copper are relatively low.

The heavy metal content of reed and periphyton was especially high at places which were impacted by large anthropogenic load. On the basis of our results, the biofilter role of reed-periphyton complex can be confirmed.

Since the main reasons of reed decline have not been revealed yet, therefore future research has to focus on it and involve reed and reed-periphyton studies, which may provide a good basis to find the most appropriate ways to save and restore the reed communities of shallow standing waters.

Acknowledgements

This study was carried out within the framework of CEU, OTKA (F 023775) and EUREED Project financial grant which is greatfully acknowledged.

References

Allen, H. L., 1971. Primary productivity, chemo-organotrophy and nutritional interactions of epiphytic algae and bacteria on macrophytes in the littoral of a lake. Ecol. Monogr. 41: 97–127.

Armstrong, J., W. Armstrong & W. H. Van Putten, 1996. *Phragmites* die-back: bud and root death, blockages within the aeration and vascular systems and the possible role of phytotoxins. New Phytol. 133: 399–414.

Dinka, M., 1986. Accumulation and distribution of elements in cattail species (*Typha latifolia* L., *T. angustifolia* L.) and reed (*Phragmites australis*/Cav./Trin. ex Steudel) living in Balaton. Proceedings EWRS/AAB 7th Symposium on Aquatic Weeds: 81–86.

Dinka, M., P. Szeglet & I. Szabó, 1995. Hungarian Group Report. In Van Putten, W. H. (ed.) Reed News Reports of EC Project EUREED-EV5V-CT92-0083, vol. 3. Heteren, The Netherlands: Netherlands Institute of Ecology, 96–107.

Dokulil, M., 1975. Planktonic primary and bacterial productivity in shallow waters within a large *Phragmites* community (Neusiedlersee, Austria). Verh. int. Ver. Limnol. 19: 1295–1304.

Dokulil, M., H. Metz & D. Jewson (eds), 1980. Shallow lakes contributions to their limnology. Dr W. Junk Publishers, The Hague: 212 pp.

Dushenkov, V., P. B. A. Nanda Kumar, H. Motto & I. Raskin, 1995. Rhizofiltration: the use of plants to remove heavy metals from aqueous streams. Envir. Sci. Technol. 29: 1239–1245.

Haslam, S. M., 1978. River plants. University Press: Cambridge.

Korponai, J., K. Mátyás, G. Paulovits, I. Tátrai & N. Kovács, 1997. The effect of different fish communities on the cladoceran plankton assemblages of the Kis-Balaton Reservoir, Hungary. Hydrobiologia 360: 211–221.

Kowalczewski, A., 1965. Changes in the periphyton biomass of Mikolaiskie Lake. Bull. Acad. Pol. Sc. 13: 395–398.

Kovács, M., G. Turcsányi, Z. Tuba, S. E. Walcsánszky, T. Vásárhelyi, A. Dely-Draskovits, S. Tóth, A. Koltay, L. Kaszab, P. Szőke & B. Jankó, 1989. The decay of reed in Hungarian lakes. In Salánki, J. & S. Herodek (eds), Conservation and Management of Lakes. Symp. biol. Hung. 38: 461–471.

Kovach, W. L., 1993. MVSP Version 2.1 Users's Manual. Kovach Computing Services, Pentraeth, Wales, U.K: ii+55 pp.

Lakatos, G., 1978. The phenomenon and significance of benthonic eutrophication in Lake Velencei, Hungary. Acta Biol. Debrecina 15: 147–168.

Lakatos, G., 1983. Accumulation of elements in biotecton forming on reed (*Phragmites australis*) in two shallow lakes in Hungary. Proc. Int. Symp. Aquat. macrophytes, Nijmegen: 117–122.

Lakatos, G., 1989. Composition of reed periphyton (biotecton) in the Hungarian part of Lake Fertő. BFB-Bericht 71: 125–134.

Lakatos, G., 1993. Studies on heavy metals of periphyton and its host plant (*Phragmites australis* (Cav.) Trin. ex Steudel) in shallow lakes. Stud. Envir. Sci. 55: 365–372.

Lakatos, G. & P. Bíró, 1991. Study on chemical composition of reed-periphyton in Lake Balaton. BFB-Bericht 77: 157–164.

Lakatos, G., I. Grigorszky & P. Biró, 1998. Reed-periphyton complex in the littoral of shallow lakes. Verh. int. Ver. Limnol. 26: 1852–1856.

Mickle, A. M. & R. G. Wetzel, 1979. Effectiveness of submerged angiosperm-epiphyte complex on exchange of nutrients and organic carbon in littoral system. III. Refractory organic carbon. Aquat. Bot. 6: 339–355.

Ostendorp, W., 1989: 'Die-back' of reeds in Europe – a critical review of literature. Aquat. Bot. 35: 5–26.

Padisák, J., 1982. The periphyton of Lake Fertő: species composition and Chlorophyll *a* content. BFB-Bericht 43: 95–105.

Pomogyi, P., 1993. Nutrient retention by the Kis-Balaton Water Protection System. Hydrobiologia 251: 309–320.

Simonian, A., I. Tátrai, P. Bíró, G. Paulovits, L. G. Tóth & G. Lakatos, 1995. Biomass of planktonic crustaceans and the food of young cyprinids in the littoral zone of Lake Balaton. Hydrobiologia 303: 39–48.

Sommer, U., 1977. Productionsanalysen am Periphyton in Schilfgürtel des Neusiedler Sees. Sitzungsber. Österr. Akad. Wiss., mathem.-naturwiss. Kl., Abt. 1 186: 219–246.

Szczepanska, W., 1970. Periphyton of several lakes of the Mazurian lakeland. Pol. Arch. Hydrobiol. 17: 397–418.

Wetzel, R. G., 1964. A comperative study of the primary productivity of higher aquatic plants, periphyton and phytoplankton in large, shallow lake. Int. Rev. ges. Hydrobiol. 49: 1–61.

Hydrobiologia **415:** 55–58, 1999.
J. M. Caffrey, P. R. F. Barrett, M. T. Ferreira, I.S. Moreira, K. J. Murphy & P. M. Wade (eds), Biology, Ecology and Management of Aquatic Plants.
© 1999 *Kluwer Academic Publishers. Printed in the Netherlands.*

Gas transfer in wetland plants controlled by Graham's law of diffusion

W. Grosse* & H.-J. Frick
Botanical Institute, University of Cologne, Gyrhofstrasse 15–17, D-50923 Köln, Germany
E-mail: wgrosse@novell.biolan.uni-koeln.de

Key words: anoxia, oxygen shortage, *Phragmites australis*, pressurized ventilation, waterlogged soil

Abstract

In wetland plants, internal transfer of gases is vital for growth on permanently water-saturated or waterlogged soils, being required to supply oxygen to the rhizomes and roots. Over the past decade, various physical mechanisms have been proposed to be responsible for pressurized ventilation in plants which are adapted to wetland and freshwater habitats. It has now become clear that transfer effects that can be described by Graham's law of diffusion are involved in this process, in addition to thermal transpiration, humidity-induced and venturi-induced gas transport in swamp grasses. Gas flow in accordance with Graham's law has also been reported for roots of *Alnus* (alder) trees. We suggest that this effect may contribute significantly to oxygen supply to the rhizosphere and to the remineralization of organic compounds in soil, thus improving plant nutrition.

Introduction

Accumulating evidence suggests that the floating leaves of many aquatic plant species, fresh or dry culms of grasses, and even trunks of wetland trees that contain abundant lenticels, operate essentially as gas flow generators (reviewed by Grosse et al., 1996). Physical effects such as thermal transpiration (Grosse, 1996), differences in humidity and Bernoulli forces (Armstrong et al., 1996), together with adaptations such as the porous partitions that separate the aerenchyma from the ambient air, can account for the gas flow phenomena observed. While gas flow based on these purely physical effects does not depend on any physiological activity of the plants, gas flow resulting from processes governed by Graham's law of diffusion may be controlled by metabolic activities of plants and rhizospherial micro-organisms. Gas flow resulting from differences in mean molar masses of gas mixtures inside and outside the plants has been reported for the surfaces of floating leaves of *Nymphaea alba* (Schiwinski et al., 1996) and in roots of *Alnus glutinosa* (Frick et al., 1997). The present studies explain how gas flow phenomena that follow Graham's law of diffusion may contribute to internal ventilation in wetland vegetation, aerating submerged organs and facilitating oxygen input into anoxic soils for stimulation of remineralization processes and bacterial growth.

* Author for correspondence

Materials and methods

Plant material

Culms of the two wetland grass species *Phragmites australis* (Cav.) Trin. ex Steud. and *Ischaemum aristatum* var. *glaucum* (Honda), cultivated outdoors under swamp conditions in the experimental garden of the Botanical Institute at Cologne, were used for the studies.

Experimental methods

Excised culms were placed in polyvinyl tubes of appropriate length and diameter, so as to establish two gas-filled spaces, one within the culm and the other between the culm surface and the wall of the tube. The open end of each tube was immersed in a water-filled vessel. To measure internal pressurization and gas flow, a pressure transducer (Type 8260, Model CP capacity 0–25 mm H_2O, Sensotec, Columbus, U.S.A.) and a mass flowmeter (Type FMA 1802, Omega Engineering, Stanford, U.S.A.; sensitivity: 0–10 cm^3

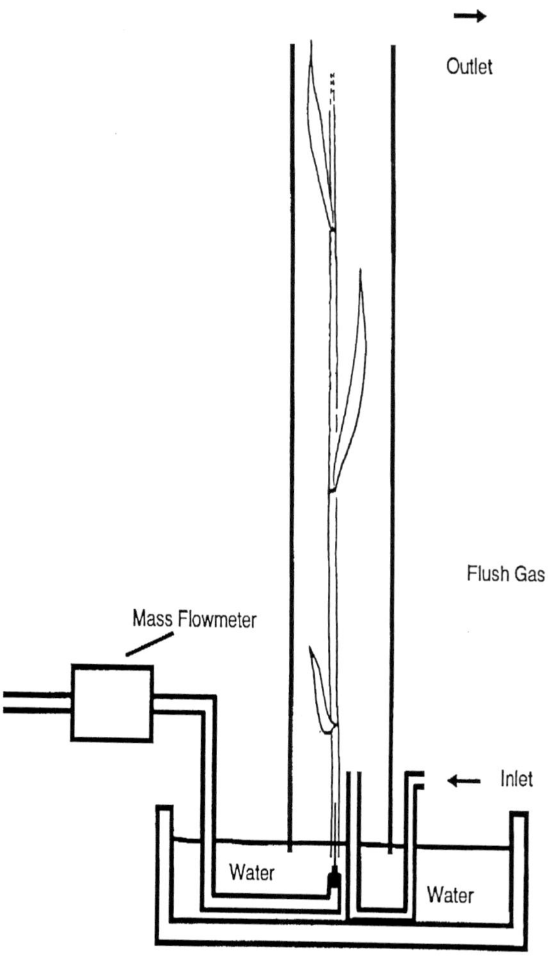

Figure 1. Experimental set up for pressurization and gas flow studies on excised culms. Arrows indicate flow direction of the flush gas.

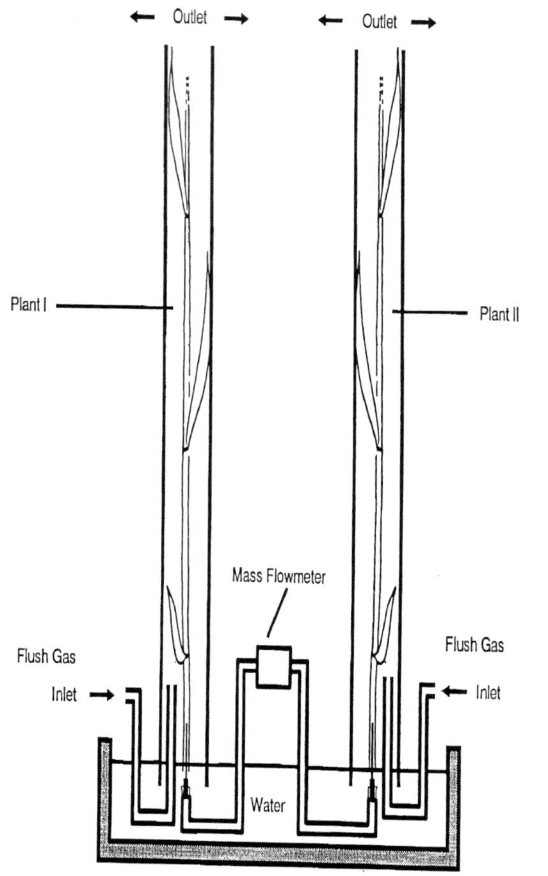

Figure 2. Experimental set up for gas flow studies on simplified plants. Arrows indicate flow direction of the flush gas.

min^{-1}), respectively, were connected to the internal gas space by inserting hypodermic needles into the basal end of the leafy culm (Figure 1). An experimental set up as shown by Figure 2 was used to study gas exchange phenomena in a simplified plant. Gas was supplied via a separate input line, at a constant flow rate (300 cm^3 min^{-1}) controlled by a manostat (Type Fa-149, Wallace & Tiernan, Guenzburg, Germany). After the system had been allowed to reach equilibrium (with the same gas in both spaces), gas flow in accordance with Graham's law was induced by flushing the tube with a different gas species. Besides air (M = 28.96 g mol^{-1}), helium (He, M = 4.00 g mol^{-1}), methane (CH_4, M = 16.04 g mol^{-1}) and nitrogen (N_2, M = 28.01 g mol^{-1}) were used as flush gases. All gases were first saturated with water by passage through a water column. An artificially rapid transient was recorded when the flush gas was changed, especially to methane. The experiments were carried out at room temperature ($T \approx 293$ K).

Results

If gas flow studies are performed on single culms of either grass species (cf. Figure 1), a transient efflux of gas from the aerenchyma through the attached flowmeter is observed both with *Ischaemum aristatum* (Figure 3, above) and with *Phragmites australis* (Figure 3, below) when the molar mass of the flush gas is lower than that of the gas used for equilibration (i.e. on shifting from N_2 to He, N_2 to CH_4, CH_4 to He). Conversely, transient gas uptake by the culm is observed

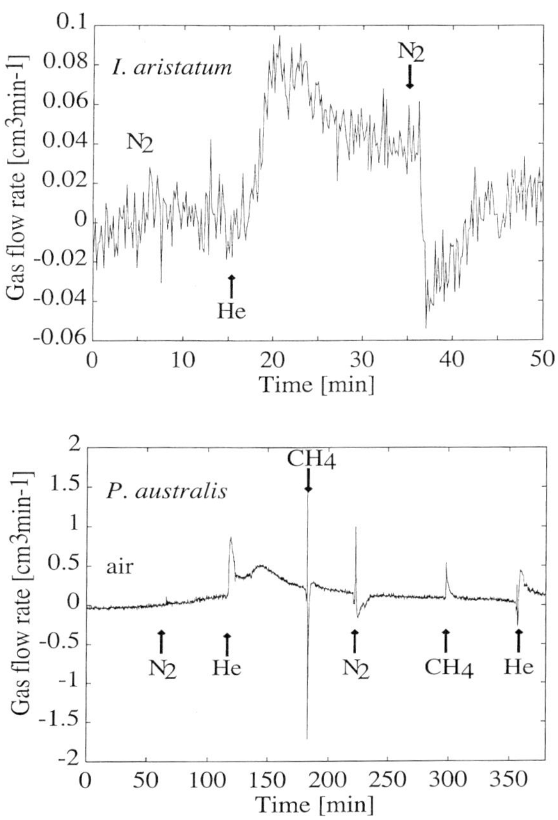

Figure 3. Gas flow through the mass flowmeter into and out of aerenchyma observed on flushing the gas space around grass culm (arrow) of *Ischaemum aristatum* (above) and *Phragmites australis* (below) with gas of molar mass different from that of the gas used for equilibration. Negative sign indicates gas suction into the culm aerenchyma.

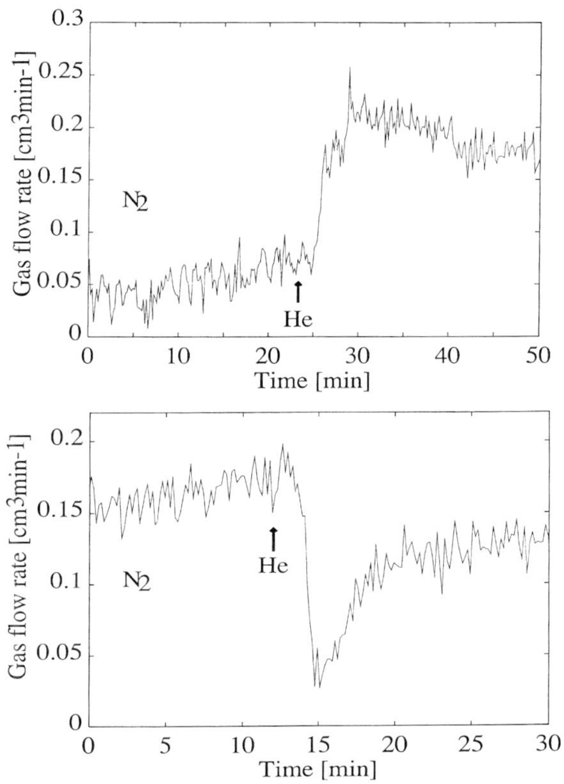

Figure 5. Convective gas flow between two culms of *Phragmites australis* attached to each other as shown in Figure 2 with flushing plant I (above) and plant II (below), respectively. Experiments are performed as in Figure 3.

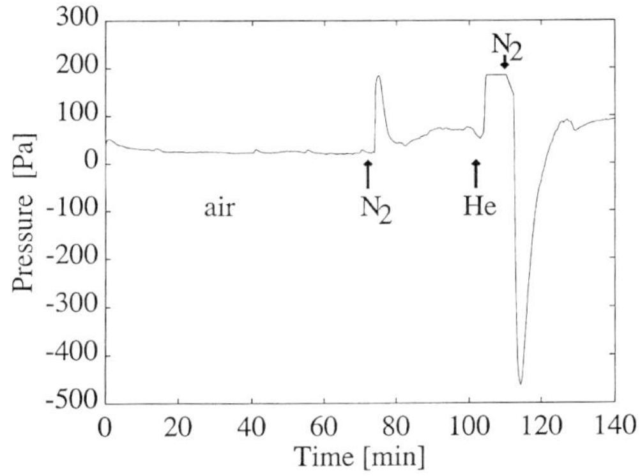

Figure 4. Transient pressure differences between the aerenchyma of *Phragmites australis* and the atmosphere as the result of the physical effect described by Graham's law of diffusion. High pressure is not recorded for the asymmetric readings of the pressure transducer. Experiments were performed as in Figure 3.

when a flush gas of higher molar mass is used (He to CH_4, CH_4 to N_2). Gas efflux and gas uptake follow on a transient increase or decrease, respectively, in the net pressure in the aerenchyma, as shown for *P. australis* in Figure 4, in which pressure changes recorded by the pressure transducer in response to alteration of the flush gas are plotted.

If two excised culms of *P. australis* are connected to each other via a flowmeter as shown in Figure 2, convective gas flow through the system, which mimics a natural stand of the reed, can be manipulated. If the tube containing plant I (cf. Figure 2) is flushed with He, a gas of lower molar mass than that of the N_2 used for equilibration, convective gas flow occurs from plant I to plant II (Figure 5, above). If plant II instead of plant I is flushed with He, the direction of net flow is reversed (Figure 5, below).

Discussion

Graham's law of diffusion describes exchange between two gases that differ in molecular mass and are confined in neighboring compartments separated by a porous partition with pore diameters in the range of the mean free path lengths of the gas molecules. Since more of the smaller molecules than the larger ones cross the partition in a given time, a net flow of gas occurs into the compartment containing the larger molecular species. This causes a transient increase in pressure in that compartment and a corresponding suction pressure in the other, that lasts until the mean molar masses of the two bulk phases reach equilibrium (Schiwinski et al., 1996). If the compartments are open, the net result of the pressure difference is an efflux from the pressurized compartment and uptake into the compartment containing the species of lower mass.

The present experimental results, of both the one- and two-culm trials, are fully in accord with the situation described above. Graham's law of diffusion accounts for the pressurization of the culm (Figure 4) and the subsequent gas efflux (Figure 3) observed when nitrogen is replaced by helium as the flushing gas (helium was chosen for these experiments because its molecular mass is markedly different from that of nitrogen, and it diffuses at high rates). The difference in flow rates between *P. australis* and *I. aristatum* may reflect the much lower surface area of the leaf sheaths in the latter, which provide the main entrance route for gases into the leafy culms of grasses (Armstrong & Armstrong, 1990).

With regard to grass tussocks, high respiratory activity in one culm will increase the mean molar mass of the gas mixture as a result of CO_2 production and O_2 consumption. The result will be a net flow of air into the culm, followed by gas flow through the rhizomes to less active culms where the gas phase may be enriched with methane taken up from the anoxic soil. Since higher concentrations of methane decrease the mean molar mass of the gas mixture, a net outflow occurs from these culms. Based on Graham's law of diffusion, plant metabolism may contribute, therefore, to convective gas throughput and root aeration. On the other hand, high rates of CO_2 production by soil bacteria in the rhizosphere of well aerated roots may initiate an efflux of oxygen out of the roots. By generating oxygenated zones in anoxic soils that permit growth of aerobic bacteria which are involved in remineralization of organic material, the process will improve the nutrient supply and facilitate nutrient uptake by plants.

Acknowledgments

These studies are part of the EC research project no ERBIC18CT960059 in the INCO-DC programme of the European Community. The authors are grateful to Dr P. A. Hardy, University of Cologne, for assistance with some linguistic problems.

References

Armstrong, J. & W. Armstrong, 1990. Pathway and mechanisms of oxygen transport in *Phragmites australis*. In Cooper, P. F. & B. C. Findlater (eds), The Use of Constructed Wetlands in Water Pollution Control. Pergamon, Oxford: 529–533.

Armstrong, J., W. Armstrong, P. M. Beckett, J. E Halder, S. Lythe, R. Holt & A. Sinclair, 1996. Pathways of aeration and the mechanism and beneficial effects of humidity- and Venturi-induced convections in *Phragmites australis* (Cav.) Trin. Ex Steud. Aquat. Bot. 54: 121–135.

Frick, H. J., D. Woermann & W. Grosse, 1997. How tree roots gather oxygen from soil for energy supply by respiration. Z. Naturforsch. 52c: 824–827.

Grosse, W., 1996. The mechanism of thermal transpiration (= thermal osmosis). Aquat. Bot. 54: 101–110.

Grosse, W., J. Armstrong & W. Armstrong, 1996. A history of pressurized gas-flow studies in plants. Aquat. Bot. 54: 87–100.

Schiwinski, K., W. Grosse & D. Woermann, 1996. Convective gas flow in plant aeration and Graham's law of diffusion. Z. Naturforsch. 51c: 681–690.

Hydrobiologia **415**: 59–65, 1999.
J. M. Caffrey, P. R. F. Barrett, M. T. Ferreira, I.S. Moreira, K. J. Murphy & P. M. Wade (eds),
Biology, Ecology and Management of Aquatic Plants.
© 1999 *Kluwer Academic Publishers. Printed in the Netherlands.*

Aquatic plants in the rice fields of the Tagus Valley, Portugal

T. Vasconcelos[1], M. Tavares[2] & N. Gaspar[2]
[1]*Departamento de Protecção das Plantas e de Fitoecologia, Instituto Superior de Agronomia, Tapada da Ajuda, P-1349-017 Lisboa, Portugal*
[2]*Escola Superior Agrária de Santarém, S. Pedro, P-2002-655 Santarém, Portugal*

Key words: ecological profiles, rice weeds, neophytes, *Oryza sativa*, *Heteranthera limosa*, *Heteranthera reniformis*

Abstract

In 1997, in 50 surveys of irrigated rice fields (*Oryza sativa* L.), 47 species of vascular plants were recorded in 19 families. Most of the species belonged to the *Gramineae* and *Cyperaceae* families. Seven species of algae were found, mostly *Hydrodictyon reticulatum* (42%) and *Chara vulgaris* (38%). *Echinochloa crus-galli* and *Paspalum paspalodes* showed a relative frequency of 98% and 82%, respectively. An interpretation of the major factors is presented influencing weed distribution. Amongst the weeds recorded, 13 were alien species. For these alien species the probable date of introduction into Portugal and into rice fields was determined. The recently introduced neophytes *Heteranthera limosa* and *Heteranthera reniformis* and the increasing presence of red rice (*Oryza sativa*), represent a potential danger to rice culture.

Introduction

The main objectives of this work were the characterisation of the rice weed flora in the Tagus Valley the relative species-factors determining ecological species behaviour and the importance of ecological factors in floral composition.

Rice growing was started in the reign of D. Dinis (1279–1325) (Vasconcellos, 1963). The major weeds were introduced with the rice crop and foreign rice varieties. Records of location and date of introduction are from national literature and annotations of *Herbarium* (Botanical Institute Gonçalo Sampaio, of Porto (PO); Botanical Institute Júlio Henriques, of Coimbra (COI); Faculdade de Ciências, of Lisbon (LISU); Instituto Superior de Agronomia, of Lisbon (LISI); Comissão Reguladora do Comércio do Arroz, of Lisbon (LISA) and Estação Agronómica Nacional (LISE), of Oeiras). For the species found in 1997, in the rice fields of Tagus and Sorraia Valley, the probable date of introduction into Portugal and into rice fields was determined.

Materials and methods

Between August and September 1997, 50 rice fields were surveyed and phytotechnical investigations were carried out. Soil samples were taken at 20 cm depth for further physico-chemical analysis. An abundance index was attributed to the surveyed species, corresponding to the number of plants per square metre, using the Barralis abundance scale (1975). This enabled the calculation of the median abundance of each species, using the Barralis formula (1976). The degree of infestation was determined as in Michez & Guillerm (1984). The following factors were also studied: discontinuous (region, rice varieties, method of sowing and weed control) and continuous (texture, pH (H_2O), assimilable phosphorus, assimilable potassium, organic matter and salt).

The method of ecological profiles and information shared between species and ecological factors (Daget & Godron, 1982) uses the following calculations:

(i) The comprehensive profile for each factor L with K classes gives the distribution of absolute frequency. If the factor has been well sampled it will be equally distributed through the classes, with high indetermination and entropy.

(ii) Entropy of factor (H_L), maximum entropy (H_{maxL}) and sampling quality (Q_L), $H_L = \sum_i^K \frac{R_i}{N} \log_2$, $H_{maxL} = \log_2 K$ and $Q_L = \frac{H_L}{H_{maxL}}$, where N is the total of surveys and R_i the number of surveys in class K.

(iii) A corrected frequency profile for species in K classes of factor is found by multiplying relative frequency by the inverse of average relative frequency over the whole survey.

(iv) Mutual information between species and factors, allowing the determination of those factors which play an important role in the distribution of species, $I = \sum_i^K \frac{U_i}{N} \log_2 \left(\frac{U_i}{R_i} \cdot \frac{N}{U_t}\right) + \sum_i^K \frac{V_i}{N} \log_2 \left(\frac{V_i}{R_i} \cdot \frac{N}{V_t}\right)$ where U is the number of surveys of the class where a species is present; U_t the total number of surveys where a species is present; V the number of surveys of the class where a species is absent and V_t the total number of surveys where a species is absent.

(v) The mean of the mutual information determining active factors which play an important role in the distribution of species.

(vi) Determined ecological preference and groups. For the continuous factors, the species were grouped in scale-imbricated groups (the species are grouped with respect to amplitude of ecological profile, e.g. 301 present in class 1, 2 and 3; 302 present in class 2 and 3) and their ecological preference determined by calculation of the center of gravity of the corrected frequency profile (Daget et al., 1971; Daget, 1976). For the discontinuous factors, species with similar profiles were marked with + (corrected frequency $\leq$ 100) and - (corrected frequency $<$ 100), according to Guillerm (1969).

Bibliographical records were consulted to determine which of the weed communities were alien and introduced into the rice fields. The origin and nomenclature of the species were in those "New Flora of Portugal" (Franco, 1971, 1984; Franco & Afonso, 1994, 1998), "Flora Europaea" (Tutin et al., 1964–1980) and "Flora of Italy" (Pignatti, 1982). In the case of the *Echinochloa* spp., the work of Carretero (1981) was used. Where necessary, the nomenclature of the species was updated.

Results and discussion

A total of 56 taxa were identified: 47 vascular plant species representing 19 families (*Gramineae*–21.3%, *Cyperaceae*–14.9%, *Lythraceae*–10.6%, *Compositae*–6.4%, *Polygonaceae*–6.4%) and 7 species of algae in four families. *Echinochloa crus-galli* and *Lemna minor* presented a high degree of infestation. The following species *Alisma lanceolatum, Alisma plantago-aquatica, Ammannia × coccinea, Glyceria declinata, Lemna gibba, Lindernia dubia, Paspalum paspalodes* and *Scirpus maritimus* presented a moderate degree of infestation. Therophytes and helophytes were predominant (36.1% and 21.3%, respectively).

The ecological groups and ecological preferences for the species with relative frequency ≥6% are shown in Table 1. The other species, *Apium nodiflorum, Cyperus fuscus* for. *virescens, Echinochloa colonum, Galega officinalis, Leersia oryzoides, Lotus pedunculatus, Lythrum junceum, Myriophyllum aquaticum, Scirpus lacustris* ssp. *tabernaemontani, Zannichellia palustris* and *Bidens frondosa, Eclipta prostrata, Eleocharis palustris, Lythrum salicaria, Panicum repens, Polypogon monspeliensis. Ranunculus sceleratus* ssp. *sceleratus, Rumex conglomeratus* and *Verbena officinalis*, showed a relative frequency of 2% and 4% respectively.

Echinochloa spp. was the greatest problem. The weeds of paddy fields differed, due to both changes in agricultural technology and also to the accidental introduction of new weeds such as *Oryza sativa* (red rice).

The mean of mutual information between species-factors for 28 species, for the factors studied and equitability of sampling, is shown in Figure 1.

The survey demonstrated that the main weeds in rice ($*$ in Table 1) were able to colonise in all classes of factor studied, but had an ecological preference for one class (indicated with +, - and number). The factor soil pH was not equitably sampled and is the most active factor (Figure 1). The most remarkable trend is shown by *Cyperus eragrostis, Lythrum thymifolia, Phragmites australis* and *Polygonum aviculare* (ecological group 303), which are restricted to alkaline soils. The factor soil texture is a active factor and was better sampled than soil pH. Eleven species have been found in all classes of soil texture (ecological group 401) and with an ecological preference for class 3, but *Echinochloa oryzicola* have an ecological preference for class 2 (between 1.5 and 2.5).

Table 1. Ecological groups (scale imbricated - si) and ecological preference (g) for the factors: geographical situation (gs), method of sowing (sk), weed control (wc), texture (t), $pH(H_2O)$ (pH), salt (s), organic matter (om), assimilable phosphorus (p), assimilable potassium (k)

Taxa	Relative frequency	gs		sk		wc		t		pH		s		om		p		k	
		1	2	1	2	1	2	si	g	si	g	si	g	si	g	si	g	si	g
Alisma lanceolatum	28	+	–	+	–	+	–	301	1.55	301	2.28	201	1.35	201	1.68	201	1.55	201	1.82
**Alisma plantago-aquatica*	46	–	+	–	+	+	–	401	3.09	301	1.61	201	1.55	201	1.34	201	1.29	201	1.21
**Ammannia×coccinea*	36	+	–	+	–	+	–	401	2.53	301	2.04	201	1.30	201	1.48	201	1.63	201	1.48
Aster squamatus	16		+	+	–	–	+	201	1.58	302	2.49	201	1.78	201	1.66	202	2.00	202	2.00
Azolla filiculoides	6		+	–	+	–	+	301	1.70	301	1.96	202	2.00	101	1.00	201	1.57	202	2.00
**Callitriche stagnalis*	16	–	+	–	+	+	–	401	3.37	301	1.48	201	1.48	201	1.28	201	1.29	201	1.10
**Cyperus difformis*	26	–	+	–	+	+	–	401	2.55	301	1.87	201	1.29	201	1.50	201	1.44	201	1.44
Cyperus eragrostis	14		–	+	+	–	+	301	2.18	303	3.00	201	1.35	201	1.32	201	1.80	202	2.00
**Echinochloa crus-galli*	98	–	+	–	+	–	+	401	2.53	301	2.02	201	1.51	201	1.49	201	1.49	201	1.49
**Echinochloa oryzicola*	32	–	+	+	–	–	+	401	2.27	301	2.19	201	1.45	201	1.48	201	1.46	201	1.71
Eichhornia crassipes	16	–	+	+	–	–	+	301	2.16	301	2.34	201	1.30	201	1.41	201	1.67	201	1.71
Glyceria declinata	14	–	+	–	+	+	–	301	2.73	201	1.19	201	1.49	201	1.32	201	1.21	201	1.06
Heteranthera limosa	6		+		+		+	302	2.72	201	1.42	201	1.27	101	1.00	101	1.00	101	1.00
Lemna gibba	10	+	–	+		+	–	201	1.17	302	2.56	101	1.00	202	2.00	201	1.73	202	2.00
**Lemna minor*	58	–	+	–	+	+	–	401	2.49	301	1.84	201	1.60	201	1.42	201	1.45	201	1.40
**Lindernia dubia*	46	–	+	–	+	+	–	401	3.09	301	1.62	201	1.59	201	1.39	201	1.34	201	1.28
Lycopus europaeus	6		+	–	+	–	+	302	2.39	302	2.39	202	2.00	101	1.00	201	1.25	201	1.41
Lythrum hyssopifolia	12		+	+	–	–	+	301	2.38	301	2.40	201	1.78	201	1.19	201	1.57	202	2.00
Lythrum thymifolia	6		+	+		–	+	302	2.39	303	3.00	201	1.59	201	1.70	202	2.00	202	2.00
**Oryza sativa*	36	+	–	+	–	–	+	401	2.77	301	2.18	201	1.42	201	1.54	201	1.57	201	1.64
**Paspalum paspalodes*	82	+	–	–	+	+	–	401	2.56	301	1.92	201	1.47	201	1.53	201	1.46	201	1.43
Phragmites australis	16	+	–	+		–	+	302	2.30	303	3.00	201	1.30	201	1.41	202	2.00	202	2.00
Polygonum aviculare	10		+	+		–	+	301	1.80	303	3.00	202	2.00	201	1.23	201	1.73	202	2.00
Polygonum salicifolium	20		+	+	–	–	+	301	1.91	302	2.74	201	1.72	201	1.33	201	1.73	202	2.00
Ranunculus trilobus	6	+	–	+		–	+	201	1.29	302	2.14	201	1.27	201	1.70	201	1.57	202	2.00
Scirpus maritimus ssp. *maritimus*	24	+	–	+		–	+	301	1.75	302	2.49	201	1.46	201	1.62	201	1.77	202	2.00
**Scirpus mucronatus*	16	–	+	+	–	–	+	401	3.04	301	1.98	201	1.68	201	1.54	201	1.67	201	1.71
Typha domingensis	8		+	+	–	–	+	201	1.45	302	2.49	201	1.42	201	1.54	202	2.00	202	2.00

Classes and comprehensive profiles for the factors: discontinuous (geographical situation) (1) north of Tagus = 20 (2) south of Tagus = 30; method of sowing (1) on flooded fields = 35 (2) on dry field = 15; weed control (1) unmixed = 23 (2) mixture = 27; + (corrected frequency profile $\leq$ 100); – (corrected frequency profile < 100) and continuous (texture) (1) silty clay = 15 (2) silty clay loam = 18 (3) silt. silt loam. loam = 14 (4) sandy loam. sandy = 3; $pH(H_2O)$ (1) acid = 13 (2) neutral = 9 (3) alkaline = 28; salt (1) with = 21 (2) without = 29; organic matter (1) $\leq 2.0\%$ = 27 (2) > 2.0% = 23; assimilable phosphorus (1) $\leq$ 200 mg P_2O_5/1000g of soil = 20 (2) > 200 mg P_2O_5/1000g of soil = 30; assimilable potassium (1) $\leq$ 200 mg K_2O/1000g of soil = 13 (2) > 200 mg K_2O/1000 g of soil = 37; scale imbricated 101 - distributed in class 1; 201- distributed in classes (1, 2); 202 (2); 301 (1, 2, 3); 302 (2, 3); 303 (3); 401 (1, 2, 3, 4); 402 (2, 3, 4); 403 (3, 4); 404 (4)

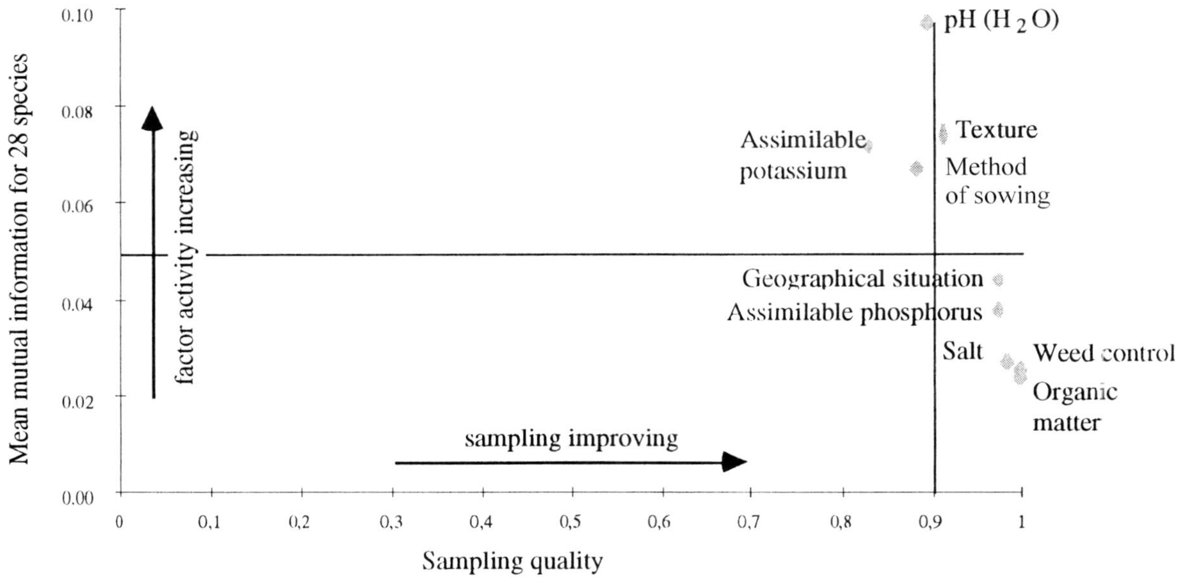

Figure 1. Activity of the factors.

In Table 2 the species of the vascular plants observed are indicated, their origin and Portuguese bibliographical references to them in this century. Many of the alien species were introduced with seeds of rice varieties (MacIntyre & Barrett, 1985). *Azolla filiculoides* was introduced before the 1930s and *Paspalum paspalodes* and *Cyperus eragrostis* were referred to in the bibliography for Portugal. In the 1940s, the species introduced were: *Ammannia×coccinea*, *Bidens frondosa*, *Eclipta prostrata*, *Eichhornia crassipes* and *Myriophyllum aquaticum*. In the 1950s, the authors registered *Aster squamatus*, *Echinochloa colonum*, *Echinochloa oryzicola* and *Lindernia dubia*. More recently, the species introduced were two species of the genera *Heteranthera* – *H. limosa* (Swartz) Willd. and *H. reniformis* Ruiz & Pavón. The latter species was not found in the study area. *Heteranthera limosa*, found in 1997 in three rice fields in the study zone, is one of the more recent introductions into Portugal. It was not recorded in rice fields in Portugal before 1980 (Espírito-Santo & Rosa, 1981). *H. limosa* was common in 1994, in the region of Alcácer do Sal (LISI). Desfayes (1996) refers to the presence of *Heteranthera rotundifolia* (Kunth) Griseb in one rice field in Coruche. This species was imported from South America into Italy in 1968–1969 together with the seeds of commercial varieties of rice. Water, birds and mechanical means aided the spread of *H. limosa* and *H. reniformis*, which became the 4th and 6th most important weeds in rice in Italy (Sgattoni et al., 1989). *H. reniformis* was not found in the study area but was introduced in 1988 in the Montemor-o-Novo area and described in the "Flora" by Franco & Afonso (1994). It is expected to became widely distributed in Portugal.

Lindernia dubia was noted for the first time in rice fields in the Coimbra zone, in 1945 by J. G. Garcia and in Salvaterra, Benavente and Coruche by C. Fontes in LISE – 1947 (Vasconcellos, 1971). It was registered in 50% of the observed areas of the region of Alcácer do Sal (Vasconcellos, 1959b) and in the region of Chamusca (Espírito-Santo & Rosa, 1981). In the studied zone, it was registered with a frequency of 46.

The survey also confirmed that the following plants were still present in drainage or rising ditches: *Azolla filiculoides* (registered in 1920, LISE), *Eichhornia crassipes* (introduced in 1939) (Guerreiro, 1976) and *Myriophyllum aquaticum* (registered in 1936). The presence of these species in the rice fields was relatively easy to control. We can control the first two species by controlling the way in which water enters, while the last should be prevented from rooting in the ditches and bordering areas.

Paspalum paspalodes seems to have decreased in presence on the border of the rice fields, this being probably due to the smaller number of banks that resulted from the increase in rice fields and greater weed control on the banks.

Table 2. Plant species present in the rice fields of Tagus Valley and the regional origin of these plants

Species name	Origin	Portuguese bibliographical reference	
		General	Rice fields
Alisma lanceolatum With	E	Coutinho (1913)	Reymão (1941)
*Alisma plantago-aquatica*L.	E	Coutinho (1913)	Ribeiro et al. (1860)
Ammannia×coccinea Rottb.	NA & CA	Morais (1936)	
Apium nodiflorum (L.) Lag.	E	Coutinho (1913)	Reymão (1941)
Aster squamatus (Sprengel) Hieron	CA & SA	(1956) LISI	(1958) LISI
Azolla filiculoides Lam.	A	(1920) LISE	(1927) LISE
Bidens frondosa L.	NA	(1936) LISI	Reymão (1941)
Callitriche stagnalis Scop.	E	Coutinho (1913)	
Cyperus difformis L.	E	Coutinho (1913)	Reymão (1941)
Cyperus eragrostis L.	TA	(1881) LISI	Andrade (1949)
Cyperus fuscus L. for. *virescens* (Hoffm.) Vahl	E	Coutinho (1913)	Reymão (1941)
Echinochloa colonum (L.) Link	TR & SR	(1957) LISI	Vasconcellos (1954)
Echinochloa crus-galli (L.) Beauv.	E	Coutinho (1913)	Ribeiro et al. (1860)
Echinochloa oryzicola (Vasinger) Vasinger	?	(1940) LISI	Vasconcellos (1954)
Eclipta prostrata (L.) L.	TA & STA	Vasconcellos (1940)	
Eichhornia crassipes (C. Martius) Solms-Laub.	SA	1939 (Guerreiro, 1976)	
Eleocharis palustris (L.) Roemer & Schultes	E		Reymão (1941)
Galega officinalis L.	E	Coutinho (1913)	
Glyceria declinata Brèb.	E	Vianna (1958)	
Heteranthera limosa (S.W.) Willd.	SA	(1994) LISI	
Leersia oryzoides (L.) Swartz.	E	Coutinho (1913)	Torres (1940)
Lemna gibba L.	E	Coutinho (1913)	Reymão (1941)
Lemna minor L.	E	Coutinho (1913)	Reymão (1941)
Lindernia dubia (L.) Pennell	NA	Vasconcellos (1971)	
Lotus pedunculatus Cav.	IP	Coutinho (1913)	
Lycopus europaeus L. ssp. *europaeus*	E	Coutinho (1913)	Reymão (1941)
Lythrum hyssopifolia L.	E	Coutinho (1913)	(1938) LISA
Lythrum junceum Banks & Solander	E& MR	Coutinho (1913)	Reymão (1941)
Lythrum salicaria L.	E	Coutinho (1913)	Vasconcellos (1954)
Lythrum thymifolia L.	E & MR	Coutinho (1913)	
Myriophyllum aquaticum (Vell. Conc.) Verdc.	SA	1936 (Teles, 1975)	
Panicum repens L.	MR	Coutinho (1913)	
Paspalum paspalodes (Michx.) Scribner	TR & SR	Sampaio (1909)	Silva (1940)
Phragmites australis (Cav.) Steudel	E	Coutinho (1913)	(1938) LISI
Polygonum aviculare L.	E	Coutinho (1913)	
Polygonum salicifolium Willd.	E	(1951) LISI	
Polypogon monspeliensis (L.) Desf.	E	Coutinho (1913)	Reymão (1941)
Ranunculus sceleratus L. ssp. *sceleratus*	E	Coutinho (1913)	Vianna (1958)
Ranunculus trilobus Desf.	E	Coutinho (1913)	Reymão (1941)
Rumex conglomeratus Murray	E	Coutinho (1913)	
Scirpus lacustris L. ssp. *tabernaemontani* (C.C. Gmelin) Syme	E		Vasconcellos (1954)
Scirpus maritimus L. ssp. *maritimus*	E	Coutinho (1913)	Menéres (1935)
Scirpus mucronatus L.	E	Coutinho (1913)	Ribeiro et al. (1860)
Typha domingensis (Pers.) Steudel	E	Neto (1952)	
Verbena officinalis L.	E	Coutinho (1913)	
Zannichellia palustris L.	E	Coutinho (1913)	Neto (1952)

CA – Central America, NA – North America, SA – South America, STA – Subtropical America, TA – Tropical America, E – Europe, IP – Iberian Peninsula, MR – Mediterranean Region, TR – Tropical Region, SR – Subtropical Region, LISA – Comissão Reguladora do Comércio do Arroz of Lisbon, LISE – Estação Agronómica Nacional of Oeiras, LISI – Instituto Superior de Agronomia of Lisbon

The species *Echinochloa oryzicola* decreased in importance because of the laser levelling which allows better specific weed control, although surveys have shown a high presence in Alcácer do Sal (Vasconcellos, 1959b), Azambuja (Vasconcellos, 1959a), Rio Maior (Carvalho, 1955), Rio Maior (Bairrão, 1957), Rio Frio (Vianna, 1958), Chamusca (Vasconcellos, 1961) and Oliveira do Bairro (Melo, 1963).

Herbarium records showed *Eclipta prostrata* in the region of Vila Nova da Rainha and Azambuja (Vasconcellos, 1940), in Sado Valley (Reymão, 1941) and in the Paul de Magos (Neto, 1952). In the 1997 survey it was less frequent than in 1958 (Vasconcellos, 1959a).

Ammannia×coccinea was grown for the first time in 1938 in Taveiro (Vasconcellos, 1940), and registered in the Paul de Magos in 1952.

Aster squamatus was registered first in 1956, and in the rice field in 1958 (Vasconcellos & Franco, 1958).

Cyperus eragrostis, although it has long been present, has less importance as a weed in the rice fields than in other cultures.

Several measures should be taken to prevent the introduction and expansion of weed species: (i) Programs to eliminate introduced weeds as soon as they are detected, or measures adopted to stop their expansion (i.e. it is prohibited to purchase, sell, own or circulate *Eichhornia crassipes* (Ordinance-law 165/74 of April 22)); (ii) Irrigation control – the weed species of the rice fields are distributed by water and care should be taken to prevent their transportation in rice fields; (iii) When sowing rice clean seed only, free of weed and other seeds, should be used; (iv) Cleaning of agricultural machines, care in the transport of rice before cleaning, and care in cleaning rice of all seeds and inert materials; and (v) Cleaning the banks, channels and ditches by flood and drainage.

Acknowledgments

We would like to thank Mr Paulo Forte for his technical assistance throughout this study, and Dr Maria José Oliveira who read and corrected the manuscript.

References

Andrade, F. G. F. F., 1949. Subsídios para o conhecimento da flora infestante dos arrozais do Vale do Muge. Relatório Final do Curso de Engenheiro Agrónomo. Instituto Superior de Agronomia. Lisboa, 91 pp.

Bairrão, J. F. M., 1957. Flora infestante dos arrozais da Bacia do Rio Torto. (Subsídio para o seu estudo). Relatório Final do Curso de Engenheiro Agrónomo. Instituto Superior de Agronomia. Lisboa, 191 pp.

Barralis, G., 1975. Résultats d'une enquete sur la repartition et la densité des mauvaises herbes en France. 8ème Conference du COLUMA, 4: 1042–1058.

Barralis, G., 1976. Méthode d'étude des groupements adventices des cultures annuelles: application à la Côte d'Or. Vème Coll. Int. Ecol. Biol. Mauvaises Herbes, 1: 59–68.

Carretero, J. L., 1981. El Género *Echinochloa* Beauv. en el Suroeste de Europe. An. Jardin Bot. Madrid 38 (1): 91–108.

Carvalho, F. M. F. P., 1955. Estudo da flora infestante dos arrozais no Vale da Ribeira de Rio Maior. Relatório Final do Curso de Engenheiro Agrónomo. Instituto Superior de Agronomia. Lisboa, 412 pp.

Coutinho, A. X. P., 1913. A Flora de Portugal Bertrand, Lisboa, 777 pp.

Daget, P., I. Drdos, M. Godron, J. L. Guillerm, E. Ruzichova & E. Urvichiarova, 1971. Fytaekologicky Vyskum Liptovkep. Zhotovenie a interpretacia e Kologichych Zapisov Stanovista. Quaest. Geobiol. 9: 77–112.

Daget, P., 1976. Ordination des profils écologiques. Nat. Monsp. 26: 109–128.

Daget, P. & M. Godron, 1982. Analyse fréquentielle de l' écologie des espèces dans les communautés. Collection d'écologie 18. Masson, Paris. 173 pp.

Desfayes, M., 1996. Une adventice nouvelle pour le Portugal: *Heteranthera rotundifolia*. Annu. Soc. Brot. 62: 21.

Espírito-Santo, M. D. & M. L. Rosa, 1981. Contribuição para o conhecimento da evolução da flora infestante dos arrozais. Centro de Botânica Aplicada à Agricultura da Universidade Técnica de Lisboa, Lisboa, 18 pp.

Franco, J. A., 1971, 1984. Nova Flora de Portugal (Continente e Açores) 1, 2. Lisboa, 648 pp., 660 pp.

Franco, J. A. & M. L. Rocha Afonso, 1994, 1998. Nova Flora de Portugal (Continente e Açores). 3(1, 2). Escolar Ed. Lisboa, 181 pp., 283 pp.

Guerreiro, A. R., 1976. O Jacinto aquático [*Eichhornia crassipes* (Mart.) Solms.] em Portugal. II Simp. Nacional de Herbologia, 1: 1–17.

Guillerm, J. L., 1969. Relations entre la végétation spontanée et le milieu dans des terres cultivées du Bas-Languedoc. Thèse 3ème cycle. Fac. Sc. Montpellier. 165 pp.

Mac Intyre, S. & S. C. H. Barrett, 1985. A comparison of weeds communities of rice in Australia and California. Proc. Ecol. Soc. Aust., 14: 237–250.

Melo, A. B. T. P. C. F., 1963. Plantas vasculares infestantes dos arrozais do Concelho de Oliveira do Bairro. Relatório Final do Curso de Engenheiro Agrónomo. Instituto Superior de Agronomia. Lisboa, 84 pp.

Menéres, F., 1935. Subsídios para o estudo económico da cultura do arroz. Relatório Final do Curso de Engenheiro Agrónomo. Instituto Superior de Agronomia. Lisboa, 263 pp.

Michez, M. & J. L. Guillerm, 1984. Signalement écologique et degré d'infestation des adventices des cultures d'été en Lauragais. 7ème Coll. Int. Ecol. Biol. Mauvaises Herbes, 1: 155–162.

Morais, T., 1936. Notas sobre a flora portuguesa. Bol. Soc. Broteriana 11 (2ª sér.): 154.

Neto, J. G. d'A., 1952. Contribuição para o estudo do emprego de herbicidas na monda selectiva dos arrozais do Paul de Magos. Relatório Final do Curso de Engenheiro Agrónomo. Instituto Superior de Agronomia. Lisboa. 109 pp.

Pignatti, S., 1982. Flora d'Italia 3. Edagricola. 780 pp.

Reymão, G. M., 1941. Subsídios para o conhecimento da flora espontânea dos arrozais do Vale do Sado. Relatório Final do Curso de Engenheiro Agrónomo. Instituto Superior de Agronomia. Lisboa, 100 pp.

Ribeiro, M. J., S. B. Almeida & J. A. Corvo, 1860. Relatório sobre a cultura do arroz em Portugal e a sua influência na saúde pública. Lisboa, 11 pp.

Sampaio, G., 1909. Manual da Flora Portuguesa. Porto, 792 pp.

Sgattoni, P., V. Ticchiati, F. Arosio, P. Villani & C. Mallegni, 1989. Distribution and importance of the main rice weeds in Italy: Results from a 1988 Technical Survey. EWRS 4th Symposium on weed problems in the mediterranean climates, 2: 301–311.

Silva, A. R. P., 1940. O género *Paspalum* em Portugal. Agr. lusit. 2: 5–23.

Teles, A. N. & A. R. P. Silva, 1975. "A Pinheirinha" (*Myriophyllum aquaticum* (Vell.) Verdc.), uma agressiva infestante aquática. Agr. lusit. 36: 307–323.

Torres, A. B. Monteiro, 1940. Subsídios para a identificação das sementes de algumas *Paniceae* e *Phalarideae*. Rev. Agr. XI: 7–17.

Tutin, T. S., V. H. Heywood, N. A. Bueges, D. R. Moore, D. H. Valentine, S. M. Walters & D. A. Webb (ed.), (1964, 1968, 1972, 1976, 1980) – Flora Europaea. I–V. Cambridge Univ. Press. 464 pp., 455 pp., 370 pp., 505 pp., 452 pp.

Vasconcellos, J. C., 1940. Anotações ao Herbário do Instituto Superior de Agronomia. Anais do Instituto Superior de Agronomia 11: 7–17.

Vasconcellos, J. C., 1954. Plantas vasculares infestantes dos arrozais. Comissão Reguladora do Comércio de Arroz, Lisboa. 188 pp.

Vasconcellos, J. C. & J. A. Franco, 1958. Novas plantas para a flora portuguesa. Bol. Soc. Broteriana 32 (2ę sér.): 237–251.

Vasconcellos, J. C., 1959a. Evolução da vegetação infestante dos arrozais no concelho da Azambuja. Comissão Reguladora do Comércio de Arroz, Lisboa. 74 pp.

Vasconcellos, J. C., 1959b. Evolução da vegetação infestante dos arrozais no concelho de Alcácer do Sal. Comissão Reguladora do Comércio de Arroz, Lisboa, 70 pp.

Vasconcellos, J. C., 1961. Evolução da vegetação infestante dos arrozais no concelho da Chamusca. Comissão Reguladora do Comércio de Arroz, Lisboa, 63 pp.

Vasconcellos, J. C., 1963. O Arroz. Comissão Reguladora do Comércio de Arroz. Lisboa, 300 pp.

Vasconcellos, J. C., 1971. Recente introdução de infestantes exóticas. 1° Simp. Nac. Herbologia, 1: 1–5.

Vianna, A. M. L. F., 1958. Contribuição para o estudo da flora infestante dos arrozais das zona orizícola de Rio Frio. Relatório Final do Curso de Engenheiro Agrónomo. Instituto Superior de Agronomia. Lisboa, 190 pp.

Hydrobiologia **415:** 67–75, 1999.
J. M. Caffrey, P. R. F. Barrett, M. T. Ferreira, I.S. Moreira, K. J. Murphy & P. M. Wade (eds), Biology, Ecology and Management of Aquatic Plants.
© 1999 *Kluwer Academic Publishers. Printed in the Netherlands.*

Plant communities of the lagoons of the Portuguese Coastal Superdistrict – a multivariate approach

J. C. Costa[1,*], J. Capelo[2], M. D. Espírito Santo[1], M. Lousã[1], A. Monteiro[1], S. Mesquita[2], M. T. Vasconcelos[1] & I. Moreira[1]

[1]*Departamento de Protecção das Plantas e de Fitoecologia, Instituto Superior de Agronomia, Tapada da Ajuda, P-1349-017 Lisboa, Portugal*

[2]*Departamento de Conservação de Recursos Naturais, Estação Florestal Nacional, Tapada da Ajuda, P-1350 Lisboa, Portugal*

Key words: CORINE, biotops, phytosociology, habitats, UPGMA, vegetation

Abstract

The Portuguese Coastal Superdistrict (Western Portugal) is a transitional biogeographical unit between the Mediterranean and Eurosiberian Regions. Due to the multiple influences which impact on the region, a wealth of taxa are found there making large areas suited for biodiversity conservation. In this study, the flora and vegetation of the lagoons and the marshes of this Superdistrict are characterized, and the exotic weed taxa causing problems in natural communities are emphasised.

Introduction

Within the European Habitat Project (Directive 92/42/EEC), previous studies included in the CORINE project (Costa & Lousã, 1989) were extended to the whole western coast of Portugal. In terms of phytogeographical typology and following Costa et al. (1998), this area belongs to the Portuguese Coastal Superdistrict: Portuguese Divisory Sector; Gaditano-Onubo-Algarvish Province; Ibero-Atlantic Superprovince; Western Mediterranean Subregion; Mediterranean Region. It ranges approximately from Sintra (near Lisbon) to Aveiro where it meets the Cantabro-Atlantic Province (Eurosiberian Region). Both the climatic influence and palaeo-historical nature of this unit are peculiar due to its transitional character and interface position on important plant migratory pathways (Izco, 1989). Between Sintra and Nazaré, the limestone or sandstone cliffs are sometimes interrupted by small beaches. North of Nazaré, the coast is dominated by more or less altered dune systems, with a few calcareous outcrops (e.g. Cape Mondego). Most of the area belongs to the mesomediterranean bioclimatic thermotype and the ombroclimate is of the sub-humid ombrotype (Rivas-Martínez, 1997; Rivas-Martínez et al., 1998). The vegetation and flora found in the wet habitats of this Superdistrict belong both to the Eurosiberian and Mediterranean Regions. The southernmost limit of the Eurosiberian taxa *Salix arenaria* (= *S. repens* ssp. *argentea*), *Carex serotina* ssp. *serotina* and *Schoenoplectus pungens* (= *Scirpus pungens*) is found in this territory. Since the Middle Ages, the area has experienced a strong human influence. Extensive plantations of maritime pine (*Pinus pinaster*), mostly during the 19th century, also shape the landscape from Nazaré to Aveiro. These pine stands were planted for timber and for preventing the inland advance of the dunes. Sometimes small lagoons, pools and marshes can be found within the pinewoods. The aim of this work is to extend the knowledge of the flora of the lagoons, pools and marshy depressions, and to annotate the aggressive species invading the natural flora of the Portuguese Coastal Superdistrict.

Materials and methods

The following freshwaters lagoons, pools and marshes were studied: Mira (Mira), Salgueira (Cantanhede),

* Author for correspondence

Table 1. Plant communities of lagoons

	1	2	3	4	5	6	7	8	9	10	11	12	13	14	15
Wolffia arrhiza	x	.	.	.	.	.	.	.	.	.	.	.	.	.	.
Lemna minor	x	.	.	.	.	.	.	.	.	.	.	.	.	.	.
Elodea canadensis	x	.	x	.	.	.	.	.	.	.	.	.	.	.	.
Nymphaea alba	.	x	.	.	.	.	.	.	.	.	.	.	.	.	.
Nuphar lutea	.	x	.	.	.	.	.	.	.	.	.	.	.	.	.
Myriophyllum spicatum	.	x	.	.	.	.	.	.	.	.	.	.	.	.	.
Myriophyllum alterniflorum	.	x	x	.	.	.	.	.	.	.	.	.	.	.	.
Potamogeton perfoliatus		x	x	.	.	.	.	.	.	.	.	.	.	.	.
Potamogeton crispus	.	x	x	x	.	.	.	.	.	.	.	.	.	.	.
Potamogeton gramineus	.	x	x	x	.	.	.	.	.	.	.	.	.	.	.
Potamogeton polygonifolius	.	x	x	x	x	.	.	.	.	.	.	.	.	.	.
Myriophyllum aquaticum	.	x	x	x	x	.	.	.	.	.	.	.	.	.	.
Utricularia australis	.	.	x	x	.	.	.	.	.	.	.	.	.	.	.
Ludwiga palustris	.	x	x	.	.	.	x	.	.	.	.	.	.	.	.
Baldellia ranunculoides	.	.	.	x	x	x	x	x	.	.	.	.	.	.	.
Myosotis lusitanica	.	.	.	.	x	.	.	.	.	.	.	.	.	.	.
Rhynchospora rugosa	.	.	.	.	x	.	.	.	.	.	.	.	.	.	.
Eleocharis multicaulis	.	.	.	.	x	.	.	.	.	.	.	.	.	.	.
Hypericum elodes	.	.	.	.	x	x	.	.	.	.	.	.	.	.	.
Ranunculus flammula	.	.	.	.	x	x	.	.	.	.	.	.	.	.	.
Anagallis tenella	.	.	.	.	x	x	.	.	.	.	.	.	.	.	.
Hydrocotyle vulgaris	.	.	.	x	x	x	.	.	.	.	.	.	.	.	.
Juncus bulbosus	.	.	.	x	x	x	.	.	.	.	.	.	.	.	.
Juncus articulatus	.	.	.	.	x	x	x	.	.	x	.	.	.	.	.
Carex serotina	.	.	.	.	.	x	.	.	.	.	.	.	.	.	.
Glyceria spicata	.	.	.	.	.	.	x	.	.	.	.	.	.	.	.
Equisetum palustre	.	.	.	.	.	.	x	.	.	.	.	.	.	.	.
Glyceria declinata	.	.	.	.	.	.	x	x	.	.	.	.	.	.	.
Eleocharis palustris	.	.	.	.	.	.	x	x	.	.	.	.	.	.	.
Scirpus pungens	.	.	.	.	.	.	x	x	.	.	.	.	.	.	.
Scirpus lacustris	.	.	.	.	.	.	x	x	.	.	.	.	.	.	.
Apium nodiflorum	.	.	.	x	.	.	x	x	.	.	.	.	.	.	.
Galium palustre	.	.	.	.	x	.	x	x	.	x	.	.	.	.	.
Alisma plantago-aquatica	.	.	.	.	x	x	x	x	x	.	.	.	.	.	.
Mentha aquatica	.	.	.	.	.	.	x	x	x	.	.	.	.	.	.
Lycopus europaeus	.	.	.	.	.	.	x	x	x	.	.	.	.	.	.
Lythrum salicaria	.	.	.	.	.	.	x	x	x	.	.	.	.	.	.
Oenanthe crocata	.	.	.	.	.	.	x	x	x	.	.	.	.	.	.
Phragmites australis	.	.	.	.	.	.	x	x	x	.	.	.	.	.	.
Phalaris arundinacea	.	.	.	.	.	.	.	x	.	.	.	.	.	.	.
Typha angustifolia	.	.	.	.	.	.	.	x	.	.	.	.	.	.	.
Typha latifolia	.	x	.	.	.	.	.	x	.	.	.	.	.	.	.
Typha domingensis	.	x	.	.	.	.	.	x	x	.	.	.	.	.	.
Carex hispida	.	.	.	.	.	.	.	x	x	.	.	.	.	.	.
Cladium mariscus	.	.	.	.	.	.	.	.	x	.	.	.	.	.	.
Juncus effusus	.	.	.	.	.	x	.	.	.	x	x	x	.	.	.
Juncus acutus	.	.	.	.	.	.	.	.	.	x	.	.	.	.	.
Holoschoenus romanus ssp. *australis*	.	.	.	.	.	.	.	.	.	x	.	.	.	.	.
Schoenus nigricans	.	.	.	.	.	x	.	.	.	x	x	.	.	.	.
Rumex crispus	.	.	.	.	.	.	.	.	.	x	x	.	.	.	.
Hypericum undulatum	.	.	.	.	x	.	.	.	.	x	.	.	.	.	.

Table 1. Continued

	1	2	3	4	5	6	7	8	9	10	11	12	13	14	15
Lythrum junceum	.	.	.	.	x	.	.	.	x	x	x	.	.	.	.
Trifolium repens	.	.	.	.	.	.	.	.	.	.	x	x	x	.	.
Holcus lanatus	.	.	.	.	.	.	.	.	.	x	x	x	.	.	.
Lolium multiflorum	.	.	.	.	.	.	.	.	.	x	x	x	.	.	.
Mentha suaveolens	.	.	.	.	.	.	.	.	.	x	x	x	.	.	.
Agrostis stolonifera	.	.	.	.	x	.	.	.	.	x	x	x	.	.	.
Cynodon dactylon	.	.	.	.	x	.	.	.	.	x	x	x	.	.	.
Lobelia urens	.	.	.	.	x	.	.	.	.	x	x	x	.	.	.
Lotus pedunculatus	.	.	.	.	x	.	.	.	.	x	x	x	.	.	.
Cyperus longus	.	.	.	.	x	.	.	.	.	x	x	x	x	.	.
Poa annua	.	.	.	.	.	.	.	.	.	x	x	.	x	x	.
Trifolium resupinatum	.	.	.	.	.	.	.	.	.	.	x	.	.	.	.
Carex divisa	.	.	.	.	.	.	.	.	.	.	x	.	.	.	.
Prunella vulgaris	.	.	.	.	x	.	x	.	.	.	x	x	.	.	.
Panicum repens	.	.	.	.	x	.	x	x	x	x	x	x	.	.	.
Paspalum distichum	.	.	.	.	.	.	x	x	.	x	x	x	x	.	.
Paspalum dilatatum	.	.	.	.	.	.	.	.	.	.	x	x	.	.	.
Ranunculus repens	.	.	.	.	x	x	x	.	.	.	.	x	.	.	.
Taraxacum adamii	.	.	.	.	.	.	.	.	.	.	.	x	.	.	.
Cerastium fontanum ssp. *triviale*	.	.	.	.	.	.	.	.	.	.	.	x	.	.	.
Potentilla erecta	.	.	.	.	.	.	.	.	.	.	.	x	.	.	.
Plantago coronopus	.	.	.	.	.	.	.	.	.	.	x	x	.	.	.
Bidens frondosa	.	.	.	.	.	x	x	x	x	.	.	x	x	.	x
Polygonum lapathifolium	.	.	.	.	.	.	.	.	.	.	.	.	x	.	.
Polygonum persicaria	.	.	.	.	.	.	.	.	.	.	.	.	x	.	.
Polygonum hydropiper	.	.	.	.	.	.	.	.	.	.	.	.	x	.	.
Atriplex prostrata	.	.	.	.	.	.	.	.	.	.	.	.	x	.	.
Xanthium strumarium	.	.	.	.	.	.	.	.	.	.	.	.	x	.	.
Solanum luteum	.	.	.	.	.	.	.	.	.	.	.	.	x	.	.
Chaetapogon fasciculatus	.	.	.	.	.	.	.	.	.	.	.	.	.	x	.
Lotus castellanus ssp. *subbiflorus*	.	.	.	.	.	.	.	.	.	.	.	.	.	x	.
Lythrum portula	.	.	.	.	.	.	.	.	.	.	.	.	.	x	.
Pulicaria paludosa	.	.	.	.	.	.	.	.	.	.	.	.	.	x	.
Scirpus cernuus	.	.	.	.	.	x	.	.	.	.	.	.	.	x	x
Juncus minutulus	.	.	.	.	.	.	.	.	.	.	.	.	.	x	x
Juncus bufonius	.	.	.	.	.	.	.	.	.	.	.	.	x	x	x
Pseudognaphalium luteo-album	.	.	.	.	.	.	.	.	.	.	.	.	x	.	x
Cyperus flavescens	.	.	.	.	.	x	.	.	.	.	.	.	.	.	x
Scirpus juncoides	.	.	.	.	x	x	.	.	.	.	.	.	.	.	x
Hydrocotyle bonariensis	.	.	.	.	.	x	.	.	x	x	x	x	x	.	.
Cotula coronopifolia	.	.	.	.	.	x	.	.	x	.	x	x	x	.	x
Rorippa sylvestris	.	.	.	.	.	x	.	x	.	.	.	.	x	.	.
Calystegia sepium	.	.	.	.	.	.	x	x	.	.	x	.	.	.	.
Stellaria media	.	.	.	.	.	.	.	x	.	.	x	.	x	.	.
Epilobium obscurum	.	.	.	.	.	.	.	.	x	.	x	x	.	x	.
Aster squamatus	.	.	.	.	.	.	.	.	.	x	x	x	x	x	.

(1) Community of *Wolffia arrhiza* and *Lemna minor*, (2) *Nymphaeetum albi-lutei*, (3) *Ludwigo palustris-Potametum polygonifolii*, (4) *Utricalarietum exoleto-australis*, (5) *Hyperico elodis-Rhynschosporetum rugosi*, (6) *Anagallido-Juncetum bulbosi*, (7) *Typho angustifoliae-Phragmitetum australis*, (8) *Glycerio declinatae-Eleocharidetum palustris*, (9) *Cladietum marisci*, (10) *Holoschoeno-Juncetum acuti*, (11) *Trifolio resupinati-Caricetum chaetophyllae*, (12) *Paspaletum dilatato-distachi*, (13) Community of *Bidens frondosa*, (14) Loto *subbiflori-Chaetopogonetum fasciculati*, (15) Community of *Cyperus flavescens*.

Vela (Figueira da Foz), Paeia-Camarção (Figueira da Foz), Braças (Figueira da Foz), Ervideira (Leiria), Pataias (Alcobaça), Coelhas (Alcobaça), Clementina (Nazaré) and Saloio (Nazaré). Pateira de Fermentelos (Águeda) and Barrinha de Esmoriz (Ovar/Espinho) are similar to those studied but outside the area. The first is very eutrophic and the second suffers from industrial pollution (Farinha & Trindade, 1994).

The vegetation relevees were made according to the sigmatist and landscape phytosociology approach (Braun-Blanquet, 1932; Rivas-Martínez, 1976; Géhu & Rivas-Martínez, 1980). In order to describe the circumscription of plant communities, ordination and clustering methods were used. Ordination by Correspondence Analysis (Hill, 1973) – CA – was performed in two steps. A first analysis (not shown) separated the *Lemnetea* communities as outliers and, therefore, a second analysis was carried out. Clustering by UPGMA (Unweigthed Pair-Group Method using Aritmetic Averages) using the Bray-Curtis coefficient (Figure 1) (Ludwig & Reynolds, 1988) and TWINSPAN (Two-Way Indicator Species Analysis) (Hill, 1979) were used as classification procedures. Relevee ordination was performed using the UPGMA order and species ordination used the first CA axis. Table refinement was made according to the classical Braun-Blanquet method (Muller-Dombois & Ellenberg, 1974).

Results

The classification of plant communities yielded similar results for both classification methods and CA. The groupings at lower similarity levels match those of the vegetation classes concepts. The description of communities at the association or basal community level was also possible. Vegetation classes summarize clear-cut floristic groupings of large ecological scope associated with definite habitat conditions, biogeographical areas and dominant physiognomical plant types. Associations are models of plant communities that show a statistical repetitiveness associated with uniform habitat conditions. Syntaxa at these levels could be easily identified and match those defined by the classical phytosociological approach. A nomenclatural model consistent with the numeric results is established (Barkman et al., 1986).

Description of communities

Lemnetea minoris

The *Lemnetea minoris* class includes vegetation of fresh waters of floating non-rooted small aquatic plants (acropleustophytes). The most eutrophic lagoons in the study are usually covered by *Wolffia arrhiza, Lemna minor* and *Elodea canadensis* which can be accounted for agricultural activity (Table 1) (Díaz-González & Fernández-Prieto, 1994; Loidi et al., 1997). The European Commisission (1996) considered this community as an European Habitat of natural eutrophic lakes with 3150 NATURA 2000 code (CORINE code 22.411).

Potametea

The *Potametea* vegetation class is composed of rooted hydrophytes and large pleustophytes on fresh or slightly sodium-rich waters (Díaz-González & Fernández-Prieto, 1994; Loidi et al., 1997).

The *Nymphaeetum albo-lutei* is a community of rooted aquatic plants with floating leaves having long petioles, occurring in deeper fresh waters than the previous class (Loidi et al., 1997) and includes: *Nymphaea alba, Myriophyllum alterniflcrum, Potamogeton crispus, P. gramineus, P. polygonifolius, Nuphar luteo* and the exotic *Myriophyllum aquaticum* (Table 1). That association is only considered habitat of CORINE biotopes of the rooted floating waterlily carpets with 22.4311 code (Comission of European Communities, 1991).

The association *Ludwigio palustris-Potametum polygonifolii* is composed of rooted plants in meso-oligotrophic still waters (lentic habitats). It is typical of the Luso-Extremadurense Province (Cano et al., 1993) but reaches the study area. It includes *Potamogeton polygonifolius, P. gramineus, P. nodosus, Ludwigia palustris, Myriophyllum aquaticum, M. alterniflorum* and *Elodea canadensis* (Table 1). This community must be insert in oligotrophic water containing very few minerals of sandy plains habitat with 3110 Natura 2000 code (22.433 CORINE biotope code).

During the summer, the carnivorous-plant with emerged flowers and fruits community of *Utricularia australis* (*Utricularietum exoleto-australis*) also appears (Table 1). This community has a somewhat pioneer and transitional character to the *Littorelletea* and *Scheuchzerio-Caricetea*. It is found in mud-rich waters near the shore (Loidi et al., 1997). That associ-

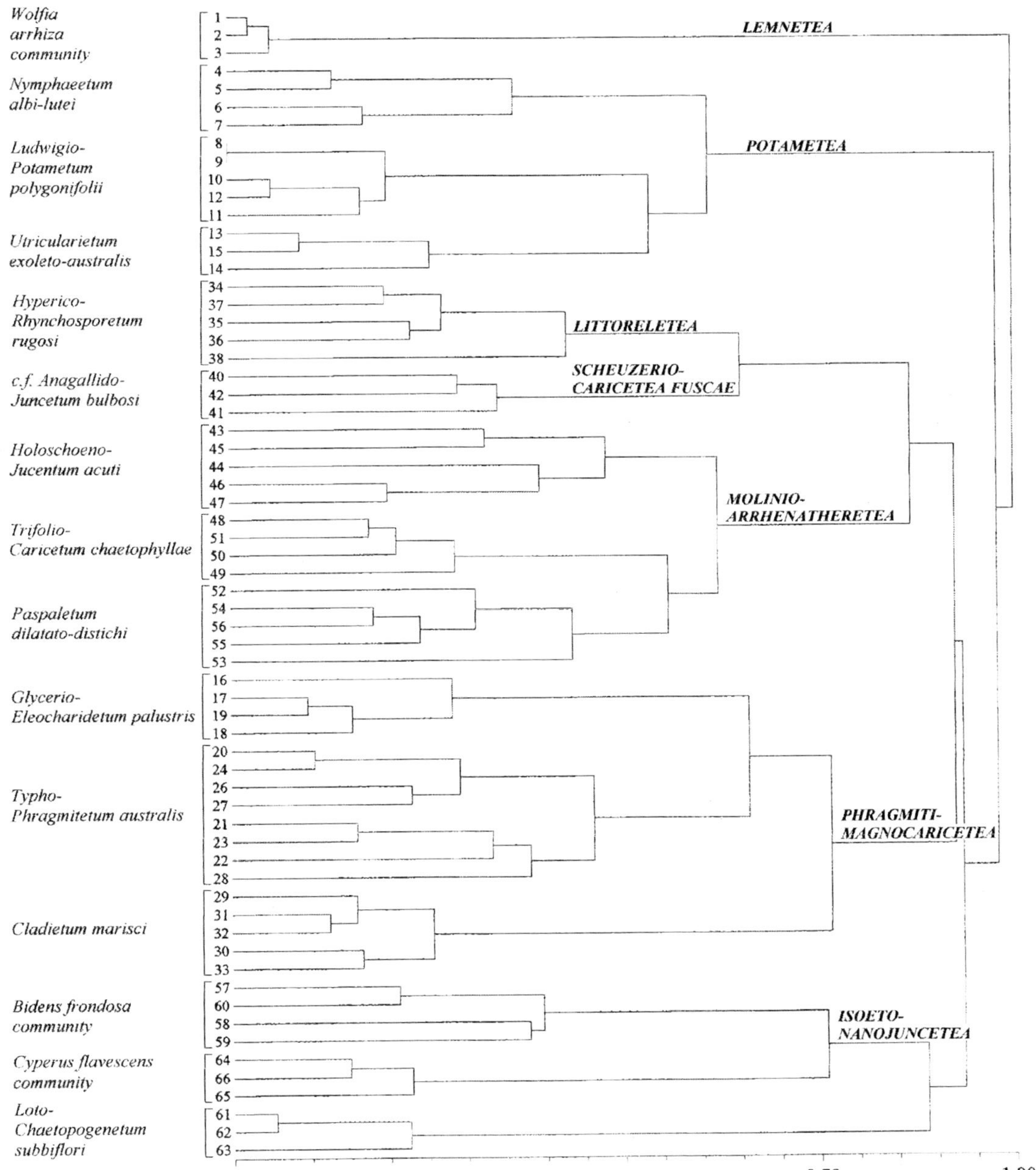

Figure 1. UPGMA Cluster Analysis of relevees.

ation belongs to dystrophic lakes habitat (3160 Natura 2000 code and 22. 414 CORNE biotopes code).

Littorelletea uniflorae *and* Scheuchzerio-Caricetea fuscae *classes*

Bog vegetation of lagoons, formed by amphibian helophytes in oligotrophic fresh waters with variable shallow water level. During the summer, the *Hyperico elodis-Rhynchosporetum rugosi* community can be observed in places where some slight degradation by animal trampling occurs (Neto, 1997). This community is endemic to Portugal and is composed of *Hypericum elodes, Juncus bulbosus* (var. *bulbosus* and var. *fluitans*), *Eleocharis multicaulis, Baldellia ranunculoides, Hydrocotyle vulgaris*, and *Rhynchospora rugosa* (Table 1). In some localities, an impoverished facies of *Anagallis tenella, Juncus bulbosus, Juncus articulatus* and *Carex serotina* ssp. *serotina* is present and may represent the undisturbed sedge bog vegetation (Table 1). These elements are similar to those of the cantabro-atlantic dystrophic-oligotrophic low peat bogs association, *Anagallido-Juncetum bulbosi* (*Scheuchzerio-Caricetea*) (Díaz-González & Fernández-Prieto, 1994). These two communities belong to the habitat: oligotrophic waters with amphibious vegetation (3130 NATURA 2000 code or 22.213 CORINE biotope code).

Phragmito-Magnocaricetea

Hydrophytic, cosmopolitan class from the margins of the rivers, bog areas and lagoons wet throughout the year. It is, in general, dominated by tall sedges, rushes and reeds.

In shallower still waters, the *Glycerio declinatae-Eleocharidetum palustris* (Table 1) is present (*Eleocharis palustris, Apium nodiflorum, Mentha aquatica, Glyceria declinata* and *Equisetum palustre*). The small reed beds is the 53.4 CORINE biotopes code.

Also, a community of *Scirpus lacustris* and *Phragmites australis* (*Typho angustifoliae-Phragmitetum australis*), together with other helophytes, occurs in deeper waters and areas flooded most of the year. Frequent taxa are: *Typha angustifolia, T. domingensis, T. latifolia, Alisma plantago-aquatica, Lycopus europaeus, Lythrum salicaria, Schoenoplectus pungens, Bidens frondosa* and *Baldelia ranunculoides* (Table 1). The *Typha* spp. are dominant due to the lentic character of the habitat. That association belongs to the reed beds 53.1 CORINE biotopes code.

The *Cladietum marisci* is present in the meso-oligotrophic margins of some lagoons (Rivas-Martínez et al., 1980; Loidi et al., 1997) and is dominated by *Cladium mariscus*. The fen-sedge beds are a 7210 NATURA 2000 priority habitat code (53.3 CORINE biotopes code).

Molinio-Arrhenatheretea

Perennial grass and rush dominated vegetation in wet soils not suffering permanent flooding with slight or no drought during the summer. It can occur in hydromorphic soils as well as in normal soils of the temperate climatic regions. In mediterranean climates, it has an azonal character occurring in soils with permanent water beds (Rivas-Martínez et al., 1980).

The rush community, *Holoschoeno-Juncetum acuti*, is a mediterranean association common in the shallow depressions of dunes and also in meso-oligotrophic lagoons. In these depressions, the water bed is near the soil surface during the winter and a pseudo-gley (due to reducing conditions) horizon is always present (Rivas-Martínez et al., 1980). Frequent species are *Holoschoenus romanus* ssp. *australis, Juncus acutus, Agrostis stolonifera, Cynodon dactylon, Galium palustre, Juncus articulatus* and *Hydrocotyle bonariensis* (Table 1). Grazing of this community by cattle maintains the *Trifolio resupinatae-Caricetum chaetophyllae*, which is a rich meadow community with *Trifolium resupinatum, Poa annua, Cynodon dactylon, Agrostis stolonifera, Carex divisa* (= *C. chaetophylla*) and *Trifolium repens* (Table 1). Those two communities are mediterranean tall-herb and rush meadows a NATURA 2000 habitat (6420 code, 37.4 CORINE biotope code).

In silt-rich margins which are flooded most of the year, a grass-dominated community appears: *Paspalum distichum* (= *P. paspalodes*), *Paspalum dilatatum, Panicum repens, Agrostis stolonifera, Trifolium repens* and *Lotus pedunculatus*. It belongs to the summer association *Paspaletum dilatato-distichi* (Table 1). Nitrophilous annual and perennial grass and sedges formations of alluvial banks are a NATURA 2000 habitat with 3229 code (24.53 CORINE biotopes code).

Bidentetea tripartitae

Annual vegetation class on riparian soils rich in nitrogen. In late summer months, on nitrophilous wet sandy biotopes, a community of *Bidens frondosa* with *Polygonum lapathifolium, P. persicaria, P. hydropiper, Atriplex prostrata, Xanthium strumarium*, etc. (Table 1) appears. That pioneer formation of annual on nitrogen rich muds is 3270 NATURA 2000 habitat code (54.52 CORINE biotopes code).

Isoeto-Nanojuncetea

Ephemeral, pioneer annual vegetation on soils with temporary flooding in the winter and early spring (Rivas-Martínez et al., 1980). In places flooded during the winter but dried out in spring, the *Loto subbiflori-Chaetopogonetum fasciculati* appears (*Chaetopogon fasciculatus, Lotus subbiflorus, Juncus minutulus, Juncus buffonius, Lythrum portula* and *Pulicaria palustris*) (Table 1). This thermomediterranean and inferior mesomediterranean community is mostly found both in the Gaditano-Onubo-Algarvish and Luso-Estremadurense Province (Costa et al., 1997). The very shallow temporary ponds which exist only in winter or late spring are a priority habitat with 3170 NATURA 2000 code (22.34 CORINE biotopes code). During the summer, a community of *Cyperus flavescens, Pseudognaphalium luteo-album* and *Juncus bufonius* can also be found in similar environments (Table 1). That community belongs to the same NATURA 2000 code of European Union Habitat.

The sequence of these communities can be see in Figure 2.

In the edges of the lagoons it is common a community of *Salix arenaria* and *S. atrocinerea* (*Salicion arenariae*) that is also an habitat of wet dune slacks with 2170 NATURA 2000 code (16.26 CORINE biotopes code).

Syntaxonomical scheme

MOLINIO-ARRHENATHERETEA Tüxen 1937 *em.* 1970
Holoschoenetalia Br.-Bl. *ex* Tchou 1948
Molinio-Holoschoenion Br.-Bl. *ex* Tchou 1948
Holoschoeno-Juncetum acuti Rivas-Martínez & Costa *in* Rivas-Martínez, Costa, Castroviejo & Valdés 1980
Plantaginetalia majoris Tüxen & Preising *in* Tüxen 1950
Trifolio fragiferi-Cynodontion dactyli Br.-Bl. & O. Bolós 1958
Trifolio resupinati-Caricetum chaetophyllae Rivas-Martínez & Costa *in* Rivas-Martínez, Costa, Castroviejo & Valdés 1980
Paspalo-Polygonion semiverticillati Br.-Bl. *in* Br.-Bl. et al. 1952
Paspaletum dilatato-distichi Herrera & F.Prieto *in* Díaz & F.Prieto 1994
PHRAGMITO-MAGNOCARICETEA Klika *in* Klika & Novak1941
Phragmitetalia W. Koch 1926 *em.* Pignati 1953
Phragmition communis W. Koch 1926
Phragmitenion communis
Typho angustifoliae-Phragmitetum australis (Tüxen & Preising 1942) Rivas-Martínez, Báscones, Díaz, Fernandez-González & Loidi 1991
Nasturtio-Glycerietalia Pignatii 1953
Glicerio-Sparganion Br.-Bl. & Sissingh *in* Böer 1942
Glicerenion fluitans (Géhu & Géhu-Franck 1987) J. A. Molina 1996
Glycerio declinatae-Eleocharidetum palustris Rivas-Martínez & Costa *in* Rivas-Martínez, Costa, Castroviejo & Valdés 1980
Magnocaricetalia Pignatti 1953
Magnocaricion elatae W. Koch 1926
Cladietum marisci (Allorge 1922) Zobrist 1939
LITTORELLETEA Tüxen 1947
Littorelletalia W. Koch *ex* Tüxen 1937
Hyperico elodis-Sparganion Br.-Bl. & Tüxen *ex* Oberdorfer 1957
Hyperico elodis-Rhynchosporetum rugosi Neto, Capelo, J. C. Costa & Lousã 1997 *in* Neto 1997
SCHEUCHZERIO-CARICETEA FUSCAE (Nordehagen 1936) Tüxen 1937
Caricetalia fuscae W. Koch 1926 *em.* Klika 1934
Anagallido-Juncion bulbosi Br.-Bl. 1967
Anagallido-Juncetum bulbosi Br.-Bl. 1967
BIDENTETEA TRIPARTITAE Tüxen, Lohenmeyer & Preising *in* Tüxen 1950
Bidentetalia tripartitae Br.-Bl. & Tüxen *ex* Klika & Hadac 1944
Chenopodium rubri (Tüxen *ex* Poli & J. Tüxen 1960) Kopecky 1960
Community of *Bidens frondosa*
ISOETO-NANOJUNCETEA Br.-Bl. & Tüxen 1943
Isoetetalia Br.-Bl. 1935 *em.* Rivas Goday 1970
Cicendion (Rivas Goday *in* Rivas Goday & Borja 1961) Br.-Bl. 1967
Loto subbiflori-Chaetopogenetum fasciculati Rivas-Martínez & Costa *in* Rivas-Martínez, Costa, Castroviejo & Valdés 1980
Nanocyperetalia Klika 1935
Nanocyperion W. Koch *ex* Libbert 1932
Community of *Cyperus flavescens*
POTAMETEA Tüxen & Preising 1942
Potametalia W. Koch 1926
Potamion W. Koch 1926
Ludwigio palustris-Potametum polygonifolii Peinado & Velasco 1983
Nymphaeion albae Oberdofer 1957
Nymphaeetum albi-lutei Nowinsky 1928

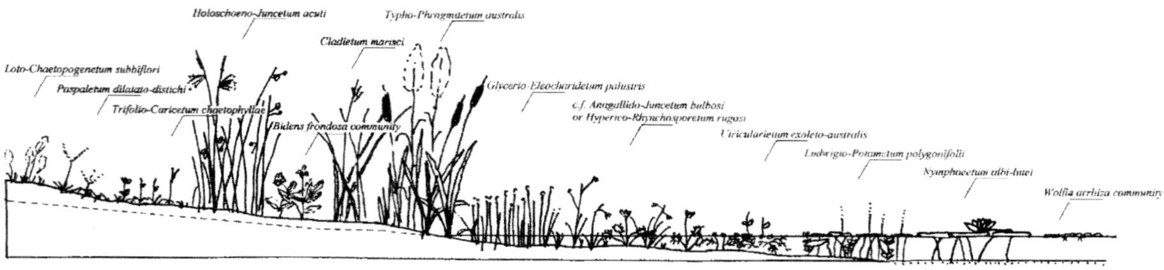

Figure 2. Theoretical vegetation transect of a lagoon.

Utricularietalia minoris Den Hartog & Segal 1964
Utricularion vulgaris Passarge 1964
Utricularietum exoleto-australis Rivas-Martínez, Costa, Castroviejo & Valdés 1980
LEMNETEA O.Bolòs & Masclans 1955
Lemnetalia O.Bolòs & Masclans 1955
Lemnion minoris O.Bolòs & Masclans 1955
Community of *Wolffia arrhiza* and *Lemna minor*

Discussion and conclusions

Some exotic invaders are starting to establish. The South American *Myriophyllum aquaticum* is a weed expand in the lagoons that menaces the vegetation conservation. In Portugal, this weed is well established in irrigation and drainage systems and rice fields (Moreira et al., 1998). The tropical *Schoenoplectus juncoides* (= *Scirpus juncoides*) is a weed in rice fields of the Mondego, Sado and Tagus basins. The South American *Bidens frondosa* is a weed of irrigation systems. The South African *Cotula coronopifolia* is well adapted in Portugal and is found in wet places of fresh and brackish waters and forms almost monospecific communities. The South American *Hydrocotyle bonariensis* has spread to coastal sandy habitats in the northwest of Portugal.

Monitoring of vegetation communities with a high degree of ecological preservation is needed for the sake of future comparisons as a reference frame. Comparisons of disturbed situations can be done with those described, when management plans or environmental impact assessment are needed in the study area. Most problems of habitat disturbance are due to eutrophication and sediment filling of the lagoons with origin in agricultural practices and touristic activities. As a consequence, the *Bidens frondosa* and *Wolffia arrhiza* communities area in the lagoons is growing and, therefore, replacing others, with loss of phytocoenological diversity. The characterization of the natural undisturbed vegetation gradient (*microgeosigmetum sensu* Rivas-Martínez, 1996) can contribute to establish an objective vegetation typology to be used in biological conservation and site restauration practices.

The extant menace of exotic species invasion can, therefore, be monitored by the degree of global floristic and phytocoenotical dissimilarity with the reference scheme now proposed.

References

Barkman, J. J., J. Moravec & S. Rauschert, 1986. Code of phytosociological nomenclature. 2nd edn. Vegetatio 67: 145–195.

Braun-Blanquet, J., 1932. Plant sociology. Mc Graw Hill Book. New York: 439 pp.

Cano, E., A. Garcia-Fuentes & N. Sánchez-Pascual. 1993. Vegetación de la Cordillera Mariánican y Cuenca del Guadiana. Departamento Biología Vegetal Universidad de Jaen: 122 pp.

Commission of the European Communities, 1991. CORINE biotopes manual Habitats of the European Community. Office for Official Publication of the European Communities. Luxembourg. 300 pp.

Costa, J. C. & M. Lousã, 1989. Flora e Vegetação da Costa de Quiaios. Centro de Botânica Aplicada à Agricultura Instituto Superior Agronomia Lisboa: 35 pp.

Costa, J. C., M. Lousã & M. D. Espírito Santo, 1997. A vegetação Natural do Parque da Ria Formosa. Studia Bot. 15: 69–157.

Costa, J. C., C. Aguiar, J. Capelo, M. Lousã & C. Neto, 1998. Biogeografia de Portugal Continental. Quercetea 0: 5–56.

Díaz-González, T. & J. A. Fernández-Prieto, 1994. La vegetación de Asturias. Itinera Geobot. 8: 243–528.

European Commission, 1996. Interpretation Manual of European Union Habitats. Version Eur 15. Dg XI, Enverioment, Nuclear Safety and Civil Protection. Brussels: 102 pp.

Farinha, J. C. & A. Trindade, 1994. Contribuição para o inventário e caracterização de zonas húmidas em Portugal Continental. Intituto da Conservação da Natureza. Lisboa: 211 pp.

Géhu, J. M. & S. Rivas-Martínez, 1980. Notions fondamentales de phytosociologie. In Diersschke, H. (ed.), Syntaxonomie. J. Cramer. Vaduz: 5–33.

Hill, M. O., 1973. Reciprocal averaging: an eigenvector method of ordination. J. Ecol. 61: 237–249.

Hill, M. O., 1979. TWINSPAN – A FORTRAN program for arranging Multivariate data in an Ordered Two-Way Table, by Classification of the individuals and attributes. Cornell University. Ithaca.

Izco, J., 1989. El Miño, barrera y camino en la migración de las plantas. In Caixa Ourense (ed.), Conferencias Sobre el Rio Miñ. Ourense. 87–97.

Loidi, J., I. Biurrun & M. Herrera, 1997. La vegetación del centro-sptentrional de España. Itinera Geobotanica 9: 161–618.

Ludwig, J. A. & J. F. Reynolds, 1988. Statistical Ecology. Wiley & Sons. New York.

Moreira, I., A. Monteiro & T. Ferreira, 1998. Biology and control of parrotfeather (*Myriophyllum aquaticum* (Vell. Conc.) Verdc.) in Portugal. Ecol., Envir. Conserv. (in press).

Muller-Dombois, D. & H. Ellenberg, 1974. Aims & methods of Vegetation Ecology. John Wiley & Sons Inc., New York: 547 pp.

Neto, C., 1997. A flora e vegetação dos meios palustres do Superdistrito Sadense. Centro de Estudos Geográficos. ISA, ICN. Lisboa: 96 pp.

Rivas-Martínez, S., 1976. Sinfitosociologia, una nueva metodologia para el estudio del paisaje vegetal. An. Inst. Bot. Cavanilles 30: 69–87.

Rivas-Martínez, S., M. Costa, S. Castroviejo & E. Valdés, 1980. Vegetación de Doñana (Huelva, España). Lazaroa 2: 5–189.

Rivas-Martínez, S., 1996. Geobotanica in Discursos pronunciados en acto de investidura de Doctor 'Honoris Causa' del Excelentísimo Senhor D. Salvador Rivas-Martínez. Universidad de Granada: 23–98.

Rivas-Martínez, S., 1997. Syntaxonomical synopsis of the North America natural potential vegetation communities. Itinera Geobot. 10: 5–148.

Rivas-Martínez, S., M. Lousã, A. Penas & L. Herreros, 1998. Aproximación à bioclimatologia de Portugal. Silva Lusitana. (in press).

Hydrobiologia **415:** 77–85, 1999.
J. M. Caffrey, P. R. F. Barrett, M. T. Ferreira, I.S. Moreira, K. J. Murphy & P. M. Wade (eds),
Biology, Ecology and Management of Aquatic Plants.
© 1999 *Kluwer Academic Publishers. Printed in the Netherlands.*

The use of isozymes and PCR-based DNA polymorphism in aquatic weed management: a case-study on introduced and native clones of a hybrid complex

Ludwig Triest* & Bart De Greef
General Botany and Nature Management, Vrije Universiteit Brussel, Pleinlaan 2, B-1050 Brussels, Belgium
E-mail: ltriest@vub.ac.be

Key words: aquatic plants, RAPD, hybridization, genetic diversity, Scirpus

Abstract

Enzyme polymorphisms have been widely used in aquatic plants since the 1980s. Studies on DNA polymorphisms are less numerous and a case-study using both methods on Scirpus is worked out. Along the unique freshwater tidal zone of the River Schelde (Belgium), clumps of *Scirpus* species are mostly scattered in small and fragmented locations on the dikes and mud flats. Most of these taxa are native *S. triqueter*, *S. tabernaemontani* or intermediate morphological forms. However, several cultivated strains of *S. tabernaemontani* have been introduced in recent years. Such 'exotic' strains have been planted to stabilize the muddy riverbanks and became well established and may perform better than the native hybrid complex. In order to determine the existing genetic diversity among these species and the possibility for genetic pollution, stems of 30 clumps from a series of locations along the tidal river were investigated for seven enzymes (SDH, PGM, EST, MNR, GOT, 6PGD and ME) and for markers at DNA level using random amplified polymorphic DNA's (RAPD) of 22 decanucleotides. Data analysis of the allozymes and of the amplified DNA fragments enabled us to classify unambiguously the different *Scirpus* taxa. Direct evidence of hybridization between *S. triqueter* and *S. tabernaemontani* could not be obtained, but the putative hybrids are genetically intermediate or close to *S. triqueter* when considering the DNA polymorphism. The introduced clones of *S. tabernaemontani* consisted of at least three groups of genotypes of which one was very related to the native ones. The escaped clumps could be assigned to a third introduced but less-related strain.

Introduction

When one is dealing with an aquatic weed problem, the first obstacle is the accurate identification of the plant in question. It is a necessity before an appropriate control and management programme can be worked out. This has much to do with different survival strategies that each species has developed. A more profound identification of a water plant is also crucial to know whether it has to be regarded from the beginning as a 'plant' or as a 'weed' (Triest, 1993). Accurate identifications in water plants may present problems in those groups that show vegetative plasticity or need dissection of detailed flowering parts. Therefore, molecular methods at enzyme and DNA level can be helpful in broadening the number of markers at species and eventually at population level. At the same time, the variability of the markers at molecular level may indicate aspects of the breeding system and allow to infer whether a species is an outcrosser, inbreeder, apomictic or even has a mixed mating system. Electrophoretic surveys of enzyme polymorphism's in water plants remained poor in the 1970s, but became more widely applied in the 1980s and beginning of the 1990s. During the second half of the 1990s, less papers appeared on enzyme polymorphism's in water plants, but first attempts were made to introduce the application of PCR-based techniques (Polymerase Chain Reaction) for *in vitro* amplification of DNA fragments in an individual sample. An overview of the most important findings about the level and dis-

* Author for correspondence

tribution of genetic diversity in water plants shows a number of generalizing trends (Triest, 1992).

The breeding system of the studied macrophytes clearly is a primary factor in determining the level and the distribution of genetic variability in populations. However, the definition of what exactly has to be regarded as a single population in aquatic plants is not straightforward and needs to be redefined for each taxon, subsequent to any molecular monitoring programme. Sometimes a population may refer to a single lake or oxbow river, but this might also be in other cases an entire river basin or even multiple catchment areas. Since PCR-based DNA polymorphisms may provide reliable markers for the identification of species, populations, their clonality and eventually events of hybridization, and we have choosen this as a model for the different native and introduced *Scirpus* L. species of the River Schelde.

Brief description of the model species and habitat

The genus *Scirpus* L. (Cyperaceae) is cosmopolitan and includes about 200 species. The cultivation of semi-aquatic *Scirpus* on a large scale allowed the development of local industries in the 18th century. Stalks were used for several purposes such as the weaving of mats and the sealing of seams during the construction of barrels. More recently, *Scirpus* can be used for macrophyte based water treatment projects, since their transported root oxygen promotes the growth of aerobic bacteria that enable the breakdown of organic material (Van Oorschot, 1990). *Scirpus* species are also (re)introduced to campshed riverbanks (Weeda et al., 1994). The freshwater tidal zone of the river Schelde (Belgium) is a unique habitat providing temporary flooded habitats along the inner dikes. Tidal daily differences are still up to 4 m and there is a strong current. *Scirpus* populations are sparsely distributed along the river banks and on the mudflats ('Schorre' in Dutch) but consist of several species and a hybrid complex. *S. triqueter* L. (2x = 40, 42) is characterized by trigonous stems with leafless sheats below. Flowers have two stigmas and nuts between 2.5 and -3.0 mm. *S. tabernaemontani* C.C. Gmelin (2x = 38, 42, 44) has terete stems, ususally two stigmas and nuts of 2.0–2.5 mm long (DeFilipps, 1980). The hybrids (*S. x scheuchzeri* Brügger) resemble more the larger phenotypes of *S. tabernaemontani*. In recent years, *S. tabernaemontani* clones, originating from Dutch companies, have been introduced at few, but well-known localities along the river Schelde. These 'exotic' clones perform very well and stems can be harvested for traditional usages. Presumed 'escaped' clones of the recent introductions could be monitored. Clumps of emergent vegetation may detach since wave action caused by navigation results in faster erosion of the muddy soft soil nearby the zone of the lower water levels (Coops, 1996).

The objective of this study was to estimate the genetic relationship between *Scirpus triqueter*, *S. tabernaemontani* and their putative hybrids by means of enzyme polymorphism and RAPD (Random Amplified Polymorphic DNA). Furthermore, the genetic identity of the different introduced and escaped *S. tabernaemontani* clones was checked against the native remnant clumps. PCR-based techniques will also be evaluated as a novel method to identify the events of genetic pollution and to trace the origin of escaped clones of aquatic macrophytes. The value of molecular methods and their role in relevant topics on aquatic weed management shall be discussed.

Material and methods

Plant material

Samples were taken from clumps of 30 *Scirpus* stands along the River Schelde, Belgium (provided and identified morphologically by M. Hoffmann). A total of 10 clones of *S. triqueter*, nine of the putative hybrids and 11 of *S. tabernaemontani* (four native, five introduced and two escaped) were considered (Table 1). The introduced *S. tabernaemontani* samples were grouped according to their provenance (Figure 1): 'Intro 1' (Appels, Vlassenbroek, two clones), 'Intro 2' (Zwijndrecht, one clone), 'Intro 3' (Burcht, Kallo, two clones), 'Escaped' (Hoboken, two clones). Plants were rinsed with water and cleaned to remove epiphyton, then transferred to liquid nitrogen and further stored at −80 °C until use for DNA extraction (Enzyme extraction was mostly done with fresh plant material).

Extraction procedures, enzyme electrophoresis and staining procedures

Mature leaves were used for enzyme analysis. The buffer conditions and the procedures for electrophoresis were as used in Triest (1989). The buds or leaves were used fresh or after freezing in liquid nitrogen and storage at −80 °C. Best results were obtained with fresh material. When using stored leaf

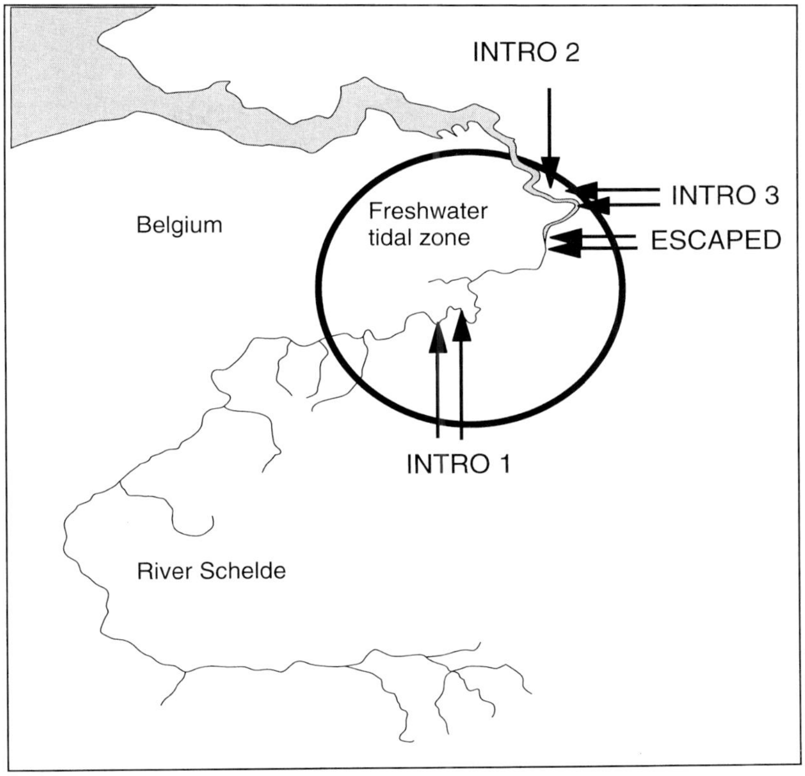

Figure 1. Thirty *Scirpus* clones of three taxa from the freshwater tidal zone of the River Schelde (Belgium), were investigated for allozymes and random amplified polymorphic DNA's. Escaped clumps were compared to the introduced and native clones.

material, the tissues were extracted immediately in cold conditions. A Tris-Cl buffer was used throughout the electrophoresis procedure. Poly-acrylamide gels of 7.5% were used for vertical slab gel electrophoresis. The staining of enzymes was according to the procedures mentioned in Triest (1989). Twenty-two enzyme systems had been tested, only the following were retained because of their clear and consistent activity: alcohol dehydrogenase (ADH, EC 1.1.1.1), β-esterase (β-EST, EC 3.1.1.1), diaphorase/menadione reductase (DIA/MNR, EC 1.6.99.2), glutamate oxaloacetate transaminase (GOT, EC 2.6.1.1), leucine-amino-peptidase (LAP, EC 3.4.11.1), malic enzyme (ME, EC 1.1.1.40), 6-phosphogluconate dehydrogenase (6PGD, EC 1.1.1.44), phosphoglucomutase (PGM, EC 2.7.5.1), shikimate dehydrogenase (SIDH, EC 1.1.1.25).

DNA extraction and PCR conditions

Total plant DNA for PCR was isolated from *Scirpus* stems, according to a protocol modified from Murray & Thompson (1980). The DNA concentration was determined spectrophotometrically and monitored by subjecting samples to 1% agarose gel electrophoresis in TBE buffer (Sambrook et al., 1989) and by visual assessment of band intensities compared with DNA standards. A set of 22 decanucleotides of arbitrary sequence (Operon Technologies Inc., Alameda, California: OPE2, OPE3, OPE4, OPE6, OPE19, OPJ4, OPJ11, OPJ12, OPJ14, OPJ15, OPJ16, OPJ19, OPO19, OPO20, OPT1, OPT4, OPT5, OPT6, OPT7, OPT8, OPT15, OPT20) out of 100 was selected and further tested on the putative hybrid complex, introduced clones and escaped clones. Amplification of RAPD markers from the different *Scirpus* samples was achieved as follows. The 25 μl reaction mixture contained approximately 25 ng template DNA, five picamoles of a single decanucleotide, 100 micromoles of dNTP's (Pharmacia) and one unit Taq polymerase in the reaction buffer provided by the manufacturer (Boehringer Mannheim). The mixture was overlaid by two drops of mineral oil (Sigma). Amplification was achieved in a Thermojet thermocycler (Eurogentec)

Table 1. Locations, origin and morphological characterization of the *Scirpus* clumps from the River Schelde used for allozyme diversity and for the assessment of hybridization between *S. tabernaemontani* and *S. triqueter* using species specific RAPD markers

No.	Location	Origin	Taxon or putative hybrid
1	Burcht	introduced (3)	*S. tabernaemontani*
2	Zwijndrecht	introduced (2)	*S. tabernaemontani*
3	Weert	native	hybrid
4	Weert	native	*S. tabernaemontani*
5	Uitbergen	native	*S. triqueter*
6	Uitbergen	native	hybrid
7	Uitbergen	native	*S. triqueter*
8	Hoboken	'escaped'	*S. tabernaemontani*
9	Hoboken	'escaped'	*S. tabernaemontani*
10	Hemiksem	native	hybrid
11	Niel	native	*S. tabernaemontani*
12	Boom	native	*S. tabernaemontani*
13	Temse	native	hybrid
14	Melle	native	hybrid
15	Schoonaarde	native	*S. triqueter*
16	Schoonaarde	native	hybrid
17	Schellebelle	native	*S. triqueter*
18	Wichelen	native	hybrid
19	Wichelen	native	*S. triqueter*
20	Berlare	native	*S. triqueter*
21	Zele	native	*S. triqueter*
22	Vlassenbroek	introduced (1)	*S. tabernaemontani*/ hybrid?
23	Appels	native	hybrid
24	Appels	native	*S. triqueter*
25	Appels	introduced (1)	*S. tabernaemontani*
26	Zele	native	*S. tabernaemontani*
27	Zele	native	hybrid
28	Zele	native	*S. triqueter*
29	Schellebelle	native	*S. triqueter*
30	Kallo	introduced (3)	*S. tabernaemontani*

programmed as follows: one cycle of 2 min 94 °C; 40 cycles of 45 s 92 °C, 1 min 36 °C, 2 min 72 °C; one cycle of 5 min 72 °C. Amplification products were subjected to electrophoresis in 1% agarose gels in 1 × TBE and stained with ethidium bromide, viewed under UV light, and photographed.

Data analysis

Levels of allozyme variation were estimated at the level of the species by the mean number of alleles per gene (A) and percent polymorphic loci (P). The genetic variability, standard genetic distances (Rogers, 1972; Nei, 1978) and fixation indices were calculated with the BIOSYS program (Swofford & Selander, 1981). The homozygotes (scored as 1 or 0) and heterozygotes (scored as 0.5) for each enzyme locus in the different samples were compared to each other. These multi-enzyme genotypes were then analysed to obtain an estimate of similarity for each pair of individuals. The similarity coefficient employed was the simple matching coefficient thereby emphasizing the different status of the heterozygotes which is important for hybrid detection. The resulting similarity matrix was 'double centered' prior to the principal coordinate analysis (PCO) using the NTSYS-PC program (Exeter Software). The presence / absence status of the selected amplification products obtained with RAPD for each clone were scored. A principal component analysis (PCA) using STATVIEW was based on the sum of all clones observed per species or considered group (native, introduced, escaped) for each diagnostic RAPD marker.

Results

Enzyme polymorphism

In a preliminary test, the clonality of each taxon within a locality could be confirmed on the basis of enzyme polymorphism, using 10–25 individual clumps, depending on the dimensions of the vegetated river bank. The *Scirpus* plants within a locality thus most likely corresponded to a single individual. Consequently, a population has been regarded as the total of all localities of a species within the freshwater tidal zone of the river. Even this might be an overestimation, due to the possible clonality between localities.

A total of nine putative loci (sdh-1, me-1, 6pg-1, lap-1, lap-2, adh-1, pgm-1, pgm-2 and pgm-3) could be revealed using six polymorphic enzymes. The mean number of alleles per locus was 1.3 in *S. triqueter* and 1.7 in *S. tabernaemontani* and the hybrids. Only 33% of the loci were polymorphic in *S. triqueter*, whereas 67% in both other taxa. The mean heterozygosity (direct count) was 0.160 for *S. tabernaemontani*, 0.222 for *S. triqueter* and 0.242 for the hybrids. The expected mean heterozygosity was 0.138 for *S. triqueter*, 0.211 for the hybrids and 0.253 for *S. tabernaemontani*. The fixation indices (Table 2) for three loci within *S. triqueter* are −1 or +1. In *S. tabernaemontani*,

Table 2. Frequencies of the most common allele for nine loci (above) and fixation indices for seven loci (below) in three *Scirpus* taxa

Locus	*S. triqueter*	hybrids	*S. tabernae-montani*
sdh-1	1.000	0.909	0.778
me-1	1.000	0.909	1.000
6pg-1	0.500	0.500	0.389
lap-2	1.000	1.000	1.000
lap-1	1.000	1.000	0.889
adh-1	0.900	0.773	0.889
pgm-1	1.000	0.909	0.333
pgm-2	1.000	1.000	1.000
pgm-3	0.500	0.636	0.444
sdh-1	mono	−0.100	+1.000
me-1	mono	+1.000	mono
6pg-1	−1.000	−1.000	−0.636
lap-1	mono	mono	+1.000
adh-1	+1.000	+0.224	+1.000
pgm-1	mono	+1.000	+1.000
pgm-3	−1.000	−0.571	−0.350

the fixation indices mostly are +1, except for 6pg-1 and pgm-3. The hybrids show more various fixation indices ranging from −1 to +1.

The genetic distances calculated after Rogers (1972) indicated that the hybrids are more closely related to *S. triqueter* (6%) than to *S. tabernaemontani* (15%). The Nei (1978) unbiased genetic distances of the hybrids with the other taxa were even lower and were calculated, respectively, as 0% and 6%.

Individual allozyme genotypes of the 30 clumps were used in a PCO to separate the different multi-enzyme genotypes within and among the three taxa. A total of 32% of the allozyme variation could be explained by the first ordinate and another 23% along the second ordinate (Figure 2). A total of 12 different multi-enzyme genotypes were detected, of which one was very common for *S. triqueter*. Only 2 /10 clones were different in *S. triqueter*, 6/9 in the hybrids, 4/4 in the native *S. tabernaemontani* and 3/7 in the introduced *S. tabernaemontani*. Most of the variation was noted in the *S. tabernaemontani* clones due to the presence of a rarer allele in sdh-1 (loc. 4 and 26). Variation in the hybrid genotypes was mostly due to the presence of heterozygotes for sdh-1 (loc. 14 and 16). The only variation in *S. triqueter* was a rarer allele in adh-1 (loc. 28). Localities 26, 27 and 28 were in close vicinity and show different genotypes for adh-1 (BB, AA, BB) and sdh-1 (BB, AA, AA), respectively. There were no other situations investigated with all three taxa in such a close vicinity (few hundreds of meters only).

DNA polymorphism (RAPD)

The selected set of 22 decanucleotides that were tested on clones from 30 localities yielded 30 clearly scorable, repeatable, polymorphic banding patterns at species or clone level. The PCA plots illustrate that the specific DNA fragments were useful for identification of the species and their putative hybrids (Figures 3 and 4). The genotype composition of the hybrid clones was positioned intermediate between the two putative parental species, especially when considering the first factor. *S. triqueter* and the hybrids were grouped closely. The genetically different groups of introduced and escaped clones of *S. tabernaemontani* (Native, Intro 1, Intro 2, Intro 3 and Escaped) were separated along the first two factors in the PCA (Figure 3). The *S. tabernaemontani* clones that escaped from an artificial plantation, clearly were related to the genotype of the 'Intro 3' group. In the PCA, both were grouped separately from the other *S. tabernaemontani* clones, according to the second and third factor (Figure 4). The 30 selected RAPD markers generated twelve specific fragments of the native *S. tabernaemontani* and nine fragments of *S. triqueter*. Clones of 'Intro 1' and 'Intro 2' had more RAPD bands in common with the native hybrid complex than 'Intro 3' (Figure 5). The escaped clones could be qualitatively assigned to "Intro 3" on the basis of 15 shared bands, of which two bands (I6d and I6c) were not present in the native hybrid complex or in the clones of 'Intro 1 and 2'.

A careful selection of molecular markers generated by RAPD at the within-species level thus can be useful to identify native and introduced clones of the presumed *S. tabernaemontani* and even can be applied to distinguish among the introduced clones.

Discussion

Allozymes

S. triqueter was very uniformic in its allozymes, but showed fairly high values of heterozygosity, due to the fixed heterozygotes in 6pg-1 and pgm-3. The hybrids showed a higher amount of alleles per locus and also were more heterozygous than the parental *S. triqueter*. Heterozygotes in the hybrids were observed for sdh-1,

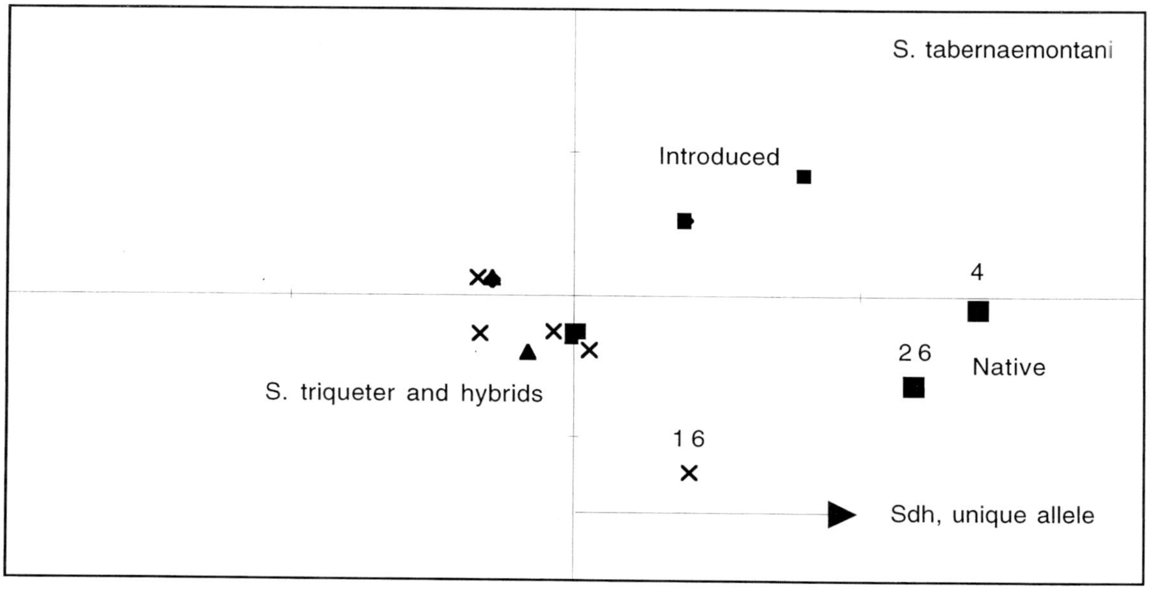

Figure 2. Multi-enzyme genotypes in three *Scirpus* taxa from the river Schelde. The presence of an unique allele in Sdh caused most of the variation among the native *S. tabernaemontani* (large squares). The introduced *S. tabernaemontani* (small squares), *S. triqueter* (triangles) and putative hybrids (crosses) are only slightly separated along the vertical axis.

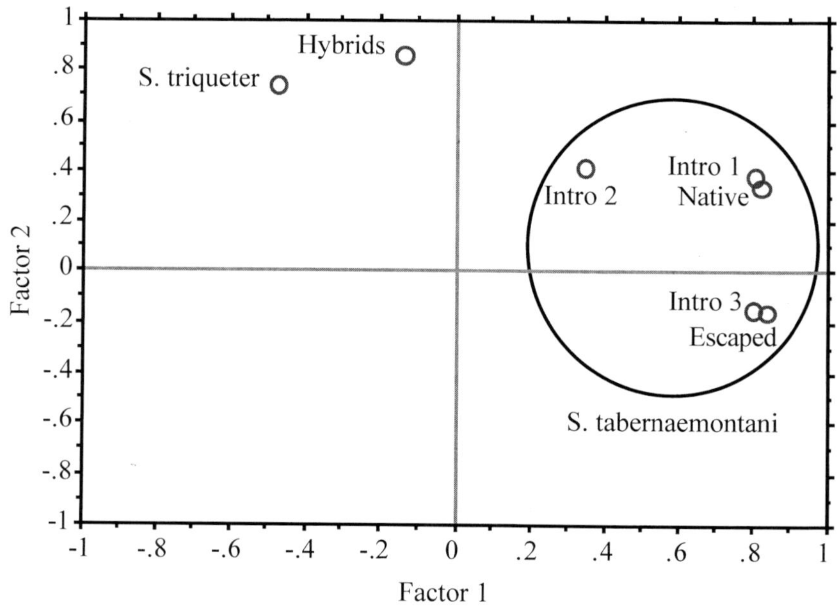

Figure 3. Three *Scirpus* taxa from the freshwater tidal zone of the River Schelde (Belgium) were separated along factor 1 of the PCA based on 30 RAPD markers.

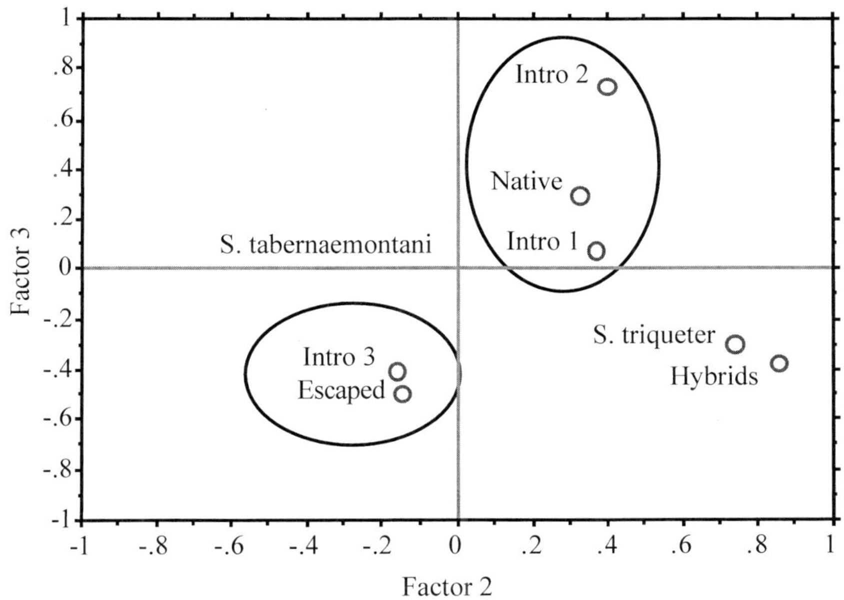

Figure 4. Different introductions of *S. tabernaemontani* from the river Schelde (Belgium) were separated along factor 2 and 3 of the PCA based on 30 RAPD markers.

adh-1 and pgm-3. The higher values of heterozygosity were in concordance with the concept of a hybrid complex. However, not every hybrid clone had the same multi-enzyme genotype, indicating either multiple origins of the hybrids or further introgressive hybridization. The PCO of individual allozyme genotypes made clear that there is no direct indication of an F1 hybrid. The observation that there were less heterozygotes observed than could be expected in *S. tabernaemontani* can be explained by the ocurrence of the different introduced clones. These recently introduced clones could not yet have caused an intraspecific gene flow or eventually take part in a putative hybrid complex.

Fixation indices in the three taxa showed that the hybrids are more variable, whereas clumps of *S. triqueter* were totally fixed for their polymorphic genes and were even fixed as heterozygotes for 6pg-1 and pgm-3. The latter values could be indicative of a more widespread and among-locality clonality of *S. triqueter*. *S. tabernaemontani* also had most of its allelic variation fixed, whereas the hybrids may show loci that are more close to equilibrium. The fixation indices of +1.0 and of −1.0 indicate that most likely clonal growth is responsible here. Locus designations have taken into account the possibility of fixed heterozygosities.

The genetic distances between the hybrids and both taxa were low and mainly indicated the close relationship for allozymes with *S. triqueter*. The more heterogeneous nature of the input data for *S. tabernaemontani* (native clumps combined with introduced clumps) made the interpretation more difficult. Therefore, a PCO of the individual multi-enzyme genotypes shed more light on that heterogeneity of *S. tabernaemontani*. Since most of the variation could be noted in the *S. tabernaemontani* clones due to the presence of a rarer allele in sdh-1 or 6pg-1 (loc. 4 and 26), it becomes clear that several introduced clones ('intro 2 and 3') are similar in their allozymes, but that the native remnants each are unique in their multi-enzyme genotype (Figure 2). Sdh-1 could be a reliable marker to indicate 'pure' or 'native' *S. tabernaemontani* of the River Schelde. The value of this marker can also be demonstrated by the detected variation in the hybrid genotypes that was mostly due to the presence of heterozygotes for sdh-1 (loc. 14 and 16). If more clumps from additional localities become available, it should be checked whether such heterozygotes eventually are F1 hybrids. The same samples from loc 14 and 16 were also investigated for RAPD (De Greef & Triest, 1999) and could be assigned as true intermediate hybrids (in that paper, these two samples are numbered as 28 and 30 in the cluster and the ordina-

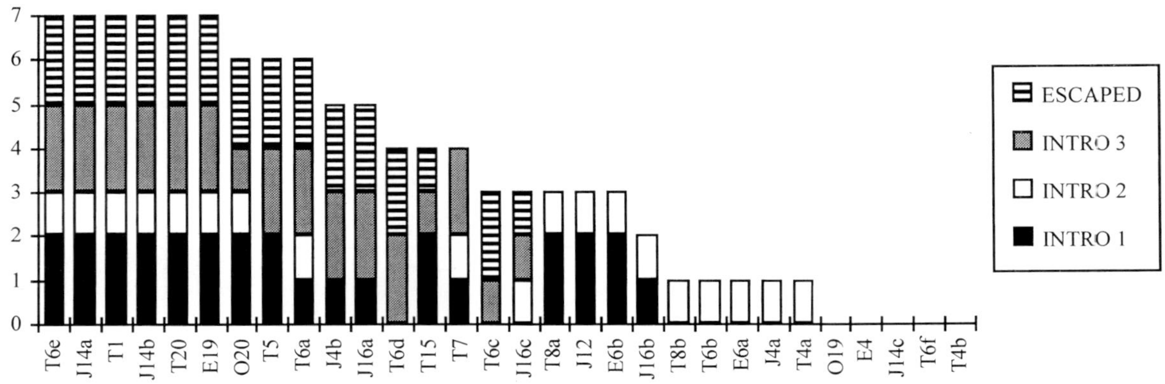

Figure 5. The distribution of 30 RAPD markers in introduced and escaped clones of the river Schelde (Belgium) indicated that the escaped clones most likely originated from the genotype 'Intro 3'. The different origin of 'intro 1' and 'intro 2' could be demonstrated with nine amplified DNA fragments. Five bands were absent in the *S. tabernaemontani* introductions.

tion). The only allozyme variation found in *S. triqueter* was a rarer allele for adh-1 (loc. 28). The observation that clumps from the three taxa (localities 26, 27 and 28 in Zele) were in close vicinity and had different genotypes for adh-1 (BB, AA, BB) and sdh-1 (BB, AA, AA), respectively, could not indicate a hybridisation event at that site, since the hybrids lack the parental alleles of both enzymes. It seems very unlikely that the hybrid clump of this locality in Zele has originated from the adjacent *S. triqueter* and *S. tabernaemontani* clumps. Though differences within *S. tabernaemontani* were obtained, enzymes failed to unambiguously distinguish among most of the clones.

RAPD

RAPD analysis allowed us to obtain unambiguously molecular markers at the level of *Scirpus* species. Though it is a species complex with hybrids, the actual populations from the freshwater tidal zone of the River Schelde consist of clumps of clonal origin. An exact assignment of the hybrid origin could not be given on locality basis, but the putative hybrids shared bands of both presumed parental species *S. triqueter* and *S. tabernaemontani*. The use of DNA based markers in problems of hybridization could be demonstrated also in riverine willow species (Triest et al., 1999). In water plants, isozymes have mainly been used to study population genetic structure (Triest, 1991a), though more recently DNA based studies on e.g. *Hydrilla* (Madeira et al., 1997) and seagrasses (Waycott, 1995) could shed a light on the usefulness of these markers at the level of strains and populations. RAPD could also be used effectively in comparing native versus introduced strains in other organisms (Novak & Mack, 1993; Largiader & Scholl, 1996). Combined studies using codominant allozyme markers as well as dominant PCR-based DNA fragments have the advantage to be a direct comparison and evaluation of both techniques. This way, it could be demonstrated that DNA fingerprinting in French populations of *Sparganium erectum* L. was again much in concordance with the allozyme genotypes (Piquot et al., 1996). When clonal for DNA fingerprints, then the allozymes also were indicating clonality, but a low DNA polymorphism could still be detected in two populations that were still uniformic in allozyme genotypes. Surprisingly, the clones or clonal composition was different among each population, indicating a poor gene flow and rare events of sexual reproduction in *Sparganium*.

In our study on *Scirpus*, RAPD analysis could be used as well to identify the individual introduced clones and to demonstrate the relatedness of escaped clones with 'Intro 3'. Since the introduced clones were not native to the freshwater tidal zone of the River Schelde, it may be questioned whether these strains might cause genetic pollution and eventually disturb the native hybrid complex in the future. At least 'Intro 3' could be regarded as an outlier in that particular ecosystem. The introduced clones obviously may spread further along the river banks through detached and escaped clumps. Navigation and associated wave action results in the removal of the planted clumps that are closest to the low tide water level. Since the use of, for example, emergent plants and willows in stabilization of river banks will gain more attention in the near future for river management in Flanders (Belgium), it should be recommended that the genetic identity and diversity of the native target species is examined before implementing a planting scheme with unknown and potential exotic clones.

An appropriate use of different molecular screening methods in aquatic plants undoubtly will become an integral part of future management plans for restoration of aquatic habitats or tackling weed problems.

Acknowledgements

We gratefully acknowledge the Fund for Scientific Research - Flanders for financial support (mandate of senior research associate for the first author and grant no. S2/5-ID-E53), the OZR-funding of the Vrije Universiteit Brussel and the VLINA 96/01 (Vlaams Impulsprogramma voor Natuurontwikkeling). We also thank Prof. Dr Maurice Hoffmann for the morphological identification of the *Scirpus* species and for providing us with the samples. Mylène D'haeseleer (Technician), Ian Charlette (Vesalius College) and William Fulton (Vesalius College) kindly helped with conducting enzyme electrophoreses.

References

Coops, H., 1996. Helophyte zonation: impact of water depth and wave exposure. PhD thesis, Katholieke Universiteit Nijmegen: 150 pp.

De Greef, B. & L. Triest, 1999. The use or RAPD for hybrid detection in *Scirpus* from the river Schelde (Belgium). Molecul. Ecol. 8: 379–386.

DeFilipps, R., 1980. *Scirpus* L. In Tutin et al. (eds), Flora Europaea 5. Cambridge Univ. Press: 277–280.

Largiader, C. & A. Scholl, 1996. Genetic introgression between native and introduced brown trout *Salmo trutta* L. populations in the Rhone river basin. Molecul. Ecol. 5: 417–426.

Madeira, P., T. Van, K. Steward & R. Schnell, 1997. Random amplified polymorphic DNA analysis of the phenetic relationships among world-wide accessions of *Hydrilla verticillata*. Aquat. Bot. 59: 217–236.

Murray, M. & W. Thompson, 1980. Rapid isolation of high molecular weight plant DNA, Nucleic Acids Res. 8: 4321–4325.

Nei, M., 1978. Estimation of average heterozygosity and genetic distance from a small number of individuals. Genetics 89: 583–590.

Novak, S. & R. Mack, 1993. Genetic variation in *Bromus tectorum* (Poaceae): comparison between native and introduced populations. Heredity 71: 167–176.

Piquot, Y., P. Saumitou-Laprade, D. Petit, P. Vernet & J. T. Epplen, 1996. Gentopyic diversity revealed by allozymes and oligonucleotide DNA fingerprinting in French populations of the aquatic macrophyte *Sparganium erectum*. Molecul. Ecol. 5: 251–258.

Rogers, J. S., 1972. Measures of genetic similarity and genetic distances. Studies in Genetics. Univ. Texas Public. 7213: 145–153.

Sambrook, J., E. F. Fritsch & T. Maniatis, 1989. Molecular Cloning: a Laboratory Manual, 2nd edn. Cold Spring Harbor Laboratory Press, New York.

Swofford, D. & R. B. Selander, 1981. BIOSYS-1: a Fortran program for the comprehensive analysis of electrophoretic data in population genetics and systematics. J. Heredity 72: 281–283.

Triest, L., 1989. Electrophoretic polymorphism and divergence in *Najas marina* L. (*Najadaceae*): molecular markers for individuals, hybrids, cytodemes, ecodemes and conservation of genetic diversity. Aquat. Bot. 33: 301–380.

Triest, L., 1991. Isozymes in water plants. Opera Botanica Belgica 4, Nat. Botanic Garden, Meise: 269 pp.

Triest, L., 1992. The role of isozymes in studies of plant populations: several considerations of data obtained in water plants. Belg. J. Bot. 125: 262–269.

Triest, L., 1993. Problems created by excessive macrophyte growth in tropical freshwater ecosystems and strategies for control. In Tilzer, M. & M. Khondker (eds), Hypertrophic and Polluted Freshwater Ecosystems: Ecological Basis for Water Resource Management. Proc. Int. Symp. Limnol., 25–28 November 1991. Dep. Bot. , Dhaka Univ., Bangladesh: 137–153.

Triest, L., B. De Greef, S. Vermeersch, J. Van Slycken & E. Coart, 1999. Genetic variation and putative hybridization in *Salix alba* and *Salix fragilis*: evidence from allozyme data. Plant Syst. Evolut. 215: 169–187.

Van Oorschoot, M. & J. Verhoeven, 1990. Wetlands for the purification of water. The Utrecht Plant Ecology News Report 11, Postacad. Onderwijs Natuurwetenschappen, Univ. Utrecht: 269 pp.

Waycott, M., 1995. Assessment of genetic variation and clonality in the seagrass *Posidonia australis* using RAPD and allozyme analysis. Mar. Ecol. Prog. Ser. 116: 289–295.

Weeda, E., R. Westra, C. Westra & T. Westra, 1994. Nederlandse Oecologische Flora: wilde planten en hun relaties 5, Lecturis BV, Eindhoven: 400 pp.

Hydrobiologia **415**: 87–91, 1999.
J. M. Caffrey, P. R. F. Barrett, M. T. Ferreira, I.S. Moreira, K. J. Murphy & P. M. Wade (eds),
Biology, Ecology and Management of Aquatic Plants.
© 1999 Kluwer Academic Publishers. Printed in the Netherlands.

Methodologies for surveying plant communities in artificial channels

David Hatcher[1]*, John Eaton[1], Mary Gibson[2] & Rick Leah[1]
[1]*School of Biological Sciences, Jones Building, University of Liverpool, P.O. Box 147, Liverpool L69 3BX, U.K.*
[2]*English Nature, Northminster House, Peterborough PE1 1UA, U.K.*

Key words: canal, methods, survey, vegetation

Abstract

The gathering of quantitative information on aquatic macrophyte communities in artificial drainage and navigation channels presents a number of methodological and analytical problems. These include: subjectivity of plant abundance estimation, the conflict between standardisation and adaptation of methods to specific purposes, and the concepts of randomness and homogeneity in linear surveying. As a result, much of the currently available information is highly subjective and difficult to use in any comparative way, either temporally or spatially. More standardised procedures should be developed which minimise these shortcomings and permit later re-use of data in comparative studies.

Introduction

There is a superficial resemblance between rivers and man-made linear waterbodies such as canals, ditches and drainage channels. Clearly, there are some similarities, but there are also important differences which affect the choice of methods that can be used to survey their plant communities. While ecological and geomorphological differences exist along and between rivers, these are far less pronounced in artificial channels, offering opportunities for sampling aquatic macrophytes with increased standardisation and reduced subjectivity. In contrast, difficulties of access to such vegetation can be a constraint to achieving this. This paper discusses the influence of these features on the design of practical survey methods

Materials and methods

Purpose

Aquatic plant communities are surveyed for different reasons. For example, a baseline for an environmental impact assessment may demand a complete inventory of species and mapping, at least, of rarities. For a national, comparative survey, rapid, standardised, semi-quantitative techniques may be appropriate, while to decide specifications for a weed control programme, a simple visual assessment may suffice. In each of these circumstances, survey methods are tailored to their objectives. The value of this clearly targeted approach is often emphasised (Catchpole & Wheeler, 1992; Jongman et al. 1995), but the former also describe the difficulties encountered when trying to compare data collected over long timescales by a variety of individually adapted methods and using the dataset for new purposes for which they were not initially intended.

It is always worth considering that once the plant community at one location has been assessed, it is very likely that there will be a desire to compare it with that at another site or to re-examine the same site at some later date. Such comparisons are made more difficult and perhaps less credible by the use of un-standardised, subjective and/or differing field methods. While practicalities must to a large extent govern methods used, it should be possible to use at least a 'base-line' level of standardisation and if data from some previous study exists for the field sites or areas concerned, consideration should be given to the possibility of producing comparable data.

Difficulties will vary according to the scale, type of survey being undertaken and the nature of the water-

* Author for correspondence

way. Below are some of the considerations that must be taken into account when surveying a large number of field sites, within a network of slow flowing channels that are distributed over a wide geographical area.

Access

Channels may have very limited road access, so the cost of transporting boats or equipment to survey sites may be high. Many navigation canals have good pedestrian access via a towing path on one bank, but the other side is often only accessible by boat. In drainage channels, high, steep banks may severely impede direct access to vegetation. Slow flow rates in artificial channels make them depositional environments with soft substrates. This, sometimes allied with deep water, may make wading difficult and even dangerous.

Turbidity

Often, eutrophic conditions lead to high densities of phytoplankton and the suspension of sediment from disturbance by wind, boat traffic (in navigation channels) or wading, further increases turbidity, making visual assessment of submersed vegetation difficult or impossible.

Such considerations have led researchers to survey different components of plant communities separately, e.g. emergent plants on each side, floating-leaved vegetation and finally submersed plants (Murphy & Eaton, 1983). The accessible emergents are sampled by means similar to those for terrestrial areas, the floating leaved vegetation and inaccessible side emergents by viewing remotely from the accessible side, and the submersed plants from material collected with a grapnel. The effectiveness of these methods probably declines as channel width increases because it becomes more difficult to see across the channel and there is a limit to the amount of vegetation a grapnel can recover and a limit to the distance to which it can be thrown.

Subjectivity and standardisation

Data collected using objective survey methods have 'statistical power' (Jongman et al., 1995) that is lacking in those that are subjective. Some benefits of standardisation have already been discussed, but the advantages must be balanced against any additional time and/or cost that may be incurred in achieving them. Standardisation may impose a methodology not ideally tailored to the objectives of any one particular study and may impede further development of methods.

These issues are currently being addressed by The Environmental Change Network, though only procedures for surveying rivers and lakes are being developed at present (ECN, in press). However, as the draft text states that sampling at one of the many points required to survey the vegetation of one lake is "likely to take approximately four hours", it seems probable that the approaches being developed will mainly be suited to the production of localised, detailed inventories.

To quantify the biomass of emergent species directly requires the removal of all plant material in a quadrat. This is destructive and very time consuming. Quicker, indirect alternatives are 'DAFOR' scales (Dominant, Abundant, Frequent Occasional or Rare, as in Kendall, 1987), relative abundance scales (Murphy & Eaton, 1983) and estimates of percentage cover in conjunction with published biomass/area conversions (e.g. Bradbury & Grace, 1983).

Submersed plants extracted with a grapnel can be drained for a set period, then weighed; again calibration against sampling of areas of known vegetation density yields conversion factors which provide estimates of biomass. However, samples thus obtained often consist of a tangled mixture of species and the time needed for physical separation can be considerable. A semi-quantitative alternative is to visually estimate species proportions in intact samples. Dry weights are more reliable measures of biomass than wet weights, but amounts of material to be transported from the field may be large (10 grapnel samples from a moderately vegetated channel can yield over 50 kg of material).

Field site selection

When long lengths of channel are surveyed, such statements as "a representative 20 m section was selected" (Alcock & Palmer, 1985) are common, but it is unclear what 'representative' actually means. It could be interpreted as 'typical of a certain longer section' in which case that longer section should be clearly defined, with reasons for its selection. Alternatively, 'representative' may mean 'has a flora that appears to be typical', 'one in which most species present in the whole area being studied are represented' or even one that is chosen at random. In practice, it is difficult to exclude subjective influences such as ease of access and presence of species of special interest to the surveyor.

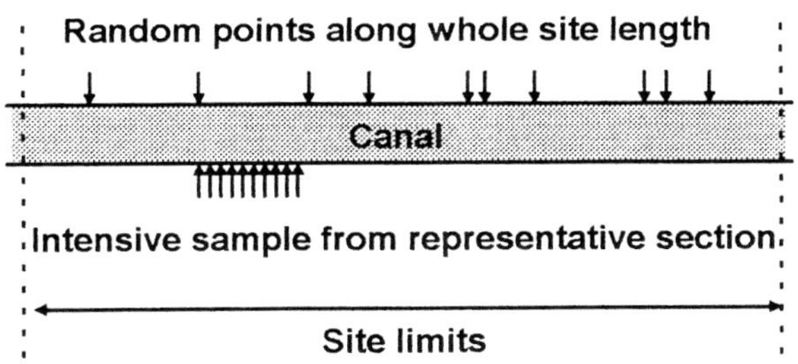

Figure 1. Alternative sub-sampling patterns applied to a large, contiguous test set of data.

Table 1. Accumulation of species detected within a 450 m test survey length on the Leeds and Liverpool Canal location, using random (r) and contiguous (c) sampling strategies at two levels of effort (10 and 20 points per sample). Total species present in test survey length = 23. Significant differences, r to c, 10 to 20 (Square root transformed, $P > 0.05$)

No. of Sample Sets	Number of species detected			
	10r	10c	20r	20c
1	14	9	17	12
2	17	14	19	16
3	18	16	20	17
4	20	17	21	19
5	20	19	22	21

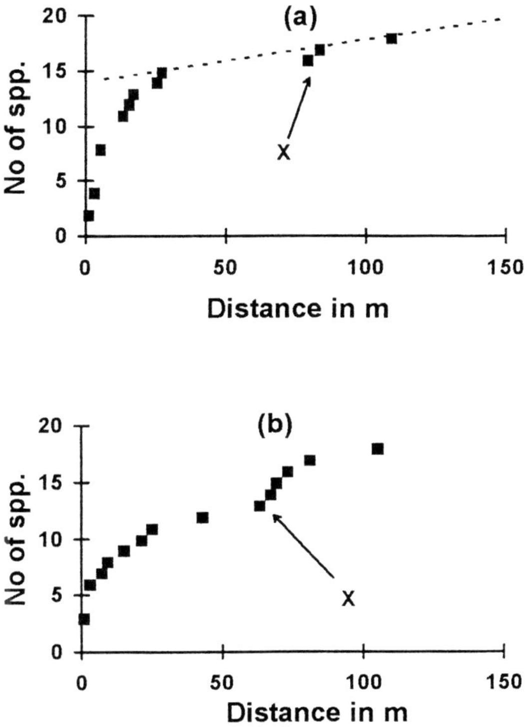

Figure 2. Species accumulation curves compiled from a linear data set from the Leeds and Liverpool Canal. Plotted (a) north to south, (b) south to north. X shows apparent change in plant community.

The terms typical, characteristic and homogeneous are similarly inexact. Presumably the ideal is a site where the plant community is wholly uniform, but as such a site does not really exist in nature, a decision must be made as to how different any part of the site has to be before it ceases to be homogeneous. For example, in a test survey on the Rufford Branch of the Leeds and Liverpool Canal, it proved very difficult to locate a 300 m long homogeneous section to use as a survey site. Initially short segments where young alder trees existed were excluded as clearly non-homogeneous, but gradually the subjective limits of homogeneity were mentally adjusted until what was judged to be the 'most' homogeneous section was finally selected.

It is preferable to choose sites randomly (e.g. as in the study of hedgerows by Hegarty & Cooper, 1994) and to sample randomly within them, as the quality of information yielded by a given survey effort must in part depend upon the distribution of the effort over the length being surveyed. This is illustrated by the differing outcomes from a range of sub-sampling patterns applied to the same large test set of data. The test set used was data from a length of the Leeds and Liverpool Canal, collected contiguously at 2 m intervals for 450 m. At two levels of effort from this dataset, random, extensively distributed sub-sets of samples were extracted and the accumulation of species compared with that from intensive, contiguous samples taken from randomly chosen starting points (see Figure 1). Table 1 shows that random sampling was always more efficient than intensive contiguous sampling at detecting species, at all levels of sampling effort.

Species accumulation curves from continuous sampling are sometimes used to determine site length as distinct from sampling effort within a predetermined length. A difficulty described by Kershaw & Looney (1985) is that in reality such curves do not reach an asymptote, and that the location of any 'break' of curve is dependant on the ratio of the axes. Figure 2(a), shows this failure to 'level out', while Figure 2(a) & (b), show another potential difficulty, where depending on the direction of sampling, a change in the plant community can be masked or exposed.

The purpose of the survey will dictate the degree of precision required for any estimates of mean biomass per unit length of channel. The sampling effort needed to achieve this can readily be determined statistically.

Linearity and independence

Underwood (1994) notes that in a linear, flowing water system, independence between sites is inevitably compromised by the continuous transfer of propagules, whole floating plants and physio-chemical effects on water quality from upstream to downstream locations. This must also be true in slow flowing artificial channels, whilst in those that are navigated there is the added potential for transfers by boat movement, even against the direction of flow.

Pattern

If vegetation occurred in a truly random manner, sample or site size could be decided purely on the basis of practicality and from species accumulation curves. However Kershaw & Looney (1985) state that non-randomness may be demonstrated in vegetation even "...where a close visual inspection does not reveal any obvious sign of clumping" and that as quadrat size is increased, a contagiously distributed population can be shown to be successively random, contagious or regular. Also, Rapson et al. (1997) suggest that interactions between species richness and biomass make richness values sensitive to sample size.

A number of recent studies have compared the discernment of pattern in species occurrence and abundance with that predicted by certain models (e.g. Spencer & Ksander, 1993; Brummer et al., 1994), including fractal models that predict clumping on a series of spatial scales (Li & Reynolds, 1994). When data are pooled from emergent, floating and submersed components to give overall community estimates (Murphy & Eaton, 1983; Kendall, 1987), within-site pattern may be masked, preventing any retrospective analysis of pattern in old data. Figure 3 shows how this may happen in the common situation when reed growth on one margin of a channel inhibits its growth on the other by concentrating flow, and boats if present, into the remaining open part of the cross-section. Greater understanding of plant distribution and community structure is achieved by analysing such data independently.

Ultimately, practicality must be a major factor in selecting the size and shape of field sites and/or sampling units. This is noted by Catchpole & Wheeler

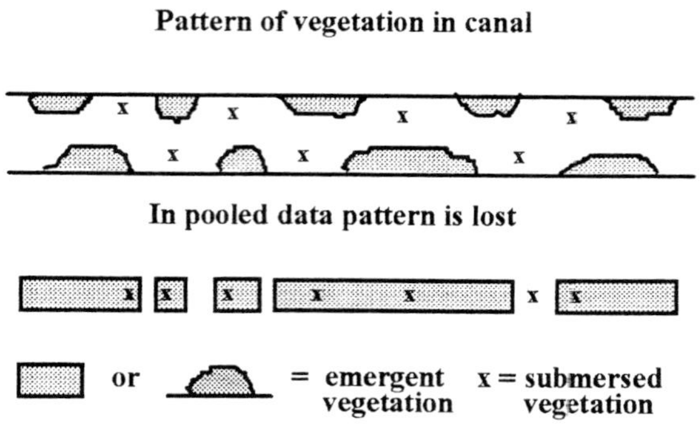

Figure 3. The effect of pooling plant distribution data on the discernment of community pattern.

(1992), who state that "...work per unit area will generally increase as quadrat size decreases" while "the sample size required will increase with the spatial variability". These interactions may lead to false assumptions being made regarding the distribution characteristics of species.

Conclusions

When undertaking surveys of vegetation in artificial channels, a tension exists between the need for widely acceptable standard procedures and the competing need for methods which are closely adapted to their specific purposes in terms of level, type of information required, and cost-effective use of the available resources. Before the procedures to be used are finalised, the following important aspects should be considered.

Precise tailoring of methods to meet objectives appears to be efficient on a within survey basis, but immediate efficiency savings may ultimately result in efficiency losses when the almost inevitable need to make temporal and spatial comparisons arises. In view of this, it would seem logical to introduce some form of standardised methodology, even if only at some base line level, to surveys at all locations including those for which no future re-assessment or monitoring is immediately planned.

The physical properties of some artificial channels may limit the choice of practical survey methods available for vegetation sampling. However, their physical uniformity provides an opportunity to use more standardised field methods.

Procedures based upon extensive and random sampling of sites are both less open to misinterpretation than those using intensive sampling of selected sub-sections, and also more efficient in terms of the information yielded per amount of sampling effort.

Multiple sub-samplings on test lengths yield data which can be used to optimise further sampling effort aimed at determining species and/or biomass characteristics of the channel. Thus species accumulation curves provide a quick indication of the sampling effort needed to detect the range of taxa present. However, the value of this approach is limited as impressions gained on the homogeneity of vegetation and the spatial distribution of particular species can easily include artefacts of the size of sample unit and the direction in which the surveyor has worked, or can miss interactive patterns within communities. Multiple sub-samplings can, secondly, provide estimates of biomass homogeneity, and continuous, ratio scale data such as these are easier to test statistically.

The aggregation of data may obscure implicit spatial patterns. Therefore, to maximise the potential for re-use, data should be retained in its original, separate categories.

Acknowledgements

D. Hatcher was supported by the Economic & Social Research Council and English Nature.

References

Alcock, M. R. & M. A. Palmer, 1985. A standard method for the survey of ditch vegetation. Chief Scientist's Team Note 37, Nature Conservancy Council, Peterborough, England.

Bradbury, I. K. & J. Grace, 1983. Primary production in wetlands and mires: swamp, bog, fen and moor. Gore, A. J. P. (ed.), Elsevier, Amsterdam.

Brummer, J. E., J. T. Nichols, K. Engel, & K. M. Eskridge, 1994. Efficiency of different quadrat sizes and shapes for sampling standing crop. J. Range Mgmt 47: 84–89.

Catchpole, W. R. & C. J. Wheeler, 1992. Estimating plant biomass: a review of techniques. Aust. J. Ecol. 17: 121–131.

ECN (in press) Monitoring aquatic macrophytes within the ECN program, Environmental Change Network. A project based at The Institute of Terrestrial Ecology, Merlewood England.

Hegarty, C. A. & A. Cooper, 1994. Regional variation of hedge structure and composition in Northern Ireland in relation to management. Proc. r. Irish Acad. 94B: 223–236.

Jongman, R. H. G., C. J. F. Ter Braak & O. F. R. Van Tongeren, (eds), 1995. Data Analysis in Community and Landscape Ecology. 2nd edn. Cambridge University Press, Cambridge.

Kendall, S., 1987. A Botanical Survey of The Leven Canal: a report produced for the Nature Conservancy Council, North Eastern Region, Peterborough, England.

Kershaw, K. A. & J. H. H. Loony, 1985. Quantitative and dynamic plant ecology. 3rd edn. Edward Arnold. London.

Li, H. & J. F. Reynolds, 1994. A simulation experiment to quantify spatial heterogeneity in categorical maps. Ecology 75: 2446–2453.

Murphy, K. J & J. W. Eaton 1983. Effects of pleasure-boat traffic on macrophyte growth in canals, J. appl. Ecol. 20: 713–729.

Rapson, G. L., K. Thompson & J. D. Hodgson, 1997. The humped relationship between species richness and biomass – testing its sensitivity to quadrat size. J. Ecol. 85: 99–100.

Spencer, D. F. & G. G. Ksander, 1993. Spatial pattern analysis for underground propagules of *Potamogeton gramineus* L. in two Northern California irrigation canals. J. Freshwat. Ecol. 8: 297–333.

Underwood, A. J., 1994. Spatial and temporal problems with monitoring. In Calow, P. & G. E. Petts (eds), The Rivers Handbook, Vol. 2. Blackwell Scientific Publications, Oxford: 101–123.

Hydrobiologia **415**: 93–100, 1999.
J. M. Caffrey, P. R. F. Barrett, M. T. Ferreira, I.S. Moreira, K. J. Murphy & P. M. Wade (eds),
Biology, Ecology and Management of Aquatic Plants.
© 1999 Kluwer Academic Publishers. Printed in the Netherlands.

Interrelations of river ship traffic with aquatic plants in the River Nile, Upper Egypt

Magdi M. Ali[1,*], Kevin J. Murphy[2] & Jérôme Langendorff[3]
[1] *Department of Botany, Faculty of Science at Aswan, South Valley University, Aswan 81528, Egypt*
[2] *Institute of Biomedical & Life Science, Division of Environmental & Evolutionary Biology, University of Glasgow, Glasgow G12 8QQ, Scotland (k.murphy@bio.gla.ac.uk)*
[3] *Department of Civil Engineering, University of Glasgow, Glasgow G12 8QQ, Scotland*

Key words: wave action, boats, tourist ships, aquatic plants, water chemistry, pollution, River Nile, Egypt

Abstract

Multivariate analysis of data from a 1997 survey of submerged vegetation at 3 locations on the River Nile in Upper Egypt showed that hydrochemical factors (total oxidised nitrogen, sulphate and sodium) were the most important environmental variables explaining the distribution and the abundance of submerged plant species. The kinetic energy of waves produced by tourist ship wash (measured using an index incorporating measured wave height at shore, depth of sampling site and distance between ship channel and sampling site) was next in importance, and explained more of the variation in submerged plant abundance and distribution than did river flow rate. Classification using TWINSPAN identified 5 major groups of sites characterised by differing plant community composition (and having significant inter-group differences in wave index and flow rate). Ship wave effects are likely to be an important factor contributing to differences observed in hydrosoil texture and organic matter content at different depths at one study location (Aswan) heavily used by ship traffic. Coarse sand and gravel fragments with low organic matter content, were predominant in the shallow water zone, most heavily impacted by ship waves; while finer sand with high organic matter content formed the hydrosoil in deeper water. Such variation in sediment characteristics is also likely to influence macrophyte growth. The environmental disturbance effects of ship traffic in the Nile in Upper Egypt currently appear to play a significant role in the ecology of the river's submerged vegetation. Given the long-term trend towards tourism use of the Nile for cruise holidays, it is likely that ship traffic intensity will increase over the next decade, adding to the impacts on aquatic vegetation.

Introduction

The tourist industry is a major source of income for Egypt, with many large tourist ships cruising the stretch of the River Nile between the Luxor and Aswan in Upper Egypt. Starting in the 1940s with steam-powered craft, the number of tourist ships on this stretch of river had increased to 12 by 1972, with diesel-powered craft replacing the old steam vessels. Numbers increased sharply thereafter, reaching 110 in 1989. By 1997, some 240 large pleasure ships (up to 70 m in length, with up to 4 decks) cruised between Aswan and Luxor (Erian, 1989).

* Author for correspondence

Effect of tourist ship traffic on the aquatic habitat

Although there is extensive literature on the effects of different abiotic factors, including boat-traffic, on the ecology and distribution of river and canal macrophyte communities, little examination has been made of the impact of large tourist ships on the plant ecology of large navigable rivers, such as the River Nile. Most investigations have been concerned with smaller waterways such as navigation canals (e.g. Murphy & Eaton, 1981; 1983). Nile tourist ships not only stress the river environment by chemical pollution (from fuel, exhaust and sewage discharge, which exacerbates the effects of pollution from industrial sources: Ali & Soltan, 1996) but also increase the intensity of habitat disturbance by inputs of kinetic energy from waves produced by propeller and hull movement. Some information is available on the impact of wind-generated waves on littoral vegetation (e.g. in lakes: Keddy, 1982). In general, exposure to wave action affects macrophytes both directly, through removal, and indirectly, by producing a gradient of substrate particle size, organic content and nutrient concentration (Wilson & Keddy, 1985; 1986). Ships produce additional specific impact on the aquatic vegetation (Murphy et al., 1995) in the following ways:

(i) direct physical damage is caused by propellers and contact with moving hulls (Cragg et al., 1980; Liddle & Scorgie, 1980);

(ii) ship-generated waves and currents cause physical damage and uprooting of the submerged vegetation (Schloesser & Manny, 1989; Vermaat & De Bruyne, 1993). Intermittent pulsed disturbance due to ship passage may lead to wash-out of free-floating plants, uprooting of lightly anchored non-rhizomatous species and mechanical damage by drag and tearing on more firmly-rooted species. Marginal reedbeds have a limited resilience to boat wash, beyond which they are damaged by stem breakage and erosion and fragmentation of the root mat (Bonham, 1980; Garrad & Heny, 1988);

(iii) eroded and re-suspended sediment may shade submerged plants (Westlake, 1966; Tanner et al., 1993) and at very high levels may cause abrasion damage to plant tissues (Edwards, 1969). Shading effects caused by boats moored for long periods in shallow water areas may also restrict the development of aquatic vegetation (Vermaat & De Bruyne, 1993); and

(iv) macrophyte establishment and spread are also likely to be inhibited by soft, accreting, unstable and periodically re-suspended layers of sediment (Murphy et al., 1995).

This study sets out to examine the importance of ship traffic in the context of a suite of physico-chemical factors likely to regulate submerged aquatic plant growth and hydrosoil particle size distribution in the River Nile in Upper Egypt.

Description of study sites

Three stations were selected, Aswan (24° 05′ N; 32° 55′ E: downstream of the Old Aswan Dam, 262 km upstream of Luxor and 982 km from Cairo), Kom Ombo (40 km N of Aswan), and Edfu (86 km N of Aswan) (Figure 1). At each station, submerged macrophyte stands were present around recently-formed islands within the river channel and along the river banks. Island and main bank shorelines bordering the main ship routes at each site (Figure 1) were influenced by wave action produced by the ships.

Materials and methods

Submerged aquatic macrophytes

Submerged macrophyte samples were collected (from 1 m depth-zone intervals to the limit of plant colonisation) at different depths along transects located perpendicular to the shore at four to six sites per station. Sites were located to include both shorelines exposed to and those sheltered from wave action, and at different distances from the main ship routes within the river channel (Figure 1). Within each depth zone a 1 m^2 area of sediment was sampled. A grapnel screwed to a metal pole was used to collect standing crop samples (five grapnel hauls per sampling area). This method gives good comparative values of abundance, but does not estimate standing crop per unit area of substrate. Plants were separated into different taxa, identified after Täckholm (1974), Moore (1986) and Triest (1989). The dry weight standing crop (DWSC) of each species per grapnel haul was determined after air drying (ambient temperature 45 °C). The average standing crop was calculated for each species per depth zone.

Water physico-chemistry

In total, 13 water physico-chemistry variables were measured. In the field, 1-litre surface water samples were collected and measurements made (during daytime *in situ*) of water temperature and pH (using an Extech Instruments Oyster pH meter) and of dissolved oxygen (DO: using a YSI model 50 dissolved oxygen meter). Total dissolved solids (TDS mg l^{-1}) and electrical conductivity (EC) were measured (using an ESD model 76 combined TDS - conductivity meter). Photosynthetically active radiation (PAR) net downwelling attenuation coefficient was recorded (using a Skye PAR meter, model SKP 200 with SKP 215 sensors). A simple estimate of flow velocity was made using a floating 1 m-long reed stick and a stop-watch. In the laboratory, water samples were analyzed using standard methods (APHA, 1985) for K and Na (Jenway flame photometer, model PFP7); total oxidised nitrogen (TON), as the sum of nitrate (chromotropic acid

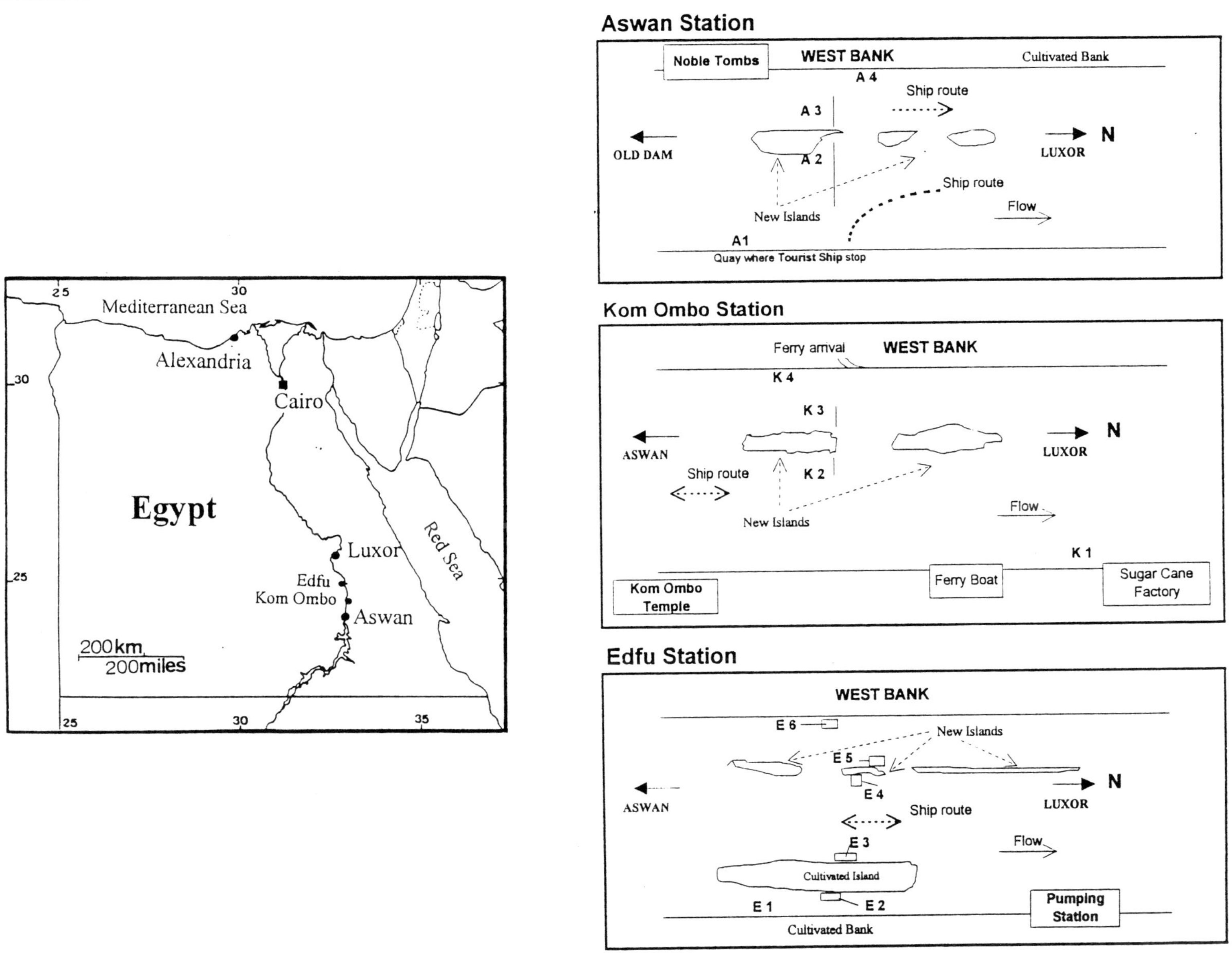

Figure 1. Location map of the River Nile showing the sites surveyed within each of the three stations in Upper Egypt.

method) and nitrite (Griess-Ilosvay method) concentrations; total soluble phosphate (molybdenum blue method); and sulphate (turbidimetric method).

Hydrosoil physical characteristics

Hydrosoil texture was determined in samples collected from sites along transect A-3 (Aswan) at 1 m, 2 m and 3 m depth) by separation of the different size particles using the pipette method of Kilmer & Alexander (1949). Hydrosoil organic matter was measured by the loss-on-ignition method of Allen et al. (1986).

Wave energy

A metre stick placed with one end on the sediment surface in the water at the shoreline was held vertical during the recording period following movement of a ship past the sampling point, and the maximum wave height produced by the wake was read. The error estimated was no more than half a centimetre. This method allowed quick movement from one transect to another. Wave height at shoreline (W m), depth of macrophyte sample (D m) and distance from ship to metre stick (R m) were used to produce a wave index (I), which is a function of the kinetic energy released in the form of the ship waves:

$$I = \frac{W}{D \times R} \times 10000\,\mathrm{m}^{-1},$$

Data analysis

The data were drawn up in the form of matrices of plant species DWSC values × samples, and environmental variables × samples. Multivariate classification of the first data matrix was undertaken using Two-way Indicator Species Analysis (TWINSPAN: Hill 1979). Ordination analysis for the two matrices was carried out using Canonical Correspondence Analysis (CCA), release 3.12 of CANOCO (Ter Braak, 1988). One-way analysis of variance was applied to examine the significance of differences in mean values of measured environmental variables between groups of sites identified by TWINSPAN.

Results

Aquatic macrophytes

In total, ten species of submerged macrophytes were found. No submerged macrophytes were observed at sites located adjacent to ship or boat moorings (sites A1, A2-1 and A2-3 at Aswan; K1 and K2-2 at Kom Ombo). There was no significant correlation ($r = -0.107$, $p > 0.05$, NS) between wave index and total standing crop of submerged vegetation. However, there was evidence of localised impact of waves produced by moving ships. Transects located in the channels heavily used by tourist ships (e.g. approaching mooring sites) were characterised by lower abundance of submerged vegetation (e.g. A2-1 [0.7 g sample^{-1}] where *Potamogeton pectinatus* was the dominant and A3-1 [25.3 g sample^{-1}] where *Potamogeton trichoides* dominated the submerged vegetation). At Edfu, in transects located on either side of the main channel used by ships, no submerged aquatic macrophytes were present in water deeper than 2 m. On the same transects, only very low macrophyte abundance occurred in water shallower than 2 m (7.6 g sample^{-1}: dominated by *Potamogeton pectinatus* at E3-2, and 2.61 g sample^{-1}: dominated by *Myriophyllum spicatum* at E4-2). At Kom Ombo no submerged vegetation was recorded in deep water (< 2 m). In shallower water, vegetation was abundant (77.3 g sample^{-1} at K2-1, mainly comprising *Chara* sp.).

Water

Water samples collected from all 30 sampling sites had low concentrations of phosphate (below the limit of detection < 0.01 mg l^{-1}). Narrow ranges were recorded for temperature (20.6–21.7 °C), pH (7.4–8.4), TDS (235–408 mg l^{-1}), and EC (352–408 μS cm^{-1}). Wider ranges were found for TON (0.01–4.9 mg l^{-1}), sulphate (14.8–22.3 mg l^{-1}), DO (6.1–16.5 mg l^{-1}), flow rate (0.01–0.5 m s^{-1}), and PAR attenuation coefficient, k (1.4–3.3 m^{-1}). Wave index ranged from zero in sheltered sites to 78.1 in sites exposed to strong ship waves.

Hydrosoil characteristics

Hydrosoil samples were collected from three depth zones (0–1, 1–2 and 2–3 m) along transect A3 at Aswan. Coarse sandand gravel fragments, with low organic matter content, predominated in the shallow water zone (0–1 m depth). In deeper water, the hydrosoil was made up primarily of finer sand, with high organic matter content, and low gravel percentages (Table 1).

Table 1. Distribution of particle size and organic matter content percentages along transect A3 in Aswan station

%	A3-1	A3-2	A3-3
Gravel	17.57	1.11	0.40
Coarse sand	80.66	36.06	32.53
Fine sand	0.67	11.08	29.33
Silt	0.01	0.57	0.41
Clay	0.28	0.73	0.48
Organic matter	0.82	50.45	36.85

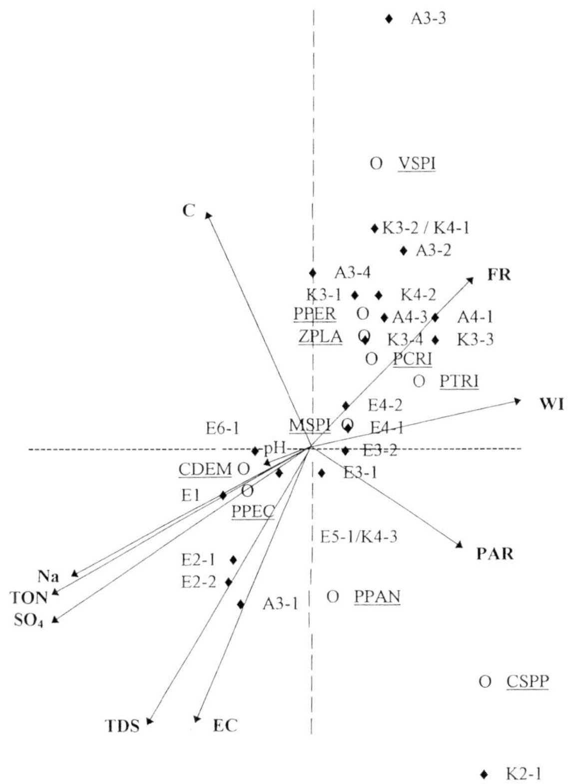

Figure 2. CCA ordination diagram illustrates the distribution of 10 submerged species (O) VSPI = *Vallisneria spiralis*, *PPER* = *Potamogeton perfoliatus*, ZPLA = *Zannichellia palustris*, CDEM = *Ceratophyllum demersum*, PPEC = *Potamogeton pectinatus*, PCRI = *Potamogeton crispus*, PPAN = *Potamogeton panormitanus*, PTRI = *Potamogeton trichoides*, MSPI = *Myriophyllum spicatum*, CSPP = *Chara* sp.; from 25 sampling sites: (♦). First is the initial letter of the station, followed by the sampling site number and depth (m) (See Figure 1 for site locations) in relation to environmental factors (arrows) WI = wave index, FR = flow rate, PAR = underwater photosynthetically active radiation, C = water temperature °C, pH = water pH, EC = water electric conductivity, Na = water sodium, TON = water total oxidised nitrogen, SO_4 = water sulphate, TDS = water total dissolved solids.

Vegetation – environment relationships

Classification of the plant samples using TWINSPAN identified five major groups of sites, characterised by differing plant community composition (and having significant inter-group differences in wave index and flow rate). Group I sites, indicated by *Potamogeton pectinatus* but with only very sparse submerged vegetation, had the highest wave index, with intermediate flow. Groups II (indicated by abundant *Potamogeton trichoides*, together with *Vallisneria spiralis* and lower abundance of *Potamogeton perfoliatus*) and IV (indicated by *Potamogeton pectinatus* in high abundance) had intermediate wave index values, but differed in flow rate: being intermediate in II and very low in IV. The wave index values for the other two site groups, III and V, were low and very low, respectively, with quite different flow characteristics. The mean flow rate for the Group III sites was the highest of any site group. However, flow rates were very low at Group V sites. These differences in flow are likely to have strongly influenced the observed differences in the submerged plant community. Group III was indicated by abundant *Myriophyllum spicatum*, while V had the highest abundance of *Ceratophyllum demersum* recorded for any group of sites.

The CCA ordination (Figure 2) accounted for 35.6% of the total variance of species data. The successive eigenvalues of the first four axes of the canonical ordination decreased rapidly (0.82–0.16), suggesting a well-structured data set. The analysis omitted potassium, phosphate and dissolved oxygen from the ordination because they have negligible variance. Sodium showed a high collinearity with TON and sulphate. The environmental vectors represented by the arrows in the ordination diagram suggest that two opposing trends are recognizable. A set of partly intercorrelated hydrochemical factors (total oxidised nitrogen, sodium, sulphate TDS and conductivity: broadly representing ionic strength of the water) were opposed to vectors representing environmental disturbance (wave index and flow rate). Gradients of water temperature and clarity ran in different directions through the data set. The interset correlations for the environmental factors suggested that TON, sulphate and sodium were likely to be of the most importance in explaining the observed variability in the dataset.

However, the variance inflation factors[1] (VIF) of TON and sulphate (together with those of temperature, pH, conductivity, underwater light availability, and TDS) were > 20. The canonical coefficients of such variables are unstable and do not usually merit detailed interpretation (Ter Braak, 1986). For example, in this dataset the actual range of temperature variation encompassed by the temperature vector is only 1.1 °C. Such a gradient is highly unlikely to be of environmental significance, and the observed vector is most likely due to correlation with some other factor. Removing from consideration the factors exhibiting both high IF values and low interset correlations leaves TON, sulphate (and their close correlate sodium), plus the physical disturbance variables, flow and wave index, as the most likely candidates for explaining the observed sample and species variation in the dataset.

Sites located in the top right of the CCA diagram (Figure 2) were those adjacent to ship routes, and most susceptible to wave disturbance. The species characteristic of these sites included *Vallisneria spiralis, Potamogeton trichoides* and *Chara* spp. In the opposite corner of the diagram, *Ceratophyllum demersum* and *Potamogeton pectinatus* were found in abundance. Sites with these plants were in backwaters at Edfu (E1, E2-1 and E2-2), sheltered from wave action, where high concentrations of total oxidised nitrogen, sulphate, sodium and total dissolved solids were recorded.

Discussion

Wave action increases habitat disturbance through direct damage to plant biomass, and may also influence the intensity of environmental stress influencing plant growth, via the creation of gradients in sediment organic content, nutrient concentration, and fine particle sizes (Wilson & Keddy, 1986). There was some evidence for such a sediment gradient along the transect at Aswan (which was heavily affected by ship traffic, with the highest WI value of the whole data set recorded from theshallow site A3-1). Table 1 shows that coarse fragment particles predominated in the shallow end of the transect, while finer particles, and a much higher level of organic matter, characterised the deeper sediments.

Looking at the strategies of the plants present at these sites, there is little evidence to suggest much change in the incidence of stress-tolerance attributes in those plants present at sites with differing wave index (WI) values. All species present have a reasonably strong element of Disturbance-tolerance (D) in their strategy (all are competitive-disturbance-tolerance (CD), except *Chara* spp.: competitive-stress-disturbance-tolerance (CSD) according to Murphy et al., 1990). Physical disturbance appeared to be associated more with a change in community composition than with a change in abundance. The four sites with very high WI (> 36) all had a low abundance of submerged vegetation, but so did many sites with lower wave index values.

Ranking species along the wave index axis (WI) on the ordination diagram, and comparing the results with those of Sabbatini & Murphy (1996), we found close similarities to rankings of the same species on a management disturbance axis in British drainage channels. This suggests that (within the CD strategy type) there are degrees of disturbance tolerance which occur consistently across different habitats and different sources of disturbance, namely (in this study) waves produced by ships and (in drainage channels) water level fluctuation and channel weed control operations.

Results of this study (species occurring in both Egyptian and UK studies in **bold**) were: High WI →*Vallisneria spiralis* / *Potamogeton trichoides* / *Chara* spp. > ***Potamogeton crispus* / *Zannichellia palustris* / *Potamogeton perfoliatus*** > ***Myriophyllum spicatum*** > ***Potamogeton panormitanus*** > *Ceratophyllum. demersum* / *Potamogeton pectinatus* → Low WI.

Results of Sabbatini & Murphy (1996) were: High D → ***Potamogeton crispus*** > ***Zannichellia palustris*** > ***Potamogeton panormitanus*** > ***Myriophyllum spicatum*** > ***Potamogeton perfoliatus* / *Ceratophyllum demersum* / *Potamogeton pectinatus*** → Low D.

There is much evidence to suggest that *Ceratophyllum demersum* is a poor disturbance tolerator. Murphy & Eaton (1983) noted that *C. demersum* was absent from UK canals where boat traffic densities were high, suggesting that fragile stems were vulnerable to mechanical damage in more heavily used channels. Of the species recorded in non-saline waters, in Argentine channel systems, Sabbatini et al. (1998) found that *C. demersum* ranked lowest, along an integrated disturbance gradient (*Ruppia* and *Enteromorpha*, tolerant of

[1] If the **Variance Inflation Factor (VIF)** of a variable is large, say if VIF > 20, then the variable is almost perfectly correlated with the other variables and therefore has no unique contribution to the regression equation. As a consequence, its regression coefficient (or its canonical coefficient in canonical ordination) is unstable and does not merit interpretation (Ter Braak, 1986).

the severe stress produced by very high salinity were even poorer D-tolerators but did not occur in low salinity channels). Sabbatini et al. (1998), also found evidence to suggest a low-to-intermediate D-tolerance status for three of the other species which also occur in low-to-intermediate disturbance conditions in UK and Egyptian waters (*Potamogeton panormitanus*, *Potamogeton pectinatus*, *Zannichellia palustris*).

Results of Sabbatini et al. (1998) (species also found in UK and Egyptian studies in **bold**) were: High D → *Potamogeton illinoensis* > *Elodea callitrichoides* > *Chara contraria* > ***Potamogeton panormitanus*** / *Myriophyllum aquaticum* > ***Potamogeton pectinatus*** > ***Zannichellia palustris*** / *Azolla filiculoides* > ***Ceratophyllum demersum*** > *Ruppia maritima* > *Enteromorpha flexuosa* → Low D.

In this study, nutrient availability (measured as TON) appeared to be more important than physical disturbance as a predictor of submerged macrophyte. A previous study of Nile sites (all located well downstream of the study sites examined here) influenced by severe nutrient pollution (Ali & Soltan, 1996) confirmed that nutrient regime was also a primary predictor of submerged vegetation type in polluted areas (though here hydrosoil phosphate was particularly important). Under present conditions, it thus seems likely that physical disturbance produced by flow and ship waves is probably the second most important influence on submerged vegetation in the Nile in Upper Egypt, after nutrient availability. Given the long-term trend towards increasing tourism use of the Nile for cruise holidays, it is quite likely that ship traffic intensity will increase over the next decade, adding to the impact on aquatic vegetation.

Acknowledgements

We thank the University of Glasgow for a contribution to the reasearch costs of J.L. which allowed him to visit Egypt to undertake fieldwork for this study, and South Valley University for provision of laboratory, boat and field equipment. Data on ship traffic in the River Nile was kindly provided by the Planning Centre, Aswan Governorate.

References

Ali, M. M. & M. E.Soltan, 1996. The impact of three industrial effluents on submerged aquatic plants in the River Nile, Egypt. Hydrobiologia 340: 77–83.

Allen, S. E. H., H. M. Grimshaw & A. P. Rowland, 1986. Chemical analysis. In P. D. Moore & S. B. Chapman (eds), Methods in Plant Ecology. Blackwell, Oxford: 285–344.

APHA (American Public Health Association), 1985. Standard methods for the examination of water and wastewater. 16th edn. American Public Health Association, New York, 1268 pp.

Bonham, A. J., 1980. Bank protection using emergent plants against boat wash in rivers and canals. Report No. IT 206, Hydraulics Research Station, Wallingford, U.K.

Cragg, B. A., J. C. Fry, Z. Bacchus & S. S. Thyrley, 1980. The aquatic vegetation of Llangorse Lake, Wales. Aquat. Bot. 8: 187–196.

Edwards, D., 1969. Some effects of siltation upon aquatic macrophyte vegetation in rivers. Hydrobiologia 34: 29–37.

Erian, F. Z., 1989. Symposium on the future of Nile tourism in Egypt. Part III. Aswan Governorate.

Garrad, P. N. & R. D. Heny, 1988. The effect of boat traffic on river regime. In White, W. R. (ed.), Conference on River Regime. Wiley & Sons, England: 395–409.

Healthwaite, A. L., 1993. Nitrogen cycling in surface waters and lakes. In Burt, T. P., A. L. Healthwaite & S. T. Trudgil (eds), Nitrate: Processes, Patterns and Management. Wiley & Sons, England: 99–140.

Hill, M. O., 1979. TWINSPAN, a FORTRAN Program. Ecology & Systematics. Cornell University.

Keddy, P. A., 1982. Quantifying within-lake gradients of wave-energy: interrelationships of wave-energy, substrate particle size and shoreline plants in Axe Lake, Ontario. Aquat. Bot. 14: 41–58.

Keddy, P. A., 1985. Wave disturbance on lakeshores and within lake distribution of Ontario's Atlantic coastal plain flora. Can. J. Bot. 63: 656–660.

Kilmer, V. J. & L. T. Alexander, 1949. Methods for making mechanical analysis of soils. Soil Sci. 68: 15.

Liddle, M. J. & H. R. A. Scorgie, 1980. The effects of recreation on freshwater animals and plants: a review. Biol. Conserv. 17: 183–206.

Moore, J. A., 1986. Charophytes of Great Britain and Ireland. BSBI Handbook No. 5. 141 pp.

Murphy, K. J. & J. W. Eaton, 1981. Water plants, boat traffic and angling in navigable canals. Proceedings of the 2nd British Freshwater Fisheries Conference, Liverpool: 173–187.

Murphy, K. J. & J. W. Eaton, 1983. Effects of pleasure-boat traffic on macrophyte growth in canals. J. Appl. Ecol. 20: 713–729.

Murphy, K. J., B. Rørslett & I. Springuel, 1990. Strategy analysis of submerged lake macrophyte communities: an international example. Aquat. Bot. 36: 303–323.

Murphy, K. J., N. J. Willby & J. W. Eaton, 1995. Ecological impacts and management of boat traffic on navigable inland waterways. In Harper, D. M. & A. J. D. Ferguson (eds), The Ecological Basis for River Management. Wiley & Sons, England: 427–442.

Sabbatini, M. R. & K. J. Murphy, 1996. Submerged plant survival strategy in relation to management and environmental pressures in drainage channel habitats. Hydrobiologia 340: 191–195.

Sabbatini, M. R., K. J. Murphy & J. H. Irigoyen, 1998. Vegetation-environmental relationships in irrigation channel systems in southern Argentina. Aquat. Bot. 60: 119–133.

Schloesser, D. W. & B. A. Manny, 1989. Potential effects of shipping on submersed macrophytes in the St. Clair and Detroit River of the Great Lakes. Michigan Acad. 21: 101–108.

Täckholm, V., 1974. Student's Flora of Egypt. Cairo University Press, Cairo, 888 pp.

Tanner, C. C., J. S. Clayton & R. D. S. Wells, 1993. Effects of suspended solids on the establishment and growth of *Egeria densa*. Aquat. Bot. 45: 299–310.

Ter Braak, C. J. F., 1986. Canonical correspondence analysis: a new eignvector technique for multivariate direct gradient analysis. Ecology 67: 1167–1179.

Ter Braak, C. J. F., 1988. CANOCO – a FORTRAN program for canonical community ordination by [partial] [detrended] [canonical] correspondence analysis, principal components analysis and redundancy analysis (version 3.12). Wageninen. Technical Report. LWA-88-02.

Triest, L., 1989. Flora of Tropical East Africa – Najadaceae. Royal Botanic Gardens, Kew, London. 1–9.

Vermaat, J. E. & R. J. de Bruyne, 1993. Factors limiting the distribution of submerged waterplants in the lowland River Vecht (The Netherlands). Freshwat. Biol. 30: 147–158.

Westlake, D. F., 1966. The light climate for plants in river. In Bainbridge, R., C. Clifford-Evans & Racckham (eds), Light as an Ecological Factor. Blackwell, Oxford: 99–119.

Wilson, S. D. & P. A. Keddy, 1985. Plant zonation on a shoreline gradient: physiological response curves of component species. J. Ecol. 73: 851–860.

Wilson, S. D. & P. A. Keddy, 1986. Species competitive ability and position along a natural stress/disturbance gradient. Ecology 67: 1236–1242.

Hydrobiologia **415:** 101–107, 1999.
J. M. Caffrey, P. R. F. Barrett, M. T. Ferreira, I.S. Moreira, K. J. Murphy & P. M. Wade (eds), Biology, Ecology and Management of Aquatic Plants.
© 1999 *Kluwer Academic Publishers. Printed in the Netherlands.*

River plants from an Iberian basin and environmental factors influencing their distribution

M. T. Ferreira* & I. S. Moreira
Forestry Department, Instituto Superior de Agronomia, Tapada da Ajuda, 1349-017 Lisboa, Portugal
E-mail: terferreira@mail.telepac.pt

Key words: fluvial macrophytes, anthropogenic factors, Portugal, Mediterranean rivers, canonical correspondence analysis

Abstract

River plants were studied at 119 sites from a southern Iberian river basin. A site length of 50–100 m was found to be adequate for characterising local plant richness in these rivers. Average number of river plant species per site was 48 (range 21–76). Species richness showed a maximum at mid-course, with considerable variation in the upper course and a decreasing trend towards the floodplain. Sixteen environmental variables were initially considered to explain the biological variability. Nine, retained by the forward selection procedure, were significant predictors of the species distribution. Canonical correspondence analysis was used to directly relate plant species composition to these environmental variables. Altitude, conductivity, river width, pH, percentage of hard substrates and fine particulate organic matter on the river bed, average rainfall and temperature, and human-related disturbance were significantly related to the species distribution. Species positioned at the axes extremes could be arranged into species groups related to different geographical areas of the river system.

Introduction

Many of the river plant classifications in Europe are based on the phytosociological approach (e.g. Rodwell et al., 1995). However, many authors have emphasised the difficulty of recognising consistent river plant associations to fit the phytosociological scheme or its capacity to relate plants to the river abiotic environment in a reliable manner (Wiegleb, 1988). Another current approach to identify macrophyte assemblages, or types, has been to survey sets of sites within particular regions or river basins, and further relate these assemblages to the characteristics of the basin, the river system or the instream and water environment. Butcher (1933) was a pioneer of such studies, relating plant groups to such variables as geology, channel gradient, altitude, substrate and current velocity. The fundamental traits of this working classification can be recognised in the 10 river community types and their 56 end groups presently established for the U.K. (Holmes et al., 1998).

When there is enough extensive and consistent data on river plants, and about their habitat requirements and interactions with the river environment, this can be used to determine species strategy and functional types (Hills et al., 1994), to develop indices based on type-specific bioindicators (Haury et al., 1996), to study human-induced disturbances (Nilsson & Jansson, 1995) or to establish predictive quality systems based on reference conditions for aquatic plant types (Holmes et al., 1998). Large scale studies of macrophytes are, however, rare in Europe. Two exceptions are the survey of Western European river plants by Haslam (1987) and a country-wide survey of British rivers (Holmes et al., 1998). Similar studies can be found, but at a smaller geographical scale (e.g. Tabacchi et al., 1990).

In Iberia, a few studies of river macrophytes have adopted a biotypological approach (e.g. Penuelas & Sabater, 1987). The river basins studied were located in northern parts of Spain and differed considerably in their environmental characteristics. In addition, few attempts were made to directly relate the plant distri-

* Author for correspondence

bution with the abiotic variables considered. Southern Iberian rivers present special features, such as a high average annual temperature and an extreme flow irregularity, accentuated by water abstraction for irrigation purposes. Some studies point to an homogenisation of plant assemblages due to eutrophication and agriculture-induced disturbances (Ferreira, 1994) or the dominance of a few widespread species and of a few and highly overlapped plant groups (Ferreira et al., 1998). Clearly, there is a need to conduct more surveys in order to establish reference community types for river plants and to determine their leading environmental factors.

Study area

The Sorraia river is the major Portuguese tributary of the Tagus basin and has a total drainage area of 7652 km^2 , occupying a large part of Central Portugal. The main axis of the river system is 198 km long, up to the 7th order. The Sorraia basin has typical Mediterranean features: 80% of the rain falls during three winter months, it has a mean annual temperature of 16–18 °C and a natural absence of flow during an average of four months in many low order streams. The basin spreads throughout a series of plains of low altitude (up to 300 m), mostly occupied by coarse and fine tertiary deposits and quaternary alluvia. However, ancient geological materials related to the 'hercinic basement' of the Iberian Peninsule, are also found in the headwaters and some middle course areas. About 60% of the basin area is used for forestry activities related to cork-oak, pine and eucalyptus production, as well as game. Gently sloping headwaters and lowland valleys are used for agriculture (mainly rice, maize, orchards and tomatoes).

Most of these rivers are to some degree regulated. Four large reservoirs distribute most of the outflow to the system of irrigation channels crossing the main river valley. Other smaller reservoirs and weirs are used for local irrigation activities, affecting the river flow and increasing the period of summer drought. River margins are also locally straightened, embanked, dredged and subject to weed cutting and cattle grazing. River water and sediments show high values of cations and nutrients and heavy siltation occurs in many sites under direct agricultural influence (Ferreira, 1992). Human settlements are small and scattered; only a few have more than 2000 inhabitants. The industries are few and also small, most of them associated with the washing and processing of farm animal products.

Materials and methods

Between 1987 and 1993, a total of 119 river sites with about 250 m in length were visited one to three times between May and July. This included 34 sites on the main course (rivers Tera-Raia-Sorraia) and 85 sites on its direct tributaries (Erra, Sor, Divor). Studied sites presented an average distance of 4.6 km from one another (range 2.3–9.4 km). To assess the suitability of the sampled river length, the cumulative plant richness was determined in successive 10 m long river sections in six sites having a different position in the river hierarchy.

Because of the climatic dryness and their intermittent flow, the banks of Mediterranean rivers are colonised by terrestrial adventitious species in various degrees during the year. River plants were considered to be as those normally growing in, on or close to the water, in waterlogged or moist environments. Terrestrial plants also found were separated from these in data treatment to avoid the skewness of site grouping being dominated by non-riverine species (Ferreira, 1994; Ferreira et al., 1998).

The following environmental variables were recorded from each site: average width (m) and depth (m) of the river (averaged from 5 points, at 5 random transects); water velocity (m s^{-1}, R.OTT Instrument); conductivity (μ cm^{-1}) and pH (WTW probes). Percentage cover of hard substrates (rocks and boulders, >20 cm diameter), of medium coarse substrates (pebbles, gravel and sand, 0.2–20 cm) and abundance of fine particulate organic matter (on a scale 1–5), were assessed visually. Altitude (m), distance from the source (km), drainage area (km^2) and order number were obtained from topographic maps (1:50 000 scale). Mean annual rainfall and the air temperature were obtained from the weather station nearest to each site. The degree of direct human disturbance of each site was visually evaluated (on a scale 1–5), including bank alterations, recreation, angling, riparian cuttings, cattle grazing and local inputs of organic matter. Dominant geological formations were also given a relative scale (1–5) of increasing texture and geological age, as: 1. quaternary alluvia and fine cenozoic deposits; 2. coarse and fine cenozoic materials; 3. as in 2, but together with other older materials; 4. metasedimentary rocks; 5. granitic rocks.

Canonical correspondence analysis, CCA (ter Braak, 1990) was used to relate plant species composition to environmental variables. All species were considered in the analysis, but rare species were down-weighted. Environmental variables were log transformed and standardised by row centering. From the 16 variables initially considered, only the subset of best predictors was retained for the final analysis, following application of a forward selection procedure similar to a step-wise regression (available in CANOCO version 3.1). A cut-off point of $P>0.10$ was used in this routine. A Monte Carlo test was used to test the significance of both the first axis eigenvalue and the trace, the sum of all eigenvalues.

Results

In all studied sites, the cumulative value of total plant richness along serial sections of 10 m reached a stable value at about 50 m, regardless of the fact that sites presented very different positions along the river course (i.e. different widths, altitudes or drainage areas, Figure 1). Hence, the sampling length used of 250 m was deemed to be representative. The average number of river plant species per site was 38 (range 21–76) and river plant richness presented a maximum at mid-course (order number 4–6), with considerable variation at the upper course and a decreasing trend towards the floodplain.

Of the 16 environmental variables initially considered, the order number, the distance from the source, the site drainage area and the percentage of medium size substrate were discarded because they presented variance inflation factors higher than 20. Medium substrate was negatively correlated with hard substrate, while the others were positively correlated with altitude. The 12 remaining variables presented pair-wise correlations smaller than 0.5 ($p<0.01$) and together explained 21.4% of the total biotic variation, that is, the quotient between the trace and the sum of all unconstrained eigenvalues (ter Braak, 1990).

Only nine of these variables were retained by the forward selection procedure because depth, geology and water velocity presented P-values greater than 0.1 (Table 1). Plant species composition and the selected environmental variables were significantly related ($P<0.01$) for both the first eigenvalues and the trace (Monte Carlo simulation with 1000 permutations, ter Braak, 1990). These variables were: altitude (and related variables river width, order number, distance to source and the site drainage area, hence, the hierarchical position of the sites on the river axis), substrate and geoclimatic regime (rainfall and temperature), mineral and nutrient enrichment (conductivity, pH and FPOM) and site disturbance.

The biplot of the first two axes (eigenvalues 0.229 and 0.182, respectively; sum of all canonical eigenvalues: 0.817) is shown in Figures 2 and 3. The first axis opposed small upstream sites with higher altitude and rainfall, lower conductivity and pH, and a higher proportion of rocks and boulders on the river bed, to drier downstream areas with larger river channels, having a higher conductivity and pH, and more disturbed. The second axis emphasises these gradients and is further related to substrate texture and pollution, opposing sites dominated by rocks and boulders to heavily silted sites. The species more related to these variables and positioned on the axes extremes are indicated in Figure 2. Species such as *Anagallis tenella*, *Hydrocotyle vulgaris* and *Hypericum elodes* typically occur in small lowland streams, having acidic and low mineralised waters, with silted beds. Species like *Callitriche brutia*, *Myosotis stolonifera* and *Myriophyllum verticillatum* were found at small upstream sites, with river beds dominated by rocks and boulders, little human disturbance, and more or less intermittent flow. *Typha angustifolia*, *Lemna gibba*, *Sonchus oleraceus* and *Geranium lucidum* were abundant in small to medium size lowland streams, that were highly enriched. *Myriophyllum aquaticum*, *Lysimachia vulgaris*, *Rumex crispus* and *Polygonum persicaria* were found in large, highly disturbed, lowland reaches, with well developed riparian formations. These four species assemblages are characteristic in reaches from different geographic areas, in fact, from each river sub-system studied (Figure 3).

Discussion

Many studies have addressed the relationship between environmental variables and river plant distribution. However, only recently has direct multivariate analysis become more popular (e.g. Haury et al., 1996) and enabled more reliable information on which environmental factors are important. In this study, for example, nine environmental factors were retained to be related to plant distribution, although 16 were initially considered.

Water velocity and depth were not selected as significant variables influencing species distribution,

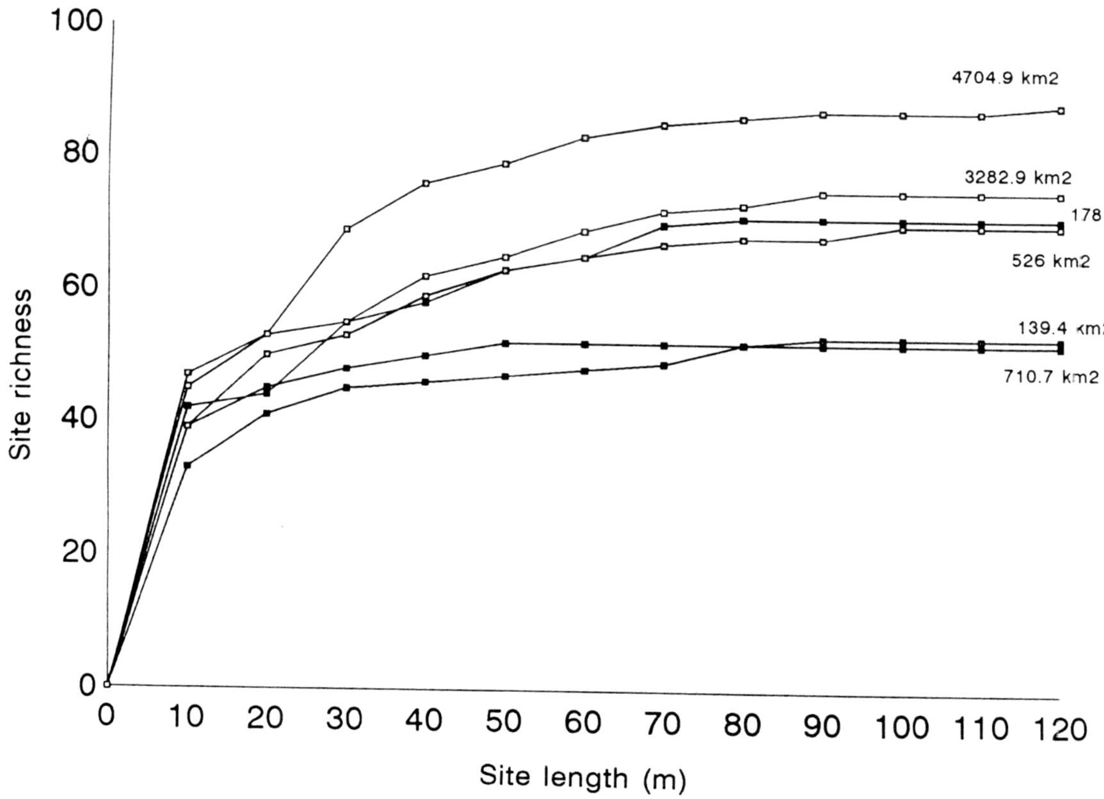

Figure 1. Changes in river plant richness in serial sections of 10 m along six selected studied sites. Also indicated is the drainage area of each site.

Table 1. Summary statistics for the canonical correspondence analysis of river macrophytes and environmental variables of the Sorraia basin. Variance explained when considering all 12 variables: 0.93. *significantly related at $P<0.05$. Also shown are the acronyms used in the Figures

Environmental variables	Forward selection of variables			Correlation with canonical axis	
	Extra-fit (%variance)	*F*-ratio	*P*-value	Axis 1	Axis 2
Altitude (ALTI)	19.3	7.43	0.01	0.5388*	0.6433*
Conductivity (COND)	17.2	6.85	0.01	-0.6997*	0.2778*
Human disturbance (HUSE)	10.7	3.94	0.01	-0.5888*	-0.4902*
Organic matter (FPOM)	9.6	2.68	0.01	0.2272	-0.2753*
Substrate (HARDSUBST)	6.4	2.27	0.01	0.1309	0.7415*
Annual rainfall (RAIN)	5.3	2.43	0.01	0.7302*	-0.0526
River width (WIDTH)	5.3	1.79	0.03	-0.6181*	-0.3189*
Annual temperature (TEMP)	7.5	1.58	0.03	-0.6021*	-0.0822
pH (PH)	5.3	1.33	0.04	-0.7465*	0.1472
Depth	4.3	1.20	0.14	-	-
Geology	3.2	1.21	0.16	-	-
Water velocity	3.2	1.19	0.14	-	-

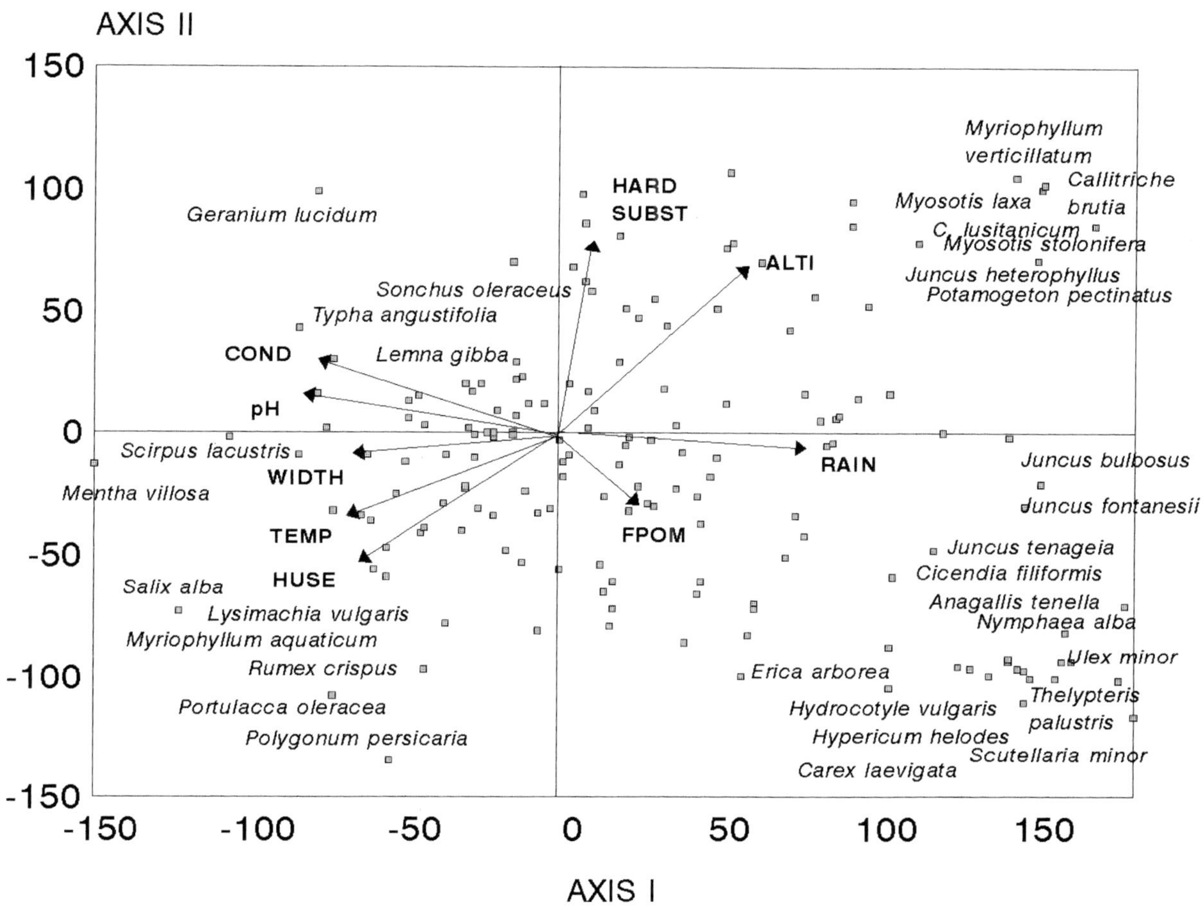

Figure 2. Biplot of canonical correspondence analysis showing plant species positions and the vectors of the nine environmental variables selected. The species indicated are the most important contributions to the canonical axes.

in spite of its generally recognised importance for river plants (Wiegleb, 1988). Water velocity and average depth were highly dependent on the valley morphology and the geological materials, and varied irregularly along the river axis, with frequent situations of flowless reaches and deep pools in upstream reaches. The upstream and middle courses areas are on the borders of the Iberian hercinic basement, and thus geological background is spatially heterogeneous, including small and entangled areas of granites, quartzites, shales and vulcanites, together with more recent coarse and medium sized detritic materials. This heterogeneity resulted in small areas of hard substrate and different gradients occurring scattered along the river courses and not following a more common upstream-downstream river gradient. The little structural importance of these variables on river plant distribution may indeed be one of the more typical features of southern Iberian rivers.

This study was conducted in a relatively large area (nearly 8000 km^2) and the two major environmental gradients detected were river hierarchy and substrate type. In smaller scale studies of the same area (Ferreira & Smeding, 1992; Ferreira, 1994), species turnover was very short and environmental gradients explaining plant grouping were naturally harder to distinguish, and usually related to local variations of shade or substrate. However, other studies on a similar scale in more southern areas were nearly unable to establish plant groups and plant-environment relationships (Ferreira et al., 1998). Further studies are needed to clarify the relative influence of scale and disturbance on plant grouping.

Other than basin scale, plant grouping is also dependent on methodological issues such as what river

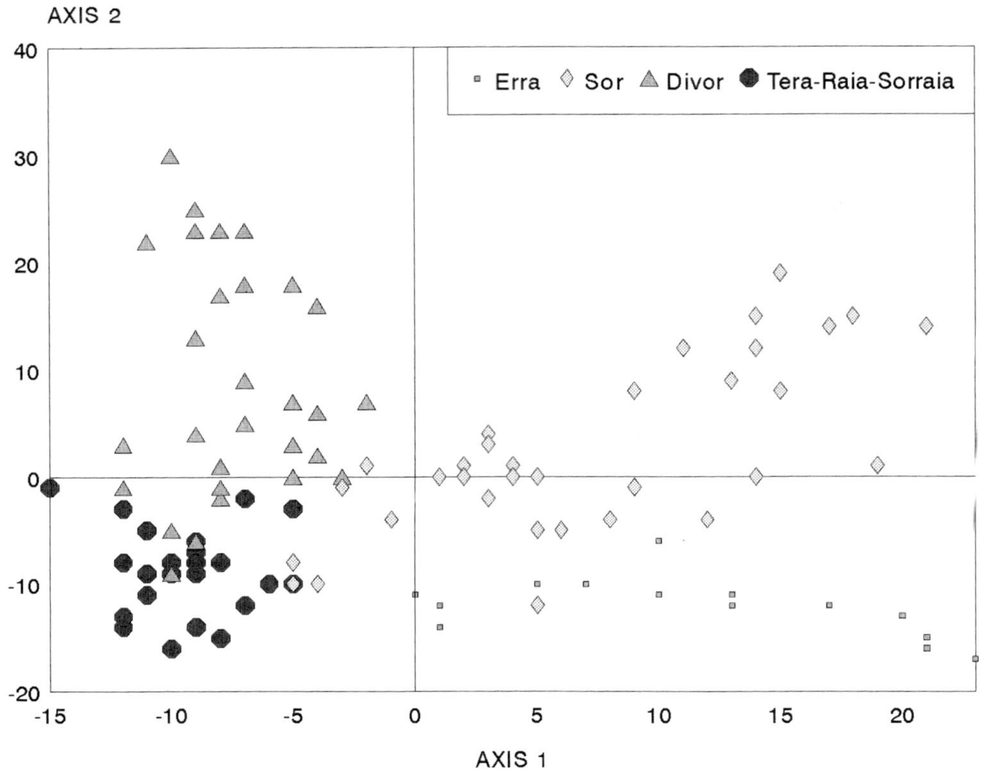

Figure 3. Biplot of canonical correspondence analysis showing site positions.

length is studied, the exclusion of terrestrial plants and the width of the river corridor considered for survey. Different authors have used different site lengths for plant surveys, for example 50 m (Haury et al., 1996), 100 m (Penuelas & Sabater, 1987), 200 m (Nilsson et al., 1988); 500 m (Demars & Harper, 1998) or 1000 m (Holmes et al., 1998). Though hydrophytes can present a very patchy distribution, with locally important cover, this study indicated that a river length of 50 m is enough for performing surveys in southern Iberian rivers.

Terrestrial adventitious plants can make up more than 50% of instream species in intermittent rivers during summer. These are mostly short-lived ruderals whose capacity of colonisation is favoured by natural disturbance. The proportions of plant traits such as life-span, lateral location and ruderality can vary considerably within and between river systems (e.g. Nilsson et al., 1994; Tabacchi et al., 1996) and these last authors found that semi-arid rivers were more influenced by local patterns than rivers with a more regular flow. In these rivers, the lateral ecotone is very thin and terrestrial systems strongly constrain the river corridor.

In southern Iberian rivers, the abiotic variables frequently explain only a small part of the species variability (Ferreira et al., 1998) and a clear separation of river plant groups is difficult. Human disturbance, enrichment and silting also tend to reduce the differences between groups (Ferreira, 1994). To date, the value of river plant assemblages as indicator of aquatic regions or river conditions has failed to be developed into a successful indicator system.

References

Butcher, R. W., 1933. Studies on the ecology of rivers. I. On the distribution of macrophytic vegetation in the rivers of Britain. J. Ecol. 21: 58–91.

Demars, B. O. L. & D. H. Harper, 1998. The aquatic macrophytes of an English lowland river system: assessing response to nutrient enrichment. Hydrobiologia 384: 75–88.

Ferreira, M. T., 1992. Estrutura e dinâmica das comunidades de macrófitos lóticos da bacia hidrográfica do Sorraia. PhD Thesis. Agronomy Institute of Lisbon: 402 pp (in Portuguese).

Ferreira, M. T., 1994. Aquatic and marginal vegetation of the river Divor and its relation to land use. Verh. int. Ver. Limnol. 25: 2309–2315.

Ferreira, M. T. & F. Smeding, 1990. Distribution and patterns of aquatic and riparian vegetation in Erra river, central Portugal. Scientia Gerundensis 16: 116–125.

Ferreira, M. T., F. N. Godinho & R. M. Cortes, 1998. Macrophytes in a southern Iberian river. Verh. int. Ver. Limnol. 26: 1835–1841.

Haslam, S. M., 1987. River plants of western Europe: the macrophytic vegetation of watercourses of the European economic community. Cambridge University Press, Cambridge: 512 pp.

Haury, J., M. C. Peltre, S. Muller, M. Tremolieres, J. Barbe, A. Dutartre & M. Guerlesquin, 1996. Des indices macrophytiques pour estimer la qualite des cours d'eau français: premieres propositions. Ecologie 27: 233–244 (in French).

Hills, J. M., K. J. Murphy, I. D. Pulford & T. H. Flowers, 1994. A method for classifying European riverine wetland ecosystems using functional vegetation groups. Funct. Ecol. 8: 242–252.

Holmes, N. T. H., P. J. Boon & T. A. Rodwell, 1988. A revised system for British rivers based on their aquatic communities. Aquat. Conserv.: Mar. Freshwat. Ecosyst. 8: 555–578.

Nilsson, C. & R. Jansson, 1995. Floristic differences between riparian corridors of regulated and free-flowing boreal rivers. Regul. Riv., Res. Mgmt 11: 55–66.

Nilsson, C., A. Ekblad, M. Dynesius, S. Backe, M. Gardfjell, B. Carlberg, S. Hellqvist & R. Jansson, 1994. A comparison between species richness and traits or riparian plants between a main river channel and its tributaries. J. Ecol. 82: 281–295.

Nilsson, C., G. Grelsson, M. Johansson & U. Sperens, 1988. Can rarity and diversity be predicted in vegetation along river banks? Biol. Conserv. 44: 201–212.

Penuelas, J. & F. Sabater, 1987. Distribution of aquatic macrophytes in relation to environmental factors in the Ter river, N.E. Spain. Int. Rev. ges. Hydrobiol. 78: 41–58.

Rodwell, J. S., C. D. Pygott, D. A. Ratcliffe, A. J. Malloch, H. J. Birks, M. C. Proctor, D. W. Shimwell, J. P. Huntley, E. Radford, M. J. Wigginton & P. Wilkins, 1995. British plant communities. Volume 4. Cambridge University Press: 283 pp.

Tabacchi, E., A. M. Planty-Tabacchi, M. J. Salinas & H. Decamps, 1996. Landscape structure and diversity in riparian plant communities: a longitudinal comparative study. Regul. Riv., Res. Mgmt 12: 367–390.

Ter Braak, C. J. F. (1990) CANOCO version 3.1. Update notes. Agricultural Mathematics group, Wageningen: 35 pp.

Wiegleb, G., 1988. Analysis of flora and vegetation in rivers: concepts and applications. In Symoens, J. J. (ed)., Vegetation of Inland Waters. Handbook of Vegetation Science no 15. Kluwer Academic Publishers, Dordrecht, The Netherlands: 311–340.

Hydrobiologia **415**: 109–115, 1999.
J. M. Caffrey, P. R. F. Barrett, M. T. Ferreira, I.S. Moreira, K. J. Murphy & P. M. Wade (eds),
Biology, Ecology and Management of Aquatic Plants.
© 1999 *Kluwer Academic Publishers. Printed in the Netherlands.*

Macrophyte cover and standing crop in the River Scorff and its tributaries (Brittany, northwestern France): scale, patterns and process

Jacques Haury[1] & Lanciné Gouesse Aïdara
[1]*Ecole Nationale Supérieure Agronomique Ecologie et Sciences phytosanitaires & I.N.R.A. Ecologie aquatique, 65 rue de Saint Brieuc, F- 35042 Rennes Cedex, France*
Tel: [+33] 2 99 28 75 39; Fax: [+33] 2 99 28 54 40; E-mail: haury@epi.roazhon.inra.fr

Key words: macrophytes, river, cover, biomass, ecomorphology, fish habitat

Abstract

In order to quantify macrophyte cover and standing crop, studies were performed in a main river and two tributaries in Brittany. Species and total plant cover were assessed using a pin-point method over a period of nearly three years. Spring and summer standing crops in the smaller tributary were collected along 1-m-long stretches, while, in the main river and the larger tributary, summer mean biomass was obtained from quadrats chosen from previous sampling on average macrophyte cover. Seasonal changes were found to be dependent on available light. They were more important in the tributary than in the main river. In the tributary, the species seasonal succession was: *Ranunculus penicillatus, Oenanthe crocata, Apium nodiflorum*. In the river, the following succession was observed: Bryophytes, *R. penicillatus, Callitriche* spp. (*C. hamulata*, *C. platycarpa* and *C. obtusangula*). In sunny stretches, minimum and maximum values of macrophyte cover were 17% and 89% of the stream bed, and 37% and 95% of the river bed, respectively. In the stream, mean summer standing crops were 28 g of dry weight per m^2 (gDW/m^2) in shaded and 58 gDW/m^2 in sunny stretches. In the main river, the average biomasses (with reduced standard deviation) were 96 and 187 gDW/m^2 for shaded and sunny stretches, respectively. These results are explained with regard to plant life strategies and species morphology; the consequences for fish habitat are pointed out.

Introduction

Quantitative studies on macrophyte production in rivers became more numerous following the pioneering work of Owens & Edwards (1962), Dawson (1976) and Kern-Hansen & Dawson (1978). Later studies sought to obtain data on standing crops in order to model primary productivity, to understand its limiting factors (Chambers et al., 1991), and to assess the relationships between colonization, production and species traits (Sabbatini & Murphy, 1996). Although macrophyte proliferations increasingly damage French rivers, few data have been published (Peltre et al., 1993) or presented in symposia (Dutartre et al., 1994; Laforge & Petitdidier, 1996), and almost none are available for lowland acidic rivers. In such watercourses, macrophytes are important components of fish habitat (Baglinière & Champigneulle, 1982), but their influence has so far not been quantified.

The present study was performed in the Scorff catchment in Brittany – at the scale of both the river and the stream – in order to assess different patterns of vegetation, to obtain some measurements of the average standing biomass and to ascertain the processes of macrophyte production as well as to analyse its ecological effects.

Study areas and methods

The River Scorff has already been presented elsewhere (Haury, 1995). It is a salmonid river flowing into the Atlantic Ocean at Lorient in Southern Brittany (Figure 1). Schists and granites form the predominant substrata of the catchment area. Its waters are sightly acidic, containing much nitrate.

Ten permanent plots were studied from April–May 1980 to September 1982: six in the main river and its major tributary around an important fish-farm at Pont-

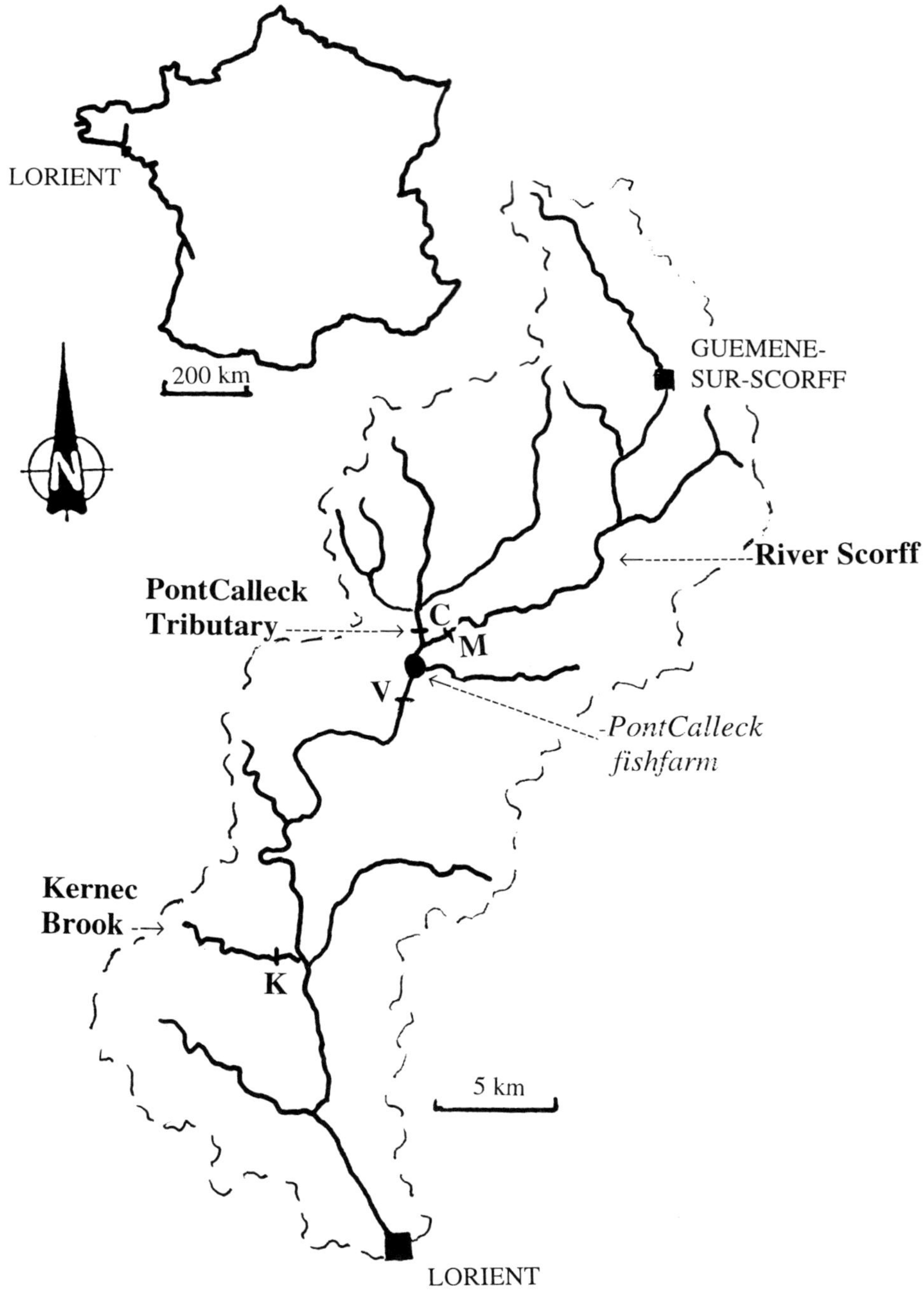

Figure 1. Study area: the River Scorff and its catchment. (Permanent plots: K (KA to KD): Kernec Brook; V (VA, VB): River Scorff, downstream; M (MA, MB): River Scorff, upstream; C (CA, CB): PontCalleck tributary)

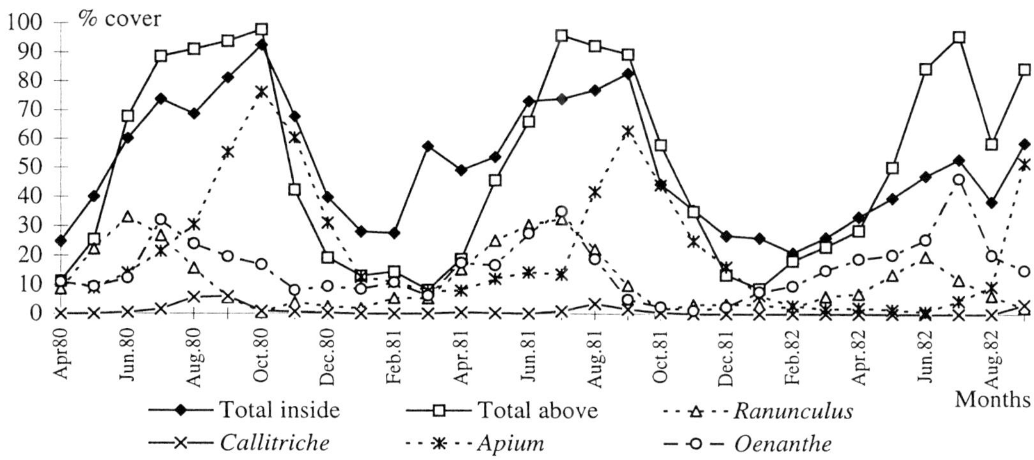

Figure 2. Seasonal changes of macrophyte cover in the sunny plot KC from April 1980 to September 1982.

Table 1. Characteristics of permanent plots (K: Kernec Brook; M: Scorff, upstream; V: Scorff, downstream; C: PontCalleck Tributary) and mean physico-chemical data (25 samples for K; 7 for M, V, C)

Permanent plots	KA	KB	KD	KC	MA	MB	CB	CA	VB	VA
Width (m)	1.9	2.1	1.3	2.0	10	9.5	11	13	14	15
Light (% of max. light)	10	48	75	85	37	37	19	17	34	51
Depth (cm)	14	12	20	16	45	46	25	31	40	38
Current velocity (cm/s)	14	26	19	17	28	31	20	16	25	29
Resistivity ($\Omega.cm^{-1}$)	5719				7490		7078		7211	
pH	6.60				6.74		6.67		6.64	
Hardness (°f)	1.3				1.5		1.6		1.5	
NO_3^- ($mg.l^{-1}$)	23.2				15.2		12.7		14.8	
NH_4^+ ($mg.l^{-1}$)	0.12				0.10		0.12		0.17	
PO_4^{3-} ($mg.l^{-1}$)	0.06				0.16		0.15		0.41	

Calleck (Daniel & Haury, 1995), as well as four in the Kernec Brook, a smaller tributary farther downstream (Haury, 1995). All the plots were 5 m in length, and 50% (KA, KB, MA, CB, VB) were located in more shaded areas (% of light measured using a luxmeter, with comparison of weighted representative measures with open area – Haury, 1985). Their characteristics, as well as some water quality features, are summarized in Table 1. Cropping was carried out downstream of the permanent plots between 1980 and 1982 in June for the brook (Haury, 1985), and in July 1985 for Pont-Calleck Tributary and the main river (Gouesse-Aïdara, 1986).

Pin-point methods within transects are now widely used to monitor vegetation cover (Ham et al., 1981). Two devices were employed: a wooden frame placed across the banks of the brook enabled observations to be made without any entrance, while a measuring tape was put across the river, with the observer wading in the river. At each point (20 cm × 20 cm for the brook, 0.5 m × 1 m for the river), the depth, bottom sediment granulometry and macrophyte species (if present) were recorded, each month between April 1980 and September 1982 (when possible), which gave from almost one hundred to more than two hundred points for each survey in one plot.

Cropping was performed along randomized 1m-long stretches in Kernec Brook. Macrophytes were cropped in the river as well as in PontCalleck Tributary using 0.5 × 0.5 m quadrats (Haury & Gouesse Aïdara, 1990), a value chosen on the basis of average macrophyte cover measurements.

Table 2. Macrophyte cover in the permanent plots, from April 1980 to September 1982. (For KA and KB, figures within brackets were observed after tree bank cutting)

Permanent plots	KA	KB	KD	KC	MA	MB	CB	CA	VB	VA
% minimum river bed cover	8	33	17	21	50	63	40	26	27	37
% maximum river bed cover										
Total cover under water	29	79	65	89	75	90	59	76	67	95
Total cover above water	39 (47)	59 (82)	97	96	12	14	34	23	12	12
Bryophytes	10	26	2	1	55	54	55	28	50	25
Algae	9 (22)	3 (38)	2	1	33	28	10	38	57	60
Total Hydrophytes	9	44	47	32	27	65	21	65	37	81
Ranunculus penicillatus	9	44	47	32	31	65	11	65	6	26
Callitriche spp.	3	2	6	6	3	8	4	8	37	72
Helophytes	8 (10)	27 (53)	38	73	6	15	1	1	5	10

Results

Macrophyte cover (Table 2)

In Kernec Brook, the general measurements obtained for the entire macrophyte vegetation, both below and above water level, show a normal seasonal pattern. Overlying cover was voluminous from spring through to autumn, due to helophytes such as *Oenanthe crocata* L., *Iris pseudacorus* L., *Sparganium erectum* L. and *Phalaris arundinacea* L., and finally *Apium nodiflorum* (L.) Lag..

A succession is apparent in the permanent plot KC (Figure 2). At first, *Ranunculus penicillatus* (Dum.) Bab. began to grow in late autumn – early winter (minimum cover in October, some growings in December), flowering in May–June and decaying during July to disappear in late summer–early autumn. *O. crocata* then began to grow in winter both from tuberous roots and from seedlings, extending first in the current before erecting up above the water surface. It flowered in June and July, then fell down into the bed, in some cases showing a second flowering period from lateral stems during August. The late-stage *A. nodiflorum* expanded in late summer and autumn, with creeping stems mainly anchored on the bank. In another sunny plot, KD, which is a narrower and deeper stretch, *R. penicillatus* exhibited the same pattern; *O. crocata* was less well developed, while *I. pseudacorus* was dominant but did not extend laterally; *A. nodiflorum* was predominant only in 1980. In both plots, some plants of *Callitriche stagnalis* appeared in late summer.

In the shaded plots KA and KB, cover was less developed. Bryophytes (mainly *Leptodictyum riparium* (Hedw.) Warnst and *Fontinalis antipyretica* Hedw., but also *Chiloscyphus polyanthus* (L.) Corda, *Fissidens pusillus* Wils. and *Platyhypnidium rusciforme* (Neck.) Fleisch.) did not show any measurable seasonal patterns. The succession was: *Lemanea* sp. (red Algae), *R. penicillatus*, *O. crocata*, *P. arundinacea*.

Differences from year to year were observed, with decreasing populations of *R. penicillatus* in sunny stretches, especially after a winter flood disturbance within the KD plot; *A. nodiflorum* also decreased, probably in relation to a greater abundance of *O. crocata* which was left without autumn cutting during the study. At the end of the study, tree bank cutting led to an autumn proliferation of green Algae in KA where the cover reached 21% of the stream bed area, and in an increase of overlying cover, up to 89% of this value in KB.

Comparing the river with Kernec Brook, general results showed less variations between sunny and shaded stretches, as well as for each taxon. In the sunny plot VA, the seasonal succession was: Bryophytes (*F. antipyretica, P. rusciforme, C. polyanthus, F. pusillus*) and *Lemanea* sp., *R. penicillatus*, at least three species of *Callitriche* (*C. hamulata* Kütz. ex Koch, *C. obtusangula* Le Gall, *C. platycarpa* Kütz.), with *Potamogeton alpinus* Balbis and *Lemna minor* L. appearing in September. Some *O. crocata* showed a later cycle than in the brook, with a flowering period in August and September, or remained in a sterilous stage, as some individuals of *Polygonum hydropiper* L.. In other sunny plots, the seasonal changes were more marked in CA, with a huge seasonal variation of *R. penicillatus* (from 6% to 61%) and a weaker one in MB. Bryophytes were abundant in both plots (with a significant population of *Scapania undulata* (L.) Dum.

in MB); the dominance of *R. penicillatus* and the presence of a few *C. hamulata* (absence *C. obtusangula* – Daniel & Haury, 1995) led to a maximal cover in June. In the shaded plots, total macrophyte cover was smaller than in sunny areas, while Bryophytes and red Algae (*Lemanea* sp., *Batrachospermum* sp.) were more abundant. The general cycle was closely similar, with a small delay (1-2 months) for most species. In all river plots, filamentous green Algae and *Melosira* sp. showed important but transient peaks probably due to temperature effects and the decay of *Ranunculus*.

Macrophyte standing biomass (Table 3), relationships between macrophyte cover and biomass.

The macrophyte standing biomass was mainly due to *Ranunculus penicillatus*, whose standing crop reached 166 g of dry weight per m^2 (gDW/m^2) in Kernec Brook. Bryophytes were present in most river quadrats (81% for V, 100% for M and 98% for C) and in shaded quadrats (100%) in the Kernec Brook, but were absent in 41% of the sunny quadrats of streams. For the dominant species (*R. penicillatus*, *F. antipyretica*), both in Kernec Brook or around Pontcalleck, linear relationships have been shown between cover and standing crop, provided depth did not change too much (Haury, 1985; Gouesse-Aïdara, 1986). Nevertheless, while significant, these relationships were not very strong, because the shape of the tufts was quite heterogeneous as was shown by Dawson & Robinson (1985).

Discussion

Analysis of macrophyte cycle

The species cycles are in good agreement with literature data from English rivers (Dawson, 1976; Ham et al., 1981; Spink et al., 1997) as well as with observations in floras. Earlier development in streams has been already described by Dawson (1980): downstream from PontCalleck Reservoir, in the PontCalleck Tributary where water can be more than 2 °C warmer, flowering of *Ranunculus* can occur about one month earlier than in the nearby River Scorff.

The observed succession and the structure of macrophyte stands can be explained by ecological tendencies, eco-morphological types (Den Hartog & Segal, 1964; Makirinta, 1978) and species traits in relation to bottom fixation, lateral spreading, vegetative organs and reproduction, as well as time of flowering. Permanent bryids are mainly found on pebbles and stones in winter; as they are small and tolerate the shade, they can maintain development under the Spermatophyta layer in the river. In the stream, however, where the shallowness leads to a closer contact between Bryophytes and hydrophytes, exclusion occurs in most cases. Furthermore, creeping bryids such as *L. riparium*, *C. polyanthus* and *P. rusciforme* form a mat allowing the anchoring of Spermatophyta. The appearance of cold-water red Algae in winter increases the species diversity of the "Bryophyte" layer. The batrachiid *Ranunculus* develops both in gravels and within the bryophytic mat from remaining rhizomes and more rarely from seedlings in the swift–flowing and sunny stretches throughout the winter, before forming broad leaves and then flowering, which determines the subsequent decay (Dawson, 1976). Dense stands of *Ranunculus* reduce the current velocity (Dawson & Robinson, 1985) and trap sediments (Haury & Baglinière, 1996), which can locally favour *Apium* development in streams, (as described with *Rorippa nasturtium-aquaticum*: Dawson et al., 1979), or *Callitriche* spp. and *Potamogeton alpinus* in rivers. The free-floating lemnid, *L. minor*, is maintained near the banks in helophyte mats and on *Ranunculus* or *Callitriche* stands reaching the surface, where the current velocity decreases. Helophytes are mainly present in the shallower stretches, close to the bank where silt and sand are predominant.

The period of active growth is very important in analysing the relationships between macrophytes and fish. Since the dominant species *R. penicillatus* begins its cycle in late autumn or early winter, and is particularly well developed in riffles, it may occupy spawning areas in gravels, thus preventing any digging of the nest. The development of *R. penicillatus* causes the trapping of fine sediments which are disliked by salmon, and also limit fish movements. Precise electrofishing measurements have yielded a density of 32.0 salmon/100 m^2 in the current vein, as against a value of 0.9 inside *Ranunculus* stands (Haury & Baglinière, 1996). Experimental work (Roussel et al., 1998) has shown that a riffle without *Ranunculus* can sustain twice as many juvenile salmon compared with areas with *Ranunculus*.

Table 3. Macrophyte standing biomass in Kernec Brook (June) and Scorff River (July). (The figures are expressed in g of dry weight per square metre (gDW/m^2). Helophytes have not been cropped in Kernec Brook.)

Macrophytes	Shaded plot K	Sunny plot K	Plot M	Plot C	Plot V
Number of crops	15 (june)	12 (june)	40 (july)	40 (july)	40 (july)
Shape	1 linear meter	1 l. meter	0.5×0.5 m	0.5×0.5 m	0.5×0.5 m
Sampling type	random		random/m.cover		
Algae			15.1 ± 1.6	2.7 ± 0.3	6.3 ± 0.4
Bryophytes	13.4 ± 8.9	1.7 ± 3.0	15.5 ± 3.1	46.3 ± 2.0	2.7 ± 0.3
Vascular Hydrophytes	14.1 ± 13.3	56.7 ± 42.0	167.0 ± 12.0	50.5 ± 5.0	110.0 ± 4.2
Ranunculus penicillatus	13.9 ± 13.0	56.6 ± 42.0	167.0 ± 10.4	49.5 ± 5.0	107.4 ± 4.3
Callitriche spp.	1.3 ± 1.8	0.3 ± 0.3	0.1 ± 0.0	1.0 ± 0.2	0.2 ± 0.0
Total hydrophytes	27.4 ± 14.1	58.3 ± 41.5	182.5 ± 16.7	96.8 ± 7.4	112.7 ± 4.8
Helophytes	–	–	4.0 ± 1.5	0.5 ± 0.1	2.3 ± 0.4
Total vegetation	–	–	186.6 ± 12.1	97.3 ± 5.5	115.3 ± 4.8

Comparisons of macrophyte standing crops in River Scorff with literature

The quantitative estimates of biomass given here fall within the lower range of literature values (100 to 700 gDW/m^2 in temperate rivers: Westlake, 1963 and 1975; 110 to 520 gDW/m^2 in chalk streams: Edwards & Owens, 1960; Westlake et al., 1972; 200 to 500 gDW/m^2 usually in lowland Danish streams, with extrema of 76 and 730 gDW/m^2: Kern-Hansen & Dawson, 1978; 21 to 672 gDW/m^2 in the eutrophicated River Vair: Peltre et al., 1993). These rivers are not very productive, but may be compared with the maximum crop obtained on granites and sandstones in Eastern France (respectively 204 gDW/m^2 and 381 gDW/m^2: Laforge & Petitdidier, 1996).

Conclusion

Clearly, the shape of a watercourse controls species distribution, as well as the architecture of macrophyte communities and seasonal patterns. On one hand, this means that any quantitative observation must be related to an accurate presentation of the study area to allow further extrapolation of the analysed patterns or processes. On the other hand, measurements of population traits will become indispensable to ascertain the structure of macrophyte communities and, consequently, its influence on the functioning of watercourses. Such morphological studies will supplement ecomorphological typologies which currently appear to be a general framework for interpretation but which are too imprecise as tools for detailed research.

Relationships between species remain to be investigated both in the field and experimentally involving such processes as competition, shading, protection from excessively strong current velocity, as well as response of dominated populations (e.g. Bryophytes). Such studies should also include time of colonization and rapidity of settlement for each species. The whole set of processes leads to a mosaic structure with species belonging to many ecomorphological types, a pattern which is characteristic of rivers in Brittany.

To assess average biomass, preliminar sampling on macrophyte cover seems to be a good compromise between long and tedious random cropping and imprecise data due to reduced number of cropped quadrats. In the future, studies are to be carried out on biomass variability both in space and time. Particular focus on maximum biomass would quantify optimum cover at which *Ranunculus* becomes a negative factor for fish species.

Such data would be important for assessing nutrient budgets in rivers, especially in areas of very intensive agriculture such as found in Brittany. Here, studies are already in progress to reduce nitrate fluxes into rivers using macrophyte production.

Acknowledgement

The authors thank Dr M. Carpenter (Univ. of Rennes) for correcting the English style.

References

Baglinière, J. L. & A. Champigneulle, 1982. Densité des populations de truite commune (*Salmo trutta* L.) et de juvéniles de saumon atlantique (*Salmo salar* L.) sur le cours principal du Scorff (Bretagne): préférendums physiques et variations annuelles (1976–1980). Acta Oecol., Oecol. Applic. 3: 241–256.

Chambers, P. A., E. E. Prépas, H. R. Hamilton & M. L. Bothwell, 1991. Current velocity and its effect on aquatic macrophytes in flowing waters. Ecol. Applic. 1: 249–257.

Daniel, H. & J. Haury, 1995. Effects of fish farms on phytocenoses in acidic rivers. Acta Bot. Gallica 142: 639–650.

Dawson, F. H., 1976. The annual production of the aquatic macrophyte *Ranunculus penicillatus* var. *calcareus* (R.W. Butcher) C. D. K. Cook. Aquat. Bot. 2: 51–73.

Dawson, F. H., 1980. Flowering of *Ranunculus penicillatus* (Dum.) Bab. var. *calcareus* (R.W. Butcher) C. D. K. Cook in the River Piddle (Dorset-England). Aquat. Bot. 9: 145–157.

Dawson, F. H., E. Castellano & M. Ladle, 1979. Concept of species succession in relation to river vegetation and management. Verh. Int. Verein Theor. Angew. Limnol. 20: 1429–1434.

Dawson F. H. & W. N. Robinson, 1985. Submerged macrophytes and the hydraulic roughness of a lowland chalkstream. Verh. Int. Verein Theor. Angew. Limnol. 22: 1944–1948.

Dutartre, A., H. Codhant & N. Mary, 1994. Les végétaux aquatiques et le transfert des nutriments dans le fleuve Charente: les macrophytes. Symp. Relations continent-zones côtières, La Rochelle, 13–15 Septembre 1994: 10 pp.

Edwards, R. W. & M. Owens, 1960. The effects of plants on river conditions. 1: Summer crops and estimates of net productivity of macrophytes in a chalk-stream. J. Ecol. 48: 151–160.

Gouesse-Aïdara, L., 1986. Biomasse, production, phénomètrie et composition minérale de quelques macrophytes des cours d'eau bretons. Thèse Dr-Ing. Ecologie, Univ. Rennes I: 135 pp.

Ham, S. F., Wright, A. D. Berrie, A. D. 1981 Growth and recession of aquatic macrophytes on an unshaded section of the River Lambourn, England, from 1971 to 1976. Freshwat. Biol. 11: 381–390.

Hartog, C. Den & S. Segal, 1964. A new classification of water-plants communities. Acta Bot. Neerl. 13: 67–393.

Haury, J., 1985. Etude écologique des macrophytes du Scorff (Bretagne-Sud). Thèse Dr.-Ing. Ecologie, Univ. Rennes I: 243 pp.

Haury, J., 1995. Patterns of macrophyte distribution within a Breton brook compared to other study scales. Landscape and Urban Planning 30: 239–248.

Haury, J. & J. L. Baglinière, 1996. Les macrophytes, facteurs structurant de l'habitat piscicole en rivière à salmonides. Etude de microrépartition sur un secteur végétalisé du Scorff (Bretagne-Sud). Cybium 20 (3) suppl.: 107–122.

Haury, J. & L. Gouesse Aïdara, 1990. Etude méthodologique préliminaire de la biomasse des macrophytes en rivière. 14 ème Conf. Int. COLUMA 1: 247–255.

Kern-Hansen, U. & F. H. Dawson, 1978. The standing crop of aquatic plants of lowlands in Denmark and the interrelationships of nutrients in plant, sediment and water. Proc. Int. Symp. Aquat. Weeds 5: 143–140.

Laforge, P. & D. Petitidier, 1997. Etude du phytobenthos dans différents types de rivières du bassin Rhin-Meuse. 5ème Conf. int. Limnologues Expression française, F.U.N.D.P., Namur, Belgique, 5-11/07/1997: 118 (summary).

Makirinta, H., 1988. Ein neues ökomorphologisches Lebenformen-System der aquatischen Makrophyten. Phytocoenologia 4: 446–470.

Owens, M. & R.W. Edwards, 1962. The effects of plants on river conditions. 3: Crop studies and estimates of net productivity of macrophytes in four streams in Southern England. J. Ecol. 50: 157–162.

Peltre, M. C., L. Léglize & J. L. Salleron, 1993. Végétation fixée et phosphore en petit cours d'eau. Conséquences d'une réduction des apports de phosphore. Bull. Fr. Pêche Piscic. 331: 357–371.

Roussel, J. M., A. Bardonnet, J. Haury, J. L. Baglinière & E. Prévost, 1998. Végétation aquatique et peuplement pisciaire: approche expérimentale de l'enlèvement des macrophytes dans les radiers d'un cours d'eau breton. Bull. Fr. Pêche Piscic. 350–351: 693–709.

Sabbatini M. R. & K. J. Murphy, 1996. Submerged plant survival strategies in relation to management and environmental pressures in drainage channel habitats. Hydrobiologia 340: 191–195.

Spink, A, K. J. Murphy & D. F. Westlake, 1997. Distribution and environmental regulation of species of *Ranunculus* subgenus *Batrachium* in British rivers. Arch. f. Hydrobiol. 139: 509–525.

Westlake, D. F., 1963. Comparisons of plant productivity. Biol. Rev. 63: 385–425.

Westlake, D. F., 1975. Aquatic macrophytes. In Whitton, B.A. (ed.), River Ecology. Blackwell Sci. Publ., Oxford: 106–128.

Westlake, D. F., H. Casey, F. H. Dawson, M. Ladle, R. H. K. Mann & A. F. H. Marker, 1972. The chalk-stream ecosystem. In Kajak, Z. & A. Hillbright-Ilkowska (eds.), Productivity Problems of Freshwater Ecosystems. Proc. IBP/UNESCO Symp. Krsimierz, Dolny Poland, 1970. Pol. Sci. Publ., Warsaw-Krakow: 615–635.

Hydrobiologia **415:** 117–122, 1999.
J. M. Caffrey, P. R. F. Barrett, M. T. Ferreira, I.S. Moreira, K. J. Murphy & P. M. Wade (eds),
Biology, Ecology and Management of Aquatic Plants.
© 1999 Kluwer Academic Publishers. Printed in the Netherlands.

Relationships of some ecological factors with the associations of vegetation in British rivers

F. Hugh Dawson[1,*] & Krzysztof Szoszkiewicz[2]
[1]NERC-Institute of Freshwater Ecology, River Laboratory, East Stoke, Wareham, Dorset, U.K.
[2]University of Agriculture, Department of Ecology and Agricultural Environment Protection, Wojska Polskiego 28, 60-625 Poznan, Poland

Key words: vegetation, floristic associations, habitat, enrichment, phosphorus

Abstract

During River Habitat Surveys on the physical structure of watercourses in Britain between 1995 and 98, detailed assessments of the floristic associations, physical habitat and water chemistry were made at 165 sites over a wide range of vegetation types. Botanical data from a total of 340 relevés from within these RHS were recorded. The vegetation was classified phyto-sociologically on the basis of the abundance of characteristic species, resulting in 45 aquatic, swamp and mire associations being identified together with nine other communities which could not be completely classified phyto-sociologically. The associations were assigned to 22 alliances, 13 orders and 11 classes excluding communities dominated by filamentous algae. Water enrichment (as soluble and total phosphorus) was the most significant factor for the differentiation of the communities but pH, conductivity and alkalinity were also important.

Introduction

Surveys of plant communities were undertaken at a representative set of sites from the 5700 standard or 'reference' surveys of rivers and streams, selected on a stratified random basis, for the detailed assessment of the physical structure of the watercourses in the U.K. (River Habitat Survey, RHS; Raven et al., 1997). The RHS assessment provides data for the many specialists involved in water management, fisheries, flood control etc, and its standardisation of method makes the data relatively compatible with other surveys. This paper links data from RHS and supplementary water chemistry with the phyto-sociological descriptions of aquatic, swamp and mire vegetation (e.g. Rodwell et al., 1991, 1995). Phyto-sociological data from different kinds of vegetation surveys are frequently used for environmental surveys, nature conservation, bio-indication, evaluation of agricultural land and forest, vegetation mapping and biological diversity assessment (Maarel, 1971; Haeupler, 1982; Dierschke, 1994). Phyto-sociological units are defined by floristic composition but they are also regarded as ecological realities – the complex interaction between plants and habitat is assumed to be the basic feature determining plant community in the phyto-sociological approach. Access to this large set of RHS data to which survey data is continually being added, will contribute to the understanding of the complex processes in the development of aquatic vegetation. The aim of this paper was to determine the links between ecological factors of habitat, water chemistry and, possibly, trophic status, with the phyto-sociological plant associations.

Materials and methods

Surveys of plant communities were carried out in 1995–98 at a subset of RHS sites (Figure 1). Following botanical assessment at 165 sites, some 340 relevés were surveyed to species level, using the taxonomic nomenclature of Stace 1997. A relevés is a uniform or homogeneous area of vegetation which cannot be divided or separated botanically and is clearly different from adjacent areas (Rieley & Page, 1990). A detailed RHS assessment of the physical

* Author for correspondence

Table 1. The phyto-sociological communities of the British rivers in this study

Class 1: *Fontinaletea antipyreticae*
Order: *Brachythecialia plumosi*
Alliance: *Scapanion undulatae: Scapanietum undulatae, Hygrohypnetum ochracei, Brachythecietum plumosi, Fontinalis squamosa* community
Order: *Fontinaletalia antipyreticae*
Alliance: *Rhynchostegion riparioides: Oxyrrhynchietum rusciformis*
Alliance: *Fontinalion antipyreticae: Fontinaletum antipyreticae*
Alliance: *Brachythecion rivularis: Brachythecietum rivularis*

Class 2: *Potamogetonetea*
Order: *Potamogetonetalia*
Alliance: *Potamogetion pectinati: Potamogetonetum pectinati, Potamogetonetum lucentis, Potamogetoneto – Myriophyletum spicati, Ceratophylletum demersi, Elodeetum canadensis, Parvopotamo – Zanichellietum palustris, Potamogetonetum perfoliati, Potamogetonetum obtusifolii*
Alliance: *Nymphaeion albae*: *Nupharietum lutenti, Potamogetonetum natanti, Polygonetum amphibiae*
Alliance: *Ranunculion fluitantis: Ranunculus penicillatus var. pseudofluitans* association, *Ranunculetum fluitantis, Callitrichetum stagnalis, Callitricho – Ranunculetum, Ranunculus trichophyllus* community
Alliance: *Hydrocharition: Hydrocharito - Stratriotetum aloides*

Class 3: *Lemnetea*
Order: *Lemnetalia*
Alliance: *Lemnion minoris: Lemnetum minoris*
Alliance: *Lemnion gibbae: Lemnetum gibbae*

Class 4: *Oxycocco-Sphagnetea*
Order: *Sphagnetalia magellanici*
Alliance: *Sphagnion magellanici: Narthecium ossifragum – Sphagnum papillosum* association

Class 5: *Littorelletea*
Order: *Littorelletalia*
Alliance: *Hydrocotylo-Baldellion: Hyperico – Potamogetonetum polygonifolius, Potamogeton polygonifolius-Juncus bulbosus* association, *Myriophylletum alterniflori*

Class 6: *Utricularietea intermedio-minoris*
Order: *Utricularietalia*
Alliance: *Sphagno-Utricularion: Juncetum bulbosi*

Class 7: *Scheuchzerio-Caricetea nigrae*
Order: *Caricetalia nigrae*
Alliance: *Caricion nigrae: Carex echinata – Sphagnum recurvum/auriculatum* association

Class 8: *Phragmitetea*
Order: *Phragmitetalia*
Alliance: *Phragmition: Phragmitetum australis, Phragmites australis - Urtica dioica* community, *Glycerietum maximae, Scirpetum lacustris, Eleocharietum palustris, Sparganietum erecti, Sagittaro - Sparganietum fluitanti*
Alliance: *Magnocaricon: Caricetum acutae, Caricetum ripariae, Caricetum rostratae, Caricetum vesicariae*

Table 1. Continued

Order: *Nasturtio-Glycerietalia*
Alliance: *Glycerio-Sparganion: Glycerietum fluitante, Glycerietum plicatae, Glycerietum declinatae,* other water-margin herbal vegetation
Alliance: *Apion nodiflori: Apium nodiflora* community, *Nasturtium officinale* community, other water-margin herbal vegetation
Alliance: *Phalaridion arundinaceae: Phalaridetum arundinaceae*
Class 9: *Bidentetalia tripartitae*
Order: *Bidentetalia tripartitae*
Alliance: *Bidention tripartitae: Bidens tripartita – Polygonum hydropiper* community, *Rorippa palustris – Filago uliginisum* community, *Myosotis scorpioides – Ranunculus scleratus* community, *Polygonetum persicaria – Polygonum lapathifolium* community
Class 10: *Montio-Cardaminetea*
Order: *Montio-Cardaminetalia*
Alliance: *Cardamino-Montion*
Sub-alliance: *Montion: Ranunculus omniophyllus - Montia fontana* community
Sub-alliance: *Cardaminion: Cardamino – Chrysosplenietum oppositifolium*
Class 11: *Molinio - Arrhenatheretea*
Order: *Molinietalia*
Alliance: *Molinion coeruleae: Molinio – Potentilletum erecti*

structure of the watercourses together with the results of the analysis of water samples taken at each site was available. The vegetation was classified phytosociologically on the basis of the abundance of characteristic species according to the system devised by Braun-Blanquet and modified by various authorities (British approach, Rieley & Page, 1990; Rodwell et al., 1991 and international syntheses: Matuszkiewicz, 1984; Best, 1988; Ellenberg et al., 1992; Dierschke, 1994). Simple correlation of the derived associations against the environmental parameters indicated the environmental preferences and ranges of each community. Further analysis was undertaken to derive the multi-dimensional Weber medians (see Lira, 1999) for use in determining the interaction of the most frequent communities with the following environmental parameters: acidity (pH), conductivity, 'soluble reactive'- and total-phosphorus, alkalinity, discharge category on a ten point exponential scale ($1 = <0.3$ m^3 s^{-1}, $2 = 0.3$–0.6 m^3 s^{-1}, $3 = 0.6$–1.2 m^3 s^{-1}... $10 = >80$ m^3 s^{-1}), bed slope (m km^{-1}), water quality classification on a 5-point scale (UK-WQC, 1 = excellent...5 = poor) and altitude (m).

Results

The aquatic vegetation, together with associated marginal wetland communities, was classified phytosociologically and gave 45 associations (Table 1); nine other communities were not fully classified. The associations belong to 11 classes, 13 orders, 22 alliances and 2 sub-alliances; some other communities which were dominated by filamentous algae (*Cladophora* agg. and *Vaucheria* agg.) were not included.

Very specific ecological requirements and tolerances were found for some of the most frequent plant communities (Table 2). The most specific of these relationships was for communities with a high abundance of *Juncus bulbosus*. This species was present in a variety of related habitats: the peaty channels between vegetation of *Utricularietea intermedio-minoris* class (*Juncetum bulbosi* association)and *Littorelletea* class (*Hyperico - Potamogetonetum polygonifolius, Potamogeton polygonifolius-Juncus bulbosus* associations), the shallow waters in the open peaty habitat of *Oxycocco-Sphagnetea* (*Narthecium ossifragum - Sphagnum papillosum* association), and the bog-pool vegetation of *Scheuchzerio-Caricetae nigrae (Carex echinata - Sphagnum recurvum/auriculatum* associ-

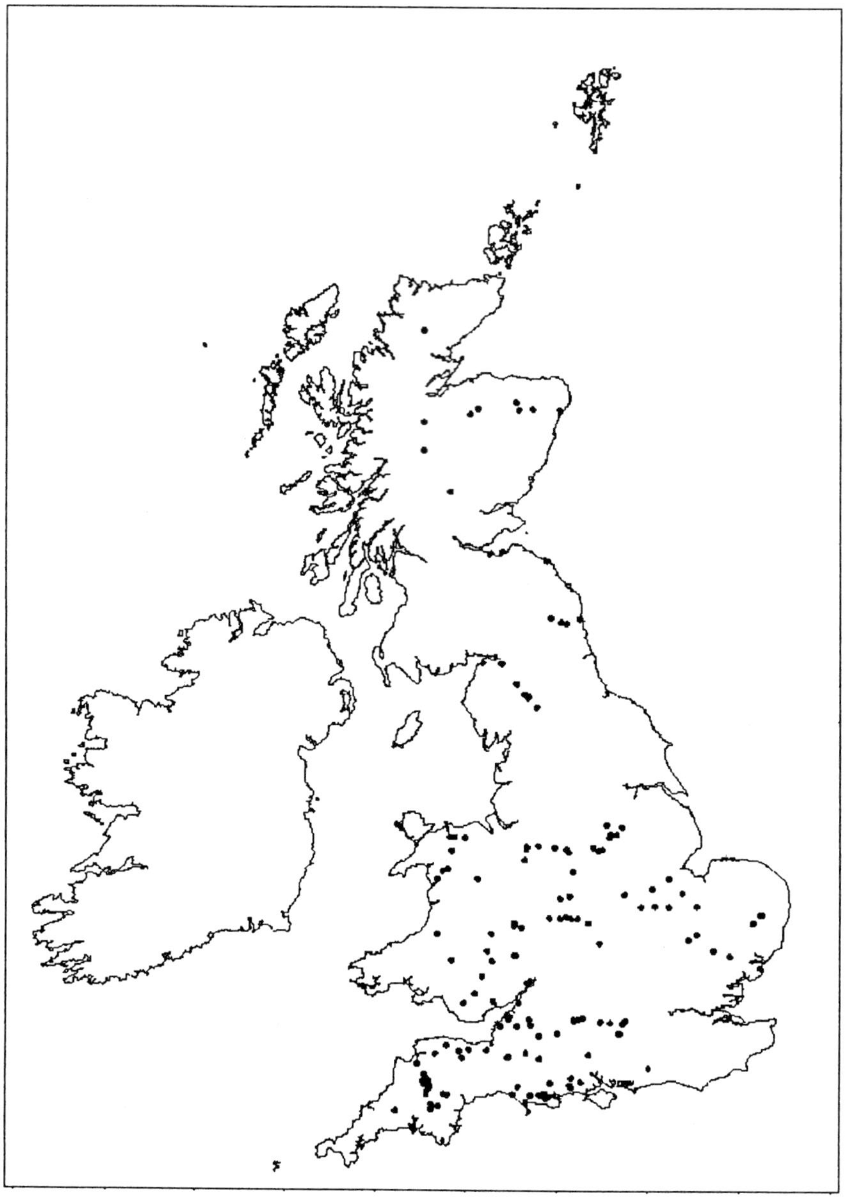

Figure 1. The distribution of surveys sites on British rivers for plant community analysis, 1995–98.

ation). The water in these areas was acid and extremely oligotrophic being low in phosphorous, conductivity and alkalinity, and was classified as excellent or good water quality. Many of these sites were located in the higher areas of Scotland, Wales and also in Southwest England on Dartmoor and Exmoor where the flow variation and general slope of the river bed is often high. Similar ecological conditions were found for the associations dominated by bryophytes, such as the *Fontinaletea antipyreticae* class, which usually develop in the central parts of shady and fast flowing rivers but always on those streams with steep bed-slopes. The *Montio-Cardaminetea* association is also

Table 2. The most frequent river phyto-sociological communities and some of the ecological parameters, arranged in order of increasing soluble reactive phosphate, given as medians from the multi-dimensional Weber analysis for water chemistry – pH, conductivity, soluble reactive phosphate (SRP), total phosphorus, alkalinity, and water quality classification, discharge category (see text), altitude and bed slope (see Table 1. for full names of communities)

Community/ Alliance	pH	Conductivity (μg S cm^{-1})	SRP (μg P l^{-1})	Total P (μg P l^{-1})	Alkalinity (meq l^{-1})	WQC	Discharge category	Altitude (m)	Slope (m/km)
Carex –Sphag. mire	5.9	41	8	116	0.1	1.0	1.0	435	6.2
Pot.polyg. -Jun. bulb.	6.7	65	8	84	0.2	1.0	0.8	263	62.2
Montion	7.1	114	8	67	0.6	1.0	0.0	168	0.4
Juncetum bulbosi	6.3	71	9	53	0.1	1.0	0.9	353	0.1
Hyperico-Pot. polyg.	6.5	71	9	24	0.2	1.0	1.0	389	14.7
Scapan. undulatae	5.4	40	11	47	0.1	1.1	0.9	452	45.0
Hygrohypnetum ochr.	6.6	64	18	22	0.1	1.1	1.9	268	18.3
Oxyrrhyn.rusciformis	7.2	177	23	106	0.8	1.1	0.7	124	44.9
*Fontinal. sqam.*comm.	7.0	135	28	32	0.6	1.3	3.9	158	8.6
Fontinaletum antipyr.	7.5	186	42	98	0.8	1.2	2.4	134	9.8
Potamoget. natantis	7.1	339	49	152	1.8	1.3	0.8	46	1.0
Ranunculet. fluitantis	8.0	402	54	86	2.8	1.2	4.9	87	1.2
Glycerietum fluitantis	7.5	361	58	189	1.9	1.7	0.9	130	11.5
Callitricho-Ranuncul.	7.8	599	83	203	3.2	1.6	1.0	73	4.7
Apium nodif. comm.	8.0	784	134	159	4.7	1.2	3.0	53	4.2
*Bident. trip.-A.*egetat	7.8	442	147	216	3.1	1.7	4.1	40	2.3
Phalaridetum arundin.	8.0	476	205	293	3.1	1.8	4.5	44	1.9
Ranunc.penic.var.ps.	7.9	533	235	482	3.5	1.6	3.5	63	3.7
Glycerietum maxim.	8.0	655	346	474	4.6	1.4	3.9	19	1.0
Scirpetum lacustris	8.3	602	369	389	4.6	1.7	6.1	66	0.4
Sparganietum erecti	8.0	666	392	554	3.9	1.7	3.5	45	1.4
Cladophora comm.	8.2	1053	806	879	4.7	1.6	3.6	6.0	3.6
Potamoget.pectinati	8.4	1500	1262	1458	4.5	1.8	4.8	23	1.1

found in these conditions but they are not so specific in acid waters.

The other classes were much more diverse. The deeper parts of mesotrophic and eutrophic rivers were dominated by *Potamogetonetea* vegetation. The most frequent were two fine-leaved submerged plant communities, the *Ranunculion fluitantis* alliance and the variety of rooted aquatic communities of the *Potamogetion pectinati* alliance with the most common, *Potamogetonetum pectinati*, developing in highly phosphate-enriched water of high conductivity, high alkalinity and high pH. Similar eutrophic conditions also favoured algal dominated communities, especially the *Cladophora* agg. which were however not classified. The deep but slow-flowing rivers were the typical habitat of the *Nymphaeion* alliance from the *Potamogetonetea* class and of the free-floating *Lemnetea* vegetation class. The water margins were dominated by *Phragmitetea* vegetation; these communities all appear to be tolerant of elevated levels of phosphate and have a wide pH range. Some shallow and enriched margins, especially when disturbed by anthropogenic factors, were dominated by the class*Bidentetea tripartiti*.

Discussion

The large number of associations found in the sites surveyed included most of the typical aquatic, swamp and mire communities of Britain, not just those of rivers (Rieley & Page, 1990; Rodwell et al., 1995). The range of communities was not extended and did not require new descriptions although filamentous algae and some bryophytes were not fully accounted for and need further study. A number of difficulties which need further investigation were found with the British phyto-sociological system (Rodwell et al., 1995) especially with very similar types of vegetation which had to be assigned to widely separated syntaxa.

The most frequent in-stream phyto-coenoses belonged to the *Potamogetonetea* class which included communities of rooted, floating and submerged plants (Table 2). The margins of the river are mostly occupied by dense fen vegetation with only narrow channels between stands of the *Nasturtio-Glycerietalia* order and *Phragmition* alliance (*Phragmitetalia* order).

Trophic status gives the clearest relationship with the plant associations assuming that it is reflected in soluble reactive phosphate (SRP); low trophic status was assumed to be indicated by low phosphate. The vegetation which developed in these very low or oligotrophic conditions is characterised by a high abundance of *Juncus bulbosus (Oxycocco-Sphagnetea, Scheuchzerio-Caricetae nigrae, Utricularietea intermedio-minoris* and *Littorelletea*), bryophyte associations of *Fontinaletea antipyreticae* and *Montio-Cardaminetea* vegetation. The most trophic habitats were dominated by various *Potamogetonetea* vegetation, especially *Potamogetonetum pectinati.* High trophic status was also associated with algal dominated communities, especially from the *Cladophora* agg. but which were not investigated in detail. Conductivity, pH, soluble- or total-phosphate, and alkalinity were quite successful in describing ecological requirements of the particular communities, whereas the map-based parameters such as slope or flow category were not so useful. The full interpretation of the results of these kind of parameter is worthy of more detailed study.

Acknowledgements

We are grateful to the RHS surveyors supervised by Paul Henville (PH) and who accompanied (KS), to PH and Darren Smith who undertook the phosphate analyses and particularly to Peter Scarlett, Sally Small and Ryszard Staniszewski, for assembling the data base and to Michael Gravelle for comments on the text. A travel grant was obtained for KS from The Central European University (Soros Foundation).

References

Best, E. P. H., 1988. The phyto-sociological approach to the description and classification of aquatic macrophytic vegetation. In Symoens, J. J. (ed.), Vegetation of Inland Waters. Kluwer Academic Publishers, Dordecht: 155–181.

Dierschke, H., 1994. Pflanzensoziologie. Eugen Ulmer Verlag, Stuttgart: 683 pp.

Ellenberg, H., H. E. Weber, R. Düll, V. Wirth, W. Werner & D. Paulissen, 1992. Zeigerwerte von Pflanzen in Mitteleuropa. Scripta Geobot. 18: 258 pp.

Haeupler, H., 1982. Enenness als Ausdruck der Vielfalt in der Vegetation. Diss. Bot. 65: 1–269.

Lira, J., 1999. The application of some chosen statistical methods based on a median in agricultural experiments. [Zastosowanie wybranych metod statystycznych opartych na medianie w do wiadczalnictwie rolniczym]. PhD dissertation, Agricultural University, Poznan: 196 pp.

Matuszkiewicz, W., 1984. Przewodnik do oznaczania zbiorowisk roslinnych Polski. PWN, Warszawa: 298 pp.

Raven, P. J., P. Fox, M. Everard, N. T. H. Holmes & F. H. Dawson, 1997. River Habitat Survey: a new system to classify rivers according to their habitat quality. In Boon, P. J. (ed.), Freshwater Quality: Defining the Indefinable. HMSO, London: 215–234.

Rieley, J. O. & S. Page, 1990. Ecology of plant communities. Longman Scientific & Technical., Burnt Mill: 163 pp.

Rodwell, J. S. (ed.), C. D. Pigott, D. A. Ratcliffe, A. J. C. Malloch, H. J. B. Birks, M. C. F. Proctor, D. W. Shimwell, J. P. Huntley, E. Radford, M. J. Wiggington & P. Wilkins, 1991. British Plant Communities. Vol. 2. Cambridge University Press, Cambridge: 628 pp.

Rodwell, J. S. (ed.), C. D. Pigott, D. A. Ratcliffe, A. J. C. Malloch, H. J. B. Birks, M. C. F. Proctor, D. W. Shimwell, J. P. Huntley, E. Radford, M. J. Wiggington & P. Wilkins, 1991. British Plant Communities. Vol. 4. Cambridge University Press, Cambridge: 295 pp.

Stace, C., 1997. New Flora of the British Isles. Cambridge University Press, Cambridge: 1130 pp.

Van der Maarel, E., 1971. Plant species diversity in relation to management. In Duffey, E. & A. S. Watt (eds), The Scientific Management of Animal and Plant Communities for Conservation. Blackwell Scientific Publications, Oxford: 45–63.

Hydrobiologia **415:** 123–130, 1999.
J. M. Caffrey, P. R. F. Barrett, M. T. Ferreira, I.S. Moreira, K. J. Murphy & P. M. Wade (eds),
Biology, Ecology and Management of Aquatic Plants.
© 1999 Kluwer Academic Publishers. Printed in the Netherlands.

Distribution of the morphological groups of aquatic plants for rivers in the U.K.

F. Hugh Dawson[1,*], Paul J. Raven[2] & Michael J. Gravelle[1]
[1]NERC-Institute of Freshwater Ecology, River Laboratory, Wareham, Dorset BH20 6BB, U.K.
[2]Environment Agency, Rio House, Waterside Drive, Almondsbury, Bristol BS12 4UD, U.K.

Key words: morphological groups, aquatic vegetation, distribution, physical habitat

Abstract

During River Habitat Surveys on the physical structure of the watercourses in the U.K., channel vegetation types were recorded as present or extensive in 10 m wide belt transects at 50 m intervals over 500 m reaches for some 2435 stratified random sites in 1995–96. Data on attached, submerged, emergent and free-floating aquatic vegetation were grouped into nine broad morphological groups such as submerged broad-, linear- and fine-leaved, free-floating, emergent, etc. Vegetation data are supported by assessments of physical parameters at each site including dimension, flow types, channel substrate and the slope of the channel. The general habitat requirements for each group are shown to agree with assumed requirements but with a broader distribution of habitats. Eight percent of channels had excessive growths of vegetation and would normally be considered for weed control; this was a minimum estimate.

Introduction

The presence and extent of aquatic vegetation is a fundamental part of the riverine habitat whether as a source of organic material for biota, a habitat, an influence on physical stream structure or simply in its own right (Whitcomb, 1968; Marshal & Westlake, 1978). Surveys of channel vegetation were undertaken in 1995–96 as part of the detailed assessment of the physical structure of the watercourses in the U.K. (River Habitat Surveys, RHS; Raven et al., 1997, 1998). The RHS assessment provides data for the many and varied specialists involved in water management, fisheries, flood control etc, and its standardisation of method makes the data relatively compatible with other surveys. Botanical surveys often include aquatic vegetation but data are then presented as distributions of species, genera or even management, within defined areas e.g. squares of 1, 2 or 10 km without reference to the actual plants habitat (e.g. Preston & Croft, 1997).

This preliminary study therefore aims to show 1. the distribution of aquatic plants by major morphological group, 2. some detail of their respective characteristic or preferred habitats, and 3. the distribution of channels choked with vegetation and potentially needing management by removal or dredging, in the rivers and streams of the U.K.

* Author for correspondence

Materials and methods

The 1995–96 River Habitat Survey was carried out to collect data on a range of features describing the physical habitat of 500 m lengths of river in the U.K. RHS sites were selected on a stratified random basis from the 10 × 10 km squares of Ordnance Survey national grid system, and provided a representative distribution of sites throughout the regions of the U.K. Channel vegetation was recorded from 10 m wide belt transects across the watercourse at 50 m intervals in the 500 m reaches for 2435 sites. Attached, submerged, emergent and free-floating aquatic vegetation was recorded as 'absent', 'present' or 'extensive' for each of nine broad morphological groups (Table 1). The RHS Guidance manual defines 'present' as at least 1% of the channel area within the 10 m transect, and 'extensive' as one third or more of this area. The definition of 'choked channel', which was also recorded, is that

Table 1. RHS-channel vegetation types with examples of the species

	RHS vegetation type	Example genera, species or comment	Occurrences, present (%)	Occurrences, extensive (%)
a	Liverworts, mosses & lichens	any exposed or submerged bryophyte e.g. *Scapania, Fontinalis*	32	3
b	Emergent broad-leaved herbaceous	*Apium, Rorippa*	17	2
c	Emergent reeds, sedges, or rushes	*Sparganium erectum, Typha, Phragmites, Glyceria maxima, Juncus, Carex*	22	4
d	Floating-leaved (rooted)	*Nuphar lutea, Potamogeton natans, S. emersum*	3	1
e	Free-floating	*Lemna, Ceratophyllum, Stratiotes*	2	0.3
f	Amphibious	*Polygonum amphibium, Agrostis stolonifera*	8	0.4
g	Submerged broad-leaved	*Nuphar, Elodea, Callitriche*	5	1
h	Submerged linear-leaved or Submerged fine-leaved	submerged *Sparganium erectum, Typha* *Ranunculus, Myriophyllum*	12	3
i	Filamentous algae	*Cladophora, Enteromorpha*	38	8

a third or more of the whole 500 m has dense vegetation of any type which is causing, or could cause, a significant impediment to water flow.

The typical habitat was determined by analysis of RHS data for these sites for predominant flow type, predominant channel substrate and channel dimensions (measured at a straight section with well-defined banks and with a riffle if possible) and slope of the channel which was derived from 1:50 000 scale maps (Environment Agency, 1997). The predominant substrate types are: artificial, bedrock, boulders, cobbles, gravel-pebble, sand, silt, clay and peat; turbid or deeper water sites in which the bed was not generally visible were recorded as 'not visible'. The predominant flow types are: 'free-fall' as in waterfalls, 'chute-flow' for cascades, 'chaotic' for mixtures of high energy flows, 'broken standing-waves' of rapids, 'unbroken standing-waves' of riffles, 'upwelling' of boils, 'rippled' of runs, or 'smooth' of glides and 'no perceptible flow' of pools or 'no flow' when dry.

Results

This national survey of the U.K. has clarified the respective characteristics or preferred habitats of the nine morphologically-similar groups of aquatic plants. The geographic distribution of each RHS channel vegetation type when it is 'present' or 'extensive' reinforces the generally-held opinion of the distribution of the various types of water plant. The proportion within vegetative types for each predominant flow type, channel substrate, channel dimensions and bed-slope gives data on the range and optimum habitat requirements of these groups of aquatic plants (Figures 1 and 2). Thus, bryophytes were common and particularly extensive in rock and boulder-dominated streams and those with higher slope or energies. Bryophytes were disproportionately less extensive in predominantly gravel-pebble substrates but they were still found infrequently in streams dominated by even finer substrates or in more stable flow conditions, where they were never extensive. However, they were not as common as might be expected on artificial substrates (cross hatched area in Figure 2a) which are as stable as bedrock or boulder substrates. Filamentous algae were the most common vegetation type occuring in a very wide range of conditions, though this may relate either to discrete local or diffuse areas with enrichment in smaller streams and general enrichment in rivers. Reeds and emergent herbs were also frequent and were not just limited to low gradient streams or rivers. They were disproportionately less extensive in higher energy flow such as riffles but were more extensive on predominantly silt substrates. Similarly,

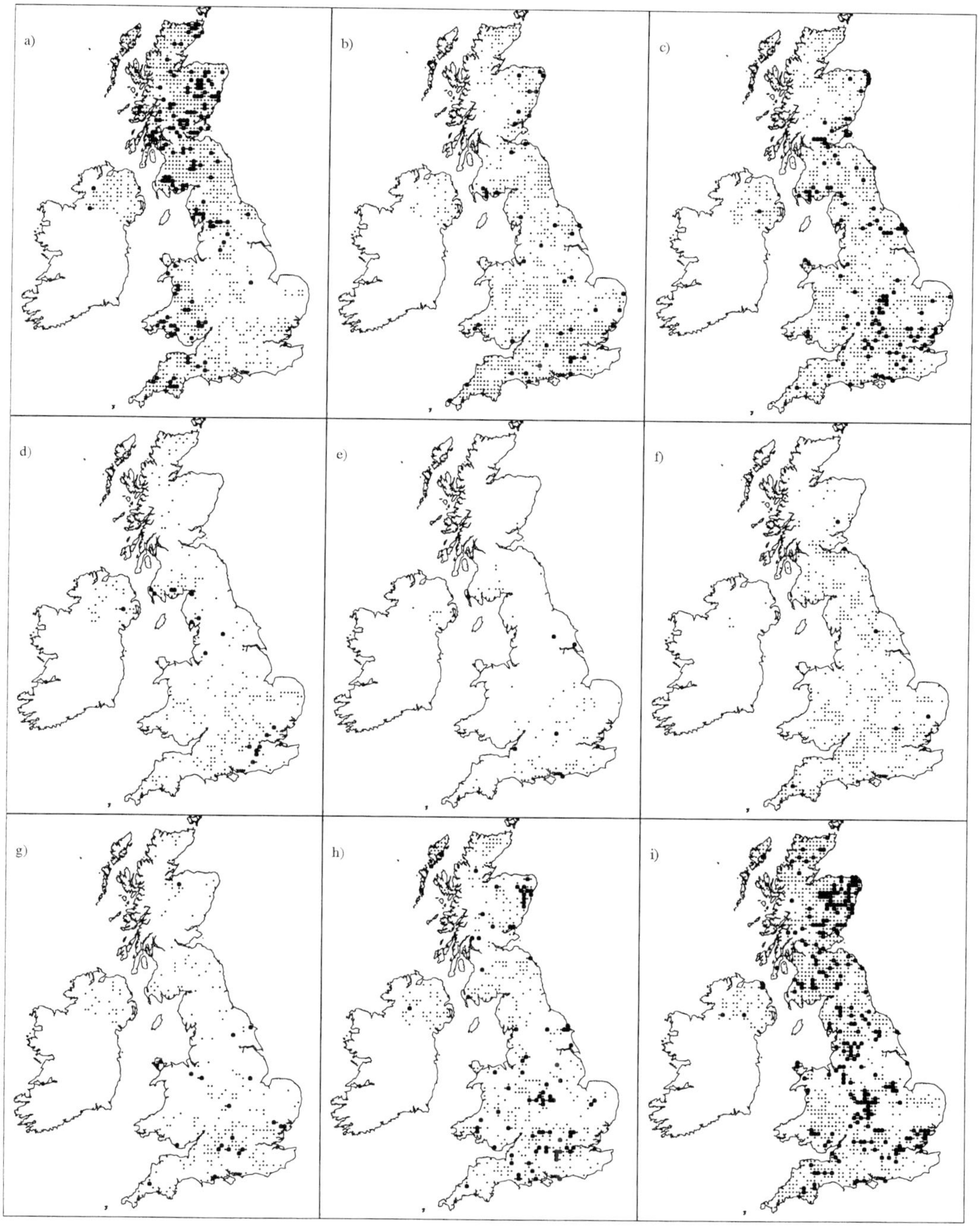

Figure 1. The distribution of channel vegetation types (a) liverworts, mosses or lichens, (b) emergent broad-leaved herbs, (c) emergent reeds, sedges or rushes, (d) rooted floating-leaved, (e) free-floating, (f) amphibious, (g) submerged broad-leaved, (h) submerged fine-leaved and (i) filamentous algae in the U.K. (Key: present –1–32% - small solid circles; ≥ 33% extensive - large solid circles).

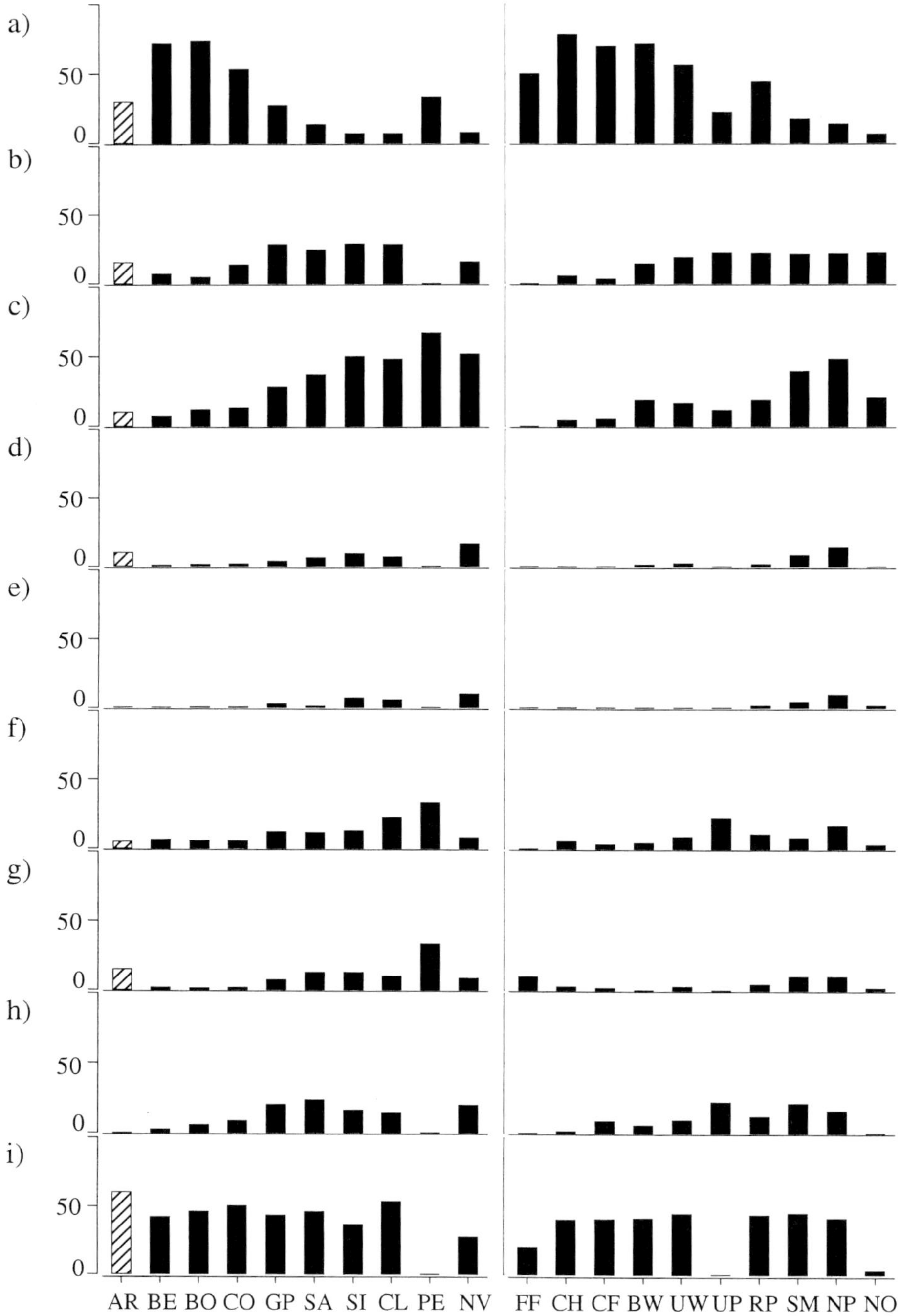

Figure 2a. The percentage within each type of predominant channel substrate and predominant water flow type, for each of the vegetation types (a) liverworts, mosses or lichens, (b) emergent broad-leaved herbs, (c) emergent reeds, sedges or rushes, (d) rooted floating-leaved, (e) free-floating, (f) amphibious, (g) submerged broad-leaved, (h) submerged fine-leaved and (i) filamentous algae in the U.K. [key: predominant substrate types: artificial (AR), bedrock (BE), boulders (BO), cobbles (CO), gravel-pebble (GP), sand (SA), silt (SI), clay (CL), peat (PE), and not visible (NV). Predominant flow types are free-fall (FF), chute-flow (CH), chaotic (CF), broken standing-waves (BW), unbroken standing-waves (UW), upwelling (UP), rippled (RP), smooth (SM), 'no perceptible flow' (NP) and no flow (NO).]

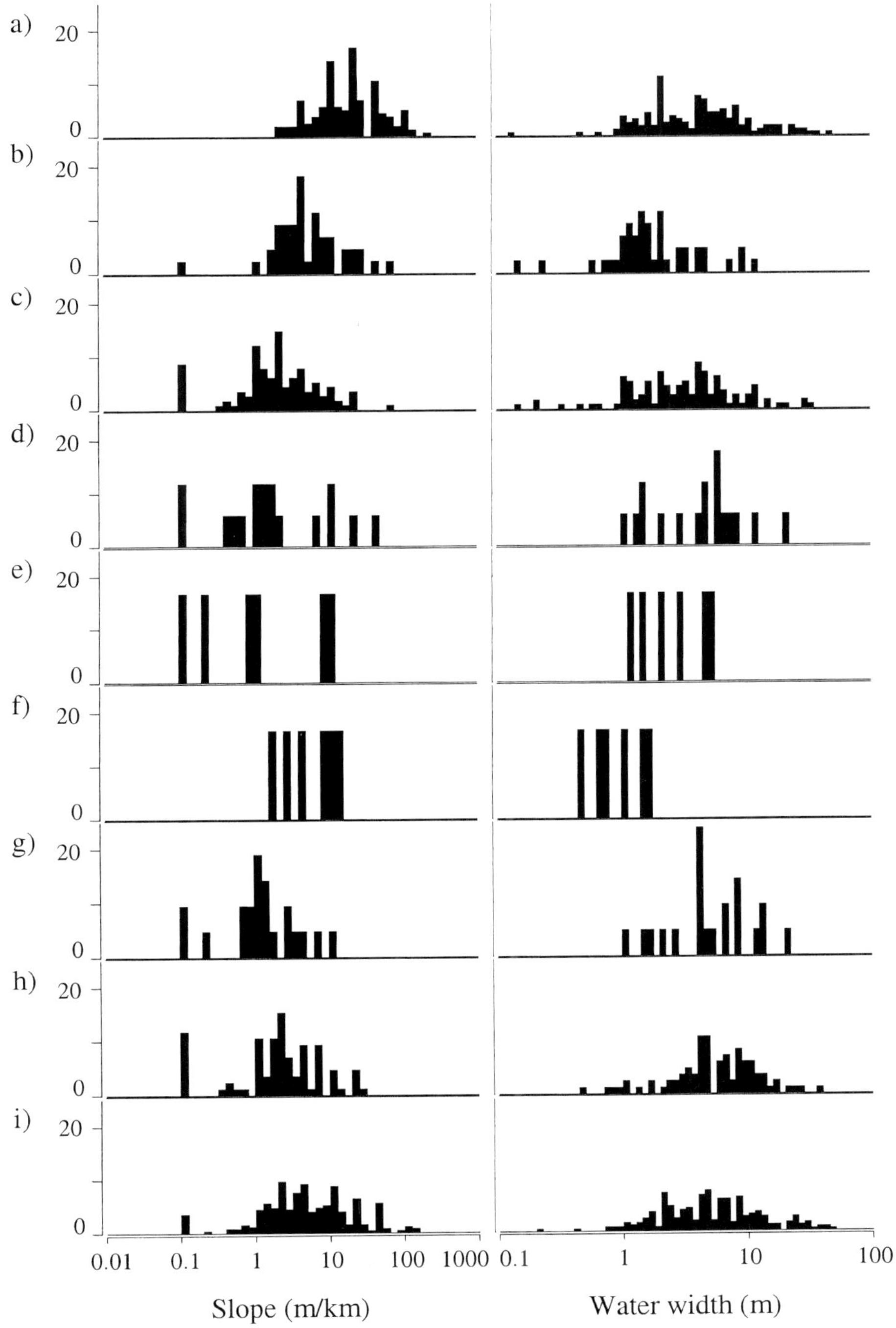

Figure 2b. The percentage within each type of bed slope (m/km) and water width (m) for each of the vegetation types (a) liverworts, mosses or lichens, (b) emergent broad-leaved herbs, (c) emergent reeds, sedges or rushes, (d) rooted floating-leaved, (e) free-floating, (f) amphibious, (g) submerged broad-leaved, (h) submerged fine-leaved and (i) filamentous algae in the U.K. [key: predominant substrate types: artificial (AR), bedrock (BE), boulders (BO), cobbles (CO), gravel-pebble (GP), sand (SA), silt (SI), clay (CL), peat (PE), and not visible (NV). Predominant flow types are free-fall (FF), chute-flow (CH), chaotic (CF), broken standing-waves (BW), unbroken standing waves (UW), upwelling (UP), rippled (RP), smooth (SM), 'no perceptible flow' (NP) and no flow (NO).]

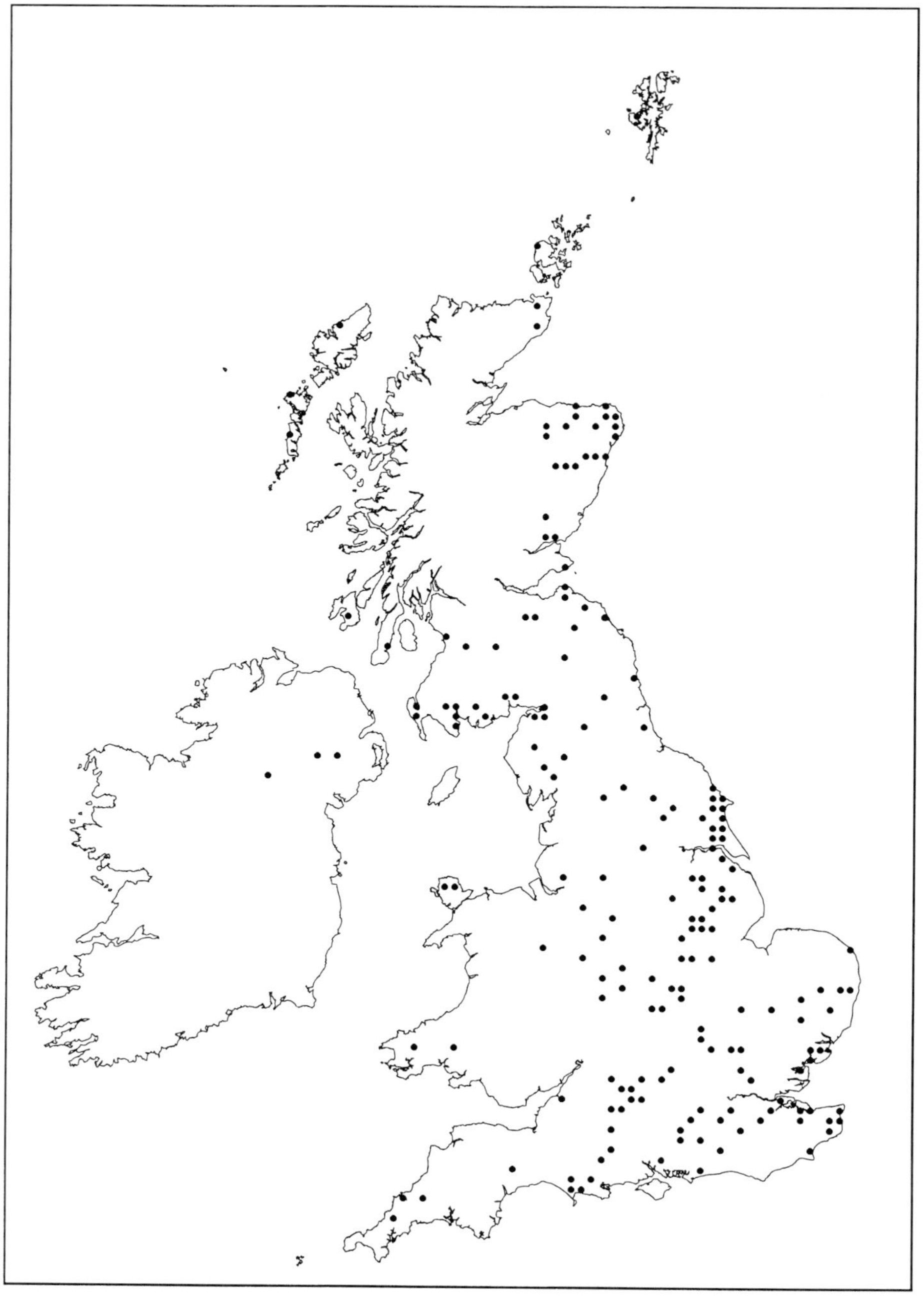

Figure 3. The geographic distribution of channels choked with vegetation in the U.K.

free-floating and floating leaved species were not limited to reaches of 'no perceptible flow' although they were, in general, uncommon, with their distribution probably related to local flow features such as the presence of pools and backwaters.

The distribution and extent of submerged, emergent and reed species appeared to exceed previous estimates of species which have traditionally been considered to need control for various river management objectives, such as land drainage and flood

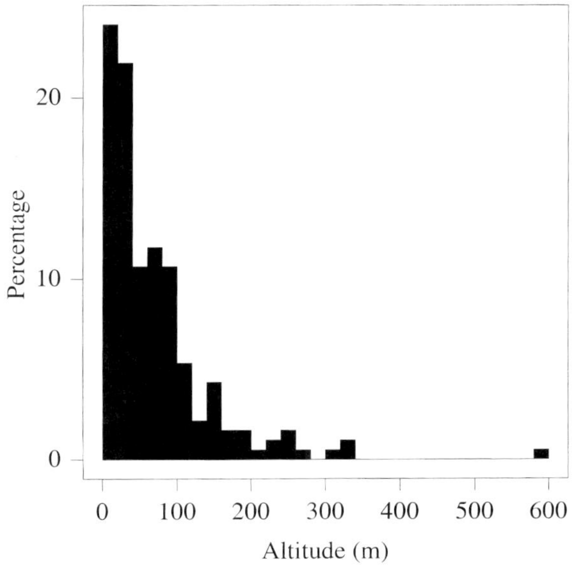

Figure 4. The distribution of vegetation-choked channels with altitude in the U.K.

control, making a combined total some half as much again as is normally considered to be present (summarised in Dawson, 1989). Eight percent of channels were considered by trained surveyors to be choked by vegetation, but this should be regarded as a minimum estimate as weed removal or dredging work may have been undertaken prior to the survey period in the year and there is also likely to be a carry-over from the results of dredging in the immediately previous years (Figure 3). The majority of the sites lie at altitudes less than 100 m with a quarter less than c. 20 m (Figure 4).

Discussion

The results show an overall pattern of habitat requirements as is generally assumed by aquatic botanists (e.g. Butcher, 1933; Dawson, 1988), but extends this knowledge to show a wider geographic distribution and a broader range of habitats for all morphological groups of plants. The additional data estimating the distribution and extent of submerged, emergent and reed species of choked channels confirms the range of previous estimates of channels needing control to meet various river management objectives, such as land drainage and flood control, but give evidence as to which sites need more examination for more environmentally-sound management techniques especially those at less than 20 m in altitude.

More analysis is needed to consider: the distribution of many of the commoner species, the variation of types of vegetation, and their combination and relationship in semi-natural and modified streams, the effects of marginal shade from trees and other vegetation, and links with studies of the effects of fencing-off rivers on marginal plant communities. In addition to this, proposals for simple measures of quality of river vegetation for habitat indices probably based on the developing framework to characterise or classify the differing types of river in terms of their driving variables including altitude, flow slope and geology is needed. So far, however, detailed initial assessments of (a) the floral communities together with their ecological preferences has been undertaken for a sub-set over a wide range of vegetation types (Dawson & Szoszkiewicz, 1999); and, (b) the occurrence of alien invasive bankside plants, Japanese Knotweed, *Fallopia japonica* (Houett.) Ronse Decr., Giant Hogweed *Heracleum mantegazzianum* Sommier & Levier, and Himalayan or Indian Balsam, *Impatiens glandulifera* Royle (Dawson & Holland, 1999). The distribution and habitat requirements of the species at these study sites will be considered elsewhere.

Acknowledgements

We are grateful to the RHS surveyors from the Environment Agency and IFE who undertook the field work including Paul Henville and Peter Scarlett, who also undertook other tasks from assembly of the database to production of the diagrams and to Nigel T. H. Holmes and the anonymous referees for their comments on the paper.

References

Butcher, R. W., 1933. Studies on the ecology of rivers. I. On the distribution of macrophytic vegetation in the rivers of Britain. J. Ecol. 21: 58–91.

Dawson, F. H., 1988. Water flow and the vegetation of running waters. In Symoens, J. J. (ed.), Handbook of Vegetation Sciences series, 15, Vegetation of Inland Waters. Dr W. Junk Publishers, The Hague: 283–309.

Dawson, F. H., 1989. Ecology and management of water plants in lowland streams. Ann. Rep. Freshwat. Biol. Ass. 57: 43–60.

Dawson, F. H. & D. Holland, 1999. The distribution in bankside habitats of three alien invasive plants in the U.K. in relation to the development of control strategies. Hydrobiologia 415 (Dev. Hydrobiol. 147): 193–201.

Dawson, F. H. & K. Szoszkiewicz, 1999. Relationships of some ecological factors with the associations of vegetation in British rivers. Hydrobiologia 415 (Dev. Hydrobiol. 147): 117–122.

Environment Agency, 1997. River Habitat Survey: 1997 Field Survey Guidance Manual including SERCON. Environment Agency, Bristol, England: 100 pp.

Marshall, E. J. P. & D. F. Westlake, 1978. Recent studies on the role of aquatic plants in their ecosystem. Proc. Eur. Weed Res. Counc. 5th Int. Symp. Aquat. Weeds: 43–51.

Preston, C. D. & J. M. Croft, 1997. Aquatic plants in Britain and Ireland. Harley Books, Colchester, U.K: 365 pp.

Raven, P. J., P. J. A. Fox, M. Everard, N. T. H. Holmes & F. H. Dawson, 1997. River Habitat Survey: a new system to classify rivers according to their habitat quality. In Boon, P. J. (ed.), Freshwater Quality: Defining the Indefinable. HMSO, London: 215–234.

Raven, P. J., N. T. H. Holmes, F. H. Dawson, P. J. A. Fox, M. Everard, I. Fozzard & K. J. Rouen, 1998. River Habitat Quality: The physical character of rivers and streams in the U.K. and Isle of Man. River Habitat Survey Report No. 2. Environment Agency: 100 pp.

Whitcomb, D. M., 1968. The fauna of aquatic plants. Proc 9th. British Weed Control Conference. British Crop Protection Council: 382–391.

Hydrobiologia **415:** 131–138, 1999.
J. M. Caffrey, P. R. F. Barrett, M. T. Ferreira, I.S. Moreira, K. J. Murphy & P. M. Wade (eds),
Biology, Ecology and Management of Aquatic Plants.
© 1999 *Kluwer Academic Publishers. Printed in the Netherlands.*

Macrophyte functional variables *versus* species assemblages as predictors of trophic status in flowing waters

M. M. Ali[1], K. J. Murphy[2,*] & V. J. Abernethy[2]
[1]*Department of Botany, Faculty of Science at Aswan, South Valley University, Aswan 81528, Egypt*
[2]*Institute of Biomedical & Life Sciences, Division of Environmental & Evolutionary Biology, University of Glasgow, Glasgow G12 8QQ, Scotland*
E-mail: k.murphy@bio.gla.ac.uk

Key words: phosphate, modelling, eutrophication, aquatic plants, rivers, irrigation channels

Abstract

A series of models was developed using functionally-derived variables (mainly based on morphological attributes of freshwater macrophytes) to predict the trophic status of river and associated channel systems. The models were compared with an existing species-assemblage based procedure for predicting British river trophic conditions (the Macrophyte Trophic Ranking scheme, MTR). We compared sites in cooler temperate conditions (in Scotland) and warmer, sub-tropical conditions (in Egypt). In total, we made measurements of 13 traits from >600 individual plant specimens of 33 species growing at 42 sites (divided into independent input and test site datasets). N status (as annual mean concentration in water of total oxidised nitrogen, TON) was only very poorly predicted by this approach. However, P (as annual mean concentration in water of soluble reactive phosphate, SRP) was better predicted: both by a model based on MTR ($r = -0.585$, $p<0.001$), and by models using functional attributes of the macrophyte vegetation. River Trophic Status Indicator (RTSI) models based on ranked plant functional group relationship to river water P concentrations ($RTSI_{FG}$), or field-measured trait sets of the plants ($RTSI_{TR}$) could also individually explain up to about 34% of the variation in P, both for the total dataset and for subsets from Egypt or Scotland alone or for high v. low-flow sites. Combining both types of RTSI measure produced the most powerful predictive model ($r = 0.72$, $p<0.001$), explaining just over half the variability in P.

Introduction

Unlike fish and most aquatic invertebrates, the low mobility of macrophytes means that they cannot easily escape the combination of flow, nutrient availability and other physico-chemical conditions influencing their survival in the lotic habitat. There is good evidence that aquatic macrophytes respond in a predictable manner to the trophic and physical characteristics of flowing-water habitats in both temperate waters (e.g. Haslam & Wolseley, 1981; Kohler & Schiele, 1984; Haury, 1996; Haury & Peltre, 1993. Husak & Vorechovska, 1996; Robach et al., 1996; Spink & Murphy, 1997) and in warmer-water systems (e.g. Agami, 1989; Springuel & Murphy, 1991; Ali et al., 1995; Ali & Soltan, 1996; Bini et al., 1999). In addition, most macrophytes are easy to see, readily sampled and reasonably simple to identify to species level. As a result, macrophytes have often been suggested as potential bioindicators of environmental conditions in flowing waters.

The assemblage of macrophyte species in a river or channel might be an effective indicator of the integrated combination of stress and disturbance pressures influencing the plant habitat. Several schemes have utilised this approach in recent years, in the U.K. and elsewhere (e.g. Ellenberg, 1973; Kohler, 1975; Holmes, 1983; De Lange & Van Zon, 1983; Caffrey, 1985, 1986; Schmedtje & Kohmann, 1987; Newbold & Holmes, 1987; Haslam et al., 1987; Daniel & Haury, 1996; Haury et al., 1996). Such schemes usually attempt to use macrophyte species-assemblage either as predictors of organic pollution or of trophic status (N or P or both). However, the functional plas-

* Author for correspondence

ticity of macrophyte populations often increases the trophic range of many species in rivers. Schemes which rely on the presence or absence of individual (or sets of) species thus tend to have low sensitivity to change. In practice, what is often found is that macrophyte species may persist across a rather wide range of trophic conditions and their response may be strongly influenced by other prevailing conditions – especially flow regime.

The approach used here was based on the ideas, originating from plant strategy theory (Grime et al.,1988) and subsequently applied by others to a wide range of vegetation types (e.g. Murphy et al., 1990; Leishmann & Westoby, 1992; Abernethy, 1994; Hills et al., 1994; McIntyre et al., 1995), that: (a) functionally-defined groups of plants tend to occupy discrete sections of environmental gradients. By identifying the group members we can use their occurrence in the environment to predict the existence of pre-defined ranges of those gradients at individual locations; and (b) functionally-defined groups of plants are so-defined on their possession of certain quantifiable sets of traits (often morphological). If these traits can be measured from plants in the field, it might be possible first to find consistent relationships between traits relating to the size or shape of plants, and the environmental variables which affect those plants; and then to use these relationships as predictors of variation in environmental variables, such as N and P.

The aim of the study was to gather evidence to test the hypothesis that variation in plant functional group presence and/or sets of functional traits of river and channel plants may provide an indication of trophic status in flowing waters. The models developed were tested for success in predicting trophic status against an existing macrophyte assemblage-based bioindicator procedure (Environment Agency, 1996).

Materials and methods

Field sampling to acquire model development data was undertaken during two sampling campaigns, in Scotland (July–September, 1996) and Egypt (December 1996–April 1997). Test data were acquired during a third sampling period, from an independent set of Scottish river sites, in August 1997. Each site comprised a 50 m length of channel. Sampling regime was identical at all 42 sites sampled. Submerged and floating macrophyte species were assessed using standard quadrat techniques to give % frequency and % cover estimates. Samples (whole plant or ramet, as appropriate) were collected for subsequent measurement of traits. Great care was taken to retrieve both leaf/shoot and root systems as intact as possible (for example by careful teasing-out of the intact root mat by hand). After collection, plants were kept cool and moist in sealed polybags in cool boxes, and morphological analysis (a suite of 13 traits: Table 1) was undertaken as quickly as possible after collection. Root lengths and submerged and/or floating leaf areas were measured using a flatbed optical scanner in Scotland and a portable scanner in Egypt downloading images to a PC. Customised DeltaT Scan plant-analysis software was used to analyse the images to produce the required data. Painstaking care was taken to divide up ramets or individual plants (stained where necessary using methyl violet) to obtain high quality scanned images. In the case of plants with finely-divided leaves (e.g. *Ceratophyllum demersum*) or strongly-developed root systems (e.g. *Eichhornia crassipes*) and consequently very large surface areas, we took sub-samples of the individual ramets from top, mid and lower portions of the plant and multiplied up by aliquots based on total dry weight of the plant parts concerned. Total above-ground (or above water surface) and below-ground (or below water surface) dry weight per ramet was recorded from a maximum of 10 (minimum 3) replicate ramets of each species sampled per site (fewer replicates were taken from rare species throughout; and

Table 1. Plant traits measured (mean value per ramet)

SHL: total shoot length (m)
BN: number of lateral branches
LAS: submerged leaf area (cm^2)
LAF: floating leaf area (cm^2)
RL: total root length (m)
RL/SL: ratio of root:shoot length
RON: number of reproductive structures (total individual flowering heads, flowering and/or fruiting structures or other relevant structures)
SB: stem biomass (mg)
LSB: submerged leaf biomass (mg)
LFB: floating leaf biomass (mg)
RB: root biomass (mg)
SHBROB: ratio of shoot:root biomass
ROB: biomass of reproductive structures (mg)

in the 1997 surveys, after analysis of the 1996 data had shown that intra-site variability was lower than expected).

Water samples were collected to give snapshot indications of conditions at each site. The data were combined with longer-term information provided by the Scottish Environmental Protection Agency for nearest upstream routine sampling point to the Scottish sites or held by the authors for Egyptian sites. Physico-chemical data measured were (*in situ*) pH; underwater light attenuation, k (using a twin-sensor Skye PAR meter); conductivity; dissolved oxygen; and water flow (using a SENSA electromagnetic flow meter to record flow at 1 m intervals across the channel at the Scottish sites, but recorded only by visual estimation at the Egyptian sites, on a three point scale: (approximately: slow <0.2 ms^{-1}; moderate c. 0.2–0.4 ms^{-1}; fast >0.4 ms^{-1}). Other variables measured were soluble reactive phosphate, total oxidised nitrogen and potassium, determined using standard methods (APHA, 1980).

Data were analysed using GENSTAT for WINDOWS (Payne et al., 1993), CANOCO, MINITAB and TWINSPAN software. Datasets were tested for normality using Ryan-Joiner test procedures and appropriate transformations applied to normalize individual datasets, where required. Canonical correspondence analysis, and both non-hierarchical (clustering) and hierarchical (TWINSPAN) classification methods, were used to define functional groups of plants, and explore their relationship to environmental conditions (Jongman et al., 1995). Multiple regression procedures were used to relate trait datasets to flow and trophic variables. Standard testing of residuals for normality and randomness against fitted values was always carried out on completion of any modelling process. In circumstances where the residuals were considered inadequate, the model was culled from the process. Details are given where relevant in the 'Results'.

Results

In total, 1150 leaf and root images were scanned during the course of the study, from a total of 604 ramets of 33 macrophyte species growing at the initial and test sets of sites. Raw data on the range of plant traits measured from these populations and environmental data recorded by us at survey sites are available on request from the authors.

Table 2. Species membership of river macrophyte functional groups for Scottish/Egyptian dataset

Group 1	**Group 4**
Eichhornia crassipes	*Persicaria amphibia*
Hottonia palustris	*Potamogeton coloratus*
Najas horrida	*Potamogeton natans*
Najas marina subsp. *armata*	*Potamogeton polygonifolius*
Oenanthe fluviatile	*Ranunculus aquatilis*
Pistia stratiotes	*Ranunculus peltatus*
Potamogeton x *nitens*	*Ranunculus penicillatus*
Ranunculus circinatus	
Ranunculus fluitans	**Group 5**
Ranunculus trichophyllus	*Callitriche hamulata*
Vallisneria spiralis	*Callitriche obtusifolius*
	Callitriche platycarpa
Group 2	*Callitriche stagnalis*
Ceratophyllum demersum	
Elodea canadensis	**Group 6**
Myriophyllum alterniflorum	*Nuphar lutea*
Myriophyllum spicatum	*Nuphar pumila*
Myriophyllum verticillatum	*Nymphaea alba*
Potamogeton berchtoldii	*Sagitarria sagittifolia*
Potamogeton crispus	
Potamogeton obtusifolius	**Group 7**
Potamogeton pectinatus	*Eleocharis acicularis*
Potamogeton praelongus	*Eriocaulon septangulare*
Potamogeton pusillus	*Isoetes lacustris*
Potamogeton trichoides	*Juncus bulbosus*
Utricularia intermedia	*Littorella uniflora*
Utricularia vulgaris	*Lobelia dortmanna*
Zannicellia palustris	*Subularia aquatica*
Group 3	**Group 8**
Glyceria declinata	*Azolla filiculoides*
Glyceria fluitans	*Hydrocharis morsus-ranae*
Potamogeton gramineus	*Lemna gibba*
Potamogeton lucens	*Lemna minor*
Potamogeton nodosus	*Lemna trisulca*
Potamogeton perfoliatus	*Spirodela polyrhiza*
Potamogeton x*sparganifolius*	
Sparganium angustifolium	
Sparganium emersum	

Definition of river plant functional groups

The approach was based on the method developed by Abernethy (1994) and given in full by Demars (1997). A set of 63 macrophyte species (including those found in European river sites surveyed in 1992–94 (Abernethy, 1994), plus those additionally recorded during the 1996–97 survey) was clustered into homogeneous groups in terms of their possession or otherwise of a set of 35 fuzzy traits on morphology, physiology and life-history attributes (information derived from literature searches undertaken primarily by Abernethy

(1994) and Demars (1997), with additional information added for Egyptian species). As no assumptions were made about the relative adaptive significance of traits, non-hierarchical analysis was appropriate (Gordon, 1981; Gauch, 1982). GENSTAT was used to run the analysis with the clustering criteria set by Total Sum of Squares intra groups (S). The initial partition was improved (Payne et al., 1993) by incorporating a starting set of groups (g) derived from a prior TWINSPAN classification (cut at the fourth level of division). The stopping rule of Krzanowski & Lai (1988) was used to define the optimal number of groups, showing as peaks in the statistic Cg at individual group-numbers. The analysis revealed strong peaks of Cg at g = 5 and g = 8, suggesting that either 5 or 8 species-groups could usefully be identified within the dataset. In order to select one of them, the analysis was run again using TWINSPAN-derived initial groups (primarily resulting from the second level of division) and allowing the clustering procedure to reallocate species at g = 8 (Jongman et al., 1995). This produced a weaker Cg peak at 7 groups, with the results more suggestive of a continuum in the data set. Comparisons of values of S produced by differing starting points suggested that overall a division based on combined analysis produced the most ecologically-meaningful results (Table 2). Although it is the best currently available, we have used this functional groups classification here on a provisional basis only, as an illustration of how the approach might be used once further refined to include a wider range of species.

Modelling trophic status using macrophyte species assemblage: use of the Macrophyte Trophic Rank (MTR) procedure

The MTR system (Standing Committee of Analysts, 1987; Environment Agency, 1996) is strictly only applicable in its current form to British rivers, although very similar schemes are in use elsewhere, e.g. France: Haury et al. (1996). Given the large number of species in common between the Scottish and Egyptian datasets, we tried out the MTR scheme against the full dataset (including all sites with >2 species given in the MTR listings). MTR values for each site were calculated (Environment Agency, 1996) and correlated against mean P data. This gave a significant though noisy relationship:

Model 1. P predicted from MTR: all sites with >2 species given in the MTR listings ($n = 29$: $r = -0.585$, $p<0.01^{**}$):

$$\log_e P = 6.89 - 0.449(sqrt\text{MTR}),$$

where n = number of samples included in data set; r: correlation coefficient; p: probability.

Table 3. Egyptian and Scottish river and channel sites (Type: *R* = river; D = drainage channel; I = irrigation channel; * = test site)

Site (region)	Type	Code (see Figure 1b)
Egypt		
Balana (Aswan)	D	BA
Nasser El-Nuba (Aswan)	D	NA
Kalabsha (Aswan)	I	KA
R. Nile: El-Ramady (Aswan)	R	RA
R. Nile: Aswan West Village (Aswan)	R	RNWA
R. Nile: New Island (Aswan)	R	RNNI
Ras El-Bar - Damietta (Delta)	I	RASD
Balamon - Sinania - Damietta (Delta)	I	SIND
R. Nile: Damietta (Delta)	R	RND
Hamam - Damietta (Delta)	D	HAMD
Balamon - Sherbin - Damietta (Delta)	I	SHD
Sherbin - Damietta (Delta)	D	SHW
Kafr Tawila (Delta)	D	KTD
R. Nile: (Mansoura)	R	RNM
Scotland		
Blane Water (Strathclyde)	R	BLA
Mouse Water (Lanarkshire)	R	MOUS
Dippool Water (Lanarkshire)	R	DIPP
Gryfe (Renfrewshire)	R	GRY
R. Don (Aberdeen)	R	DON
R. Ythan (Grampian)	R	YTH
South Ugie Water (Grampian)	R	SUG
North Ugie Water (Grampian)	R	NUG
R. Spey (Strathspey)	R	SPEY
R. Lossie (Nairn)	R	LOS
Muckle Burn (Nairn)	R*	MB
Forres Burn (Nairn)	R*	FB
R. Earn (Perthshire)	R*	RE
Eddleston Water (Borders)	R*	EDW
Ettrick Water (Borders)	R*	ETW
R. Teviot (Borders)	R*	TEV
R. Tweed (Borders)	R*	TWE
R. Lonan (Argyll)	R*	LON

Modelling trophic status using macrophyte functional variables: development of River Trophic Status Indicator (RTSI) scheme

The scheme developed had two elements. The first was

based upon macrophyte functional group representation at individual river sites and the relationships of river macrophyte functional groups to P concentration in river (or channel) water. This was termed $RTSI_{FG}$. The second approach was based upon regression models of the relationships between measured functional traits of macrophyte populations present at river sites and P concentration in river (or channel) water. This was termed $RTSI_{TR}$.

$RTSI_{FG}$: development of macrophyte functional groups model for predicting river trophic status

The first element of the indicator scheme was based on the approach developed by Abernethy (1994) relating macrophyte functional group (FG) membership to environmental conditions. The functional group membership of individual species at each site was used as input data for a CCA ordination of sites × functional group composition, weighted by plant abundance score (i.e. each site was characterised by the total abundance of species from each of seven FGs represented by at least one species in this dataset (FG6 was not represented). Mean values (1995–96) for environmental variables collected for each site, including soluble reactive phosphate (P) and total oxidised nitrogen (N) concentrations in water, were used to ordinate functional groups in relation to environmental variables. Flow was input as a three-class categorical variable: FR, equivalent to category 1; MFR, category 2; HFR, category 3, as outlined above. The CCA ordination results are shown in Figure 1a for functional groups and Figure 1b for sample site ordination (site codes are listed in Table 3). Ranked positions of the seven FGs were identical along the P and N gradients, which had high affinity with Axis 1. High flow had a high affinity with Axis 2. Scores (nutrient rank score: N_r) were allocated to each FG in order of position along the gradients from high to low P or N:

FG	Rank score (N_r) on nutrient (N and P) gradient	
8	7	High nutrient affinity
2	6	
7	5	↓
3	4	
1 and 5	2	
4	1	Low nutrient affinity

The nutrient rank score (N_r) per species was weighted by % plant abundance score (A), summed, and divided by number of species (n) present to give a raw river trophic score index RTSI. Testing showed that taking account of variation in flow improved the index. This was simply achieved by dividing by flow class value (F) to give final values of $RTSI_{FG}$. To normalise this dataset a $\log_e$ transformation was required. In summary:

$$RTSI_{FG} = \log_e[\Sigma(N_r.A/n)/F].$$

This simple index, based on the occurrence of functionally-defined sets of river and channel macrophytes, takes account of relative abundance of species from those FGs present at individual river and channel sites, and also incorporates the likely influence of flow upon the functional set of species present, and their abundance. How well does it predict nutrient status?

After $\log_e$ transformation to normalise the data, a significant relationship was found between $RTSI_{FG}$ and P status at those sites where >1 macrophyte species occurred:

Model 2. P predicted from $RTSI_{FG}$: all sites with >1 macrophyte species present (n = 32: r = 0.522; p<0.01):

$$\log_e P = 0.72 + 0.858\,RTSI_{FG}$$

This is a noisy relationship, which fits the data slightly less well than using species assemblage data (Model 1).

$RTSI_{TR}$: development of macrophyte traits model for predicting river trophic status

A multiple regression approach was used to assess the relationships between sets of trait data, measured in the field for the macrophyte populations present at individual sites. The average value of each trait (see Table 1 for plant trait codes) per site was calculated for ramets of each species present at the site, and the model was calculated using these site-averaged trait data. The optimal predictive model for P (as annual mean soluble reactive phosphorus concentration in water) using macrophyte trait predictors is given below.

Model 3. P predicted from $RTSI_{TR}$ variables: ramets from all sites, averaged across species present per site (n = 31, r = 0.57, p<0.001***)

$$sqrt\,P = 0.495 - (0.0688 \log_e \text{BN}) - (0.0342 \log_e \text{LSB}) + (0.262 \sin \text{LFB})$$

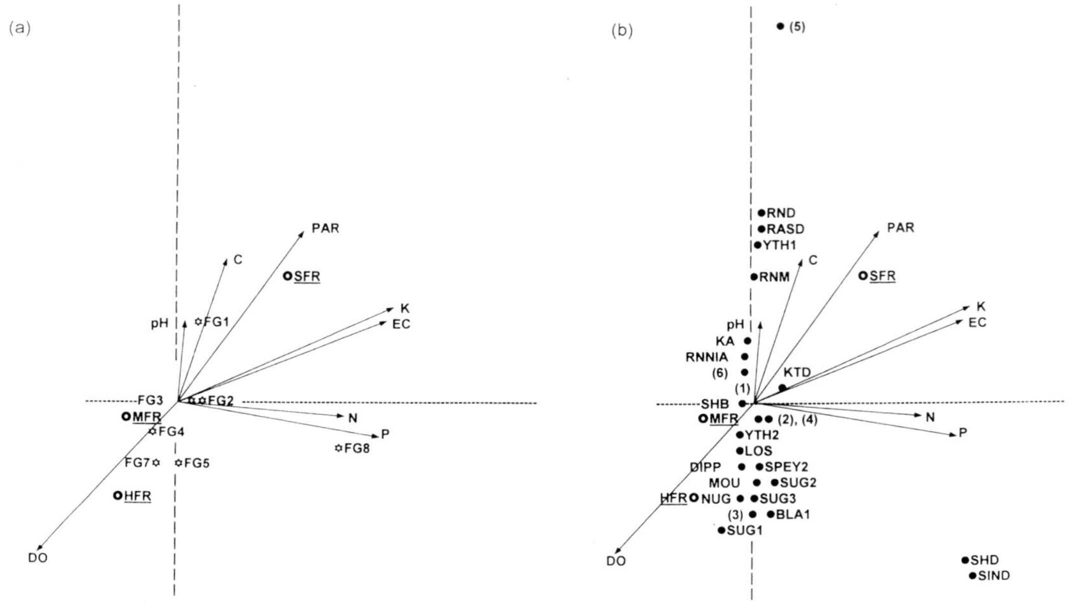

Figure 1. CCA ordination diagram shows: (a) ranking of the eight macrophyte functional groups (⊛FG) obtained from TWINSPAN classification along nutrient (N = total oxidised nitrogen and P = soluble reactive phosphate) gradient in relation to other environmental variables measured (arrows): C = water temperature, pH = hydrogen ion concentration, PAR = underwater light attenuation (k), EC = electric conductivity, DO = dissolved oxygen concentration, K = potassium concentration, and flow rate classes: ◎SFR = slow (approximately: very slow <0.2 ms^{-1}), ◎MFR = moderate (c. 0.2–0.4 ms^{-1}, or ◎HFR = fast (>0.4 ms^{-1}); (b) distribution of Egyptian and Scottish sites (●) along the environmental variables measured.

The combined RTSI model for predicting river trophic status

The final stage in development of the RTSI model was to combine the two approaches used into a single predictive model. This was done on a site-by-site basis using predictor variables from Models 2 and 3: three sites were excluded (compared with Model 2) because although $RTSI_{FG}$ data were available, traits were not measured for plant populations present at these sites so the required $RTSI_{TR}$ variables were unavailable. The size of the dataset was, therefore, reduced to n=29. Despite the reduced size of the input dataset, the final RTSI model, incorporating both functional group- and trait-based predictors, proved to be the most powerful predictive model for P of those developed in this study:

Model 4. P predicted from $RTSI_{FG}$ and $RTSI_{TR}$ variables averaged across species present per site ($n = 29$, $r = 0.72$, $p<0.001$***)

$$sqrt P = -0.167 + (0.150\,\mathrm{RTSI_{FG}}) + (0.0052 \log_e \mathrm{BN}) - (0.0399 \log_e \mathrm{LSB}) + (0.220 \sin \mathrm{LFB})$$

The final model predicts just over half the variability in P concentration in river and channel water of the Egyptian and Scottish sites used. This is a substantial improvement on the predictive success of either the MTR method (Model 1: 34%), or any of the the individual component RTSI models (Models 2 and 3) which individually account for between 27 and 32% of variation in P.

Testing of models

Soluble reactive phosphorus concentrations in water were allocated on OECD (1982) criteria for flowing water to trophic bands (oligotrophic OLIGO < 10; mesotrophic MESO 10–20; meso-eutrophic MESO-EU 21–40; eutrophic EU 41–100; hypertrophic HYP >100 $\mu g\ l^{-1}$P). Testing of Models 1 and 4 against the trait sets collected from species present at an independent set of seven Scottish river sites, sampled in August 1997, found that the trophic band (derived from mean P value calculated from chemical monitoring data) was correctly predicted by RTSI (Model 4) about half the time and when incorrect the predicted band was usually only one category out (e.g. MESO-EU instead of EU). MTR (Model 1) predicted trophic band correctly about one third of the time, and when wrong, was sometimes more than one band out. For example, both models correctly estimated the trophic band of the Ettrick Water (MESO-EU), but while Model 4 correctly assigned the Forres Burn to the EU category, Model 1 underestimated this site as MESO-EU. Both models underestimated (MESO-EU) the actual eutrophic status of the River Earn.

Discussion

For some macrophyte species, the sediment is a potentially important source of P and macrophyte-sediment interactions play an important role in the functional ecology of many freshwater ecosystems (e.g. Barko et al., 1991; Moore et al., 1994). In a previous study of the macrophyte vegetation of polluted sections of the R. Nile, we found that sediment P was more strongly associated than water P with macrophyte community composition (Ali et al., 1995). For other species, such as free-floating plants (e.g. *Lemna gibba*, *Eichhornia crassipes*) which have roots in contact only with water under normal circumstances, sediment P is probably of lesser functional importance. Recent work in France with weakly-rooted macrophytes such as *Elodea nuttallii* has provided evidence that shoot uptake of P is related to the soluble reactive phosphorus content of the water, regardless of whether the plant is rooted or not (Robach et al., 1996). In order to obtain a better model of how the trophic status of river and channel systems is related to macrophyte functional characteristics, it seems essential to investigate both water and sediment nutrient conditions. Further work is in progress to model the relationships between sediment concentrations of P and RTSI predictor variables, based on the current dataset.

The functional models developed could predict P conditions in the water, within rather noisy limits but generally to at least the same degree of precision as, or better than existing species-assemblage-based methods, such as the MTR approach adopted by the Environment Agency (1996) in the U.K. MTR is based on much larger data sets and has a much longer history of use and refinement than our new approach. We conclude that models of river and channel P concentration in water, utilising predictor variables derived from plant functional group membership ($RTSI_{FG}$) and field-measured morphological traits of macrophyte populations present in the river ($RTSI_{TR}$ variables), appear to provide a potentially useful means of assessing river trophic status.

Acknowledgements

We acknowledge with thanks the contribution of Benoit Demars and Dr Nigel Willby in developing the macrophyte functional groups classification on which this study is in part based. We also thank Dr M. E. Soltan, Dr Ricardo Sabbatini, Andrew Morrison, Benoit Demars and Margaret McCormick for their invaluable assistance of field and laboratory work. In Egypt, we are very grateful for the help and assistance provided by Dr Abdel-Hamid Khedr and Mohammed Mousa at Mansoura University in carrying out the field survey work in the Delta area. The work was funded by the Natural Environment Research Council (GR9/02474) and British Ecological Society (Grant No. 1326).

References

Abernethy, V. J. A., 1994. The functional ecology of euhydrophyte communities of European riverine wetland ecosystems. Ph.D. Thesis, Univ. Glasgow: 265 pp.

Agami, M., 1989. Effects of water pollution on plant species composition along the Amal River, Israel. Arch. Hydrobiol. 100: 445–454.

Ali, M. M. & M. E. Soltan, 1996. The impact of three industrial effluents on submerged aquatic plants in the River Nile, Egypt. Hydrobiologia 340: 77–83.

Ali, M. M., A. Hamad, I. V. Springuel & K. J. Murphy, 1995. Environmental factors affecting submerged macrophyte communities in regulated waterbodies in Egypt. Arch. Hydrobiol. 133: 107–128.

APHA (American Public Health Association), 1980. Standard Methods for the Examination of Water and Waste Water. 16th edn. American Public Health Association. New York: 1268 pp.

Barko, J. W., D. Gunnison & S. R. Carpenter, 1991. Sediment interactions with submerged macrophyte growth and community dynamics. Aquat. Bot. 41: 41–65.

Bini, L. M., S. M., Thomaz, K. J. Murphy & A. F. M. Camargo, 1999. Aquatic macrophyte distribution in relation to water and sediment conditions in the Itaipu Reservoir, Brazil. Hydrobiologia 415 (Dev. Hydrobiol. 147): 147–154.

Caffrey, J. M., 1985. A scheme for the assessment of water quality using aquatic macrophytes as indicators. J. Life Sci. R. Dubl. Soc. 5: 105–111.

Caffrey, J. M., 1986. Macrophytes as biological indicators of organic pollution in Irish rivers. In Richardson, D. H. S. (ed.), Biological Indicators of Pollution. Royal Irish Academy, Dublin: 77–87.

Daniel, H. & J. Haury, 1996. Ecology of aquatic macrophytes in an Armorican river (the River Scorff, Southern Brittany, France), application to bioindication. Ecologie (Brunoy) 27: 245–256.

De Lange, L. & J. C. J. Van Zon, 1983. A system for the evaluation of aquatic biotopes based on the composition of the macrophytic vegetation. Biol. Conserv. 25: 273–284.

Demars, B., 1997. Classification des hydrophytes sur la base des traits de l'histoire de vie. Thesis, DEA, Univ. Paris XI and Univ. Glasgow: 25 pp.

Ellenberg, H., 1973. Chemical data and aquatic vascular plant as indicator for pollution in the Moosach river system near Munich. Arch. Hydrobiol. 72: 533–549.

Environment Agency, 1996. Methodology for the assessment of freshwater riverine macrophytes for the purposes of the Urban Waste Water Treatment Directive. Version 2. Environment Agency, Bristol, U.K: 34 pp.

Gauch, H. G. 1982. Multivariate analysis in community ecology. Cambridge University Press, Cambridge, U.K: 298 pp.

Gordon, A. D., 1981. Classification: Methods for the Exploratory Analysis of Multivariate Data. Chapman & Hall, London: 193 pp.

Grime, J. P., J. G. Hodgson & R. Hunt, 1988. Comparative plant ecology. Unwin Hyman, London, 742 pp.

Haslam S.M. & P.A. Wolseley, 1981. River Vegetation: its Identification, Assessment and Management. Cambridge University Press, Cambridge.

Haslam, S. M., J. P. C. Harding & D. H. N. Spence, 1987. Methods for the use of aquatic macrophytes for assessing water quality 1985–86. In Methods for Examination of Waters and Associated Materials. HMSO, London.

Haury, J., 1996. Assessing functional typology involving water quality, physical features and macrophytes in a Normandy river. Hydrobiologia 340: 43–49.

Haury, J. & M.-C. Peltre, 1993. Intérêts et limites des 'indices macrophytes' pour qualifier la mésologie et la physico-chimie des cours d'eau: exemples armoricains, picards et lorrains. Annls. Limnol. 29: 239–253.

Haury, J., M.-C. Peltre, S. Muller, M. Tremolieres, A. Dutartre & M. Gherlesquin, 1996). Macrophyte indices for the assessment of stream water quality in France: preliminary proposals. Ecologie (Brunoy) 27: 233–244.

Hills, J. M., K. J. Murphy, I. D. Pulford & T. H. Flowers, 1994. A method for classifying European riverine wetland ecosystems using functional vegetation groups. Funct. Ecol. 8: 242–252.

Holmes, N. T. H., 1983. Focus on Nature Conservation. 4. Typing British Rivers according to their flora. Nature Conservancy Council, Shrewsbury, U.K.

Husak, S. & V. Vorechovska, 1996. Stream vegetation in different landscape types. Hydrobiologia 340: 141–145.

Jongman, R. G. H., C. J. F. Ter Braak & O. F. R. van Tongeren, 1995. Data Analysis in Community and Landscape Ecology. Cambridge University Press, Cambridge: 299 pp.

Kohler, A., 1975. Macrophytische Wasserpflanzen als Bioindikatorn für Belastungen von Fliessgewässerökosystemen. Verh. Ökologie, Wien 3: 255–276.

Kohler, A. & S. Schiele, 1984. Versauerungsresisternz submerser Makrophyten in Gewässerversauerung in der Bundesrepublik Deutschland. Matierialien 1/84. Erich Scmidt Verlag, Berlin: 353–369.

Krzanowski, W. J. & Y. T. Lai, 1988. A criterion for determining the number of groups in a data-set using sum-of-squares clustering. Biometrics 44: 23–34.

Leishman, M.R. & M. Westoby, 1992. Classifying plants into groups on the basis of associations of individual traits – evidence from Australian semi-arid woodlands. J. Ecol. 80: 417–424.

McIntyre, S., S. Lavorel & R. M. Tremont, 1995. Plant life-history attributes: their relationship to disturbance response in herbaceous vegetation. J. Ecol. 83: 31–44.

Moore, B. C., J. E. Lafer & W. H. Funk, 1994. Influence of aquatic macrophytes on phosphorus and sediment porewater chemistry in a freshwater wetland. Aquat. Bot. 49: 137–148.

Murphy, K. J., B. Rørslett & I. Springuel, 1990. Strategy analysis of submerged lake macrophyte communities: an international example. Aquat. Bot. 36: 303–323.

Newbold, C. & N. T. H. Holmes, 1987. Nature conservation: water quality criteria and plants as water quality monitors. Water Pollut. Contr. 86: 345–364.

OECD, 1982. Eutrophication of waters: monitoring, assessment and control. Organisation for Economic Cooperation and Development, Paris, France.

Payne, R. W., P. W. Lane, P. G. N. Digby, S. A. Harding, P. K. Leech, G. W. Morgan, A. D. Todd, R. Thompson, G. Tunnicliffe Wilson, S. J. Welham & R.P. White, 1993. GENSTAT 5 Release 3 – Reference Manual. Clarendon Press, London: 796 pp.

Robach, F., S. Merlin, T. Rolland & M. Tremolieres, 1996. Ecophysiological approach of water quality bioindicating using aquatic plant materials: the role of phosphorus. Ecologie (Brunoy) 27: 203–214.

Schmedtje, U. & F. Kohmann, 1987. Bioindication by macrophytes – can macrophytes indicate saprobity? Arch. Hydrobiol. 109: 455–469.

Spink, A. & K. J. Murphy, 1997. Distribution and environmental regulation of Batrachian *Ranunculus* in British rivers. Arch. Hydrobiol. 139: 509–525.

Springuel, I. & K. J. Murphy, 1991. Euhydrophyte communities of the River Nile and its impoundments in Egyptian Nubia. Hydrobiologia 218: 35–47.

Standing Committee of Analysts, 1987. Methods for the use of aquatic macrophytes for assessing water quality 1985–86. HMSO, London.

Hydrobiologia **415:** 139–145, 1999.
J. M. Caffrey, P. R. F. Barrett, M. T. Ferreira, I.S. Moreira, K. J. Murphy & P. M. Wade (eds), Biology, Ecology and Management of Aquatic Plants.
© 1999 *Kluwer Academic Publishers. Printed in the Netherlands.*

Vegetation dynamics in a fishpond littoral related to human impact

Zdenka Hroudová & Petr Zákravský
Institute of Botany, Academy of Sciences of the Czech Republic, CZ-252 43 Průhonice, Czech Republic
E-mail: hroudova@ibot.cas.cz

Key words: wetlands, vegetation mapping, reeds, trophic level, fishpond management, Czech Republic

Abstract

To determine how fishpond vegetation reflects human activities, changes of littoral vegetation of the Opatovický fishpond (South Bohemia) were studied over a period of 26 years. The littoral vegetation of this fishpond has been mapped in 1992 and 1997, and compared with a former study in 1971 together with relation to the soil chemistry of the bottom sediments. The following factors influence the vegetation dynamics: fluctuations of water level, trophic level of habitat, changes in management (mowing) and mechanical disturbances (scraping). The expansion of reed stands (*Phragmites australis*), disappearance of small sedge stands and a decrease in the area of *Glyceria maxima* stands were the main characteristics of the vegetation dynamics within the period studied; most other plant communities remained stable. In addition, successional development on scraped shores led to the re-establishment of wetland communities (positive effect) or to ruderal vegetation on islands of deposited material (negative effect). The fishpond may be characterized as meso- to slightly eutrophic and was found to be relatively resistant to the influence of local sources of pollution: the total trophic level (total N, nitrates, phosphates) did not increase at all. As the fishpond has been proposed as a nature reserve for valuable plant communities and water bird diversity, fishery management and human activities in the surrounding area need to be controlled.

Introduction

The Opatovický fishpond near Třeboň (South Bohemia, 48 °59′ N, 14 °46′ E, alt. 433 m, water area 165 ha) is an example of a reservoir influenced during recent decades by multiple human impact – fish production, camping sites and agricultural activities in the surrounding land. At the same time, valuable plant communities have persisted in its littoral providing suitable biotopes for the breeding and resting of water birds and the fishpond has been proposed as a nature reserve. Therefore human impact needs to be controlled.

The littoral vegetation was studied three times within a long period (in 1971, 1992 and 1997) and compared with habitat conditions which may have been responsible for the vegetation changes. Intensive investigations of the vegetation, especially the reed stands, were undertaken in the 1970s within the scope of the International Biological Programme (Hejný, 1973). The littoral vegetation was mapped and the soil chemistry studied in 1971 (Dykyjová & Hejný, 1975). These data provide a good basis for the evaluation of vegetation dynamics and changes of trophic level of bottom sediments in the long-term, and can be compared with our investigations undertaken in 1992 (Hroudová & Zákravský, 1995) and 1997 (Hroudová et al., 1999). The aim is: 1. to find qualitative and quantitative changes in littoral plant communities and total plant diversity and relate them to soil chemistry; 2. to evaluate the trophic level of the fishpond; 3. to find possible local sources of pollution; and 4. to evaluate the results in order to make recommendations for nature conservation and fishpond management.

Methods

The vegetation of the fishpond littoral was mapped in 1992 and 1997. The areas occupied by plant communities were recorded in the field on a map on a scale of 1:5000. Aerial photos from a height of about 1500 m were used to map precisely the shoreline and boundaries of individual plant communities. The areas occupied by plant communities were measured by planimeter. Plant communities were characterized by phytocoenological relevés using the 7-point Braun-Blanquet scale (Klika et al., 1954). The floristic composition of the vegetation was also recorded in those sites where the vegetation was not sufficiently homogeneous for phytocoenological relevés or where only fragments of the communities were developed. The nomenclature of vascular plants follows Rothmaler (1982), subspecies of *Bolboschoenus maritimus* sensu Casper & Krausch (1980). Author associations were given according to the survey of plant communities of the Czech Republic (Moravec, 1995).

Soil samples of bottom sediments were taken from the rhizosphere (0–25 cm) and from the ground layer (below 25 cm) in 1992–1993. Sampling sites were chosen in relation to the extent of main plant communities and also in these sites where plants seemed to indicate sources of pollution: in stands of *Phragmites australis* (7 sampling sites), *Glyceria maxima* (6), *Carex gracilis* (2), *Schoenoplectus lacustris* (2), in one sampling site in the stands of *Typha angustifolia*, *Calamagrostis canescens*, *Bolboschoenus maritimus*, *Acorus calamus* and in small lagoon near the shore - altogether, 22 sampling sites. In 1997 only samples from the rhizosphere could be taken due to the high water level; the samples were taken in stands of *Phragmites australis* (9 sampling sites), *Glyceria maxima* (5), *Carex gracilis* (3), *Schoenoplectus lacustris* (2), one sampling site in the stands of *Typha angustifolia* and *Calamagrostis canescens* and on three sites where former littoral stands had disappeared – altogether, 24 sampling sites. 8 of the sampling sites were the same as in 1967–1971 (Dykyjová & Hejný, 1975). The following analyses were performed: pH in the water and in 1 mol KCl – electrometrically; exchangeable Ca, Mg, K and Na by atomic absorption spectrophotometry; $N\text{-}NH_4$- colorimetrically using Nessler's reagent (Hraško et al., 1962); $P\text{-}PO_4$, colorimetrically using extraction with sodium bicarbonate (Olsen et al., 1954); total N and C – using CHN analyzer by Heraeus; and NO_3, SO_4^{2-} and Cl^- using isotachophoresis by ITP analyzer (Everaerts et al., 1976).

Results

Vegetation changes

The following plant communities were found in 1992 and 1997 in the littoral of the Opatovický fishpond (see Figure 1): All. *Phragmition communis* Koch 1926: *Scirpetum lacustris* Schmale 1939, *Phragmitetum communis* (Gams 1927) Schmale 1939, *Glycerietum maximae* Hueck 1931, *Typhetum angustifoliae* Pignatti 1953, *Typhetum latifoliae* Lang 1973, *Sparganietum erecti* Roll 1939, *Acoretum calami* Eggler 1933; All. *Caricion gracilis* Neuhäusl 1959 em. Balátová-Tuláčková 1963: *Caricetum gracilis* Almquist 1929, *Calamagrostidetum canescentis* Simon 1960, *Phalaridetum arundinaceae* Libbert 1931, unclassified units: *Bolboschoenus maritimus* subsp. *maritimus*-community, *Deschampsia caespitosa*-community; regenerating mixed stands.

Table 1. The extent of stands (in hectares) of dominant plant species in 1971 (Dykyjová & Hejný, 1975), 1992 and 1997

Stands	1971	1992	1997
Phragmites australis	5.70	13.85	18.50
Glyceria maxima	9.50	5.20	5.70
Carex gracilis	–	6.10	5.50
Calamagrostis canescens	–	0.65	1.75
Phalaris arundinacea	–	1.05	0.80
Typha angustifolia	1.60	1.45	1.60
Typha latifolia	0.21	0.10	0.15
Schoenoplectus lacustris	0.45	0.65	0.60
Sparganium erectum	0.01	–	0.02
Bolboschoenus maritimus	3.90	10.40	–
Acorus calamus	0.05	0.05	–
Deschampsia caespitosa	–	0.05	0.22
mixed regenerating stands	–	2.35	0.22
Carex gracilis + *Cariceta* div.	14.30	–	–

From 1971 to 1997 the following vegetation changes occurred (Table 1):

1. stands of small sedges disappeared;
2. stands of *Calamagrostis canescens, Deschampsia caespitosa, Phalaris arundinacea* and mixed regenerating stands were newly recorded in 1992 and 1997,
3. stands of *Phragmites australis* increased;
4. stands of *Glyceria maxima* decreased, partly due to scraping;
5. the area of *Bolboschoenus maritimus* fluctuated markedly, and *Acorus calamus* and *Sparganium erectum* stands occurred irregularly over small areas;
6. areas of other communities (*Typha angustifolia, T. latifolia, Schoenoplectus lacustris*) were relatively stable.

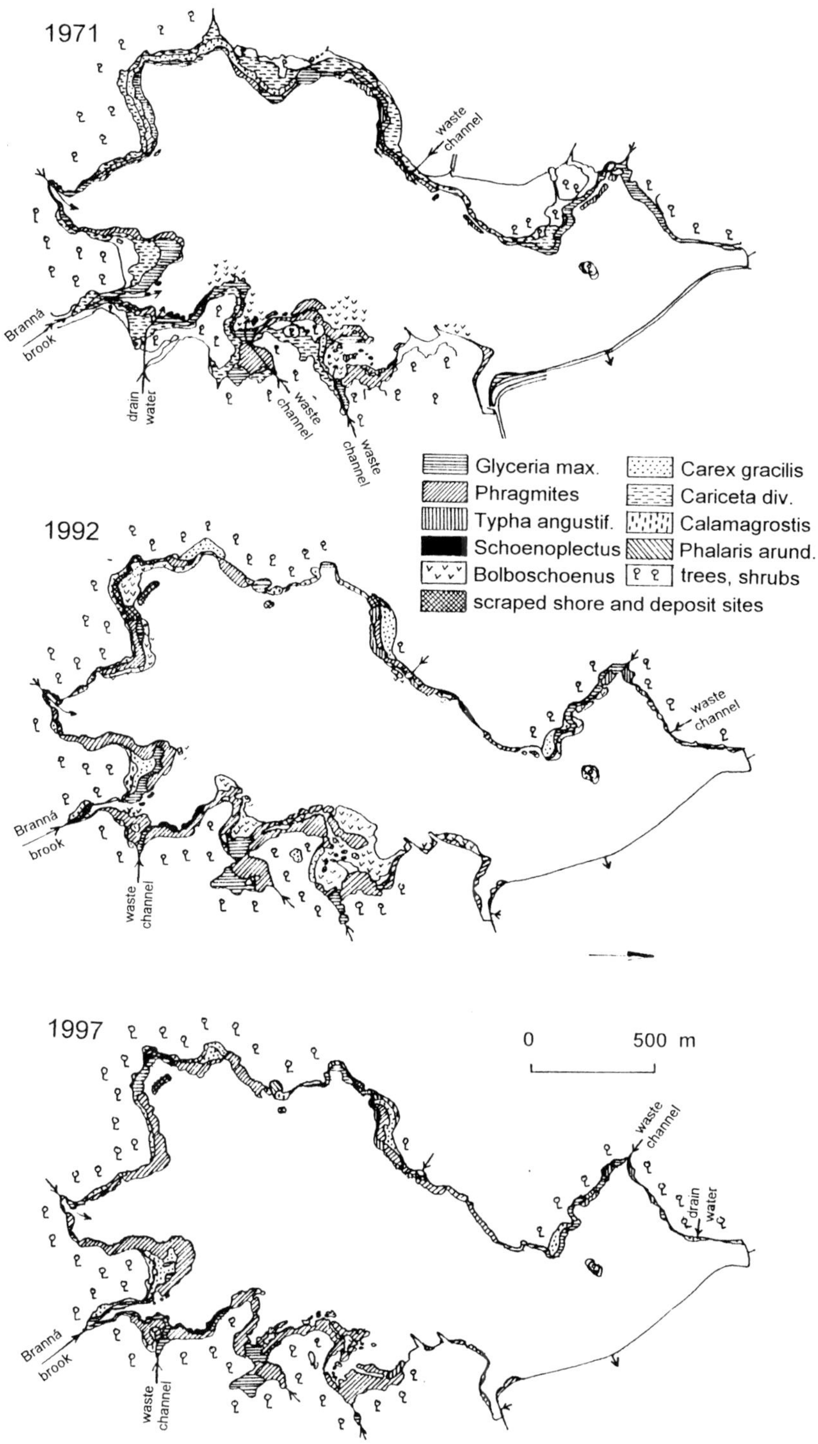

Figure 1. Vegetation map of the Opatovický fishpond in 1971 (adapted from Husák & Hejný, 1978), in 1992 and 1997. The following main vegetation units are shown: stands of *Glyceria maxima, Phragmites australis, Typha angustifolia, Schoenoplectus lacustris, Bolboschoenus maritimus* subsp. *maritimus*, *Carex gracilis, Cariceta* div., *Calamagrostis canescens, Phalaris arundinacea*, vegetation of scraped shores and deposit sites of bottom material. Local sources of pollution are marked.

Sedge stands in 1971 and in the 1990s were not fully comparable, because the unit *Cariceta* div. in 1971 contained stands of various sedges (*Carex gracilis, C. nigra, C. rostrata*) and probably also stands of *Calamagrostis canescens*. In 1992, only stands of *Carex gracilis* were found and *C. rostrata* and the small sedges had disappeared. Only small fragments of the acidophile community with *Juncus filiformis, Agrostis canina, Comarum palustre, Peucedanum palustre, Carex gracilis* and *C. vesicaria* persisted within the willow stands (*Salix aurita, S. cinerea*).

The most extensive reed-bed stands occur in the south-eastern part of the fishpond. Zoning corresponding to the slope of the shore has developed there: willow carr (*Salix aurita, S. cinerea*), *Calamagrostis canescens, Carex gracilis, Glyceria maxima, Phragmites australis* and open water. Scraping of muddy bottom sediment was performed twice: in the 1970s and in 1990. Sediments were removed by bulldozer on several parts of banks and scraped material deposited into fishpond near the shore, forming small islands (see Figure 1). Mixed regenerating stands appeared in 1992 on exposed bottoms uncovered by scraping and on islands created by deposited bottom material. Those stands developed after scraping on a flat sheltered shore (a slope of ca. 1°30′) into littoral plant communities during the following 5 years (1992 to 1997), providing favourable conditions for many species, including some rare and endangered plants (*Utricularia australis, Lysimachia thyrsiflora, Scirpus radicans*). On the other hand, on more sloping wave – exposed banks (ca. 5°) littoral stands did not regenerate. The islands created by deposited bottom material remained vegetation – free or became covered by weed species (*Urtica dioica, Cirsium palustre, C. arvense, Senecio viscosus*).

A total of 125 vascular plant species were found in 1992 and 1997; 86 of them were present in both years. The differences were a result of different water levels (presence of species of exposed bottoms in 1992). The proportion of ruderal and weed species increased from 7 in 1992 to 16 in 1997.

Soil chemistry

The changes of trophic level were small within the whole period (Table 2). However, a somewhat different analytical method was used in 1971 (Dykyjová & Hejný, 1975), resulting in higher values measured, and thus these values are not fully comparable to those from 1992 and 1997. The trophic level of the sediments had increased slightly in the 1970s, as a result of increased nutrient supply to the fishpond (Dykyjová & Hejný, 1975), and nutrient infiltration into the ground soil layer continued during the period 1971–1992 (Hroudová & Zákravský, 1995). The total trophic level of sediments did not increase at all until 1997, and the fishpond may be considered as mesotrophic to slightly eutrophic. A considerable decrease of the carbon and nitrogen content resulted from the removal of mud by scraping in 1990. Except for several terrestrial sites, the littoral sediments were low in organic matter, partly due to scraping of the bottom. Low pH values reflected acid, base-poor ground, typical of the whole Třeboň basin (Ložek, 1978) and low salt input (low content of basic cations and sulphate and chloride ions, only Ca ions were increased due to liming). A high content of nutrients and organic matter was found in the terrestrial stands of *Glyceria maxima* and *Phragmites australis*, in which plant detritus was deposited for the whole period 1971–1997. The local expansion of *Glyceria maxima* indicated local sources of pollution changing over time: a camping site, a drainage channel from the field, and a brook running from Branná village (Figure 1).

Table 2. Soil chemistry of the bottom of the Opatovický fishpond in 1971 (according to Dykyjová & Hejný, 1975), 1992 and 1997. Mean values of all sampling sites and all soil horizons are given. *n* = number of sampling sites

Year	1971	1992	1997
	n:8	22	24
pH_{H_2O}	5.48	5.09	5.13
pH_{KCl}	–	4.49	4.60
Na (ppm)	–	115	28
K (ppm)	147	129	132
Mg (ppm)	405	234	300
Ca (ppm)	1940	1713	1777
$P\text{-}PO_4$(ppm)	24	17	16
SO_4^{2-} (ppm)	–	287	412
Cl^- (ppm)	–	97	37
$N\text{-}NH_4$ (ppm)	–	31	17
NO_3^- (ppm)	–	97	56
N_{tot} (%)	0.991	0.255	0.342
C (%)	10.29	2.71	4.70
C/N	9.20	11.12	12.36

Discussion

Although annual fluctuations (or fluctuations over several years) may be observed in the fishpond vegetation

(Hroudová & Kloubec, 1997), some long-term trends appeared in the development of the littoral vegetation for the period from 1971 to 1997. All changes were influenced by the following factors: fluctuations of water level, trophic level of the habitat, mechanical disturbance (removal of mud by scraping) and other changes in management of the littoral stands (mowing).

The following main trends in vegetation development appeared during the period:

a) successional expansion of shrubs and trees on grass and sedge stands. This was initiated when seasonal mowing and grazing of sedge stands ceased (Jeník & Větvička, 1973).

b) Expansion of the reed (*Phragmites australis*), which mainly replaced sedge stands and partly stands of *Glyceria maxima*, or which spread into the open water. The increase in reeds might be supported by a slight increase of trophic level of the fishpond in the 1970s, contrary to the situation in highly eutrophic habitats (Čížková et al., 1996). The gradual replacement of sedge stands by reed stands seems to correspond to the position of the ecological optimum of *Phragmites australis* among other dominant species in relation to trophic level (Hroudová & Zákravský, 1995; Hroudová et al., 1999); a slight increase of trophic level favoured *Phragmites australis*, but strong eutrophication favoured *Glyceria maxima*.

The position of ecological optima of tall sedges and reed species in relation to trophic level is as follows: low trophic level, then increasing eutrophication, then high trophic level; *Carex rostrata* ⇒ *C. elata* ⇒ *C. gracilis* ⇒ *Phragmites australis* ⇒ *Glyceria maxima*.

A similar consequence was observed also in other South Bohemian fishponds (Hroudová, 1972; Hroudová et al., 1988). Reed decline has been observed in highly eutrophic habitats with a deep layer of continually flooded organic sediments (Čížková et al., 1996, Van den Putten, 1997). This is not the case with the Opatovický fishpond: most of the reed stands occur in a slightly eutrophic environment on mineral substrate, with a low proportion of organic matter. The only sites with a high nutrient content and a deep bottom layer of organic matter are terrestrial and are only temporarily flooded; thus anaerobic processes are limited there. The total trophic level of the bottom sediments did not increase at all since the 1970s. Sources of pollution alternated over time; the pig farm was closed after the 1970s and the inflow of waste water from the camp site stopped after 1992. The gradual infiltration of nutrients from the bottom into the ground layer might be considered to be a "buffering" process preventing nutrient accumulation in the rhizosphere layer. The main remaining sources of pollution are the brook running from Branná village and drainage ditches from fields. The decrease of nitrates and ammonium ions in 1997 (Table 2) indicates a gradual decrease of fertilization of the surrounding fields.

Glyceria maxima stands were limited when muddy sediments were removed by scraping, and regenerated only to a small extent. Well-developed stands of *Glyceria maxima* indicated local sources of pollution, corresponding to high phosphorus content or nitrates in the bottom sediments (Hroudová & Zákravský, 1995). This is also in accordance with the ecological optimum of this species in eutrophic habitats.

c) Regeneration of vegetation on scraped sites. Scraping had both positive and negative effects on vegetation. Positive effects were the removal of muddy deposits preventing anaerobic processes on the bottom; and the formation of open sites on the shores free from competition by reeds, on which species-rich littoral stands could develop during succession. Favourable conditions for the development of some rare or endangered plant species were also provided there.

Negative effects occurred on sites with a greater angle of slope – the regeneration of shore vegetation was difficult or impossible due to the high water level and the action of wind and waves. Sites created by the deposition of bottom material became centres of ruderal vegetation.

Plant succession on flat scraped sites developed during the period – in 1997 the vegetation was fully re-established on sites scraped in the 1970s, while the sites scraped in 1990–91 were at the stage of non-homogeneous stands. New reed stands regenerated relatively slowly. Total evaluation of the effects of scraping performed in 1990 will be possible after a longer period; the detailed study of successional development of vegetation on scraped sites continues.

Changes of water level may initiate changes of vegetation to a different extent; a temporary decrease of water level (at least one growing season) initiated the development of stands of *Bolboschoenus maritimus* subsp. *maritimus* over large areas along shores, while high water levels led to the disappearance of stands of this species which survived only as dormant tubers on the bottom (Hroudová & Kloubec, 1997). Long-lasting changes of water level may lead to irreversible changes in the extent of some communities. *Glyceria maxima* appeared to be least resistant to these changes – it was damaged as much by high water levels (Dykyjová & Hejný, 1975) as by drought (Hroudová & Kloubec, 1997). High water levels limited the extent of all littoral communities, while a temporary decrease of water level showed positive effects on reed stands and also on generative reproduction of many other species.

Conclusions

Comparing the effects of all factors influencing the littoral vegetation, we can conclude that:

1. The fishpond ecosystem appeared to be relatively resistant to different impacts of human activity during the whole 26-year period, and wetland plant communities persisted. An increase in reed stands at the expense of sedges may be explained by the shift in trophic habitat conditions towards more favourable conditions for *Phragmites* expansion.
2. Nutrient contents in bottom sediments did not increase since the 1970s, partly due to the exclusion of some sources of pollution. The fishpond may be considered to be meso- to slightly eutrophic.
3. While the resistance of the fishpond ecosystem may be limited, depending on the ecological amplitudes of the dominant plant species and nutrient supply in the bottom sediments, the following recommendation are made for fishpond management: The prevention of an increase of trophic level; this requires the control of fishery management (limitation of unnecessary manuring, temporary decrease of water level), and limitation of the remaining local sources of pollution. No further scraping of the shores; although this showed some positive effects, the extent of littoral stands is so limited that there are no suitable sites for further scraping. A temporary decrease of water level over an interval of 2–4 years, which would enable the renewal of littoral stands and interrupt anaerobic processes in bottom sediments; and the restriction of public areas to the scope of the proposed nature reserve in the most valuable parts of the fishpond.

Acknowledgements

Our sincere thanks are due to our collaborators from the Administration of Protected Landscape Area and Biosphere Reserve, Třeboňsko, namely B. Kloubec for help with field work and J. Ševčík for providing aerial photographs; to Dagmar Dykyjová and K. Přibáň from the Section of Plant Ecology of the Botanical Institute in Třeboň for valuable information on the investigated sites in 1971; to J.Pokorný for initiation of the study in 1992 within the scope of his project; to Eva Zamazalová for technical assistance and John Cross for language assistance. The work was supported by the Federal Committee of Environment (Grant No. RR/104/93) and the Agency of Nature and Landscape Conservation, Czech Republic (Grant No. PPZP/610/3/97 - DÚ 5).

References

Čížková, H., J. A. Strand & J. Lukavská, 1996. Factors associated with reed decline in a eutrophic fishpond, Rožmberk (South Bohemia, Czech Republic). Folia Geobot. Phytotax. 31: 73–84.

Casper, J. & H.-D. Krausch, 1980. Süßwasserflora von Mitteleuropa. 1. Teil: *Lycopodiaceae* bis *Orchidaceae*. Fischer, Jena, 403 pp.

Dykyjová, D. & S. Hejný, 1975. Produkce a podíl pobřežních rostlinných společenstev na eutrofizaci Opatovického rybníka. Sborn. Jihočes. Muz., České Budějovice, Přír. Vědy, 15: 11–66.

Everaerts, F. M., J. L. Beckers & T. P. E. M. Verheggen, 1976. Isotachophoresis. Elsevier, Amsterdam.

Hejný, S. (ed.), 1973. Ecosystem study on wetland biome in Czechoslovakia. PT-PP/IBPReport No 3, Třeboň, 262 pp.

Hraško, J., L. Červenka, Z. Facek, J. Komár, J. Němeček, F. Pospíšil & V. Sirový, 1962. Rozbory pôd. SVPL, Bratislava, 335 pp.

Hroudová, Z., 1972. Příspěvek k fytocenologickému a floristickému výzkumu rybníků v okolí Jarošova nad Nežárkou. Sborn. Jihočes. Muz., České Budějovice, Přír. Vědy, XII (3): 129–143.

Hroudová, Z., S. Hejný & P. Zákravský, 1988. Littoral vegetation of the Rožmberk fishpond. In Z. Hroudová (ed), Littoral vegetation of the Rožmberk fishpond and its mineral nutrient economy. Studie ČSAV 9, Academia, Praha,: 23–60.

Hroudová, Z. & P. Zákravský, 1995. Změny rybniční vegetace jako indikace změn prostředí. In P. Kovář & H. Härtel (eds), Využití terénní botaniky v ekologii krajiny. Zpr. Čes. Bot. Společ., Praha, 30, Mater. 12: 152–158.

Hroudová Z. & B. Kloubec, 1997. Vliv cíleného managementu na ochranu biodiversity v rybničním ekosystému. Ms., Závěrečná zpráva projektu PPZP/610/3/97, 22 pp., depon. in Inst. of Botany, Průhonice.

Hroudová, Z., B. Kloubec & P. Zákravský, 1999. Dynamika litorální vegetace a avifauna Opatovického rybníka u Třeboně. Příroda, Praha, 14: 173–98.

Husák, Š. & S. Hejný, 1978. General characteristics of the Třeboň basin and Lednice region.- In D. Dykyjová & J. Květ (eds.), Pond littoral ecosystems. Ecological Studies, Vol. 28, Berlin,: 13–22.

Jeník, J. & V. Větvička, 1973. Ecology and structure in stands of *Salix* spp. in the Třeboň basin. In S. Hejný (ed.), Ecosystem study on wetland biome in Czechoslovakia. PT-PP/IBP Report No. 3, Třeboň 1973,: 39–46.

Klika, J., V. Novák & A. Gregor, 1954. Praktikum fytocenologie, ekologie, klimatologie a půdoznalství. Nakl. ČSAV, Praha, 776 pp.

Ložek, V., 1978. Zvláštnosti kvartéru Třeboňska. In J. Jeník & S. Přibil (eds.), Ekologie a ekonomika Třeboňska. Třeboň, p. 93–95.

Moravec J. (ed.) 1995. Rostlinná společenstva České republiky a jejich ohrožení. Severočeskou přírodou, Příl. 1995, Litoměřice, 206 pp.

Olsen, S. R., C. V. Cole, F. S. Watanabe & L. A. Dean, 1954. Estimation of available phosphorus in soil by extraction with sodium bicarbonate. USDA Circ. No. 939, Washington D.C., 19 pp.

Rothmaler, W., 1982. Exkursionsflora für die Gebiete der DDR und BRD, Band 4, Kritischer Band. Volk und Wiessen Volkseigener Verlag, Berlin, 811 pp.

Van den Putten, W. H., 1997. Die-back of *Phragmites australis* in European wetlands: an overview of the European Research Programme on Reed Die-back and Progression (1993–1994). Aquat. Bot. 59: 263–275.

Hydrobiologia **415**: 147–154, 1999.
J. M. Caffrey, P. R. F. Barrett, M. T. Ferreira, I.S. Moreira, K. J. Murphy & P. M. Wade (eds),
Biology, Ecology and Management of Aquatic Plants.
© 1999 Kluwer Academic Publishers. Printed in the Netherlands.

Aquatic macrophyte distribution in relation to water and sediment conditions in the Itaipu Reservoir, Brazil

L. Mauricio Bini[1,*], Sidinei M. Thomaz[2], Kevin J. Murphy[3] & Antonio F. M. Camargo [4]
[1]*Universidade Federal de Goiás, ICB, DBG, Caixa Postal 131, CEP 74001-970, Goiânia, GO, Brazil*
E-mail: bini@icb1.ufg.br
[2]*NUPELIA (Núcleo de Pesquisas em Limnologia, Ictiologia e Aquicultura), Universidade Estadual de Maringá, Avenida Colombo 5790, CEP 87020-900, Maringá, Brazil*
E-mail: smthomaz@nupelia.uem.br
[3]*Institute of Biomedical and Life Sciences, Division of Environmental and Evolutionary Biology, Graham Kerr Building, University of Glasgow, Glasgow G12 8QQ, Scotland*
E-mail: k.murphy@bio.gla.ac.uk
[4]*Departamento de Ecologia, Instituto de Biociências, UNESP, Rio Claro, SP, Brazil*
E-mail: afmc@life.ibrc.unesp.br

Key words: freshwater macrophytes, reservoirs, South America, aquatic weeds

Abstract

Aquatic macrophyte community distribution along the eastern shoreline of the Itaipu Reservoir (one of the South America's largest impoundments) is described in relation to limnological and sedimentological factors. The central body of the reservoir is mesotrophic, while the arms (flooded influent river valleys) along the eastern shore may be oligo-mesotrophic to eutrophic, depending on time of year and sub-catchment characteristics. Macrophyte community composition and species cover were surveyed at 30 sites in four arms, in relation to sediment total P and organic matter; underwater light regime; and water total P and Kjeldahl N concentration, alkalinity, conductivity, depth and pH. Seventeen euhydrophyte and six emergent macrophyte species were recorded. Large stands of *Egeria najas* dominated the euhydrophyte vegetation, together with free-floating weed species (*Pistia stratiotes* Linn., *Salvinia auriculata* Aublet and *Eichhornia crassipes* (Mart.) Solms.). Canonical Correspondence Analysis of the data showed that two sets of variables were important predictors of aquatic macrophyte community structure. Floating macrophyte assemblage was closely related to concentration of nutrients in both water and sediment, while light penetration was the strongest predictor of submerged species occurrence. Although a large number of potential nuisance species were present, dense growths were restricted to shallower areas of the Itaipu Reservoir, causing localised problems. The possibility of increasing interference by these plants with fisheries, recreational use, transport and hydroelectricity generation suggests a need for continued monitoring of weed distribution and abundance, and investigation of appropriate management measures.

Introduction

Abundant growth of aquatic macrophytes is a common feature of large reservoir systems in sub-tropical and tropical regions, especially in impoundments which have complex shapes with extensive areas of sheltered, shallow water ideal for macrophyte growth (e.g. Mitchell et al., 1990; Ali et al., 1995). If unchecked by appropriate management, aquatic weeds (especially free-floating plants such as *E. crassipes*, *S. auritulata* and *P. stratiotes,* and submerged species of Hydrocharitaceae, such as *Egeria densa* Planch. and *Egeria najas* Planch.) may cause serious problems in reservoirs with suitable hydrogeomorphological, climatic and trophic characteristics. Weed species of Hydrocharitaceae for example, are notorious for their

* Author for correspondence

tendency to form large detached masses of vegetation which interfere with the utilization of water resources (Pieterse & Murphy, 1990).

Such problems are increasingly common in South America. For example, Tundisi et al. (1993) reported problems of 'excessive growth' of *E. crassipes* and *Eichhornia azurea* (Swartz) Kunth in reservoirs of the Tietê River in São Paulo State, Brazil. It is a notable feature of South American reservoir systems that aquatic weed problems are usually caused by native Neotropical species (Fernández et al., 1990) unlike reservoirs elsewhere in the subtropical - tropical world where introduced species are commonly the main cause of problems. *E. najas*, *E. densa*, *E. crassipes* and *P. stratiotes* are all native to the Paraná system (Hoehne, 1948, Neiff, 1986). *E. najas* typically occurs in high transparent ponds with remote influence of the High Paraná River in Argentina (Neiff, 1986) and in backwaters of the Paraná in Brazil, upstream to Itaipu reservoir (S.M.Thomaz, personal observation). This species has recently been widespread in Itaipu and in several Brazilian reservoirs (Thomaz et al., 1999).

The objective of this work was to identify the structure of euhydrophyte assemblages present in the Itaipu Reservoir and to establish the relative importance of sets of abiotic factors as predictors of vegetation type. This study arose from concerns of the Itaipu Reservoir managers about the development of free-floating and submerged aquatic weed problems in the lake.

Materials and methods

The Itaipu Reservoir (24° 05′ – 25° 33′ S; 50° 00′ – 50° 30′ W) is a large impoundment (mean area 1350 km^2; mean depth 21.5 m; length 170 km) formed since 1982 behind the Itaipu Dam on the Rio Paraná, itself one of the largest rivers in the world, draining an area of 2.8×10^6 km^2 of South America. The reservoir has an irregular shape (Figure 1) with a large number of shallow arms representing the flooded lower valleys of small to medium-sized tributaries draining sub-catchments having spatially and seasonally varied characteristics.

A preliminary survey of the presence of aquatic macrophytes along the Brazilian shore of the Itaipu Reservoir conducted during 1995 showed substantial variation in both abundance and species composition of the macrophyte community (Thomaz et al., 1999). Results from this study were used to select four reservoir arms for further study during August 1996. This intensive survey of macrophyte abundance and ambient conditions in water and sediment was undertaken at 90 individual stations, located at 30 sites in the selected arms (Figure 1). At each site, three stations were selected at random along the 0.5, 1.0 and 1.5 m isobaths for sampling from a boat. Water levels in the reservoir were low (some 0.7 m below average) but did not vary during the sampling period. Submerged plants were sampled in deeper water using a Petersen grab (0.12 × 0.30 m), which worked well in the soft sediments of the reservoir. A floating quadrat (0.5 × 0.5 m) was used in shallow water stations. In both cases, plant species present within the sample were visually scored on a Domin-Krajina abundance scale (1 = <20; 2 = 21–40; 3 = 31–60; 4 = 61–80; 5 = 81–100% cover).

Water transparency at each site was measured by Secchi disc, and an estimate of k (light extinction coefficient) was obtained by taking three paired measures of underwater photosynthetically-active radiation (PAR) at each of 0 (sub-surface) and 0.6 m, using a twin-sensor Skye SKP210 underwater PAR meter, linked to a Skye Datahog logger and downloaded to a notebook computer on site. The depth (d in meters) was measured from the surface to the sediment. Data for k were used to calculate the proportion of the water column which stands within the euphotic zone - Z_{eu}/d (assuming 3% of surface light availability defining Z_{eu} (euphotic zone) for submerged macrophytes: Moss, 1993). Water turbidity, pH, conductivity and temperature were measured at each site using a Horiba multimeter. Water samples were collected for subsequent laboratory determination of total Kjeldahl N (Mackereth et al., 1978), total P (Golterman et al., 1978) and total alkalinity (Gran titration: Wetzel & Likens, 1991). A sediment sample was collected from each station, using the Petersen grab, for laboratory analysis of total P and organic matter content. The macrophyte environment dataset was analysed using CCA (Ter Braak, 1986) to determine the relative importance of individual environmental factors as predictors of macrophyte status in the Itaipu Reservoir. A TWINSPAN classification (Hill, 1979) of the macrophyte data established sample group composition. The significance of differences in mean values for environmental variables between these sample groups was tested using ANOVA with subsequent mean separation by Tukey's Least Significant Difference test.

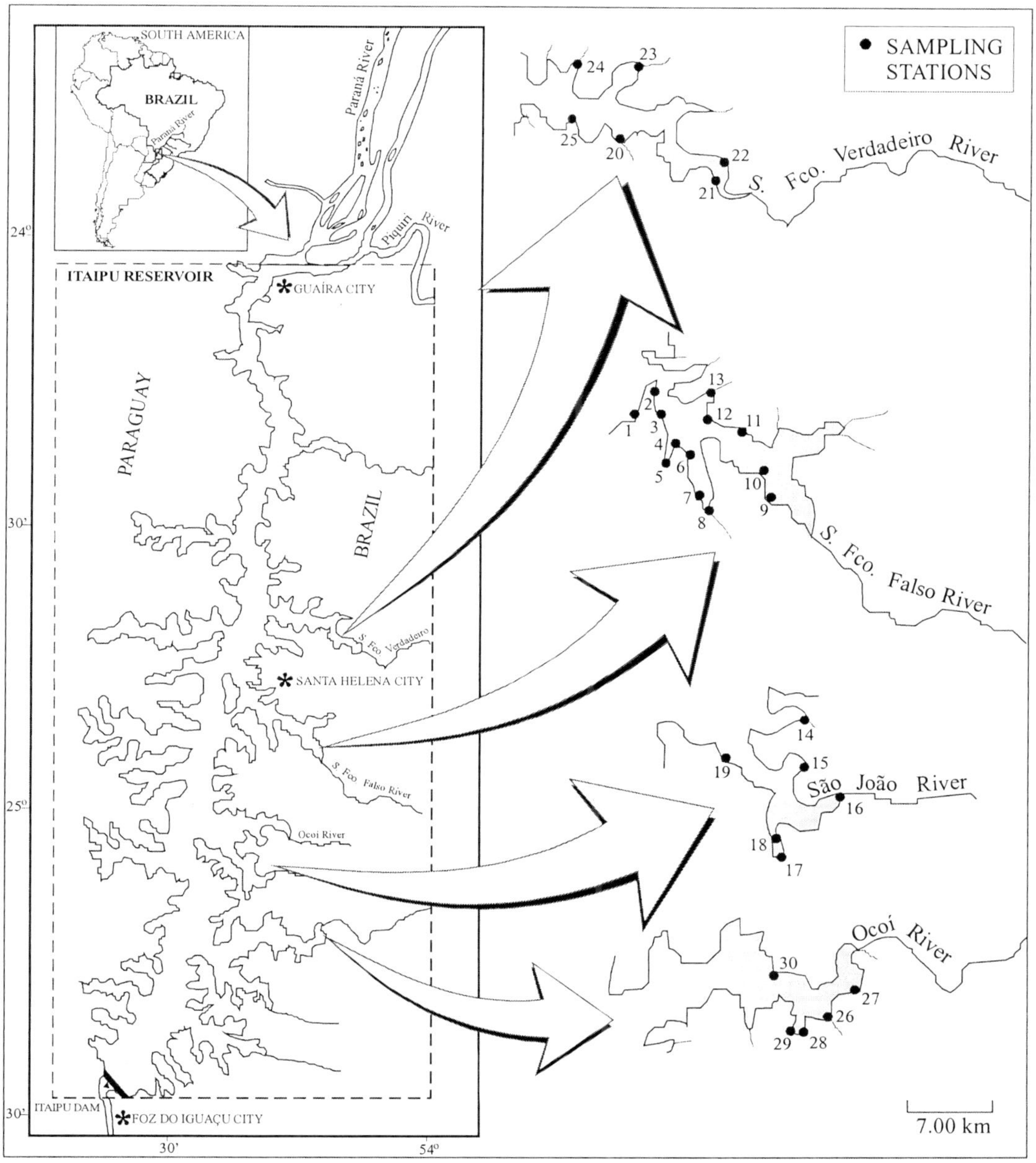

Figure 1. Map of the Itaipu Reservoir showing sampling site locations for 1996 intensive survey. Arms codes used in the text and in Table 1 are as follow: São Francisco Verdadeiro (SFV), São Francisco Falso (SFF), São João (SJ) and Ocoi (OC).

Results and discussion

Limnological characterization

Table 1 summarises data on environmental factors likely to influence aquatic macrophyte growth in the four reservoir arms studied. In general, the main differences between the arms were seen in pH, water and sediment phosphorus availability, and underwater light climate. One arm (SFV) was eutrophic, while the other three were at most mesotrophic. Arm SJ had the lowest trophic status, bordering on oligo-mesotrophic and the clearest water (measured by light extinction). This arm, and arm OC, had acidic water, while the other two had higher pH. Dense phytoplankton blooms were

Table 1. Mean and standard deviation (in parentheses) of environmental variables from four Itaipu Reservoir arms. *N*: number of sites; Secchi: Secchi disc depth (m); *k*: light extinction coefficient (m^{-1}); Z_{eu}/d: euphotic depth divided by actual water depth; Cond: conductivity ($\mu S\ cm^{-1}$); Turb: turbidity (NTU); TKN: total Kjeldahl nitrogen in water ($mg\ l^{-1}$); Alk: alkalinity ($\mu eq\ l^{-1}$); P_w: total phosphorus in water ($\mu g\ l^{-1}$); P_s: total phosphorus in sediment ($mg\ g^{-1}$); OMs: organic matter in sediment ($mg\ g^{-1}$)

Arm	*N*	Secchi	*k*	Z_{eu}/d	pH	Cond	Alk	Turb	TKN	OMs	P_s	P_w
SFF	13	1.02	2.38	1.99	7.61	42.54	443.62	17.08	0.54	12.52	0.82	25.08
		(0.24)	(0.81)	(1.16)	(0.63)	(11.16)	(127.96)	(11.31)	(0.26)	(5.31)	(0.34)	(12.15)
SJ	6	1.15	1.65	2.59	6.93	36.83	370.33	11.17	0.37	14.74	0.80	17.83
		(0.10)	(0.46)	(1.60)	(0.52)	(2.93)	(30.40)	(6.01)	(0.13)	(7.80)	(0.24)	(4.62)
SFV	6	1.00	2.32	2.19	7.33	42.17	371.83	18.33	0.42	13.97	1.20	52.83
		(0.24)	(0.81)	(1.22)	(0.99)	(16.65)	(158.85)	(10.98)	(0.33)	(3.36)	(0.41)	(39.70)
OC	5	1.20	1.91	2.42	6.70	56.40	506.60	5.80	0.47	13.76	0.90	25.40
		(0.32)	(0.60)	(1.38)	(0.20)	(15.76)	(73.16)	(7.05)	(0.10)	(3.16)	(0.28)	(15.58)

observed in SFV and to a lesser extent in SFF. The higher pH and light extinction values at these locations were consistent with the effects normally associated with the presence of algal blooms.

Macrophyte distribution in relation to environmental factors

In the extensive 1995–96 surveys of the Itaipu Reservoir, we recorded 62 taxa from 25 families and 41 genera (Thomaz et al., 1999). Emergent macrophytes had the highest diversity (39 taxa), but euhydrophytes made up one third of the aquatic flora. In this study, we considered only the euhydrophyte species (23 out of the 62 species recorded by Thomaz et al., 1999). Aquatic macrophyte communities were primarily located within the arms of the reservoir, adjacent to the mouths of the main influent streams, and in water of less than 2 m depth.

The richest macrophyte flora found from the 1996 intensive survey occurred in OC (seven species), where samples tended to have three to four times as many species present as in the other three. In general, there was a clear positive relationship between the number of sites at which species occurred and their abundance within each site (measured as cover index). *E. azurea* was an exception: this species occurred at only two sites but was a dominant species in the aquatic vegetation at both. This reflects the pattern of local abundance seen for *E. azurea* elsewhere in the Paraná system.

The CCA species-environment analysis (Figure 2a) showed that pH, P concentration in water and sediment and underwater light climate were the three most important predictors of macrophyte species distribution. Free-floating species, including *E. crassipes, P. stratiotes, S. auriculata* and Lemnaceae, were strongly associated with high nutrient conditions (total phosphorus and conductivity), and, together with emergent plants such as *Polygonum, Commelina* and *Urochloa*, were unaffected by underwater light availability. The free floating species predominantly obtain their nutrient requiriments from the water column and thus the association between this group and sediment P concentrations may be a result of the direct relationship between water and sediment nutrients concentration. Additionally, rich sediments may be an important source of P to the water column through release in anoxic conditions and turbulence (Wetzel, 1983). Other studies have also indicated that aquatic plant growth occurred in nutrient rich water and sediments. The distribution of emergent and floating species in drainage canals in the north-eastern Nile Delta, for example, was correlated with water electrical conductivity and total phosphorus (Khedr & El-Demerdash, 1997). Elsewhere in Egypt, Ali et al. (1995), found evidence that several rooted submerged species occurred at sites with high concentrations of sediment P rather than water P.

The submerged plants *Najas microdon* A.Br., *Utricularia gibba* L. and the isoetid *Ottelia brasiliensis* (Planchon) Walters, were associated with the clearest waters and with intermediate nutrient conditions, in line with the findings of other studies which have examined plants with similar functional preferences (Spence, 1982; Middelboe & Markager, 1997; Van den Berg, 1999). *E. azurea* (which has mainly emergent leaves arising from a submerged stem) was

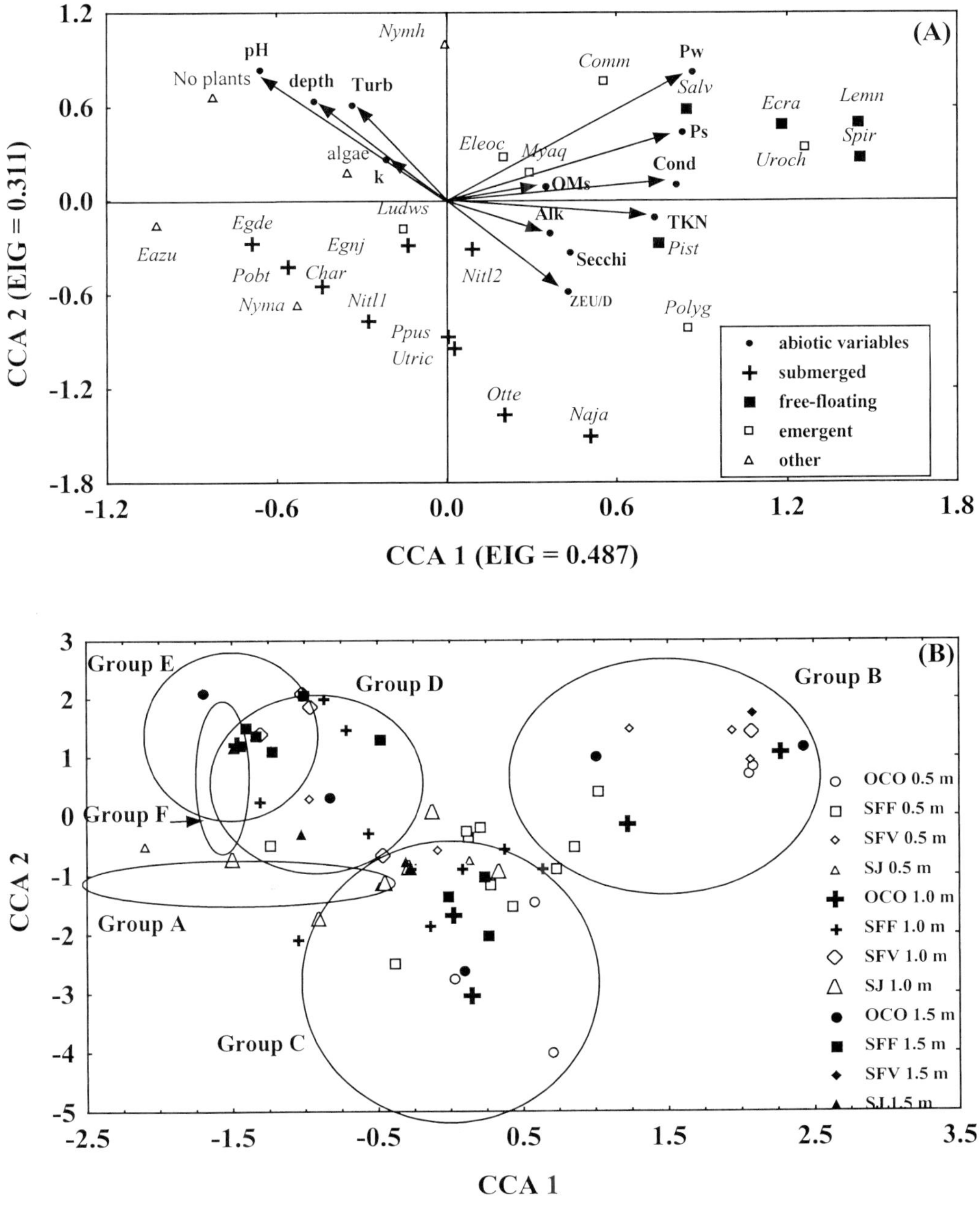

Figure 2. (a) CCA species-environment plot (EIG= axis eigenvalues) for 1996 survey data. Axes 1 and 2 explained the highest significant proportion of the variability in the data set (Monte Carlo test: F-ratio = 5.62; p<0.01). For abiotic variable codes see Table 1; except Z_{eu}/d: euphotic depth (Z_{eu}) divided by actual water depth (d). Species codes – Char: *Chara tomentosa*; Nitl1: *Nitella furcata*; Nitl2: *Nitella* spp; Comm: *Commelina nudiflora*; Eleoc: *Eleocharis* cf. *fistuloides*; Myaq: *Myriophyllum aquaticum*; Egnj: *Egeria najas*; Egde: *Egeria densa*; Otte: *Ottelia* cf. *brasiliensis*; Lemn: *Lemna minor*; Spir: *Spirodela* cf. *polyrhiza*; Utric: *Utricularia gibba*; Nymh: *Nymphoides humboldtiana*; Naja: *Najas* cf. *microdon*; Nyma: *Nymphaea* cf. *rudgeana*; Ludws: *Ludwigia suffruticosa*; Uroch: *Urochloa plantaginea*; Polyg: *Polygonum hydropiperoides*; Eazu: *Eichhornia azurea*; Ecra: *Eichhornia crassipes*; Pobt: *Potamogeton obtusifolius*; Ppus: *Potamogeton pusillus*; Salv: *Salvinia auriculata*; No plants: samples without macrophyte vegetation; algae: green filamentous macroalgae. (b) CCA sample plot (statistical details as for Table 2) showing location of samples by arm and depth. Boundaries of TWINSPAN sample groups A–F are shown overlaid on the biplot.

tolerant of high turbidity and pH. Sites occurring further along this gradient (towards the top-left part of the ordination plot) tended to have dense phytoplankton blooms, or deeper water or both, and supported no macrophytes. The remaining species formed a group tending to occur in sites with relatively low nutrient status, and usually intermediate to high underwater light. These included *Potamogeton obtusifolius* Mert. & W. D. J. Koch, *Potamogeton pusillus* L., nymphaeids, charophytes, *Ludwigia*, *E. najas* and *E. densa*. It was notable that the centroid of distribution of *E. najas* (the submerged species considered to have the highest potential for causing nuisance problems in the reservoir) was very close to the centre of the plot, suggesting that environmental conditions within much of the reservoir will support the species. This species occurred at the highest number of sites, and had the highest cover index at those sites within the four surveyed arms. On present evidence, it is only likely to be excluded from the most nutrient-poor and most nutrient-rich areas of the reservoir arms.

TWINSPAN analysis separated the sample stations ($n = 90$) into 6 groups (A–F) at the third hierarchical level of the classification. The sample-groups are delineated on the CCA sample ordination in Figure 2b. There were significant differences (Table 2) between the sample groups in terms of mean depth, total P in sediment, total P in water (all: $p<0.001$, ***); pH, total Kjeldahl nitrogen, Z_{eu}/d (all: $p<0.01$, **); and conductivity ($p<0.05$, *). There were no significant differences between sample groups ($p>0.05$, not significant) in terms of alkalinity or sediment organic matter content.

A brief summary of the characteristics of the vegetation types and environmental characteristics of each sample group is shown in Table 2. Groups A (indicated by *E. azurea*) and F (indicator: *E. densa*) were each represented at only a small set of sampling stations, located in different arms. In assemblage terms, they are completely different, with no species in common. Group A occurred in shallower water and had moderately high sediment phosphorus (similar to Groups B, C and E), but low water phosphorus and TKN concentrations. Group F had the lowest values for both Ps and Pw of any group, but had quite high TKN. Overall, both A and F can be characterised as predominating in parts of the reservoir where trophic conditions are no more than mesotrophic. The high pH of Group F sites probably reflects a high level of photosynthetic activity by the abundant stands of *Egeria densa* present.

Group E was indicated by the presence of a high proportion of samples lacking macrophyte growth ('no plants') and had the lowest values of Z_{eu}/d of any group. The poor underwater light availability is the most likely cause of macrophyte absence. The eutrophic status of these sites probably contributed to the high turbidity, by promoting algal blooms. Additionally, the deep waters of this group also contribute in producing low values of Z_{eu}/d.

Groups C and D can be considered as variants of the same vegetation type, since both were indicated by the presence of *E. najas*. There were differences in geographical representation (Group C had samples from all four arms, but Ocoi samples were not represented in Group D). Group C had the richer plant diversity, with 18 taxa recorded, while Group D had only six species in total, and a high proportion of individual samples scoring 'no plants'. Although Group C was significantly shallower than D, there was no significant difference in Z_{eu}/d, both sample-groups lying comfortably within the euphotic zone. There was no significant difference in Pw or TKN between the two groups, but Group D was significantly poorer in sediment P. This last variable is the only abiotic predictor which significantly discriminated between the two groups and may account for the lower abundance of *E. najas* and other species within the Group D vegetation type.

Finally, Group B samples were indicated by the free-floating plants *E. crassipes* and *S. auriculata*. This group had high species richness, mainly comprising free-floating species and emergent plants (the latter reflecting the rather shallow water characteristic of these samples). For this type of vegetation, water transparency is of little relevance, although Z_{eu}/d was quite high. More important as predictors of this group were trophic variables. TKN, Pw and Ps all reached maximum values at Group B sites. The absence of SJ samples from this group is notable: this arm of the reservoir had the lowest trophic status of those studied (see Table 2).

The presence of two important aquatic weeds, *E. najas* and *E. densa*, as the dominant species of three of the six open-water vegetation types present in the reservoir is a cause for concern. Hydrocharitaceae species have long been considered nuisance species in natural and man made environments (Wells & Clayton, 1988; Haramoto & Ikusima, 1988; Thomaz et al., 1999). Where local conditions in Itaipu produce nutrient-enriched conditions, the submerged vegetation is replaced by a free-floating macrophyte com-

Table 2. Number of stations per TWINSPAN group, group species indicator and species richness (*S*). Differences in limnological variables between TWINSPAN sample groups are also shown. For each limnological variable, means (standard error between brackets) sharing the same letter are not significantly different ($P>0.05$: Tukey's LSD test). *S*: species richness. For variables codes and units see Table 1

Groups	A	B	C	D	E	F
No. of stations	3	12	41	14	17	3
Indicator	*E. azurea*	*E. crassipe* *S. auriculatas*	*E. najas*	*E. najas*	no plants	*E. densa*
S	3	10	18	6		2
Depth	0.67^a (0.17)	0.88^a (0.12)	0.87^a (0.06)	1.29^b (0.09)	1.26^b (0.08)	1.0^{ab} (0.29)
Z_{eu}/d	3.81^a (0.7)	2.66^b (0.4)	2.43^b (0.2)	1.71^{bc} (0.3)	1.46^c (0.2)	1.98^{bc} (0.7)
P_s	0.86^{ab} (0.5)	1.2^a (0.12)	0.91^{ab} (0.04)	0.67^b (0.08)	0.96^a (0.07)	0.35^c (0.08)
P_w	14.0^c (1.0)	55.6^a (9.9)	25.2^b (2.3)	25.0^b (4.8)	29.8^b (6.6)	16.0^c (0)
pH	7.46^a (0.24)	7.07^b (0.23)	7.00^b (0.09)	7.38^{ab} (0.21)	7.66^a (0.17)	8.27^a (0)
TKN	0.21^b (0.08)	0.59^a (0.08)	0.51^a (0.03)	0.44^{ab} (0.06)	0.34^b (0.05)	0.53^a (0)
Cond	35^b (0)	54^a (6)	43^b (2)	41^b (3)	43^b (3)	34^b (0)

munity, also with a high potential for causing nuisance (Mitchell & Pieterse, 1990).

In 1996, dense growths of *Egeria* spp. and free-floating weed species were restricted to shallow areas in certain arms of the Itaipu Reservoir, causing only localised problems. However, the possibility of future larger-scale interference by these aquatic weeds with fisheries, recreational use, transport and hydroelectricity generation suggests a need for continued monitoring of weed distribution and abundance, and investigation of appropriate management measures.

Acknowledgements

The study was funded by Itaipu Binacional and CNPq. We acknowledge this funding with gratitude. We also thank our driver (Sr. Domingos Durante), boatman (Sandro Alves Heil), and technical assistants (Lucinéia Paes da Silva, Raul Ribeiro and Maria do Carmo Roberto) for their excellent assistance in the fieldwork and laboratory analyses.

References

Ali, M. M., A. Hamad, I. V. Springuel & K. J. Murphy, 1995. Environmental factors affecting submerged macrophyte communities in regulated waterbodies in Egypt. Arch. Hydrobiol. 133: 107–128.

Fernández, O. A., D. L. Sutton, V. H. Lallana, M. R. Sabbatini & J. H. Irigoyen, 1990. Aquatic weed problems and management in South and Central America. In Pieterse, A. H. & K. J. Murphy (eds), Aquatic Weeds. Oxford University Press, Oxford: 406–425.

Golterman H. L., R. S. Clymo & M. A. M. Ohmstad, 1978. Methods for Physical and Chemical Analysis of Freshwaters. Blackwell Scientific Publications, Oxford: 214 pp.

Haramoto, T. & I. Ikusima, 1988. Life cycle of *Egeria densa* Planch., an aquatic plant naturalized in Japan. Aquat. Bot. 30: 389–403.

Hill, M. O., 1979. TWINSPAN – a FORTRAN Program for Two-Way Indicator Species Analysis. Cornell University, Ithaca, New York: 52 pp.

Hoehne, F. C., 1948. Plantas Aquáticas. Publicação da Série D, Instituto de Botánica, São Paulo, Brazil: 167 pp.

Khedr, A. H. A. & M. A. El-Demerdash, 1997. Distribution of aquatic plants in relation to environmental factors in the Nile Delta. Aquat. Bot. 56: 75–86.

Mackereth F. Y., J. G. Heron & J. J. Talling, 1978. Water Analysis: Some Revised Methods for Limnologists. Freshwat. Biol. Ass. 17: 120 pp.

Middeboe, A. L. & S. Markager, 1997. Depth limits and minimum light requirements of freshwater macrophytes. Freshwat. Biol. 37: 553–568.

Mitchell, D. S., A. H. Pieterse & K. J. Murphy, 1990. Aquatic plant problems and management in Africa. In Pieterse, A. H. & K. J. Murphy (eds), Aquatic Weeds. Oxford University Press, Oxford: 341–354.

Moss, B., 1993. Ecology of Freshwaters: Man and Medium. Blackwell, Oxford: 417 pp.

Neiff, J. J., 1986. Aquatic plants of the Paraná system. In Davies, B. R. & K. F. Walker (eds), The Ecology of River Systems. Dr W. Junk Publishers, Dordrecht: 557–571.

Pieterse, A. H. & K. J. Murphy (eds), 1990. Aquatic Weeds. Oxford University Press, Oxford: 593 pp.

Spence, D. H. N., 1982. The zonation of plants in freshwater lakes. In Macfadyen, A. & E. D. Ford (eds), Advances in Ecological Research. Academic Press, London: 37–125.

Ter Braak, C. J. F., 1986. Canonical Correspondence Analysis: a new eigenvector technique for multivariate direct gradient analysis. Ecology 67: 1167–1179.

Thomaz, S. M., L. M. Bini, M. C. Souza, K. K. Kita & A. F. M. Camargo, 1999. Aquatic Macrophytes of Itaipu Reservoir, Brazil: survey of species and ecological considerations. Braz. Arch. Biol. Technol. 42: 15–22.

Tundisi, J. G., T. Matsumara-Tundisi & M. C. Calijuri, 1993. Limnology and management of reservoirs in Brazil. In Straskraba, M., J. G. Tundisi & A. Duncan (eds), Comparative Reservoir Limnology and Water Quality Management. Kluwer Academic Publishers, Dordrecht: 25–55.

Van den Berg, M., 1999. Charophyte colonization in shallow lakes: Processes, ecological effects and implication for management, Ph.D. thesis, University of Amsterdam, Netherlands: 137 pp.

Wells, R. D. S. & J. S. Clayton, 1991. Submerged vegetation and spread of *Egeria densa* Planchon in lake Rotorua, central North Island, New Zealand. New Zeal. J. Mar. Freshwat. 25: 63–70.

Wetzel, R. G. 1983. Limnology. Saunders, New York: 860 pp.

Wetzel, R. G. & G. E. Likens, 1991. Limnological Analysis. Springer-Verlag, New York: 391 pp.

Hydrobiologia **415:** 155–162, 1999.
J. M. Caffrey, P. R. F. Barrett, M. T. Ferreira, I.S. Moreira, K. J. Murphy & P. M. Wade (eds),
Biology, Ecology and Management of Aquatic Plants.
© 1999 *Kluwer Academic Publishers. Printed in the Netherlands.*

Influence of water level fluctuation on the structure and composition of the macrophyte vegetation in two small temporary lakes in the northwest of Spain

Camino Fernández-Aláez*, Margarita Fernández-Aláez & Eloy Bécares
Department of Ecology, University of León, 24071 León, Spain

Key words: life forms, aquatic macrophytes, phenology, zonation, shallow lakes, temporal water bodies

Abstract

The influence of water retention time and depth variation on the macrophyte community in two small shallow lakes in the northwest of Spain was studied. One lake is periodically flooded and the other deeper lake, which is usually inundated, dries up sporadically. The influence of water regime and the spatial heterogeneity of the abiotic environment on diversity values and their temporal evolution are shown. For each lake, a model of horizontal arrangement of species is established. A differential arrangement of optima for dominant species was observed along the depth gradient. This arrangement remained constant in time.

Introduction

In temporal water bodies, depth and water retention play a major role with regard to the occurrence of plant species and their distribution, as well as the temporal structure of communities. Several authors (Spence, 1982; Duarte et al., 1986; Froend & McComb, 1994) have pointed out the significant influence of the water regime, as described by water depth and seasonal variability, on the growth and distribution of wetland vegetation.

In the northwest of Spain, there is an area of wetlands including the small lakes of Chozas de Arriba, which were declared Wetlands of Special Interest in 1991. These water bodies belong, according to Cirujano's definition (1995), to the 'annual wetlands'. Flooding is mainly caused by the seasonal variation in rainfall. These water bodies are being threatened by a series of stress factors related to agricultural activities, particularly the over-exploitation of aquifers, which modify their water regime.

The main aim of this study was to identify the influence of the water regime on temporal changes in abundance, diversity and distribution of macrophytes in two small lakes having different water retention times.

* Author for correspondence

Study area

Two lentic systems located in the northwest of Spain, named Chozas A and Chozas C, were studied (Fernandez-Aláez et al.,1984). Both systems belong to a complex of three shallow natural lakes (Figure 1). Due to dike construction, the water regime in these lakes has been modified. The lakes are located in the northwest of Spain, in the the Páramo Leonés region, surrounded by plains and platforms terraced at levels ranging from 1100 to 750 m. The climate in this area is referred to as dry temperate Mediterranean, with most rainfall during the cold winters. The summers are hot and relatively dry, which leads to lake A drying up and to a marked depletion in the water volume of lake C.

During the study period, annual rainfall averaged 380 mm. This is well below the average for 1974–1988, which was 565 mm. Most rainfall occurred from October to February (60.4%) which resulted in the lakes filling up. From February to September there was a long, relatively dry period followed by increased rainfall in May. This, however, did not significantly increase the water level in the lakes (Figure 2).

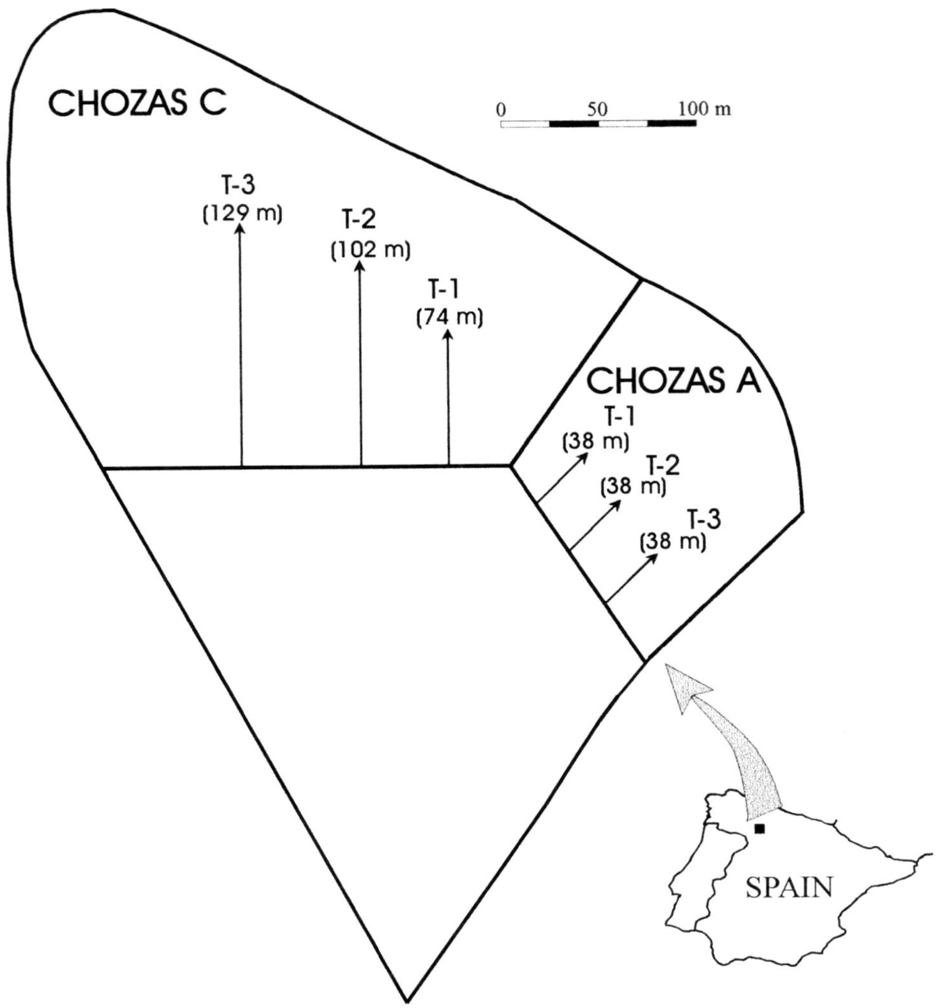

Figure 1. Geographical location of the lakes of the Chozas de Arriba complex. The position of transects is indicated.

As regards retention of water, Chozas A is a temporal environment, whereas Chozas C is a semipermanent environment which, although usually flooded, runs the risk of drying up.

The total anion concentrations are 1.87 mEq/l for lake A and 1.66 mEq/l for lake C (Fernández Aláez, 1996).

Materials and methods

Four quantitative samplings of macrophyte vegetation were carried out from March to June 1995. Sampling could not be extended to the following months because low rainfall in spring and summer caused premature withering of most vegetation. Three transects were placed in each lake starting from the bank. Their position was fixed with stakes to make it easy to locate them for subsequent samplings. A square sampling unit, measuring 1 m × 1 m, was used. The first 10 sampling units of each transect were carried out in lake A without any separations and the next ones at distances of 3 m and 5 m. The same procedure was carried out in lake C, although only the first 6 sampling units of each transect were carried out without any

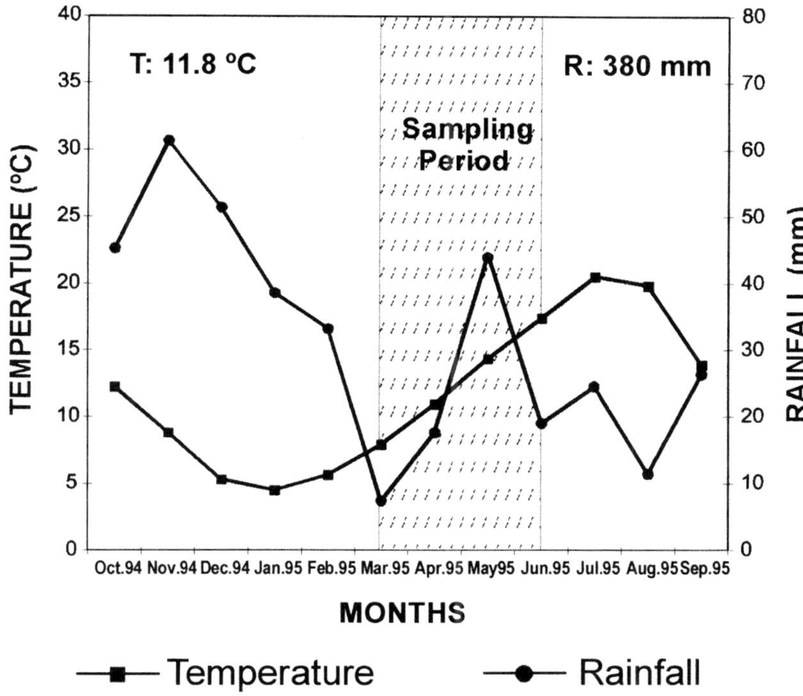

Figure 2. Rainfall and temperature conditions during the hydrological cycle in the study period. T: average temperature, R: total rainfall.

separations. Thus, 51 inventories were carried out at lake A and 68 at lake C in each sampling. The cover percentage of species was quantified for each unit. The change in depth along each transect does not correspond to a progressive increase; water depth initially increased and then diminished (Figures 4 and 5). Four semiquantitative samplings were also carried out in March, April, June and July using a scale from 1 to 7. The diversity of the macrophytic community in each lake was calculated by using the Shannon-Weaver index (1949).

Results

Life form composition

The macrophyte species in both lakes were grouped in terms of their dependence on water using the semiquantitative sampling data. The following groups have been distinguished: hydrophytes (submerged or floating), small helophytes whose juvenile stages resemble aquatic plants (helophytes* in Figure 3), helophytes without heterophylly and hygrophilous species or plants associated with humid zones.

The temporal variation in species number is very similar in both lakes. The lowest values were recorded in March and subsequently there was an increase until July. This is a result of the increase in the number of hygrophilous species, since occurrence of the other life forms remained more or less the same throughout the study period (Figure 3).

Lake C is richer in hydrophytes with a maximum of seven species in the June sampling. In lake A, only three species were registered: *Ranunculus peltatus* Schrank, *Callitriche brutia* Petagma and *Chara globularis* Thuillier, of which only the first two persist in June as terrestrial forms, because the lake dried up completely. In July, hydrophytes disappeared from this lake.

Helophytes in an aquatic juvenile stage were the most frequently found biological type in the March and April samplings. This applies especially to lake C, where few hygrophilous species occurred during this period when the level of water in the lake was still high. On the other hand, in the more temporal lake, there is greater equilibrium between both biological types. As the water volume decreased, hygrophil-

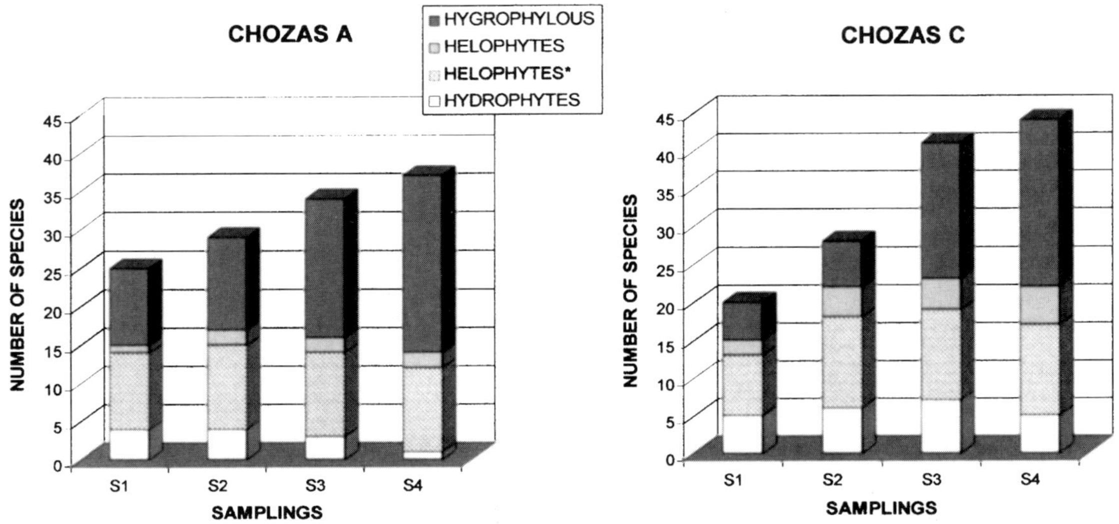

Figure 3. Life form composition in both lakes throughout the study period. The data correspond to the semi-quantitative sampling carried out in March (S1), April (S2), June (S3) and July (S4) 1995.

Table 1. Changes in diversity in both lakes throughout the study period. The values are based on total cover of species recorded in the three transects in each lake

		March	April	May	June
Chozas A	Diversity	2.194	2.392	2.677	3.145
	Richness	21	25	31	31
	Equitability	0.499	0.515	0.540	0.635
Chozas C	Diversity	3.178	3.350	3.307	3.064
	Richness	19	24	29	26
	Equitability	0.748	0.731	0.681	0.652

ous species increased in both lakes, becoming the predominant biological type.

Characteristics of diversity

The structural analysis of the macrophytic community in each lake, based on the calculation of diversity, was achieved by considering the value of species importance for three transects as a whole. In lake A, diversity fluctuated between 2.19 and 3.14 bits, while in lake C, it was always above 3 bits, with a maximum of 3.35 (Table 1). The temporal change in these values showed differences between the two lakes. In lake A, there was a continuous increase from March to June, and differences in diversity between the beginning and end of the study period were significant. This increase was linked to an increment in both richness and uniformity. The rise in specific richness, especially in May and June, was essentially based on hygrophilous species, which are favoured by the removal of water, although in most cases the contribution of these species to the plant cover is very insignificant. The increase in uniformity was due to a decrease in *Littorella uniflora* (L). Ascherson, which covers a great part of the lake during flooding. This decrease in June was due to desiccation causing rapid plant withering.

Diversity was greater in Chozas C and the temporal change was also less apparent than in Chozas A. This greater diversity was, to a large extent, connected with increased uniformity, since the number of species recorded in the transects was always smaller than in lake A. Dominance of *L. uniflora* decreased considerably and the fact that water retention was greater in this lake favours large scale development of species such as *Myriophyllum alterniflorum* DC. in Lam. & DC., *Potamogeton natans* L. and *Juncus heterophyllus* Dufour. All the above contributed to greater equitability within the community in lake C. However, at the end of the study period, the water level had drastically reduced and this caused a decline in hydrophytes. Some species disappeared, such as *R. peltatus*, *C. brutia* or *Potamogeton trichoides* Cham. Schlecht. At the same time, emergent vegetation, mainly represented by *Eleocharis palustris* (L.) Roemer & Schultes, became predominant.

Zonation and phenological rhythm of the species

The spatial disposition and importance of dominant species in every transect, as well as the change in this sequence through time, were analysed (Figures 4 and 5). Taking into account that each transect corresponds to a depth gradient, species will tend to occupy preferential positions, which have been represented by the centre of gravity of abundance distribution (cover).

In lake A, the species developing earliest were the hydrophytes *R. peltatus* and *C. brutia*, which reached maximum cover in March and April, respectively (Figure 4). *R. peltatus* is more ephemeral, since it practically disappeared in May, whereas *C. brutia* in its terrestrial form was still important. These species overlap with an optimum in the area of greatest depth (approximately 50 cm).

The helophytes *Antinoria agrostidea* (DC). Parl. and *J. heterophyllus* have an important aquatic juvenile stage. This allowed them to reach high dominance in March and April, occupying the same deep area as the hydrophytes. In May, and even in June, their cover was marked although initially in a more restricted space. *A. agrostidea* was predominant in this ecological space when *J. heterophyllus* was absent (A3), but was displaced by *C. brutia* (A1) with a temporal development similar to that of helophytes. The dominance of *C. brutia* over helophytes also became evident in the shifting of *J. heterophyllus* to shallower zones (A1).

Baldellia ranunculoides (L) Parl. initially behaved as a hydrophyte and attained maximum cover in April. At this stage, its optimum coincided with that of hydrophytes and helophytes in juvenile aquatic stages. However, if the depth gradient is greater (A3), it does not spread to deeper areas which are more favourable to *A. agrostidea*. In May, the area covered by *B. ranunculoides* diminished but its presence throughout the profile increased.

E. palustris was the last species to appear, reaching peak growth in May and June, when the lake was dry. Obviously, it prefers areas with high edaphic humidity, which was during flooding of the deepest areas. However, as mentioned above, greater depth favoured the predominance of *A. agrostidea* in April, whereas *E. palustris* moved to the area also occupied by *B. ranunculoides*.

L. uniflora showed clear dominance over all the other species in the lake, as it occupies relatively shallow areas (less than 40 cm), which predominate in the transects. Its behaviour as a pseudo-hydrophyte (den Hartog & Van der Velde, 1988) allows it to reach peak growth in March and April and, even though the lakes dry up completely in May, it can still maintain dense growth. Withering of vegetation in June brings about a significant decline in cover, in particular near the shoreline. *Juncus bulbosus* L. and *Eryngium viviparum* Gay, which nevertheless prefer relatively shallow areas, appear along with *L. uniflora*. *J. bulbosus* also behaved as a pseudohydrophyte, attaining peak cover in March. In contrast, *E. viviparum* increased in importance when the lake dried up.

The earliest species in lake C are also hydrophytes, *R. peltatus* and *C. brutia*, reaching maximum development in March and April, respectively (Figure 5). Both species preferred the deepest areas (70–80 cm), although *R. peltatus* was also present in a shallower stretch together with the helophytes. When these hydrophytes became less important in May, the maximum cover of late-developing but very dominant hydrophytes occured: *M. alterniflorum* and *P. natans*. The optimum for these two species seemed to shift to shallower areas, especially in the case of *P. natans*. Besides, its tolerance to a wider range of depths is exhibited by forming strips at different depths and overlapping the helophytes. Helophytes dominated in lake A (usually below 50 cm). *J. heterophyllus* is also important in this lake. Owing to the capacity that this species has to adapt to flooding changes, it was recorded throughout the study period with maximum cover in March and April. The presence of another helophyte in an aquatic juvenile stage, *Scirpus fluitans* L, is significant. It was not present in lake A, and showed a preference for deeper areas than *J. heterophyllus*. It reached maximum cover in April and May and seemed to displace *A. agrostidea* to shallower areas where *J. heterophyllus* is more important. *E. palustris* dominated the space occupied by helophytes in June and increased in importance progressively throughout the study period. It coexisted partly with *S. fluitans*, but preferred less depth, like *J. heterophyllus*. As in lake A, *L. uniflora* colonized the shallowest end of the gradient, with an optimum at around 20 cm. Its cover did not change significantly from March to May, but was reduced in June when the lake almost completely dried up.

Discussion

The small lakes of Chozas de Arriba considered in this study have a well-developed macrophytic community with notable diversity, which is indicative of water

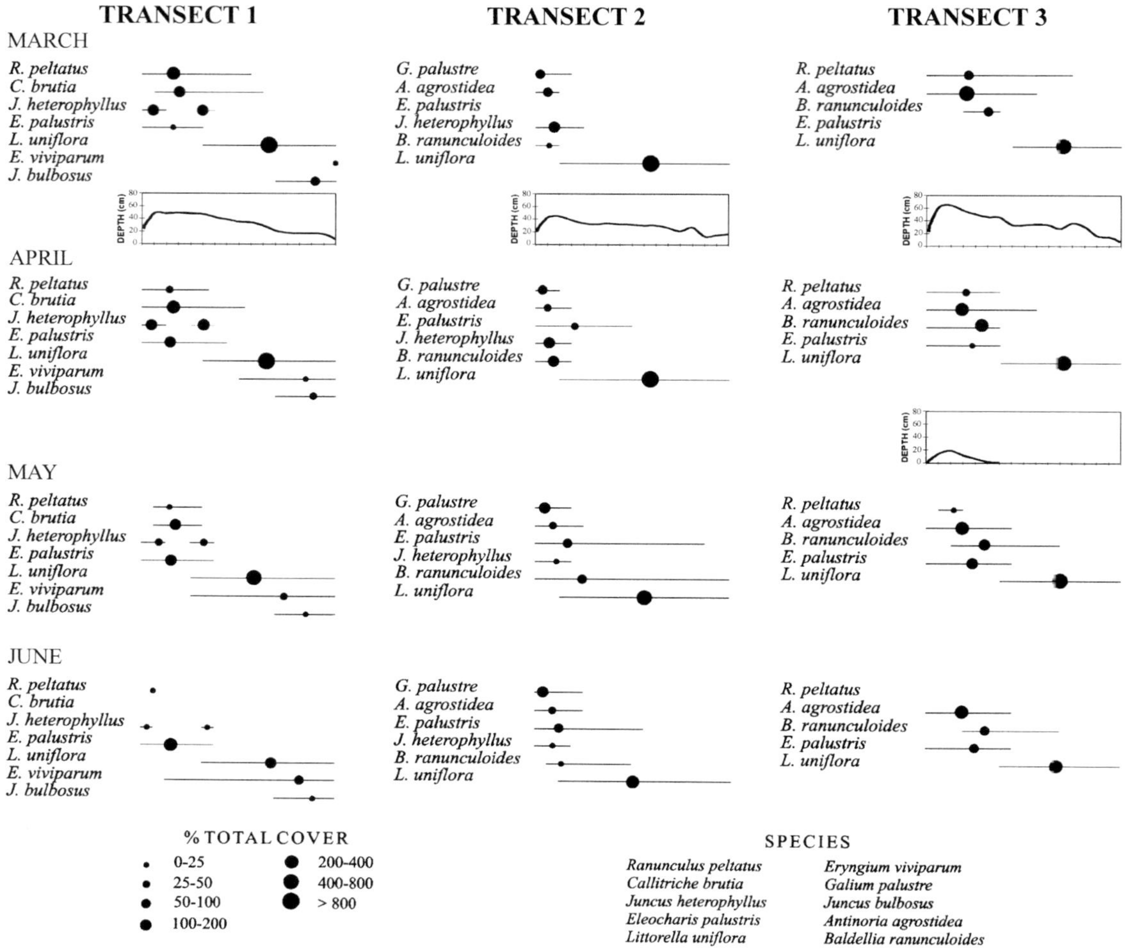

Figure 4. Patterns of spatial and temporal change for the dominant species in the transects of the Chozas A lake, showing the preferential position and importance value for every species.

bodies filled predictably and containing water for several months (Comin & Williams, 1994). However, there are fundamental structural differences based on water retention times. The greater water retention in Chozas C created higher availability of resources and higher environmental heterogeneity, contributing to a greater species diversity. In contrast, Chozas A is more temporal and environmentally more uniform, a major part of its surface being occupied by *L. uniflora* which is indicative of lesser diversity. Therefore, the diversity in the macrophytic community in the lakes of Chozas de Arriba is determined by the spatial heterogeneity of the abiotic environment. Our results concur with those obtained by Chambers & Prepas (1990) who worked on submerged plant communities. The influence of water regime and habitat availability on diversity values and their temporal evolution was evident.

The comparison between both lakes indicates the influence of the duration of the flooding period on the contribution of the life forms to community composition. The drying out of lake A in summer adversely affects aquatic species. However, and despite the differences between the two water bodies as regards the degree of temporality, the considerable number of helophytes in a juvenile aquatic stage in the two lakes emphasizes their shallowness, as does the great variation in water levels with maximums at the beginning of spring and complete drying out in summer.

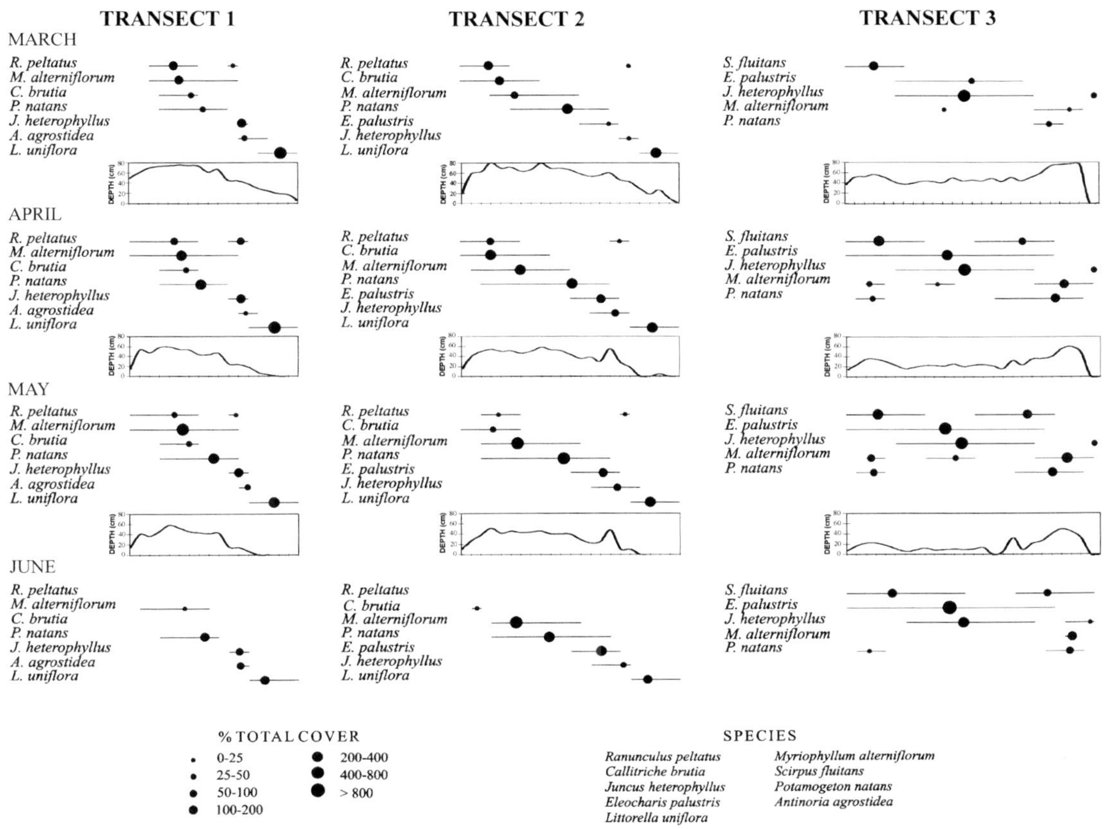

Figure 5. Patterns of spatial and temporal change for the dominant species in the transects of the Chozas C lake, showing the preferential position and importance value for every species.

In general, a clear differential disposition of species optimums is noticed in every transect. This remains constant through time and suggests spatial sharing of resources. Grubb (1977) notes that the differential use of resources may be achieved through a variation in life form or phenological spread. Spatial segregation is obviously marked among species belonging to different life forms, for example, between *P. natans* and *J. heterophyllus*. However, since the change in depth occurs on a gentle slope, widespread overlapping of various species occurs. Coexistence and spatial overlapping of dominant species of the hydrophytes (*R. peltatus*, *C. brutia*, *M. alterniflorum* and *P. natans*) may be explained by the development of community architecture, that is to say, the way in which it fills the available space (Den Hartog & Van der Velde, 1988). This architecture is closely connected to the annual activity rhythm of the species. Thus, *R. peltatus* and *C. brutia* attain their maximum biomasses in early spring and exhibit subtle temporal segregation. In constrast, *M. alterniflorum* and *P. natans* contribute to the maximum development of the hydrophyte community in late spring. The phenological rhythm of both these species prevents them from occurring in the more temporal lake, since prompt removal of water precludes the completion of their development cycle.

In lake A, the early-spring water level does not seem to be decisive for the colonization of these two species; however, this water level allows the development of early hydrophytes tolerant of depths similar to those that *P. natans* and *M. alterniflorum* can tolerate. In Chozas A, the space of these hydrophytes is occupied by helopytes in aquatic juvenile stages: *A. agrostidea* and *J. heterophyllus*. These species, owing to their morphological adaptation to the aquatic environment, are able to lengthen their growth period compared to other strict helophytes such as *E. palustris*. *A. agrostidea* seems to relegate *E. palustris* to shallower areas and is, in its turn, displaced by *C.*

brutia, a hydrophyte possessing a similar phenological rhythm, which maintains its terrestrial form after water removal.

The influence of water depth and its seasonal variability on the vegetation growth and distribution in wetlands has been stressed. It allows zonation of the species wich occupy different positions in relation to water depth (Spence, 1967, 1982; Hutchinson, 1975; Keddy, 1983). A model of horizontal ordination of the species has been established for each lake taking species preference into account. It is clear that depth and retention time condition community structure to a greater extent in these highly seasonal environments.

Acknowledgements

We would like to thank the Junta de Castilla y León and CICYT (AMB94-0292) for financing the research projects that have made this study possible.

References

Chambers, P. A. & E. E. Prepas, 1990. Competition and coexistence in submerged aquatic plant communities: the effects of species interactions versus abiotic factors. Freshwat. Biol. 23: 541–550.

Cirujano, S., 1995. Flora y vegetación de las lagunas y humedales de la provincia de Cuenca. Junta de Comunidades de Castilla-La Mancha. Consejo Superior de Investigaciones Científicas. Real Jardín Botánico de Madrid. Madrid: 224 pp.

Comín, F. A. & W. D. Williams, 1994. Parched continents: Our common future. In Margalef, R. (ed.), Limnology Now: A Paradigm of Planetary Problems. Elsevier Science, The Netherlands: 473–527.

Der Hartog, C. & G. Van der Velde, 1988. Structural aspects of aquatic plant communities. In Symoens, J. J. (ed.), Vegetation on Inland Waters. Handbook of Vegetation Science, Vol. 15/1. Kluwer Academic Publishers, Dordrecht: 113–153.

Duarte, C. M., J. Kalff & R. H. Peters, 1986. Patterns in biomass and cover of aquatic macrophytes in lakes. Can. J. Fish. aquat. Sci. 43: 1900–1908.

Fernández-Aláez, M., E. Luis & C. Fernández Aláez, 1984. Distribución y análisis de la vegetación macrofítica en las lagunas de Chozas de Arriba. León. Limnetica 1: 101–110.

Fernández Aláez, M., 1996. Estudio de la evolución de los humedales de la Meseta y Páramo leoneses. Propuesta para su recuperacion y conservación. Proyecto de Investigación subvencionado por la Junta de Castilla y León. Informe final. Referencia: LE 30.

Froend, R. H. & A. J. McComb, 1994. Distribution, productivity and reproductive phenology of emergent macrophytes in relation to water regimes at wetlands of South-western Australia. Aust. J. mar. Freshwat. Res. 45: 1491–1508.

Grubb, P. J., 1977. The maintenance of species richness in plant communities: the importance of the regeneration niche. Biol. Rev. 52: 107–145.

Hutchinson, G. E., 1975. A Treatise on Limnology, 3. J. Wiley & Sons, New York: 660 pp.

Keddy, P. A., 1983. Shoreline vegetation in Axe Lake, Ontario: effects of exposure on zonation patterns. Ecology 64 331–344.

Shannon, C. E. & W. Weaver, 1949. The Mathematical Theory of Communication. Univ. Illinois Press, Urbana.

Spence, D. H. N., 1967. Factors controlling the distribution of freshwater macrophytes with particular reference to the lochs of Scotland. J. Ecol. 55: 147–170.

Spence, D. H. N., 1982. The zonation of plants in freshwater lakes. Adv. Ecol. Res. 12: 37–125.

Hydrobiologia **415:** 163–168, 1999.
J. M. Caffrey, P. R. F. Barrett, M. T. Ferreira, I.S. Moreira, K. J. Murphy & P. M. Wade (eds), Biology, Ecology and Management of Aquatic Plants.
© 1999 *Kluwer Academic Publishers. Printed in the Netherlands.*

Environmental factors related to the establishment of algal mats in concrete irrigation channels

M. T. Ferreira[1]*, A. Franco[1], L. Catarino[2], I. Moreira[1] & P. Sousa[1]
[1]*Instituto Superior de Agronomia, Tapada da Ajuda, 1349-017 Lisboa, Portugal*
E-mail: terferreira@mail.telepac.pt
[2]*Instituto de Investigação Científica Tropical, Rua Junqueira 86, 1300 Lisboa, Portugal*

Key words: macroalgae, periphyton, concrete canals, Portugal, redundancy analysis

Abstract

The establishment of algal mats in concrete irrigation channels presents a serious weed problem in Mediterranean areas. These periphytic growths can reach two meters in filament length, 20 cm in mat thickness, 100% cover and 1 kg of fresh weight per square meter. These biological characteristics of the algal mats, and also of the drifting mats, present important temporal and spatial variation in the channel system. The species richness of the algae mats was low and only two macroalgae species were present. The diatom *Biddulphia laewis* was dominant in most channels. The green *Oedogonium* sp. was predominant in one channel characterised by lower conductivity and alkalinity, but with higher nutrient levels and this species also dominated in drift samples. Redundancy analysis showed that channel width and the residence time of the water in the channels were the variables that explained most of the variability in the biological data. Some water characteristics were of secondary importance, especially pH and temperature. Biotic factors, such as grazing, were only locally determinant. Water velocity did not present enough variation to create a gradient significantly affecting the biotic variables, except at specific locations, such as the control-gates.

Introduction

In Iberian regions, characterised by long periods of summer dryness, most of the highly productive agricultural activities are dependent upon irrigation. The irrigation channels are concrete lined. They receive water from reservoirs upstream, or pumped from the main river, carrying it along the fertile river valley, distributing it to the crops and finally discharging it downstream through direct drainage or earthen channels. Concrete irrigation channels are only active during the growing season, that is, generally between April and October. During winter, they only maintain static residual water in the bottom. The channels present serious weed problems caused by the establishment of periphytic algae mats. These significantly decrease the carrying capacity of the channels, in some years by up to one third of their original volume. Detached algae continuously drift down the channels, clogging the control-gates, the channel outlets and the crop sprayers.

The weed problems posed by algae mat in concrete irrigation channels are present in most warmwater parts of Europe (Portier, 1971; Aboal et al., 1995) but also in other continental regions (Joska & Bolton, 1995; Pryfogle & Ghio, 1996). Some studies on algae mats produce proposals for weed control, others describe the taxonomy of the algae found. However, few studies detailing the environmental factors that trigger the establishment of these communities, which could be the key for their management and control, are available in the literature.

Study area

The Sorraia Valley national irrigation area was constructed between 1933 and 1959, and has an irrigated area of 16 327 ha. It has a total of 505 km of irrigation channels, mostly with a trapezoidal cross-section, the

* Author for correspondence

smaller ones rectangular shaped. The dominant crops in the irrigation area are rice, tomato, maize, melon, sunflower and orchards.

Two upstream reservoirs are used for water supply, the Maranhão and the Montargil reservoirs (Figure 1). The water is abstracted from the mid-hypolimnia of the reservoirs and is conducted to the crops through concrete channels, which continuously bend or/and ramify and become smaller in their sections. The water outlets from the crop fields are via subterranean drains.

Material and methods

Samples were collected monthly at nine sites during the crop season (May–October). Three sites were sampled having a direct inflow from the two reservoir hypolimnia (CM1 and CM2 from Montargil reservoir; and CR1 from Maranhão reservoir) plus six sites after the junction of these, and along the valley system (CPN, EES, CAZ, CFE, DE1 and DFG), whose location is shown in Figure 1. Collection of the biological material was performed with a dip net having a mouth framed area provided with a blade scraper. At least three samples of a channel surface of 4500 cm^2 were scraped at each site. The thickness of the algae mat (to the nearest cm) and the maximum length of the macroalgae filaments (to the nearest cm) were also measured, and the cover (%) was visually assessed. Temperature (°C), conductivity (μS cm^{-1}), pH and dissolved oxygen (DO, mg l^{-1}), were measured with WTW field probes, as well as the water top width (cm). The water velocity (m s^{-1}) was measured with a R.OTT instrument, at about 40 cm of the algae mat and 10 cm under the water surface. Water samples were collected for the determination of alkalinity, calcium, magnesium, sodium, potassium, phosphates, ammonia, nitrates, BOD and COD. The distance of each site from the nearest reservoir upstream from which it received its water supply (m) was also determined and used as a rough evaluation of the residence time of the water in the channels.

Fresh weight biomass of algae and macroinvertebrates was determined on site (to the nearest 0.1 g) after sorting the latter. Drifting algal biomass was collected with a net of 1 mm mesh size and its fresh weight also determined on site. In July, a more extensive survey was performed at a total of 14 sites (CR2, DE2, SBO, CBO and CBN, Figure 1) with similar data collection. A sample of the algae was collected for identification and cell counting, using a sedimentation chamber of an inverted microscope.

Redundancy analysis (RDA), a canonical form of PCA, was used to directly relate the biotic data and the environment of the 14 sites survey (Jongman et al., 1997: 144-147). RDA selects the linear combination of environmental variables that gives the smallest total residual sum of squares. Environmental and biological variables were log transformed (log x+1) and standardised by row centering (Ludwig & Reynolds, 1988) to make the canonical coefficients comparable. The CANOCO programme, version 2.1, was used to perform the RDA (ter Braak, 1987). For interpreting the variation patterns of the nine monthly surveyed sites, a cluster analysis was performed, using χ^2 as a measure of similarity and the complete linkage method for group generation. The programme NTSYS version 1.5 was used in this data treatment (Rohlf, 1986).

Results

The algal parameters showed an important temporal variation during the study period and also among sites (Figure 2), but there were no clear temporal trends. Overall, biological parameters were weakly correlated with one another (Pearson correlation <0.5, $P<0.5$). Weak correlations were also found in general between biotic and abiotic variables, namely positive relations between water velocity and biomass, and between cover and canal top width (Pearson correlation <0.5, $P<0.5$).

Abiotic parameters also showed considerable variation during study period, especially between sites, with large standard deviations. Water velocity varied according to the size of the canal and the amount of water being supplied to the farmers. The water showed in general a moderate conductivity, high dissolved oxygen content, pH and alkalinity, and a medium to low content of organic matter and nutrients. Calcium and sodium are the dominant cations. However, sites receiving a direct inflow from Montargil reservoir presented distinctive water characteristics because the drainage basin of this reservoir is dominated by eruptive and metasedimentary rocks, whilst the geological background of the basins of Raia and Sorraia is mainly composed of tertiary deposits and alluvia. Montargil water thus has lower conductivity, pH, alkalinity and cationic levels, but is richer in organic matter and nitrogen.

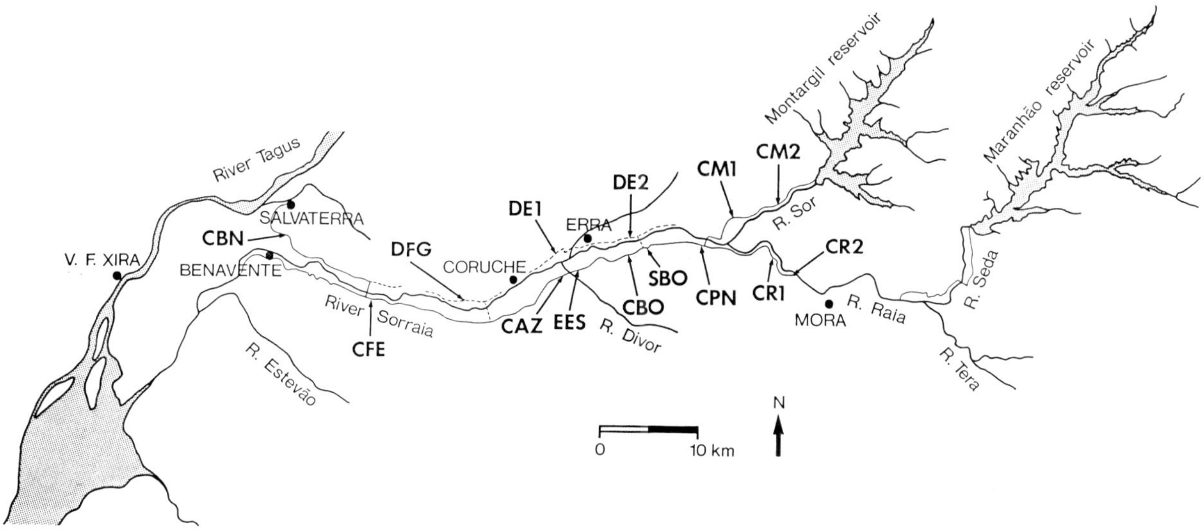

Figure 1. Principal concrete irrigation channels of the Sorraia Valley National Irrigation Area, with the location of the sampling sites. Bolder lines are the river systems, fine narrow lines are main distribution canals with up to 750 cm top width, and dotted lines are secondary channels with up to 250 cm top width.

Algae communities collected in July were quite similar, presented a low richness, and were dominated by diatoms (Table 1). Macroalgae found belonged only to two filamentous species, the diatom *Biddulphia laevis*, dominant in most sites, and the green *Oedogonium* sp., dominant in the two sites of Montargil canal and in the drift samples. The two filamentous algae behave differently. *Oedogonium* sp. produced long filaments, and thick but heterogeneous mats, whilst *B. laevis* tended to occur as less long filaments, less thick but more homogeneous and stickier algae mats. The rest of the community was composed of microscopic epiphytic algae, some of them locally displaying a considerable abundance, such as *Melosira granulata, Surirella ovata*, *Cocconeis placentula* and *Microcystis aeruginosa* ($>10^3$ cells ml^{-1}). Montargil sites and drift samples presented a smaller richness, a greater algae abundance and were dominated by chlorophytes, mostly *Oedogonium* sp. (Table 1). In both cases and in most sites, cover was nearly always more than 80% of the channel surface.

Macroinvertebrates were dominated by Chironomidae and especially Gastropoda, and nearly all individuals of this last group belonged to *Potamopyrgus jenkinsi* (Prosobranchia: Hydrobiidae), an aggressive alien snail from New-Zealand rapidly expanding in Portugal (Simões, 1988). Grazing events can be locally important and affect the algae. For example, in DE1, *P. jenkinsi* reached a fresh weight of 48 g m^{-2} and the algae mat eventually became confined to a thin layer.

The biplot of the first two axes (eigenvalues 0.54 and 0.24, respectively) for the RDA analysis is shown in Figure 3. The cumulative percentage variance explained by these first two axes accounted for 77.1% of the total variation of the biotic data, and 91.9% when the species-environment relation is considered. The variance inflation factors for all the variables selected was low (<20). Table 2 summarises the relations found between the abiotic variables and the two first canonical axes. Axis 1 is related to temperature, pH and top width. Thus, larger, colder and less alkaline channels showed the maximum values of cover, algae biomass, filament length and mat thickness (Figure 3). Larger amounts of drifting biomass were opposed to the larger filaments, cover and standing biomass and related to smaller channels. DE1, the site highly affected by grazing, is also opposite to the larger and more weed infested channels CR1 and CR2. The second canonical axis is essentially related to the distance to the water source upstream and the channel temperature, that is, the residence time of the water in the channels. Smaller channels, at the end of the irrigation system, such as CFE, DFG and CBN, are warmer and showed algal mats relatively more constant in length and biomass.

The cluster analysis performed with the biotic data from the nine monthly sampled sites showed that the

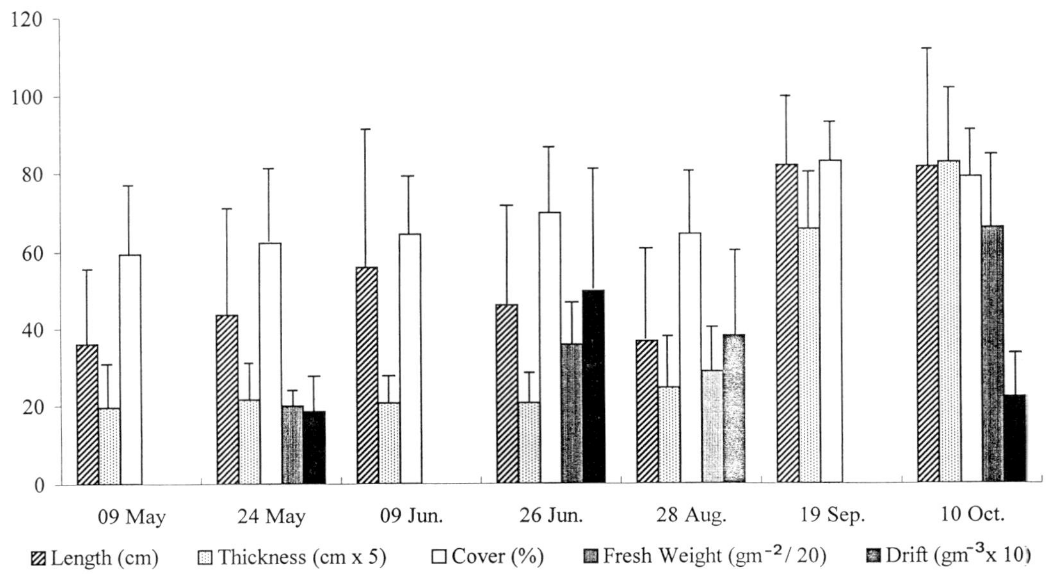

Figure 2. Monthly evolution of the average, and standard deviation, of algae mat width (cm), thickness (cm), cover(%), standing wet weight (g m^2) and drifting wet weight (g m^3).

Table 1. Summary of the periphytic composition from the July survey, including total richness, total number of cells per ml and percentile values for the algae groups found and the two dominant filamentous algae

	All sites[a]		Montargil[b]	Algae drift[c]
	Mean	Range		
Richness	28.3	22–38	13	6–9
Number cells ml^{-1}	33 808.5	16 569.5–68 584.3	105 031.3	87 953.1–99 792.0
Cyanophyta	5.2	0–29.9	–	0–0.3
Euglenophyta	–	0.1–0.2	–	–
Chlorophyta	9.8	1.9–15.8	57.3	86.4–88.7
Bacillariophyceae	85.9	61.5–97.6	42.7	11.0–12.3
Biddulphia laewis	57.9	35.4–89.1	15.6	7.1–9.2
Oedogonium sp.	3.8	0–9.4	57.3	87.8–90.9

[a]Except sites from Montargil canal, [b] mean of CM1 and CM2, [c] range of CM1, CPN and CFE.

Table 2. Redundancy analysis: canonical coefficients and intra-set correlations of environmental variables with the two axes of RDA for the algae mat data. Acronyms on Figure 3 are also indicated. Total sum of squares in species data: 59.06. *Significantly related at $P < 0.05$

Variables	Canonical coefficients		Correlation with canonical axis	
	Axis 1	Axis 2	Axis 1	Axis 2
Temperature (°C)- TEMP	0.93*	−0.79*	0.501*	−0.349*
Conductivity ($\mu S\ cm^{-1}$)- COND	−0.19	0.34*	−0.105	−0.217
pH- pH	0.62*	−0.07	0.536*	0.102
Dissolved oxygen ($mg\ l^{-1}$)- DO	−0.08	0.05	0.095	−0.190
Water velocity ($m\ s^{-1}$) - VELO	0.49*	0.07	−0.231	0.195
Top width (m)- WIDTH	−0.48*	−0.48*	−0.727*	−0.219
Distance to the reservoir (km)-DIST	−0.23	0.75*	0.232	0.588*
Invertebrate biomass ($g\ m^{-2}$) –SNAILS	−0.04	0.26	−0.201	−0.195

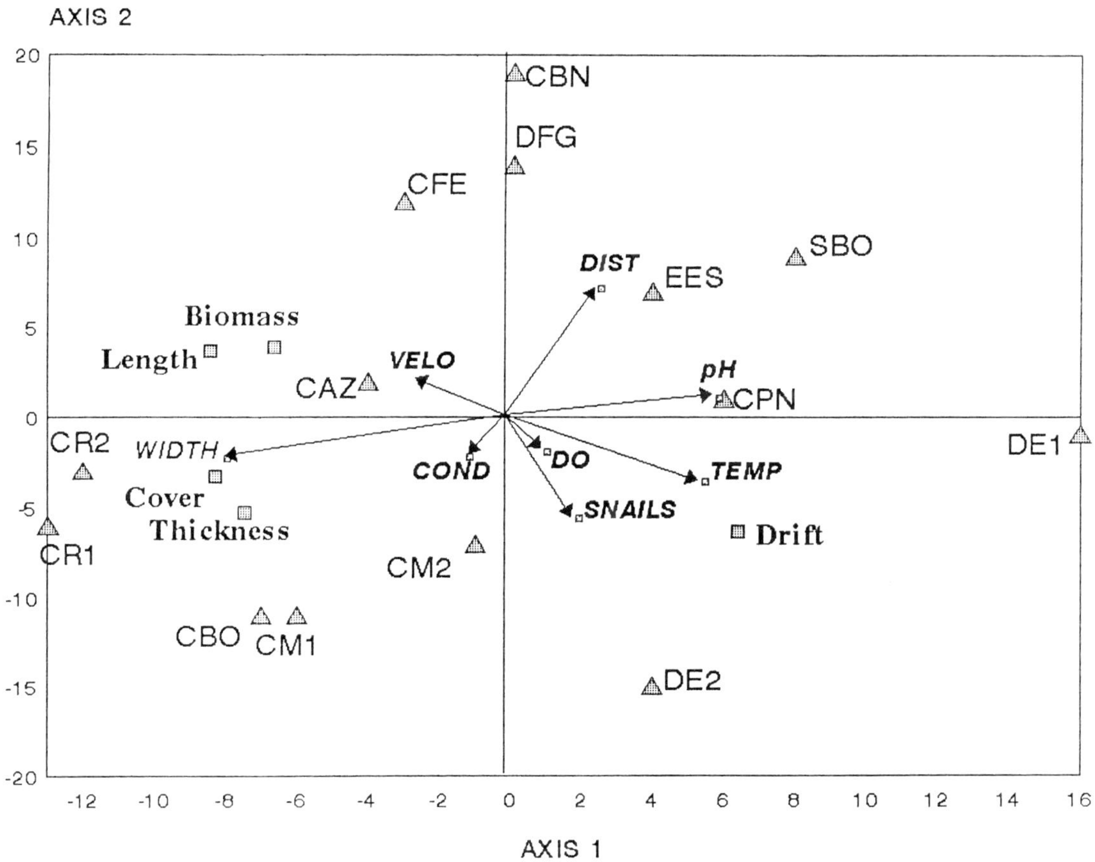

Figure 3. RDA ordination diagram of the algae mat data with environmental variables represented by arrows. Sum of all canonical eigenvalues: 0.839. Scale of the diagram: multiplier 10.

biotic temporal variation is also grouped according to the abiotic characteristics selected by the RDA based on the spatial variation displayed by the July survey. DE1 is firstly opposed to all the other sites because of the important decrease in biomass due to grazing. Two other groups of sites (χ^2 coefficient less than 0.5) were positioned along the channel system, with one of the groups presenting wide channels nearer the water entrance and the other including channels smaller and farther.

Discussion

The composition of the mat in filamentous species was very poor and similar between sites, and further it was dominated by an estuarine related species, *Biddulphia laevis*, whose occurrence had not been previously reported for Portuguese freshwaters. Montargil channel and drifting mats, however, presented a different composition, being dominated by *Oedogonium* sp. Though no systematic monthly collection was done, point observations in some months and sites seemed to indicate a temporal stability of the algae composition.

Of the 431 species of filamentous green algae found in the Iberian Peninsula, the *Oedogonium* group is the most important, accounting for 37.12% of the species, followed by *Spirogyra*, *Mougeotia* and *Ulothrix* groups (Cambra & Aboal, 1992). The taxonomy of Iberian filamentous greens is poorly Aboal et al. (1995) could not determine the species of *Oedogonium* and other filamentous greens on the channel system of Tejo-Segura, but its diversity was greater than in the Sorraia Valley system. In other channels, other species also occur abundantly, especially *Cladophora* and stoneworts (Joska & Bolton, 1995). When the channels walls become damaged or there is sediment deposits in the bottom, macrophytes are also known to occur, especially *Zanichellia palustris* and *Potamogeton pectinatus* (Penuelas, 1984; Aboal et al., 1995). Environmental variables such as the dimensional type and the age and condition of the channels, as well as the physico-chemical characteristics of the water, are likely to be involved in species richness and composition, but clearly we need a wider spatial sampling strategy, including different channels types on different systems, to create a working model.

The composition of the algae mat was an important information in terms of the weed problems caused. Diatoms produce less dense but more homogeneous mats, and in our study, drifting biomass was always dominated by green algae, while the composition of standing biomass could be different. According to Aboal et al. (1995), the gas bubbles produced by the photosynthesis that typically accumulate within the green algae filamentous mats, facilitate the detachment and drifting. Cell counting in sedimentation chambers is not very efficient for filaments and probably underestimated, so the other estimates of the quantitative expression of the algae mat used in this study, such as fresh weight and mat thickness, were more reliable for data treatment. More advanced measurement methods, such as the P/B ratio, would probably give a good account of the macroalgae community dynamics.

The amount of water distributed to the crops decreased in September, thus water velocity also decreased, and algal standing crop and other biotic features tended to increase towards the end of the irrigation period, though with large variations related to different channel types (mean 1315.9 g m^2, SD 696.92). Though suspected a priori to be an important variable influencing algal establishment, water velocity showed small correlation with the canonical axes and a weak relation with the biomass and filament lengths. Indeed, the water velocity in the different channels and throughout the irrigation season did not produce enough variation to create a gradient significantly affecting the biotic variables.

Channel width was the key-variable found in this study to be proportionally related to the algae establishment. The channels tended to decrease in top width along the irrigation system or rather the river valley axis. However, they also bend, turn in right angles or ramify, becoming rectangular shaped, thus also proportionally receiving less light, so the residence time of the water is not directly related to the channel width and was found to be the second most important environmental factor. Water characteristics are secondarily important, namely pH and temperature. Biotic factors such as grazing were only locally important.

A wider sampling of macroalgae establishment is needed, on different irrigation areas, to assess whether these abiotic variables are also important in different channel types or if other environmental factors also play a role, such as the lower summer temperatures in more northern irrigation areas or the age and condition of the channel systems.

Acknowledgements

This study was financed by the Project PAMAF 8126/97 from the Ministry of Agriculture, Fisheries and Rural Development. We would like to thank Dr Manuela Simões for the identification of the Gastropoda. Dr Isabel Andrade and Dr Fátima Brito are also thanked for helping with the algae identification and counting.

References

Aboal, M., M. Prefasi & A. D. Asencio, 1995. The aquatic microphytes and macrophytes of the transvase Tajo-Segura irrigation system, Southeastern Spain. Hydrobiologia 340: 101–107.

Cambra, J. & M. Aboal, 1992. Filamentous green algae in Spain. Limnetica 8: 213–220.

Jongman, R. H., C. J. Ter Braak & O. F. Tongeren, 1997. Data Analysis in Community and Landscape Ecology. Cambridge University Press, Cambridge: 299 pp.

Joska, M. A. & J. J. Bolton, 1995. Filamentous freshwater macroalgae in South-Africa – a literature review and perspective on the development and control of weed problems. Hydrobiologia 340: 295–300.

Ludwig, J. A. & J. F. Reynolds, 1988. Statistical Ecology. A Primer on Methods and Computing. John Wiley & Sons, Chichester: 377 pp.

Penuelas, P., 1984. Vegetatio aquatica dels canals de regar del rio Muga. Problemes i algunas soluciones. Ann. Ins. Estudis Empordanesos 24: 14 pp (in Spanish).

Portier, J., 1971. Le probléme des algues dans les canaux betonnées d'irrigation du sud-est de la France: recénts essais de traitment. Proc. 3rd Int. EWRS Symp. Aquatic Weeds. Oxford: 265–274 (in French).

Pryfogle, P. A. & G. Ghio, 1996. Algal control problems and practices. Proc. Symp. Non-Indigenous Species in Western Aquatic Ecosystems. 15th Annual Meeting of the Western Aquat. Plant Mgmt Soc.: 30–39

Rohlf, F. J., 1986. NTSYS-PC, Numerical Taxonomy and Multivariate Analysis System, version 1.5. Exeter Publ., Ltd, New-York: 253 pp.

Simões, M., 1988. *Potamopyrgus jenkinsi* (Prosobranchia: Hydrobiidae) e sua distribuição em Portugal. Iberus 8: 243–244 (in Portuguese).

Ter Braak, C. F., 1987. CANOCO – a FORTRAN programme for canonical community ordination by (partial) (detrended) (canonical) correspondence analysis (version 2.1) ITI-TNO. Institute of Applied Computer Sciences. Wageningen: 82 pp.

Hydrobiologia **415:** 169–176, 1999.
J. M. Caffrey, P. R. F. Barrett, M. T. Ferreira, I.S. Moreira, K. J. Murphy & P. M. Wade (eds),
Biology, Ecology and Management of Aquatic Plants.
© 1999 *Kluwer Academic Publishers. Printed in the Netherlands.*

Invertebrate hydraulic microhabitat and community structure in *Callitriche stagnalis* Scop. patches

Matthew T. Ó Hare & Kevin J. Murphy*
Division of Environmental and Evolutionary Biology, Institute of Biomedical and Life Sciences, University of Glasgow, Glasgow G12 8QQ, U.K.
E-mail: k.murphy@bio.gla.ac.uk

Key words: flow, macrophytes, benthic invertebrates, rivers, Scotland

Abstract

In this paper, we report the structure of the benthic invertebrate community in submerged *Callitriche stagnalis* Scop. stands in relation to velocity, in a Scottish river, the Blane Water. We compared the community within the macrophyte beds to that of adjacent unvegetated substrate. *Callitriche stagnalis* stands supported different taxa richness (no. of taxa) and abundances (no. of individuals) of benthic invertebrates than the unvegetated riffle substrate (Wilcoxon Matched-Pairs test: $p<0.01$, $n = 10$). To see if different sections of the stands supported different invertebrate communities, samples were taken from the outer, mid and root sections of the stands. The sections were separated by Detrended Correspondence Analysis (DCA) using their macroinvertebrate communities (eigenvalues: axis 1 = 0.6, axis 2 = 0.136). Abundance and taxa richness were different between the outer and mid sections, and between outer and root sections (Wilcoxon Matched-Pairs test: respectively $p<0.05$, $p<0.01$, $n = 10$). Community structure also differed between sections. The outer section community had a structure similar to that of an extreme environment where Simuliidae were dominant. *Ephemerella ignita* dominated in the mid and root sections. Differing combinations of plant structure and velocity appear to be a major factor influencing habitat structure, creating a range of stability conditions in the stands, which support the observed diversity of invertebrate assemblages present.

Introduction

When macrophytes are included in rehabilitation schemes for channellised rivers, they can potentially provide a diverse habitat for invertebrates by interacting with local flow conditions (Marshall & Westlake, 1990; Newbury & Gaboury, 1993). For practical management purposes, it is useful to know how much additional invertebrate diversity and biomass may be supported by utilising macrophytes as part of an individual river rehabilitation scheme, under specific flow conditions. We use the term diversity of invertebrates to encompass taxa richness, abundance and evenness (the relative number of individuals of each taxa). Previous work (e.g. Gregg & Rose, 1982) has demonstrated the extent to which macrophytes can alter the streamflow microenvironment. Large-scale studies (e.g. Jenkins et al., 1984; Ormerod, 1988) have demonstrated how this increased habitat diversity in turn influences the occurrence of invertebrate assemblages in vegetated and unvegetated parts of the river environment. Sand-Jensen & Mebus (1996) showed that a range of flow conditions occurred in patches of the submerged macrophyte *Callitriche cophocarpa*. Other studies (Jeppesen et al., 1984; Wright, 1992) found that invertebrate communities present in macrophyte patches (including *Callitriche*) could differ substantially from those occurring in unvegetated patches of river bed, having higher taxon richness, numerical abundance and specifically high numbers of Chironomidae and Simuliidae.

The Blane Water is a small unregulated stream which supports a *Callitriche*-dominated macrophyte flora typical of many small British rivers in which rehabilitation schemes are being considered or im-

* Author for correspondence

plemented. We investigated the relationship between flow conditions and invertebrate community structure within different parts of submerged beds of *Callitriche stagnalis* Scop., and in adjacent unvegetated patches of river substrate in the Blane Water. The results are considered in relation to the value added to river rehabilitation schemes by the inclusion of macrophytes, in terms of increased invertebrate diversity support capacity.

Materials and methods

Plants and substrate were collected from riffle sections of the Blane Water, a gravel bed river, at Blane Bridge (U.K. National Grid reference NS507852) in Scotland during June 1998. The site is downstream from a trout fishery and is classified as organically enriched by the Scottish Environmental Protection Agency, although water quality is high (Doughty & Maitland, 1994). The only noticeable effects of enrichment are higher numbers of chironomids and annelids than would otherwise be expected. Channel width in the sampling area ranged from 8 to 13 m, depth from 7 to 22 cm and velocity from 0 to 0.35 m/s.

Ten *Callitriche stagnalis* stands were chosen at random within a c.100 m stretch of river. Water velocity and depth were measured in front, above, in the middle and at the bottom of each stand using an electromagnetic velocity meter (SENSA). At the mid channel side of each stand, velocity and depth were measured and substrate type was visually estimated (% scale) using the following particle size scale: sand (0.06 mm–2 mm nominal diameter), fine gravel (2–10 mm), gravel (10–64 mm) and cobble (64–256 mm). A reference scale was used in the field to confirm particle size when necessary. Mean water column velocity measurements taken outwith the stands were measured at 0.4 depth from the river bed (Smith, 1975) as 50 second averages.

The total area of bed occupied by each stand was estimated using a quadrat with 25 cm^2 squares. Invertebrate samples were taken separately from the outer, middle and root sections of the plants by placing a net downstream, trimming the desired section of the stand with scissors and letting material float into the net. Sections were defined using stand length and depth as follows; outer = outer 20%, mid = all other material above substrate, root = material within the bed. When removing the roots, all the substrate which had been underneath the stand was disturbed and the invertebrates collected into the net placed downstream. A Surber sample was then taken from the mid channel point where substrate type had been assessed (10 Surber samples in total). We estimated the near bed velocities at the Surber sample points by using values which were 10% of those at 0.4 depth from bed as suggested by Carling (1992). All samples, and the sections of stands, were immediately placed in 70% alcohol to preserve the invertebrates. All samples were sorted on white trays and animals identified to the following taxonomic levels: Annelida to subclass; Diptera to family or genus and all other taxa to genus or species. Other studies have shown that larvae and adults of Coleoptera have differing hydraulic habitat preferences (Degani et al., 1993) and so these were treated separately.

To compare densities of invertebrates between stand sections, abundance values were standardised to estimated wet weight of stands. Because of the necessity of immediately preserving invertebrate samples in alcohol, in the field, the plant material in the samples could only be weighed after a storage period in 70% alcohol. It was necessary to see if samples of different weights were affected equally. The relationship between dry weight post storage with wet weight prior to storage was linear for log-log data as determined by regressing the wet weight of 24 control stand samples against their dry weight.

$$\text{Ln Wet weight} = 2.5926 + 0.92503 \; \text{Ln Dry weight} \quad (R^2 = 0.978)$$

This equation was used to transform dry weight values to wet weight for the experimental plants. Where direct comparisons between samples were made the Wilcoxon Matched-Pairs Signed-Ranks Test was used, which allows comparison between related samples (Sidney, 1956). As either different sections of stands are compared or stands and adjacent sections of substrate (Surber samples) it was considered that they represented related, non-independent samples and so this test was appropriate. Benthic invertebrates can have clumped distributions and it is possible that this could occur around some of the *Callitriche* stands: see Elliott (1977) for a comprehensive assessment of the possible spatial dispersions of benthic invertebrates. It was believed this could further influence the relatedness of spatially close samples and enhanced the applicability of the Wilcoxon Matched-Pairs Signed-Ranks Test.

Table 1. Total abundance, number of individuals of all taxa per sample and taxa richness (number of taxa per sample). Values for stand sections (outer, mid, root) are standardised by weight of plant material in each section (no. of individuals per g wet weight of plant material). Values for whole stands and Surber samples are adjusted for the amount of river bed covered by the stand or Surber sampler (no. of individuals per cm^2 of river bed). In all cases n =10

Sample	Total Abundance					Taxa richness				
	Outer	Mid	Roots	Stand	Surber	Outer	Mid	Roots	Stand	Surber
1	4.63	1.34	1.95	0.95	0.11	0.18	0.14	0.63	0.06	0.02
2	11.61	1.44	5.75	2.64	0.22	0.39	0.22	2.03	0.03	0.03
3	3.50	1.54	1.78	0.78	0.14	0.18	0.13	0.47	0.03	0.02
4	10.49	2.54	1.39	0.92	0.15	0.14	0.41	0.38	0.05	0.02
5	2.97	2.07	2.63	0.78	0.21	0.17	0.12	0.49	0.04	0.03
6	6.01	1.51	0.90	0.62	0.15	0.25	0.15	0.23	0.17	0.02
7	5.31	2.22	0.87	0.34	0.08	0.13	0.07	0.14	0.13	0.03
8	3.12	2.56	1.19	0.67	0.16	0.07	0.33	0.27	0.29	0.02
9	1.96	5.02	2.00	1.57	0.09	0.32	0.76	0.49	0.03	0.02
10	1.29	1.81	1.99	0.87	0.14	0.14	0.24	0.49	0.17	0.03

Results

Although composed of many individual plant fronds, all stands had a similar growth form, being wider at the upstream end and tapering downstream, similar to a classic streamlined strut as described by Vogel (1994) but wider at the upstream end and generally more ragged. The area of stream bed occupied by stands was on average 467 cm^2 (s.d. 290 cm^2). Substrate in the Surber samples was dominated by gravel (% composition of gravel always >90%). The substrate underneath the *Callitriche stagnalis* stands was mainly sand (% composition of sand always > 80%). *Callitriche* stands had higher, and significantly different total taxa abundance (no. of individuals of all taxa) and taxa richness than neighbouring substrate samples (Wilcoxon Matched-Pairs test $p<0.01$, N=10; values standardised for sample area: the area of bed occupied by each stand. P value is for both cases, see Table 1 for abundance and richness data per sample). A list of taxa occurring in the Surber samples and plant stands is given in Table 2 with abundance (no. of individuals per sample) and frequency of occurrence (no. of samples in which a taxon occurs).

Taxa richness was significantly different between outer and root sections of the stands. From the box plot, it is clear that the outer section supports the lowest taxa richness (see Figure 1). Total taxa abundance was significantly different between outer and mid sections and outer and root sections. The box plot shows that the outer section of the stand supported the greatest abundance (see Figure 2). See Table 1 for abundance and richness data per sample.

Samples taken from the outer, mid and root sections of the stands and Surber samples were separated by Detrended Correspondence Analysis (DCA) (Jongman et al., 1987) using their macroinvertebrate communities (see Figure 3). Root section and Surber samples were not separated by the DCA, but three species did show a preference for the Surber samples over the root sections (see Table 3). To standardise between Surber and stand samples, sample area was used but section weight was not taken into account. Other analysis showed that the same separation occurred if section weight was also used but separation of sites was not so clear. Samples from the outer section were most dissimilar to the Surber sample, whilst the root section was most similar. The outer section was dominated by the filter feeding Simuliidae, while in all other plant sections *Ephemerella ignita* dominated. In the Surber samples, *Baetis rhodani* was the most abundant invertebrate.

Community structure also differed between sections (see Figure 4). All three stand sections supported invertebrate communities which followed the broken stick model (Gray, 1987). The outer section, which is exposed to the highest velocities, was the least equitable, with the steepest gradient at the start, and is the closest to a geometric series. Such a series is found in communities poor in species at the earliest successional stage after colonisation, or when exposed to extreme conditions (Gray, 1987). The middle sec-

Table 2. Invertebrate taxa occurring in *Callitriche stagnalis* stands and on unvegetated substrate; Surber samples. Abundance values for stand sections are standardised by weight of plant material in each section (no. of individuals per g wet weight of plant material). Abundance values for whole stands and Surber samples are adjusted for the amount of river bed covered by the stand or Surber sampler (no. of individuals per cm^2 of river bed). All abundance figures are (means ± standard deviation) × 10^{-3}. Number of occurrences is the number of samples in which a taxon occurs. In all cases n =10. Taxa for which no code is given (-) are those with very low occurrence, omitted from ordination analysis (see Figure 3)

Taxa	Taxa codes	Abundance					No. of occurrences				
	(see Figure 3)	Outer	Mid	Root	Stand	Surber	Outer	Mid	Root	Stand	Surber
Caenis rivulorum Eaton	caeriv	9 ± 84	2 ± 367	17 ± 115	5 ± 9		2	1	2	3	1
Baetis rhodani (Pictet)	baerho	149 ± 84	203 ± 367	100 ± 115	40 ± 38	30 ± 2078	10	8	6	10	10
Ephemerella ignita (Poda)	ephign	541 ± 313	1131 ± 829	629 ± 306	175 ± 40	21 ± 2066	10	10	10	10	9
Ecdyonurus sp.	ecd	7 ± 15	34 ± 55	54 ± 64	6 ± 5	8 ± 1875	2	5	7	9	9
Rhithrogena semicolorata (Curtis)	–	0	0	0	0	0.3	0	0	0	0	2
Hydropsyche siltalai Döhler	hydsil	0	14 ± 39	9 ± 22	3 ± 7	2 ± 834	0	2	2	4	7
Athripsodes sp.	ath	2 ± 6	2 ± 5	43 ± 108	5 ± 10	1 ± 416	1	1	2	2	2
Brachycentrus subnubilus Curtis	brasub	11 ± 34	17 ± 55	0	1 ± 3	0	1	1	0	1	2
Odontocerum albicorne (Scopoli)	–	0	0	8 ± 24	0.2 ± 1	0	0	0	1	1	0
Rhyacophila sp.	rhy	8 ± 17	1 ± 4	1 ± 4	1 ± 1	2 ± 625	2	1	1	3	6
Glossosoma boltoni Curtis	globol	0 ± 0	0	2 ± 5	0	3 ± 208	0	0	1	1	5
Asellus aquaticus (L.)	aseaqu	5 ± 11	63 ± 82	28 ± 37	10 ± 16	1 ± 1667	2	7	5	8	4
Gammarus pulex (L.)	gampul	7 ± 14	42 ± 43	66 ± 76	9 ± 7	2 ± 1667	2	6	6	8	7
Ancylus fluviatilis Muller	ancflu	4 ± 13	4 ± 13	21 ± 54	3 ± 5	3 ± 833	1	1	2	4	4
Pisidium sp.	psi	0	4 ± 13	3 ± 9	1 ± 3	0	0	1	1	2	0
Limnius volckmari (Panzer) larva	limvla	0	13 ± 23	78 ± 74	7 ± 6	4 ± 1876	0	3	9	9	9
Limnius volckmari (Panzer) adult	limvad	4 ± 14	10 ± 21	3 ± 7	1 ± 1	2 ± 834	1	2	2	4	6
Elmis aenea (Muller) larvae	elmaen	4 ± 9	16 ± 36	73 ± 73	8 ± 10	2 ± 1667	2	3	8	8	5
Elmis aenea (Muller)adult	elmad	4 ± 14	0	6 ± 13	1 ± 1	0.2 ± 417	1	0	2	2	1
Oulimnius tuberculatus (Muller)larvae	oultuad	7 ± 16	6 ± 18	13 ± 29	1 ± 2	0.5 ± 417	2	1	2	2	2
Oulimnius tuberculatus (Muller)adult	oultula	0	0	15 ± 34	2 ± 3	0.2 ± 834	0	0	2	4	1
Esolus parallelpipedus (Muller) adult	esopar	0	0	4 ±13	1 ± 2	0.3 ± 208	0	0	1	1	2
Oulimnius troglodytes (Gyllenhal) adult	–	0	0	4 ± 13	0.2 ± 1	0	0	0	1	1	0
Scirtidae	–	0	0	0	0	0.2 ± 0.3	0	0	0	0	1
Leuctra fusca (L.) group	leufus	46 ± 62	114 ± 122	172 ± 207	25 ± 23	9 ± 2082	5	8	7	10	10
Capnia sp.	–	0	1 ± 4	0	0	0.3 ± 208	0	1	0	1	1
Tipulinae	tip	0	0	18 ± 26	1 ± 1	0.2 ± 834	0	0	4	4	1
Limoniinae	lim	2 ± 7	6 ± 18	90 ± 95	8 ± 9	3 ± 1875	1	1	9	9	6
Ceratopogonae	–	0	6 ± 18	0	0.2 ± 1	0	0	1	0	1	0
Simuliidae	sim	3206 ± 2447	193 ± 227	25 ± 55	508 ± 593	13 ± 2076	10	6	3	10	7
Chironomidae	chi	1072 ± 1584	303 ± 412	121 ± 188	316 ± 433	26 ± 855	4	4	4	4	4
Hirudinae	hir	2 ± 6	17 ± 20	144 ± 99	17 ± 18	4 ± 1874	1	5	9	9	10
Oligochaeta	oli	0	2 ± 8	300 ± 620	25 ± 51	8 ± 1665	0	1	8	8	10

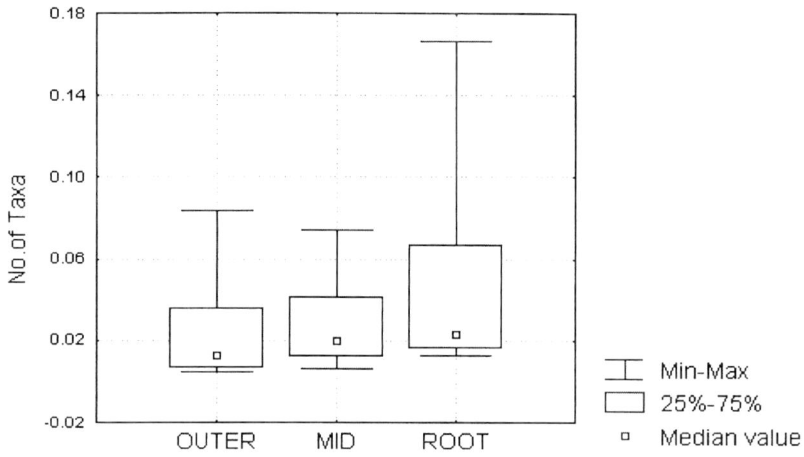

Figure 1. Number of invertebrate taxa no./g fresh weight. Outer was significantly different from root, Wilcoxon Matched-Pairs test: $p<0.05$, $n=10$. Values standardised for section weight.

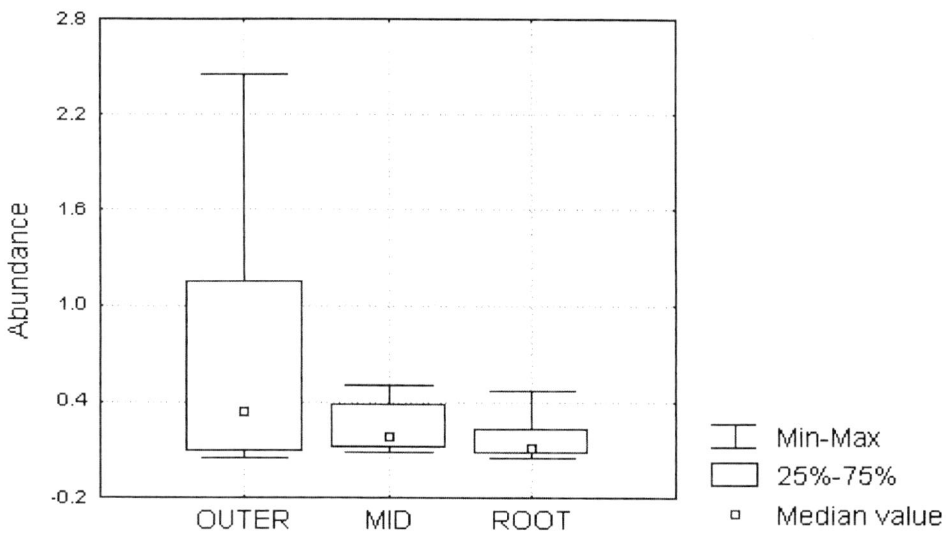

Figure 2. Relative abundance of invertebrates no./g fresh weight. Outer was significantly different from mid and root, Wilcoxon Matched-Pairs test: $p<0.05$, $n=10$. Values standardised for section weight.

tion is more equitable and the root section is the most equitable having the largest number of species which share the resource more evenly, in terms of producing individuals.

It is suggested that stand structure and velocity combine to create a range of stability conditions, forming a series of microhabitats, in, on and outwith the plant beds. Velocity was significantly different and noted to be lower over the top of the stands than directly in front of the stands or in the adjacent Surber samples (Wilcoxon Matched Pairs $p<0.05$, $n = 10$). It is suggested that the shape of the stand is deflecting water to either side and creating flow conditions which are similar to sheet flow over the top of the stand; that is turbulent but uniform in direction of flow. The velocity at the top of the stand (0.25 m/s) was higher and significantly different (Wilcoxon Matched Pairs $p<0.05$, $n= 6$) than that in the middle (0.046 m/s), but velocity at the bottom (0.040 m/s) of the stand was not significantly different from that in the middle. The near bed velocities at the Surber sample points (0.042 m/s), (estimated by adjusting velocity meas-

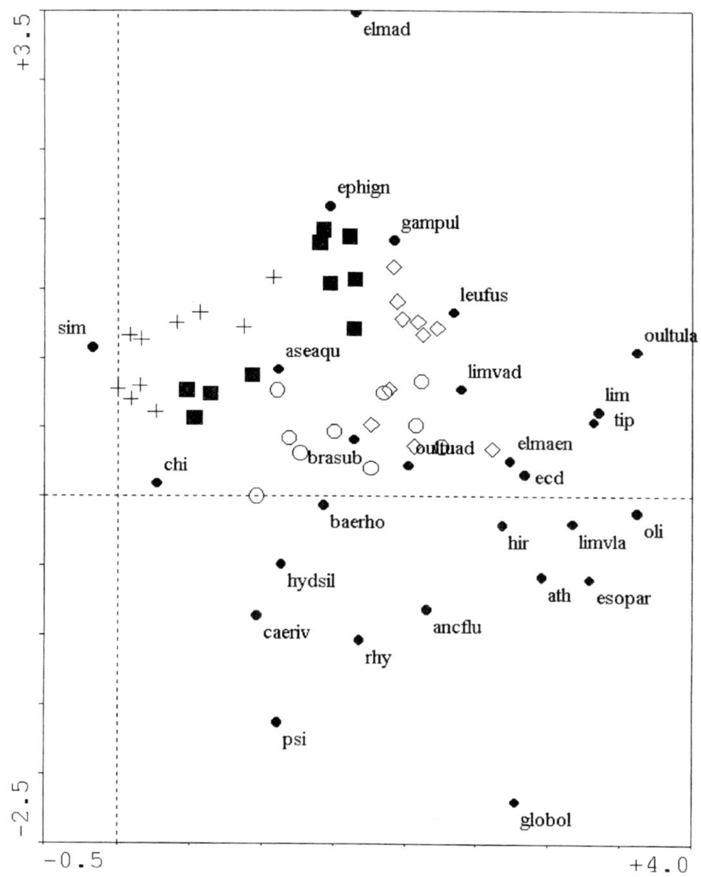

Figure 3. Detrended Correspondence Analysis site by species biplot, (eigenvalues: axis 1 = 0.6, axis 2 = 0.136) Plant section and Surber symbols: Outer +, Mid ■, Root ◊ and Surber ○. The location of each taxon is marked with ♦ and the taxa code: see Table 2. Taxa with very low occurrences are omitted (see Table 2).

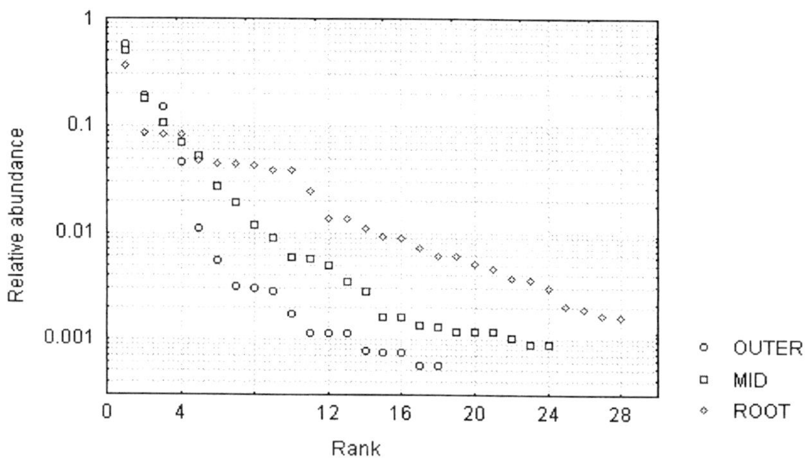

Figure 4. Rank abundance plot for the 3 plant sections. Relative abundance of invertebrates no/cm2/g fresh weight.

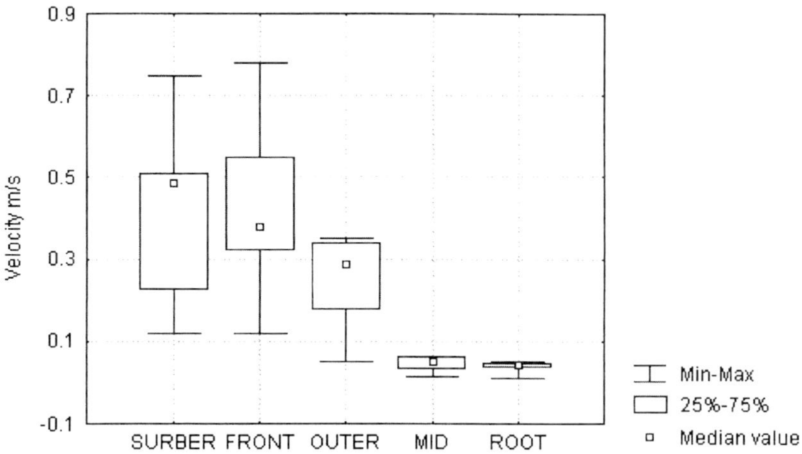

Figure 5. Velocity m/s; 50 second means. Adjusted: adjusted Surber velocity.

Table 3. Taxa with a significant preference for Surber samples over root sections (Wilcoxon matched pairs test, $N = 10$)

Taxon	z	P value
Baetis rhodani	2.80	0.01
Rhyacophila	2.20	0.05
Limnius volckmari	2.20	0.05

ured at 0.4 of depth from the bed) are not significantly different to those found in the mid and root sections of the stand. Estimates of near bed velocity over substrate were an order of magnitude lower than those measured on the outside of the plant stands. An estimate of the required velocity at 0.4 of depth above substrate flowing over unvegetated substrate that would produce a near bed velocity similar to the velocity experienced by the outer section of the stands would equate to 2.5 m/s, or near spate conditions. This figure was derived from using Carling's (1992) estimate procedure in reverse starting with the flow over the outer section of the stands (0.25 m/s).

Discussion

The presence of *Callitriche stagnalis* stands in the Blane Water increases the diversity of invertebrate habitat within the river. Taxa which occurred in the plant stands also occurred on the substrate, but some were present in higher numbers, on and within macrophyte beds. The stands extend the available substrate for invertebrate colonisation up through the water column, providing a continuum of conditions from high to low velocity. Adjacent benthic substrate does not provide the high velocities present on the outside of the plant stand in a sustained manner, that is only under near spate conditions would these high velocities be achieved. In addition, the external surface of the stands greatly increases the area of high-flow substrate available for colonisation by invertebrates adapted to high flow conditions, especially Simuliidae (Hart et al., 1996).

Although taxon richness and abundance were not significantly different between the mid and root sections of the plants, they were separable using DCA. The rank abundance plots demonstrated that the root section was more equitable than the mid section in resource distribution, hence the separation in the DCA biplot. This would suggest that conditions were more extreme in the mid section than the root section, which could be a product of longer term conditions than those measured in this study. The conditions present in these two sections allowed *E. ignita* to dominate where *B. rhodani* dominated in the Surber samples.

It has been suggested by Sand-Jensen & Madsen (1992) that *Callitriche cophocarpa* "patches form a mutually protecting structure against high flows [and] may, therefore, ... increase nutrient supply from sedimentation fine-grained particles"... He later found (Sand-Jensen, 1998) that the sediment under stands often has high levels of organic matter, which could provide food for detritivores. In this study, we did not measure the amount of organic matter in the sediment. However, despite observing that the substrate under *C. stagnalis* stands was more fine grained than that in unvegetated areas, we found no detrivore showing a preference for the root section, compared to the adjacent unvegetated sediment.

The net result of the presence of the plants is to increase the invertebrate abundance and diversity present in the Blane Water. The assemblages present can be related to the structure of the stands and how this influences the flow conditions experienced by the animals. The complexity of invertebrate community supported appears to be functionally related to position within the stand, with extreme conditions (on the external surface) supporting fewer species than within the bed. The total diversity supported is however a function of the total available set of conditions present in the river: on, in and between the stands.

From a management point of view, it is clear that the presence of the stands substantially increases the invertebrate diversity and abundance supported by the river, particularly by increasing the habitat available for organisms (e.g. Simuliidae) which would otherwise find only a limited space for colonisation in their preferred high-flow conditions.

Callitriche stagnalis is a common species in British rivers, and elsewhere in Europe, with a high tolerance of disturbance (Sabbatini & Murphy, 1996). Inclusion of transplanted material as part of rehabilitation schemes in the small gravel-bed rivers which are well-suited to this species would be likely to make a useful contribution to improving invertebrate support function of the river.

Acknowledgements

This project was funded by the Scottish Office, via a grant to the University of Glasgow. We thank Judith Milne for her excellent help with field work and Jenny Devine for useful comments.

References

Carling, P. A., 1992. The nature of the fluid-boundary layer and the selection of parameters for benthic ecology. Freshwat. Biol. 28: 273–284.

Degani, G., 1993. Relationship between current velocity, depth and the invertebrate community in a stable river system. Hydrobiologia 263: 163–172.

Doughty, C. R. & P. S. Maitland, 1994. The ecology of the River Endrick: present status and changes since 1960. Hydrobiologia 290: 131–151.

Elliot, J. M. 1977. Some Methods for the Satistical Analysis of Samples of Benthic Invertebrates. Windmere, Freshwater Biological Association.

Gray, J. S., 1987. Species-abundance patterns. In Gee, J. H. R. & P. S. Giller (eds), Organisation of Communities; Past and Present. Blackwell Scientific Publications, Oxford: 53–68.

Gregg, W. W. & F. L. Rose, 1982. The effects of aquatic macrophytes on the stream microenvironment. Aquat. Bot. 14: 309–324.

Hart, D. D., B. D. Clark & A. Jasentuliyana, 1996. Fine-scale field measurement of benthic flow environments inhabited by stream invertebrates. Limnol. Oceanogr. 41: 297–308.

Jenkins, R. A., K. R. Wade & E. Pugh, 1984. Macroinvertebrate habitat relationships in the Tefi catchment and significance to conservation. Freshwat. Biol. 14: 23–42.

Jeppesen, E., T. M. Iversen, K. Sand-Jensen & C. P. Jorgensen, 1984. Økologiske konsekvenser af reduceret vandføring i Susån., Biologiske processor og vandvalitets-forhold. Miljøstryrelsen, Copenhagen.

Jongman, R. H. G., C. J. F. Ter Braak & O. F. R. Van Tongeren, 1987. Data Analysis in Community and Landscape Ecology. Cambridge University Press, Cambridge: 299 pp.

Marshall, E. J. P. & D. F. Westlake, 1990. Water velocities around water plants in chalk streams. Folia Geobot. Phytotax. 25: 279–289.

Newbury, R. & M. Gaboury, 1993. Exploration and rehabilitation of hydraulic habitats in streams using principles of fluvial behavior. Freshwat. Biol. 29: 195–210.

Ormerod, S. J., 1988. The micro-distribution of aquatic macroinvertebrates in the Wye river system: the result of biotic or abiotic factors? Freshwat. Biol. 20: 241–247.

Sabbatini M. R. & K. J. Murphy, 1996. Response of *Callitriche stagnalis* Scop. and *Potamogeton crispus* L. to cutting, dredging and shade in English drainage channels. J. Aquat. Plant Mgmt. 33: 8–12.

Sand-Jensen, K., 1998. Influence of submerged macrophytes on sediment composition and near- bed flow in lowland streams. Freshwat. Biol. 39: 663–679.

Sand-Jensen, K. & J. R. Mebus, 1996. Fine-scale patterns of water velocity within macrophyte patches in streams. Oikos 76: 169–180.

Sand-Jensen, K. & T. V. Madsen, 1992. Patch dynamics of the stream macrophyte, *Callitriche cophocarpa*. Freshwat. Biol. 27: 277–282.

Sidney, S., 1956. Nonparametric statistics for the behavioural sciences. McGraw-Hill Book Company, Inc., Tokyo: 312 pp.

Smith, I. R., 1975. Turbulence in lakes and rivers. Scientific publication no. 29. Freshwater Biological Association, Ambleside, U.K: 79 pp.

Wright, J. F., 1992. Spatial and temporal occurrence of invertebrates in a chalk stream. Hydrobiol, 248: 11–30.

Vogel, S., 1994. Life in moving fluids The physical biology of flow. Princeton University Press, Princeton, New Jersey: 467 pp.

Hydrobiologia **415:** 177–185, 1999.
J. M. Caffrey, P. R. F. Barrett, M. T. Ferreira, I.S. Moreira, K. J. Murphy & P. M. Wade (eds), Biology, Ecology and Management of Aquatic Plants.
© 1999 *Kluwer Academic Publishers. Printed in the Netherlands.*

Habitat conditions of nymphaeid associations in Poland

Marcin Szańkowski* & Stanisław Kłosowski
Department of Environmental Botany, University of Warsaw, Al. Ujazdowskie 4, 00-478 Warsaw, Poland

Key words: ecological amplitudes, indicator value, CCA ordination

Abstract

A comparative analysis of the habitats of seven nymphaeid associations: Potamogetonetum natantis, Polygonetum natantis, Nuphar-Nymphaeetum albae, Nymphaeetum candidae, Nupharetum pumilae, Trapetum natantis and Nymphoidetum peltatae, was made within Poland. On the basis of water and substrate properties in the phytocoenoses analysed, it was demonstrated that most nymphaeid associations in Poland are distinctly associated with specific habitat conditions. Trapetum natantis occurs in neutral waters, poor in NH_4^+, but markedly rich in dissolved organic matter, K^+, PO_4^{3-}, total Fe and Ca^{2+}. It develops best on slightly acidic mineral substrates poor in Ca^{2+} and Na^+, but markedly rich in PO_4^{3-} and NO_3. Nymphoidetum peltatae occurs in alkaline waters with high concentrations of Na^+, Cl^- and NO_3^- and with relatively high concentrations of K^+ and PO_4^{3-}. Its substrates are mineral, poor in SO_4^{2-} and total N and rich in NO_3^-, PO_4^{3-} and K^+. Nupharetum pumilae develops in waters extremely poor in SO_4^{2-}, Na^+, K^+ and Ca^{2+}, but rich in NO_3^- and dissolved organic matter. It mainly occurs on organic substrates poor in Ca^{2+} and Na^+. Nymphaeetum candidae and Potamogetonetum natantis are mainly found in mesotrophic waters. Polygonetum natantis mainly inhabits mineral substrates poor in NO_3^-, Na^+ and dissolved SiO_2. Nupharo-Nymphaeetum albae occurs in a wide range of most types of habitat.

Introduction

In Europe, aquatic plants with floating leaves are typical components of the littoral vegetation of standing waters. In various types of water bodies, they form large and compact stands mainly dominated by one species (Van der Velde et al., 1986). Seven nymphaeid associations can be distinguished in Poland (Tomaszewicz, 1979): Potamogetonetum natantis, Soó, 1927; Polygonetum natantis, Soó, 1927; Nupharo-Nymphaeetum albae, Tomasz., 1977; Nymphaeetum candiae, Miljan, 1958; Nupharetum pumilae, Oberd., 1957; Trapetum natantis, Müll et Görs, 1960; Nymphoidetum peltatae (All. 1922); Bellot, 1951. They belong to the alliance Nymphaeion Oberd. 1957 and class Potamogetonetea R.Tx. et Prsg., 1942. The floristic composition and phytocoenose structure of the above associations in Poland have been thoroughly studied to date. Studies concerning their habitat conditions have been conducted since 1978. A number of works contain information on the habitat conditions of some of the associations (Kłosowski & Tomaszewicz, 1986, 1993, 1996).

The current work aims at summarizing results of ecological studies on nymphaeid associations conducted in Poland. A comparative analysis of the habitats of the associations was performed.

* Author for correspondence

Materials and methods

The survey was conducted in Poland between 1978 and 1997. All sites (nymphaeid phytocoenoses) were sampled during the period of maximal growth (July and August). In each phytocoenose single water and single substrate samples were taken for chemical analyses. Substrate samples from the rhizome-root layer were taken with a tubular bottom sampler. Each substrate sample contained a mixture of three subsamples taken randomly within a phytocoenose. Water samples were taken from a depth of 0.5–1.0 m. Altogether 12 water and substrate properties were analysed. Both analyses of water and substrate were performed immediately after the samples were transferred to the laboratory.

Water samples were anlaysed for: acidity (pH) with pH-meter; NH_4^+ by distillation and by Nessler's colorimetric method; NO_3^- colorimetrically with phenoldisulphonic acid and by cadmium reduction method; COD-$KMnO_4$ (as $KMnO_4$ consumption in acid medium), PO_4^{3-} by the colorimetric molybdate blue method; SO_4^{2-} by the nephelometric method; dissolved SiO_2 by the colorimetric molybdate method; Cl^- by Mohr's argentometric method and by mercuric thiocyanate method, total Fe by the colorimetric rhodanate method, Na^+, K^+ and Ca^{2+} by flame photometry.

In substrate samples, total Fe, Ca^{2+}, SO_4^{2-} and PO_4^{3-} were analysed in acid extracts (HCl 1+1). Concentrations of NO_3^- and dissolved SiO_2 were determined in water extracts. Hydration was determined by drying substrate samples at 105 °C to constant weight and organic matter content was assessed by ignition of dry samples at 550 °C for 1.5h. Na^+ and K^+ were determined from ignited sediment. Total nitrogen by Kjeldahl's method. The analyses from acid and water extracts, the determinations of Na^+, K^+ and acidity were conducted by the methods described above.

The associations studied, number of sampling sites (phytocoenoses) and number of water bodies from which phytocoenoses originated were as follows: Potamogetonetum natantis – 53, 41; Polygonetum natantis – 51, 32; Nupharo-Nymphaeetum albae, subass. with *Nuphar lutea* – 40, 31; Nupharo-Nymphaeetum albae, subass. with *Nymphaea alba* – 23, 20; Nymphaeetum candidae – 29, 9; Nupharetum pumilae – 28, 11; Trapetum natantis – 26, 9; Nymphodetum peltatae – 28, 8. The association of Nupharo-Nymphaeetum albae was split up into two subassociations depending on the dominance of either *Nuphar lutea* or *Nymphaea alba* in the phytocoenoses.

The vegetation-habitat relationships were determined by means of Canonical Correspondence Analysis (CCA). CCA is a multivariate technique developed to directly relate (plant) community composition to known variation in the environment (Ter Braak, 1986). The CCA ordination axes represent the main gradients of the variation in species and habitat data. The species points represent the centroids of the sites in which they occur on CCA model (Ter Braak, 1988, 1990). Since the nymphaeid associations are always dominated by their characteristic species, the ecological optima of associations studied can be estimated as the species points displayed in the diagrams. Water and substrate data were analysed separately. Both analyses were performed in two stages: first on all associations, then on associations that appeared to be very close to each other in the first stage. The computer program CANOCO 3.1 (Ter Braak, 1990) was applied.

Results

The arithmetic mean, minimum and maximum values of each habitat property for each association investigated are presented in Figures 1 and 2. (The arithmetic mean of acidity was calculated on the basis of H+ ion concentration.) Results of ordination are displayed on four ordination diagrams (Figures 3–6). Only the properties best correlated with ordination axes are shown. All associations studied, *versus* properties of water, are displayed on first ordination diagram (Figure 3). The biplot on Figure 4 shows the ordination on water properties after exclusion of Trapetum natantis and Nymphoidetum peltatae that appeared to be very distinct (see Figure 3). Results of analysis of all characteristic species *versus* substrate habitats are displayed in Figure 5. The biplot in Figure 6 displays results of ordination on the basis of substrate properties after exclusion of Nymphoidetum peltatae, Trapetum natantis and Polygonetum natantis which were best represented in the first stage (see Figure 5).

Results of ordination analyses provide insight into optimum conditions of nymphaeid associations. To describe the amplitudes of habitat properties one has to also consider the layout of minimum and maximum values (see Figures 1 and 2). On the basis of both CCA results (Figures 3–6) and the extreme values (Figures 1 and 2), it is clear that Trapetum natantis occurs in neutral waters (pH about 7), poor in NH_4^+, but markedly rich in dissolved organic matter (high COD-$KMnO_4$), K^+, PO_4^{3-}, total Fe and Ca^{2+}. Mean $Na^+$and Cl^- concentrations are also high. This association develops best on slightly acidic, mineral substrates poor in Ca^{2+} and Na^+, but markedly rich in PO_4^{3-} and NO_3^-.

Nymphoidetum peltatae mainly occurs in alkaline waters with high concentrations of Na^+, Cl^- and NO_3^-. It also indicates relatively high concentrations of K^+, PO_4^{3-} and dissolved organic matter. Substrates of Nymphoidetum peltatae are mineral, poor in SO_4^{2-} and total N and like those of Trapetum natantis rich in NO_3^-, PO_4^{3-} and K^+.

Nupharetum pumilae develops in waters extremely poor in SO_4^{2-}, Na^+, K^+ and Ca^{2+}, but relatively rich in NO_3^- and dissolved organic matter (indicated

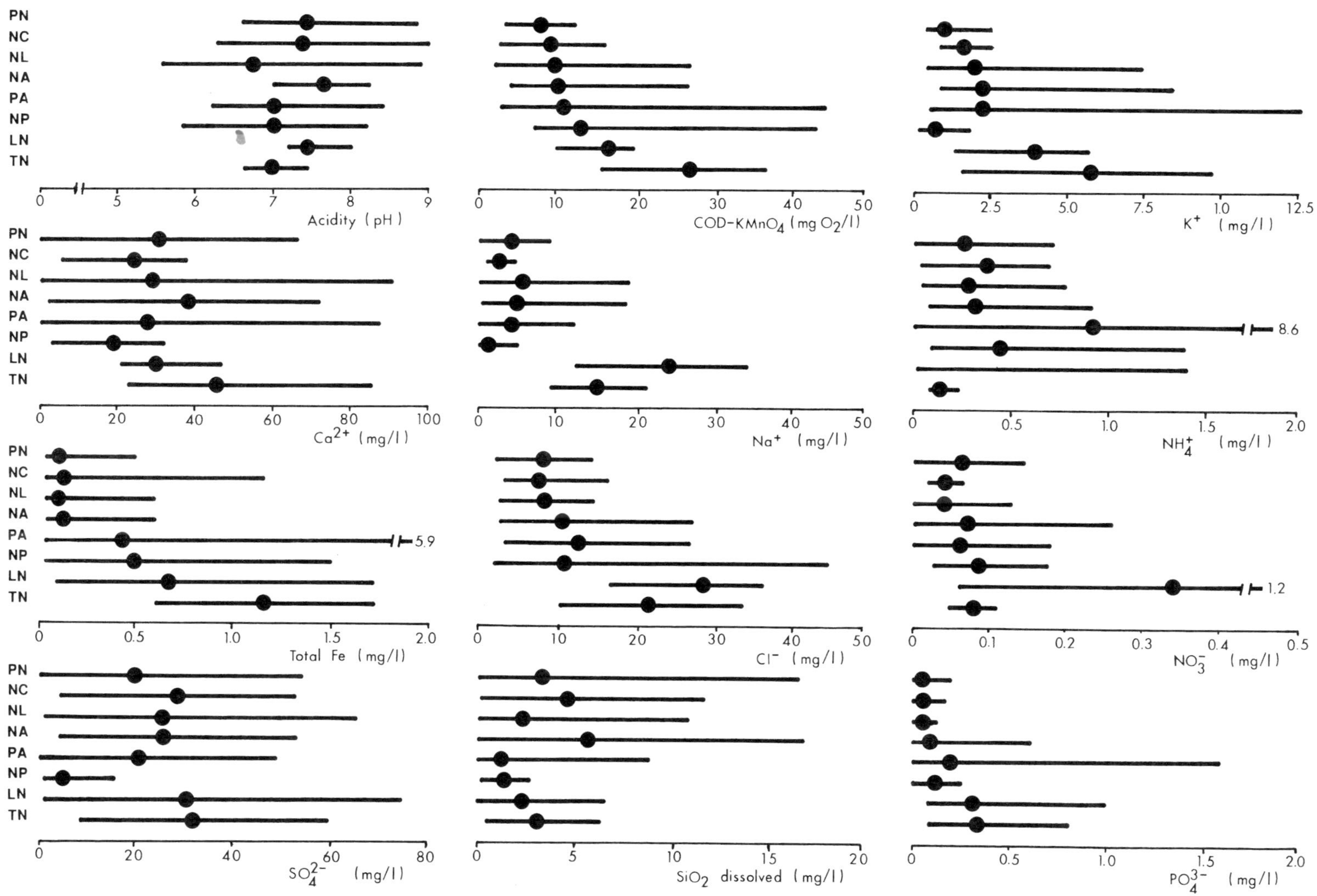

Figure 1. Water properties of nymphaeid associations studied. Ranges (minimum – maximum) and arithmetic mean. PN – Potamogetonetum natantis (n=53), PA – Polygonetum natantis (n=51), NL – Nupharo-Nymphaeetum subass. with *Nuphar lutea* (n=40), NA – Nupharo-Nymphaeetum subass. with *Nymphaea alba* (n=23), NC – Nymphaeetum candidae (n=29), NP – Nupharetum pumilae (n=28), TN – Trapetum natantis (n=26), LN – Nymphoidetum peltatae (n=28).

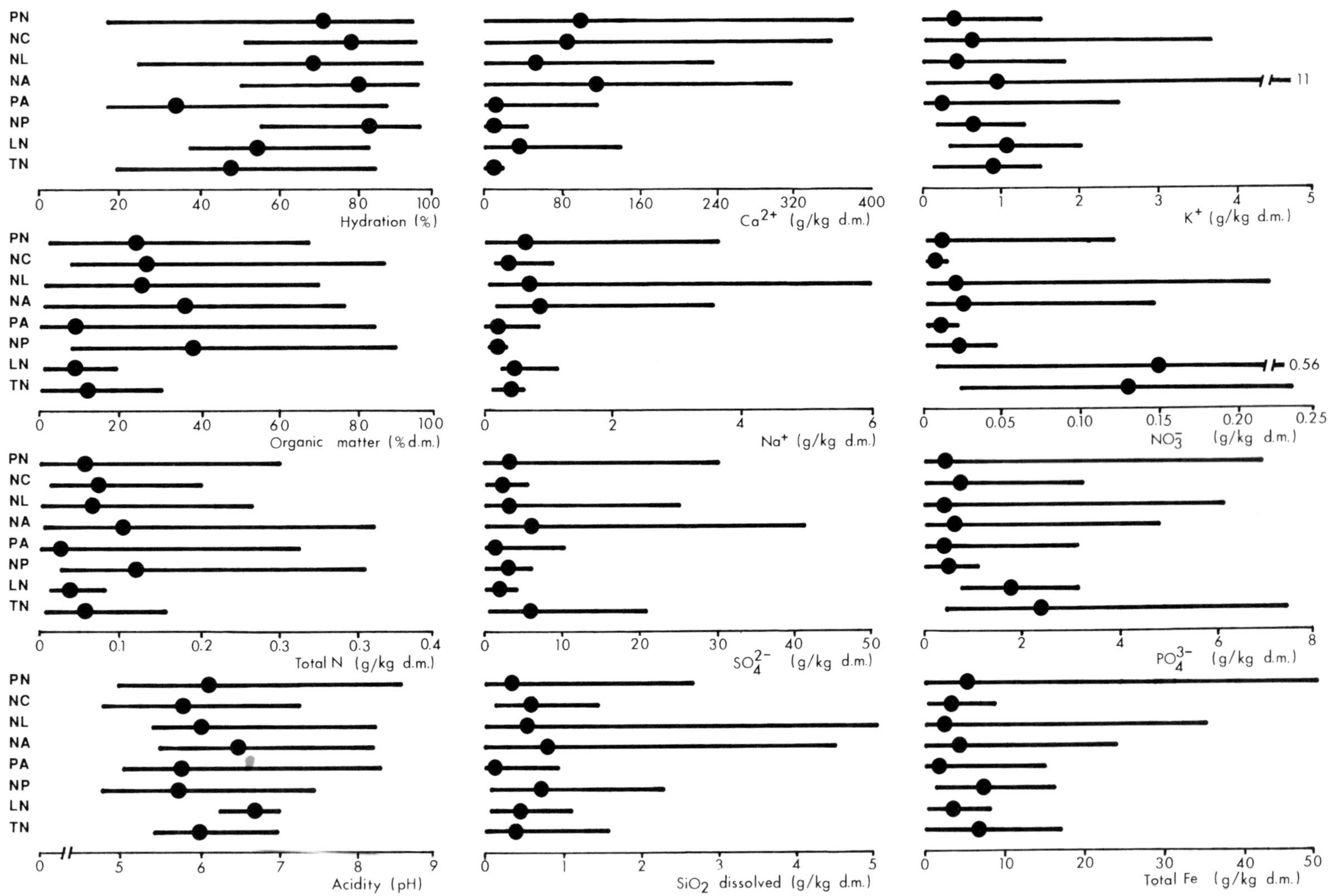

Figure 2. Substrate properties of nymphaeid associations studied. Ranges (minimum – maximum) and arithmetic mean. PN – Potamogetonetum natantis (n=53), PA – Polygonetum natantis (n=51), NL – Nupharo-Nymphaeetum subass. with Nuphar lutea (n=40), NA – Nupharo-Nymphaeetum subass. with *Nymphaea alba* (n=23), NC – Nymphaeetum candidae (n=29), NP – Nupharetum pumilae (n=28), TN – Trapetum natantis (n=26), LN – Nymphoidetum peltatae (n=28). 'd.m.' = dry matter.

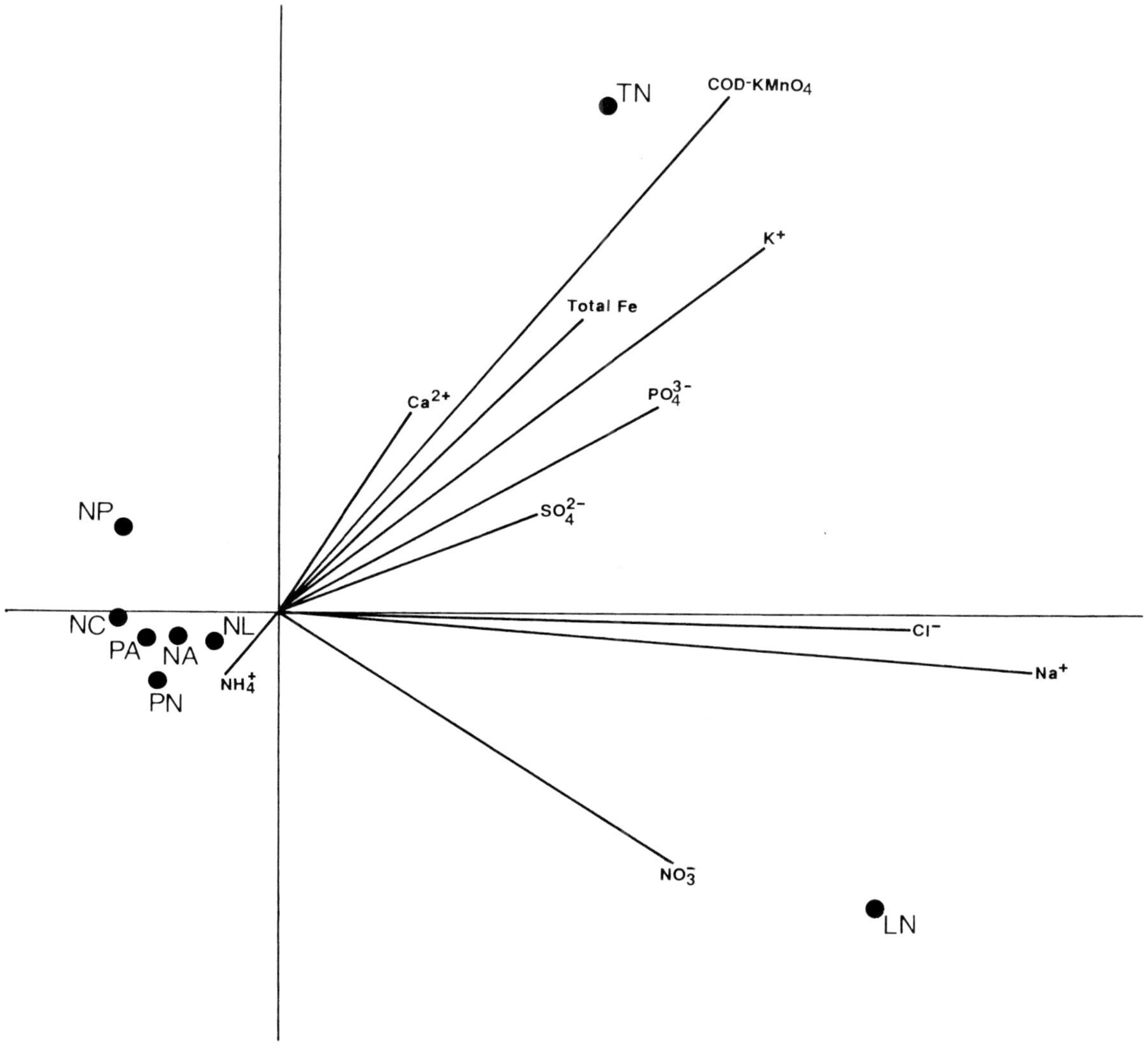

Figure 3. CCA ordination diagram of characteristic species of nymphaeid associations. Analysis was based on all associations studied and properties of water. Eigenvalues of 1st and 2nd ordination axes: L_1=0.84 and L_2=0.53; diagram displays 61% of variance in species data. Codes of associations – see Figure 1.

by COD-$KMnO_4$). Substrates of Nupharetum pumilae are highly hydrated and poor in Ca^{2+} and Na^+ but rich in organic matter, total Fe and total N.

Nymphaeetum candidae mainly occurs in mesotrophic waters (narrow ranges of nutrient concentrations) and acidic, highly hydrated, mineral-organic substrates.

Potamogetonetum natantis is distinctly associated with specific water properties. It occurs (like Nymphaeetum candidae) in mesotrophic waters.

Polygonetum natantis mainly inhabits mineral substrates poor in Na^+, NO_3^- and dissolved SiO_2, but the ranges of organic matter, hydration and total N are wide.

Nupharo-Nymphaeetum albae occurs in the wide range of most properties of habitat (water and substrate). However, when its subassociations are analysed separately, the subassociation with *Nuphar lutea* occurs in more acidic waters, poorer in PO_4^{3-}, NO_3^-, Cl^- and dissolved SiO_2 (Figures 1 and 4). Substrates of this subassociation are poorly hydrated and

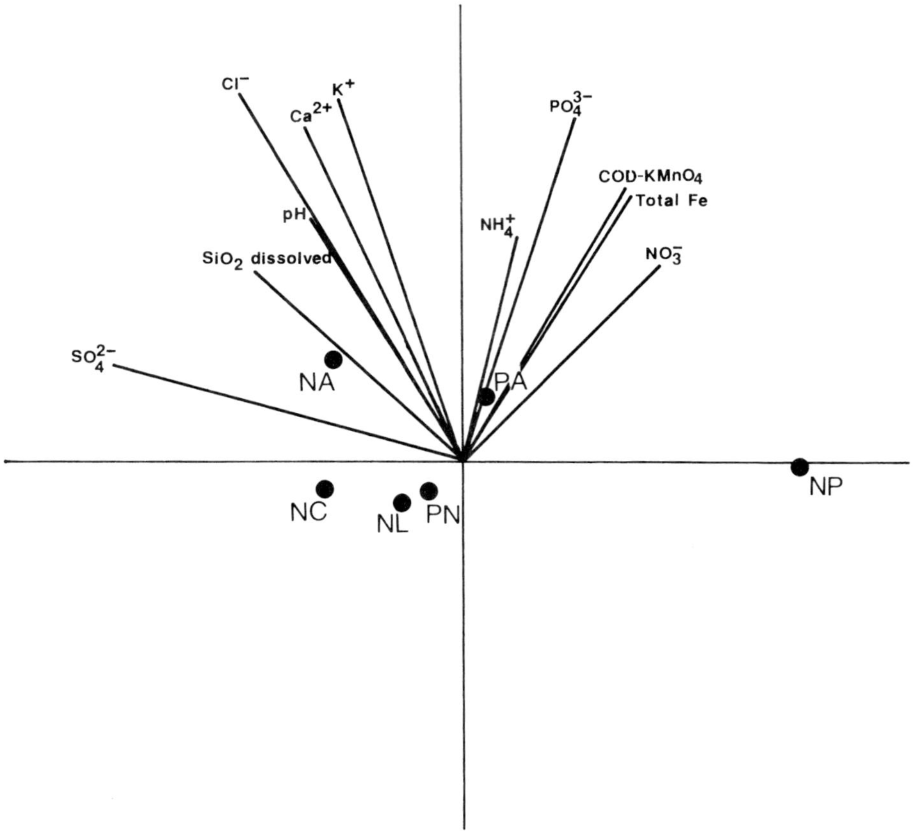

Figure 4. CCA ordination diagram of characteristic species of nymphaeid associations. Analysis was based on properties of water. The associations of TN and LN were not included in the analysis. Eigenvalues of 1st and 4th ordination axes: L_1=0.41 and L_4=0.12; diagram displays 41% of variance in species data. Codes of associations – see Figure 1.

have lower concentrations of organic matter and Ca^{2+} (Figures 2 and 6).

Discussion

Current findings indicated that the nymphaeid associations in Poland were distinct with respect to water and sediment chemistry preferences. The biggest differences between habitats were found in the case of: Trapetum natantis, Nymphoidetum peltatae, Nupharetum pumilae, Nymphaeetum candidae (water and substrate properties), Polygonetum natantis (substrates) and Potamogetonetum natantis (water properties). Results obtained confirm data from numerous works on the habitat conditions of nymphaeid species and their communities in other regions of Europe. Data of Jorga et al. (1982) and Pietsch (1982) indicated that *Trapa natans* and Trapetum natantis phytocoenoses inhabited neutral and alkaline waters, poor in NH_4^+ and relatively rich in SO_4^{2-}. Papastergiadou & Babalonas (1993a, b) demonstrated that Trapetum natantis occurred in relatively fertile waters.

In the case of Nymphoidetum peltatae, it was shown that the above association mainly developed in alkaline waters (Pott, 1980; Pietsch, 1982), rich in NO_3^- (Papastergiadou & Babalonas, 1993a, b), chlorides and Na^+ (Van der Velde et al., 1986). Data of Van der Velde's et al. (1986) confirmed that *Nymphoides peltata* was mainly associated with substrates relatively rich in K^+ and PO_4^{3-} but poor in SO_4^{2-}.

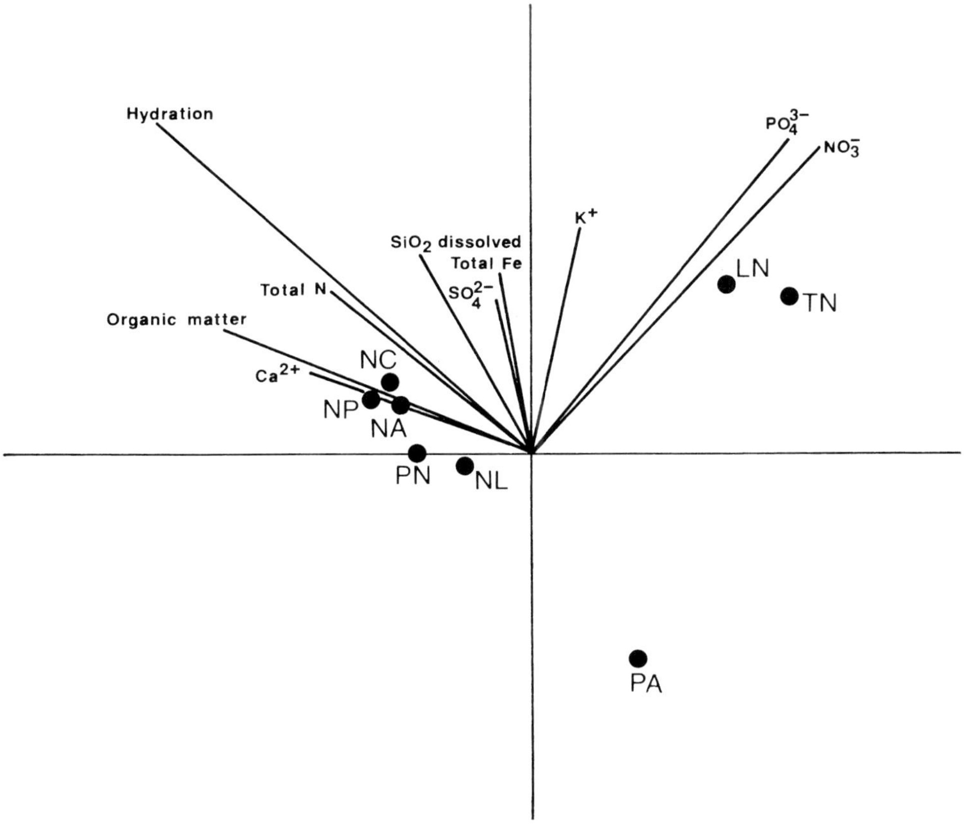

Figure 5. CCA ordintion diagram of characteristic species of nymphaeid associations. Analysis was based on properties of substrate. Eigenvalues of 1st and 2nd ordination axes: L_1=0.63 and L_2=0.46; diagram displays 58% of variance in species data. Codes of associations – see Figure 1.

The specific habitat of *Nuphar pumila* and of the phytocoenoses formed by it (waters poor in Ca^{2+}) was confirmed by Kadono (1982), Roweck (1988) and Doll (1991). Roweck's (1988) data were in agreement with our findings indicating that patches of *Nuphar pumila* were confined to organic substrates.

Pietsch (1982) reported that *Nymphaea candida* and the phytocoenoses formed by it were mainly restricted to mesotrophic waters. Moreover, results of Van der Velde et al. (1986) indicated that *Nymphaea candida* was confined to more restricted ranges of most analysed water components and some substrate properties, compared to *Nymphaea alba* and *Nuphar lutea*. According to the present data, this also applies to the phytocoenoses of Nymphaeetum candidae.

Various data are available for *Potamogeton natans* and *Polygonum amphibium* and the phytocoenoses formed by the two species. According to some authors, the aforementioned species and phytocoenoses are associated with mesotrophic waters (e.g. Felzines, 1977; Pott, 1980). Other authors (e.g. Wiegleb, 1978; Pietsch, 1982) point to the wide ecological amplitude of both species. In Poland the phytocoenoses of Potamogetonetum natantis are usually found in mesotrphic waters. In the case of Polygonetum natantis, the ranges of most properties of water are wide. The substrates inhabited by the aforementioned association are usually mineral, poor in NO_3^-, Na^+ and dissolved SiO_2.

The present studies and the aut- and synecological data in the relevant literature point to a wide ecological amplitude of *Nuphar lutea* and *Nymphaea alba* and phytocoenoses formed by the two species regarding their water (Wiegleb, 1978; Pott, 1980; Pietsch, 1982;

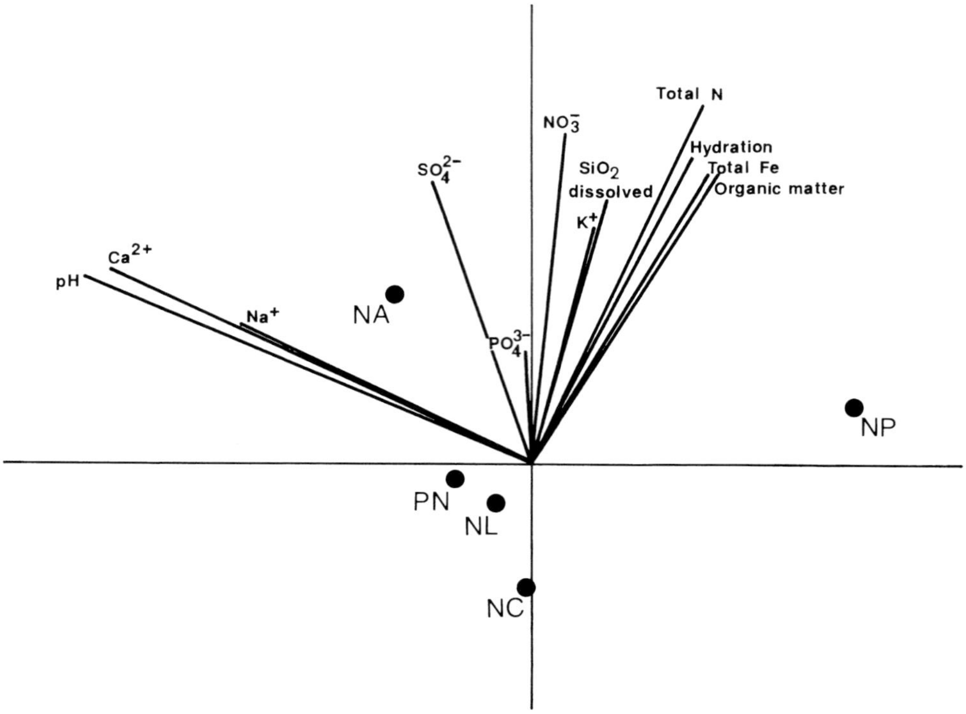

Figure 6. CCA ordination diagram of characteristic species of nymphaeid associations. Analysis was based on properties of substrate. The associations of LN, TN and PA were not included in the analysis. Eigenvalues of displayed 1st and 3rd ordination axes: L_1=0.31 and L_3=0.12; diagram displays 54% of variance in species data. Codes of associations – see Figure 1.

Van der Velde et al., 1986) and substrate (De Lyon & Roelofs, 1986a, b; Van der Velde et al., 1986; Van Katwijk & Roelofs, 1988) properties.

Our results and data of other authors indicate that considerable differences between most of the nymphaeid communities are found both with respect to their ecological amplitude and optimum habitat conditions. The structural differentiation of the associations analysed, based on floristic dominance, has been confirmed by this habitat differentiation. Hence, associations such as Trapetum natantis, Nymphoidetum peltatae, Nupharetum pumilae and Nymphaeetum candidae can be considered as indicators of various types of littoral habitats.

References

De Lyon, M. J. H. & J. G. M. Roelofs, 1986a. Waterplanten in relatie tot waterkwaliteit en bodemgesteldheid. Deel 1. Laboratorium voor Aquatische Oecologie. Katholieke Universiteit Nijmegen, Nijmegen: 106 pp.

De Lyon, M. J. H. & J. G. M. Roelofs, 1986b. Waterplanten in relatie tot waterkwaliteit en bodemgesteldheid. Deel 2. Laboratorium voor Aquatische Oecologie. Katholieke Universiteit Nijmegen, Nijmegen: 126 pp.

Doll, R., 1991. Die Pflanzengesellschaften der stehenden Gewässer in Mecklenburg-Vorpommeern. Teil I.3. Potamogetonetea Tx. et Prsg. 42- Laichkrautgesellschaften. Feddes Repertorium 102: 217–317.

Felzines, J. C., 1977. Analyse des relations entre la mineralisation des eaux stagnantes et la distribution des végétaux qui les peuplent. Ann. Sci. Nat. Bot. 18: 221–249.

Jorga, G., W. Pietsch & G. Weise, 1982. Beitrage zur Ökologie und Bioindikation von *Trapa natans* L. Limnologica 14: 385–394.

Kadono, Y., 1982. Occurrence of aquatic macrophytes in relation to pH, alkalinity, Ca^{2+}, Cl^- and conductivity. Jap. J. Ecol. 32: 39–44.

Katwijk, M. M. Van. & J. G. M. Roelofs, 1988. Vegetatis van waterplanten in relatie tot het milieu. Laboratorium voor Aquatische Oecologie. Katholieke Univeeersiteit Nijmegen, Nigmegen: 133 pp.

Kłosowski, S. & H. Tomaszewicz, 1986. Habitat requirements of *Polygonetum natantis* Soó, 1927 and *Potamogetonetum natantis* Soó, 1927 phytocoenoses in north-eastern Poland. Acta. Soc. Bot. Pol. 55: 141–157.

Kłosowski, S. & H. Tomaszewicz, 1993. Standortverhältnisse der Gesellschaften mit Dominanz einzelner Nymphaeaceen in Nordost-Polen. Tuexenia 13: 75–90.

Kłosowski, S. & H. Tomaszewicz, 1996. Zur Ökologie des Trapetum natantis und des Nymphoidetum peltatae in Polen. Tuexenia 16: 105–116.

Papastergiadou, E. & D. Babalonas, 1993a. The relationships between hudrochemical environmental factors and the aquatic macropytic vegetation in stagnant and slow flowing waters. I. Water quality and distribution of aquatic associations. Arch. Hydrobiol./Suppl. 90: 475–491.

Papastergiadou, E. & D. Babalonas, 1993b. The relationships between hydrochemical environmental factors and the aquatic macrophytic vegetation in stagnant and slow flowing waters. II. Evaluation of plant associations indicative value. Arch. Hydrobiol./Suppl. 90: 493–506.

Pietsch, W., 1982. Makrophytische Indikatoren fur okochemische Beschaffenheit der Gewässer. In: Breitig, G. & W. Tumpling (eds), Ausgewahlte Methoden der Wasseruntersuchung 2. Gustav Fischer, Jena: 67–88.

Pott, R., 1980. Die Wasser- und Sumpfvegetation eutropher Gewässer in der Westfalischen Bucht – Pflanzensoziologische und hydrochemische Untersuchungen. Abh. Landesmus. Naturkde. Münster in Westfalen 42: 1–156.

Roweck, H., 1988. Ökologische Untersuchungen an Teichrosen. Arch. Hydrobiol./Suppl. 81: 103–358.

Ter Braak, C. J. F., 1986. Canonical Correspondence Analysis; a new eigenvector technique for multivariate direct gradient analysis. Ecology 67: 1167–1179.

Ter Braak, C. J. F., 1988. CANOCO – a FORTRAN program for canonical community ordination by (partial) (detrended) (canonical) correspondence analysis, principal components analysis and reduncancy analysis (version 2.1). Report LWA-88-02. Agricultural Mathematics Group, Wageningen: 95 pp.

Ter Braak, C. J. F., 1990. Update notes: CANOCO version 3.1. Agricultural Mathematics Group, Wageningen, 35 pp.

Tomaszewicz, H., 1979. Roślinność wodna i szuwarowa Polski (klasy: Lemnetea, Charetea, Potamogetonetea, Phragmitetea) wg stanu zbadania na rok 1975. Rozprawy, Uniw. Warszawskiego 160: 1.325.

Van der Velde, G., C. P. C. Custers & M. J. H. de, Lyon, 1986. The distribution of four nymphaeid species in the Netherlands in relation to selected abiotic factors. Proceedings EWRS/AAB 7th. Symposium on Aquatic Weeds, Loughborough: 363–368.

Wiegleb, G., 1978. Untersuchungen über den Zusammenhang zwischen hydrochemischen Umweltfaktoren und Makrophytenvegetation in stehenden Gewässern. Arch. Hydrobiol. 83: 443–484.

Hydrobiologia **415:** 187–191, 1999.
J. M. Caffrey, P. R. F. Barrett, M. T. Ferreira, I.S. Moreira, K. J. Murphy & P. M. Wade (eds),
Biology, Ecology and Management of Aquatic Plants.
© 1999 *Kluwer Academic Publishers. Printed in the Netherlands.*

Classification of Danube Delta lakes based on aquatic vegetation and turbidity

Hugo Coops[1]*, Jenica Hanganu[2], Marian Tudor[2] & Willem Oosterberg[1]
[1] *Institute for Inland Water Management and Waste Water Treatment (RIZA), P.O. Box 17, NL-8200 AA Lelystad, The Netherlands*
E-mail: h.coops@riza.rws.minvenw.nl
[2]*Danube Delta Institute (DDI), 165 Babadag Str., RO-8800 Tulcea, Romania*

Key words: submerged macrophytes, lakes, classification, eutrophication, remote sensing, transparency

Abstract

The aquatic vegetation present in lakes in the delta of the River Danube was studied using field survey data and satellite image. Based on the spectral information from satellite images, three categories of lakes were distinguished: clear / macrophyte-dominated, intermediate and turbid / poorly vegetated. The satellite-based classification was consistent with vegetation cover and water transparency measured in the field. Cluster analysis of vegetation data from 235 relevés made in 22 lakes (1996–1998, each year in June) identified ten submerged vegetation types. The lakes were characterised by the frequency of occurrence of each vegetation type and ordinated, using Principal Components Analysis (PCA). A strong relationship appeared between the composition and density of aquatic vegetation and water transparency. Isolated lakes within the extensive beds of floating reed (plaur lakes) formed a separate group. Based on the ordination analysis, connectivity of lakes in the Danube Delta was concluded to be a major factor for the development of dense macrophyte vegetation.

Introduction

Shallow lakes located in the floodplains and estuaries of large rivers are often characterised by a well-developed submerged vegetation. Macrophyte composition and cover reflect the relative influence of river water in these lakes (Van den Brink et al., 1991; Janauer & Kum, 1996). Hydrology and morphology of the floodplain determine fluxes of river water to and from lakes situated within the floodplain. River water loaded with suspended matter and often rich in nutrients enters water bodies via channels, overland flow during floods, seepage or by flowing under floating reedbeds. The external input of water can be regarded as the 'driving force' of the ecosystem, the impact of which varies due to differences in access, residence time and dimensions of the lakes involved. Intact floodplains consist of a complex gradient from water bodies with an open connection to the river channel to highly isolated ones.

Light and nutrients are major factors controlling macrophyte development in lakes (Spence, 1982; Chambers, 1987). Inflow of water loaded with suspended matter and nutrients that may result in turbid conditions prevent development of a rich macrophyte community. On the other hand, submerged vegetation can have a strong positive effect on water clarity (Van den Berg et al., 1997). The Romanian part of the Danube delta is a large (4400 km^2) complex system that contains hundreds of shallow lakes. Due to nutrient loading increase (N, P fertilizers and urban discharge), eutrophication problems of these lakes have been increasing since the early 1970s (Heliotis et al., 1994). At the same time, newly-dug channels in the Delta have increased the degree of connectivity of many lake systems with the main river channels.

In the present paper, we examine the aquatic vegetation communities in relation to lake turbidity in the Danube Delta. We used satellite images to classify lakes and study the relationship between defined lake types and their vegetation characteristics.

* Author for correspondence

Materials and methods

A classification was made of the water bodies in the Danube Delta, using the Landsat TM satellite image of 6th July 1996. Terrestrial and emergent vegetation were masked using the infrared spectral band, but visual correction was necessary to account for floating vegetation. To classify the water types, the reflectance features of spectral bands 1–4 were combined in feature spaces, using the image analysis programme ERDAS-Imagine (Smith & Brown, 1997). Each individual pixel was classified into one of the categories clear water, algal blooming, suspended solids (low, intemediate and high levels), submerged vegetation and floating vegetation. Lakes were characterised on the basis of presence/absence of these categories.

Submerged vegetation was sampled between 1996 and 1998 in 235 plots situated in 22 water bodies in the Danube Delta. The samples were taken in June of each year, to avoid significant temporal variance. Each plot had a diameter of 5 m and was made from a canoe. Submerged plants were collected using a rake, and for each species the percentage cover projected on the bottom was visually estimated using a 7-point scale (<1%, 1–5%, 5–15%, 15–25%, 25–50%, 50–75%, 75–100%). Each of the total percentage covers of the vegetation and filamentous algae were determined separately. Additionally, measurements of water depth and transparency (Secchi-depth: water depth) were made at each sampling point.

A cluster analysis (TWINSPAN; Hill, 1979) was performed to classify vegetation into a number of distinct types. Frequencies of occurrence of the vegetation types within each lake were used as a characteristic of the lake's vegetation. Data collected from lakes (vegetation types, average cover of vegetation and filamentous algae, average water depth, transparency and distance from main river channel) were ordinated using Principal Components Analysis (CANOCO, Ter Braak, 1991).

Table 1. Occurrence of lake categories in 19 lakes in the Danube Delta, based on Landsat image 6th July 1996. Categories: sv = submerged vegetation, cl = clear water, ls = low suspended solids, ms = intermediate suspended solids, hs = high suspended solids, ab = algal blooming, fv = floating vegetation

Lake group		sv	cw	ls	is	hs	ab	fv
Clear lakes	Plin	x						
	Serbata	x						
	Raduculet	x	x					
	Tataru	x	x					
	Potcoava (north)	x	x					
	Raducu	x	x					
	Cuibul cu Lebede	x	x					
	Baclanesti	x	x					
	Gorgova (west)	x	x					
	Babina	x	x					
Intermediate lakes	Holbina 2 (north)	x	x		x		x	
	Isac	x	x		x		x	
	Furtuna	x	x		x		x	
	Gorgova	x	x	x	x		x	
Turbid lakes	Uzlina				x	x	x	x
	Rosulet			x	x	x	x	
	Rosu			x	x	x	x	
	Potcoava (south)				x	x	x	
	Holbina 2 (centre)				x	x	x	

Results

Three types of lakes were distinguished based on the satellite classification (Table 1): 'clear' lakes with only pixels categorised as clear water and/or submerged vegetation; 'turbid' lakes showing a pattern of high suspended solids and/or algal blooms (with patches of floating vegetation in the case of Uzlina); the 'intermediate' lakes showed a combination of clear water, submerged vegetation, suspended solids and algal blooming.

Submerged plants were present in almost all vegetation samples. The samples from 'clear' lakes showed a high average vegetation cover compared to 'turbid' lakes (81% and 32%, respectively). Cluster analysis of the samples delineated a number of vegetation types (Table 2). Characteristics of each vegetation type are listed in Table 3.

The correlation between vegetation composition (in terms of frequency of each vegetation type in the set of vegetation samples made in a lake) and environmental parameters, was evident from the Principal Components Analysis (Figure 1). The 'clear' to 'turbid' lake gradient identified from the satellite classification was clearly reflected in the PCA.

One end of the range consists of lakes where water originating from the main river branches enters via short, direct connections, showed a high turbidity and a sparse vegetation characterised by *Potamogeton crispus*. At the other end of the range were the clear

Table 2. Occurrence of macrophyte species in the vegetation types based on TWINSPAN classification. Frequency classes: 1 = 0–10%, 2 = 10–20%, ... , 10 = 90–100% of relevés. Vegetation types: Pc = *Potamogeton crispus*-type, Pp = *P. pectinatus*-type, Ms = *Myriophyllum spicatum*-type, Sa = *Stratiotes aloides*-type, Pl = *P. lucens*-type, Tn = *Trapa natans*-type, Pt = *Potamogeton trichoides*-type, Nl = *Nuphar lutea*-type, Cd = *Ceratophyllum demersum*-type, No = *Nitellopsis obtusa*-type. Species not exceeding class 1 in any vegetation type are not included

	Pc	Pp	Ms	Sa	Pl	Tn	Pt	Nl	Cd	No
Potamogeton crispus	10	4	1		7	4	5	1	2	1
Potamogeton perfoliatus	2	1	1			2	1			
Potamogeton pectinatus	2	10	7	1	3	6	6	2		3
Myriophyllum spicatum	2	1	10		3	1	1	3		1
Elodea nuttallii		2	6	2	3	6	4	5	2	1
Najas marina	1		3			1			1	1
Stratiotes aloides			1	10		1		1	1	1
Hydrocharis morsus-ranae			1	7		1	1	4	2	1
Lemna minor		1	2	5	3	5	2	5	1	1
Lemna trisulca	1	2	4	7	7	2	2	5	2	1
Spirodela polyrhiza	1	2	3	10	3	7	3	6	2	1
Salvinia natans			1	3	3	1	1	3	1	1
Potamogeton lucens			3	1	10	1				
Trapa natans						10	1	1	1	
Potamogeton trichoides		2	3		7	4	10	2		1
Potamogeton compressus		3	1			2	3	2	1	2
Potamogeton pusillus	1	3	1				4	3		1
Nuphar lutea	1			2	3	1		7		
Nymphaea alba			1	2	3	1		4		
Nymphaea candida			1	1				2		
Ceratophyllum demersum	3	7	10	10	9	8	10	10	10	10
Nitellopsis obtusa	1	1	1				1		1	10

lakes almost completely filled with vegetation (typically *P. trichoides*). At the time of observation, in June 1996–1998, the cover of filamentous algae was very high in wind-exposed parts of these lakes. Macrophytes covered by filamentous algae showed signs of decay. Observations made in August (results not presented) indicate that only few highly isolated lakes remained macrophyte-dominated throughout the summer, whereas most lakes became highly turbid as the result of phytoplankton blooms.

The group of relatively small, isolated lakes situated within floating reed beds formed a separate cluster, correlated with a long distance to the main river channels. Within these lakes, a range from hypertrophic, green-algae dominated lakes with very sparse macrophyte vegetation, to highly translucent lakes with a high density of *Ceratophyllum demersum* and the charophyte *Nitellopsis obtusa*, was observed.

Discussion and conclusions

Landsat images are not particularly applicable for a classification of surface water because of restrictions of spectral information and scale (Lehmann & Lachavanne, 1997). Climatic conditions could additionally mask the distinct features of lakes in the survey. However, by making a supervised classification (implying that ground-truth data were used for defining the categories), the information was shown to be sufficient to distinguish clear and turbid water, and to indicate the dominance of aquatic vegetation, algal blooms or high loads of suspended silt in the water.

Submersed vegetation development is strongly related to the underwater light climate (Spence, 1982). The presence of certain vegetation types thus may reflect the water quality of lakes. For example, Charophytes such as *Nitellopsis obtusa* are associated with clear-water systems. Other vegetation types are less dependent on water clarity, such as nymphaeids. *Nymphaea* and *Nuphar*-dominated vegetation in the Danube delta often indicates at the occurrence of sunken reedbeds (the so-called 'plaur'; J. Hanganu, pers. observation).

Water transparency in the lakes and the proliferation of submersed vegetation may be strongly related to the hydrology of the delta. Floods usually occur in springtime and the start of the growth season for aquatic plants typically coincides with falling water levels. As our survey was done in June, it represents this falling-water level situation. For most lakes, the residence time of water in this period is relatively short, preventing algal blooms to occur. Such blooms, that occur in summer, may cause high turbidities and as the result, the submerged plants have disappeared from many lakes in late summer (J. Hanganu, pers. observation). Dense submersed vegetation only remains present throughout the summer in highly isolated water bodies.

The degree of connectivity to the main river branches is most likely to be of major importance: the Danube river carries high amounts of suspended silt and nutrients. These can be transported through the delta via a chain of lakes connected by canals. The first lakes in the chain are often characterised by high suspended sediment concentrations and a high sedimentation. Algal blooming appears later in the chain, due to the time necessary for the algal population to grow. The lakes within the floating reedbeds form a separate category, due to the origin of the inflowing water, which has flowed under the reedbeds before

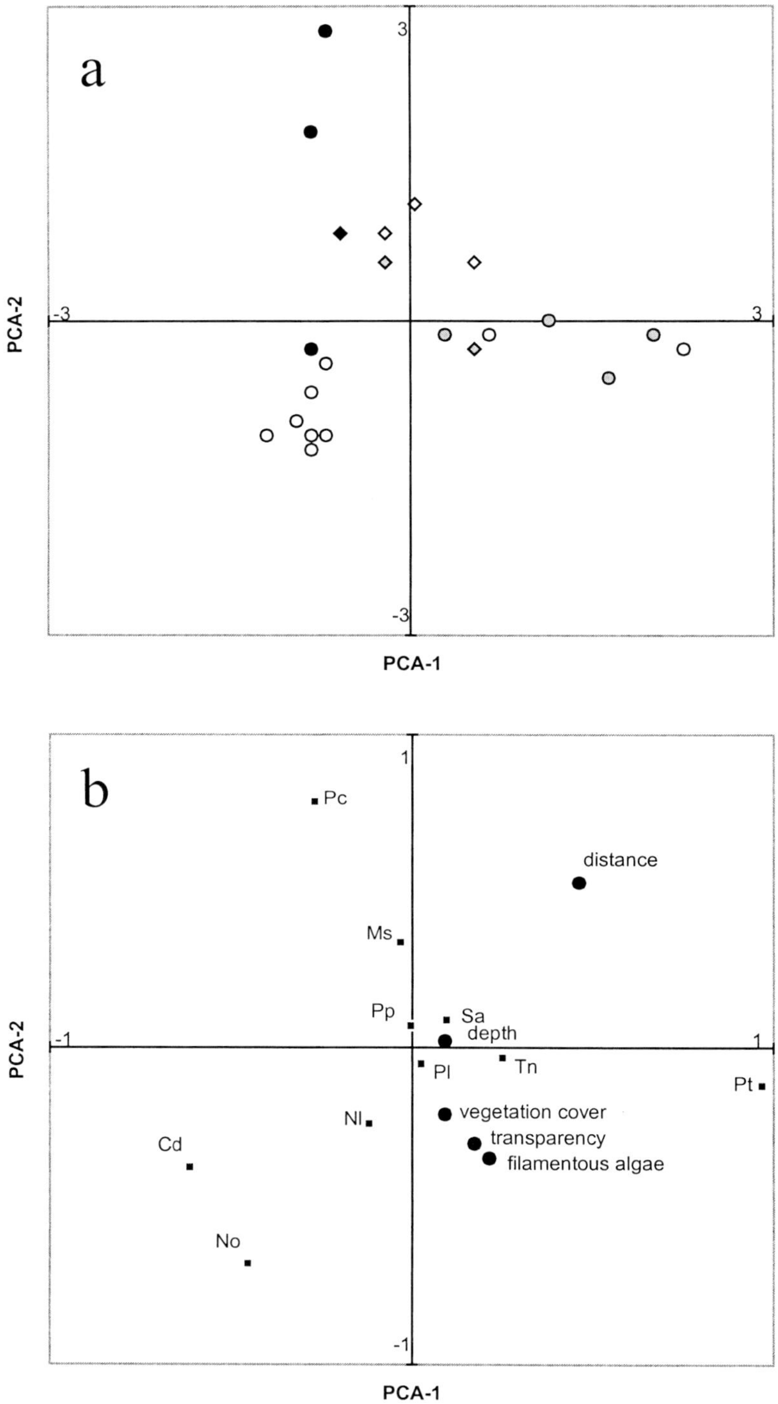

Figure 1. Ordination (Principal Components Analysis) of (a) lakes, (b) vegetation types and environmental variables of 22 lakes in the Danube Delta. The first two axes of the ordination are shown (eigenvalue of the first two axes 0.36 and 0.21, respectively). Classes of lakes derived from satellite image are indicated: black = turbid lakes, grey = intermediate lakes, and white = clear lakes. Diamond-shaped symbols indicate lakes which formerly were in use as polders or fishponds. Codes for vegetation types: see Table 2. Environmental variables: vegetation cover = lake mean vegetation cover%, filamentous algae = lake mean filamentous algae cover%, depth = mean water depth, transparency = mean transparency, distance = distance between main river channel and inflow in lake.

Table 3. Percent vegetation cover (% ± s.d.), water depth (cm ± s.d.) and number of species per relevé (n ± s.d.) of submerged vegetation types (Number of samples in brackets after vegetation type)

	Cover (%)	Water depth (cm)	Number of species
Potamogeton crispus-type (15)	19 ± 31	244 ± 99	2.3 ± 1.6
Potamogeton pectinatus-type (21)	62 ± 32	163 ± 43	3.7 ± 1.7
Myriophyllum spicatum-type (31)	77 ± 28	110 ± 49	5.7 ± 2.2
Stratiotes aloides-type (10)	90 ± 22	139 ± 24	6.1 ± 1.9
Potamogeton lucens-type (5)	64 ± 32	130 ± 51	5.6 ± 2.5
Trapa natans-type (12)	67 ± 32	136 ± 49	6.0 ± 3.3
Potamogeton trichoides-type (41)	77 ± 29	190 ± 54	4.9 ± 2.0
Nuphar lutea-type (17)	81 ± 24	133 ± 33	6.2 ± 2.1
Ceratophyllum demersum-type (37)	68 ± 35	181 ± 38	1.8 ± 1.4
Nitellopsis obtusa-type (41)	86 ± 25	175 ± 31	2.9 ± 1.4

entering the lakes. They are characterised by a high retention of autochthonous organic matter and oxygen depletion in the water. The ordination has shown that the vegetation in these lakes is different from the others (dominance of *Ceratophyllum demersum* and *Nitellopsis obtusa*).

Changes over the past decades in macrophyte dominance have been attributed to increased eutrophication . The response of floodplain lakes to increased nutrient levels in the river has been shown to relate to the degree of connectivity for instance by the studies of Van den Brink et al. (1993) and Heiler et al. (1995).

In view of the management policy for the Danube Delta, which is aimed at a sustainable exploitation of reeds and fish stocks, as well as at preserving a high biodiversity, it would be worthwhile to elucidate the possibility of sustaining intermediate levels of connectivity (e.g. by blocking particular channels).

Acknowledgements

We gratefully acknowledge Hans den Hollander (Rijkswaterstaat, Survey Department) for the spectral analysis of a Landsat image. The survey was part of the project "Eutrophication of Danube Delta lakes", within the cooperation of DDI (Romania) and RIZA (the Netherlands).

References

Chambers, P. A., 1987. Light and nutrients in the control of aquatic plant community structure: II. *In situ* observations. J. Ecol. 75: 621–628.

Heiler, G., T. Hein, F. Schiemer & G. Bornette, 1995. Hydrological connectivity and flood pulses as the control aspects for the integrity of a river - floodplain system. Regul. Riv. 11: 351–361.

Heliotis, F., G. Vellidis, D. Bandacu & C. Pringle, 1994. The Danube Delta: historical wetland drainage and potential for restoration. In Mitsch, W. J. (ed.), Global Wetlands: Old World and New. Elsevier, Amsterdam: 759–767.

Hill, M. O., 1979. TWINSPAN, a FORTRAN program for arranging multivariate data in an ordered two-way table by classification of individuals and attributes. Cornell University, Thaca, N.Y.: 90 pp.

Janauer, G. A. & G. Kum, 1996. Macrophytes and floodplain water dynamics in the river Danube ecotone research region (Austria). Hydrobiologia 340: 137–140.

Lehmann, A. & J.-B. Lachavanne, 1997. Geographic information systems and remote sensing in aquatic botany. Aquat. Bot. 58: 195–207.

Smith, C. & N. Brown, 1997. ERDAS Field Guide. 4th edn. ERDAS Inc: 655 pp.

Spence, D. H. N., 1982. The zonation of freshwater plants. Adv. ecol. Res. 12: 37–125.

Ter Braak, C. J. F., 1991. CANOCO, a FORTRAN program for canonical community ordination by (partial) (detrended) (canonical) correspondence analysis, principal components analysis and redundancy analysis. TNO Institute of Applied Computer Science, Wageningen: 95 pp.

Van den Berg, M. S., H. Coops, M. L. Meijer, M. Scheffer & J. Simons, 1997. Clear water associated with a dense Chara vegetation in the shallow and turbid Lake Veluwemeer, The Netherlands. In Jeppesen, E., M. Sondergaard, M. Sondergaard & K. Christoffersen (eds), The Structuring Role of Submerged Macrophytes in Lakes. Springer Verlag, Berlin: 339–352.

Van den Brink, F. W. B., M. M. J. Maenen, G. Van der Velde & A. Bij de Vaate, 1991. The (semi-) aquatic vegetation of still waters within the floodplains of the rivers Rhine and Meuse in The Netherlands: historical changes and the role of inundation. Ver. int. Ver. Limnol. 24: 2693–2699.

Van den Brink, F. W. B., J. P. M. De Leeuw, G. Van der Velde & G. M. Verheggen, 1993. Impact of hydrology on the chemistry and phytoplankton development in floodplain lakes along the Lower Rhine and Meuse. Biogeochemistry 19: 103–128.

Hydrobiologia **415:** 193–201, 1999.
J. M. Caffrey, P. R. F. Barrett, M. T. Ferreira, I.S. Moreira, K. J. Murphy & P. M. Wade (eds), Biology, Ecology and Management of Aquatic Plants.
© 1999 *Kluwer Academic Publishers. Printed in the Netherlands.*

The distribution in bankside habitats of three alien invasive plants in the U.K. in relation to the development of control strategies

F. Hugh Dawson[1,*] & David Holland[2]
[1]*NERC-Institute of Freshwater Ecology, River Laboratory, Wareham, Dorset, U.K.*
[2]*Formerly Environment Agency - North West Region, Sale, Cheshire, U.K. Present address: 20 Springmeadow Road, Liverpool L25 3PT, U.K.*

Key words: bankside, invasive plants, introductions, control strategies

Abstract

The national distribution of *Fallopia japonica* (Japanese Knotweed), *Heracleum mantegazzianum* (Giant Hogweed) and *Impatiens glandulifera* (Himalayan Balsam) in bankside habitats is given for 1994–96 in the U.K. and contrasted with the occurrence of large stands of other nuisance plants. The typical habitats of these plants were derived by analysis of flow, channel and bank substrates from River Habitat Surveys and shows their invasion potential across the U.K. Study of the dispersal mechanisms and control techniques shows: (a) the consequences of accidental introductions to river banks from locations away from the river, (b) their further dispersal by the downstream drift of seed or fragments with their subsequent invasion, and (c) the key areas in the development of appropriate control programmes. The concentration of effort to immediate action and to small areas, especially upstream, is recommended, in preference to widespread but incomplete control.

Introduction

The occurrence of invasive bankside plants was one of the 25 sub-sets of data recorded during River Habitat Surveys (RHS) of the general physical structure of watercourses at 5560 sites in the United Kingdom in 1994–96 (Raven et al., 1997). This sub-set had been added to the main survey following the increased interest by managers with conservation interests and the absence of a national picture of their occurrences in bankside habitats. Thus, three alien invasive plants *Fallopia japonica* (Houett.) Ronse Decr., Japanese Knotweed, *Heracleum mantegazzianum* Sommier & Levier, Giant Hogweed and *Impatiens glandulifera* Royle, Himalayan or Indian Balsam, were specifically recorded as absent, present or extensive, because these were considered to be a problem at the national level (NRA, 1994; Roblin, 1994). Other vegetation surveys of the countryside noted the presence of large stands of plants of similar habit in bankside habitats (e.g. Bunce et al., 1999) which may also be considered to have a nuisance value or adverse effect on conservation interests. Initial consideration of the data set suggested that the extensive network of river corridors provides for passage of seeds of *I. glandulifera* and *H. mantegazzianum* and propagules such as stem sections or roots of *F. japonica* by downstream movement (Pysek & Prach, 1994; Raven et al., 1998). Other aspects of the current status of research on the ecology and management of these species are considered elsewhere, general studies (de Waal et al., 1994a) and by plant (Beerling et al., 1993; de Waal et al., 1994b; Tiley et al., 1996).

This paper aims to show: 1. the national distribution of these three alien invasive species on the banksides of U.K. watercourses; 2. their characteristic or preferred habitat; 3. comparison with large stands of nuisance plants of similar habit; and 4. the results of a study of the dispersal mechanisms and control techniques applied in one of 26 Environment Agency areas of England and Wales in which these plants are all widely distributed.

* Author for correspondence

Materials and methods

The River Habitat Survey was undertaken in 1994–96 to collect data on a range of features describing the physical habitat of selected 0.5 km lengths of river in the U.K. RHS sites were selected on a stratified random basis from 10 × 10 km squares of the Ordnance Survey national grid system and provide a representative distribution of sites throughout the UK. Although three sites per grid square were chosen in England and Wales and undertaken in 1994–96 (totalling 4530 sites), two from N. Ireland for 1995–96 (246) and one from Scotland undertaken over two years 1995–96 (778), comparisons were presented on a one site per square basis. Some bias may exist in the random selection of sites, for example, in Scotland, the constraint was imposed that sites should be selected from locations within 3 km of a road or track. The species *F. japonica*, *H. mantegazzianum* and *I. glandulifera* were specifically recorded as 'absent', 'present' or 'extensive' on the bank face or bank top as defined by the RHS methodology (Environment Agency, 1997). 'Extensive' is defined as occupying more than 33% of the total bank length; the 'bank face' is the area from the edge of water to the first major break (change) in slope; and the 'bank top' is the strip of land parallel to the river, and from the bank top for a width of 5 m away from the river bank. In addition, large stands of other plants which were considered by surveyors on-site to be nuisance species, were identified and recorded for a subset of sites in England & Wales in 1995, this was extended in 1996 by the use of a list of species as a prompt to surveyors (listed in Table 2). The habitat was determined by analysis of RHS data on predominant flow type, predominant channel substrate and predominant bank material (as defined in RHS Method Manual) determined from 1 m wide transects at 50 m intervals over the 500 m sections of rivers and from general site data derived from the map-based parameters including altitude, bed slope and water quality.

Results

The three alien invasive plants *F. japonica*, *H. mantegazzianum* and *I. glandulifera* are widespread in England and Wales and present in Scotland and Northern Ireland. Each species currently has a distinct preferred area but there is also much overlap (Figure 1). Thus, *F. japonica* is more common in south Wales, central north-west and south-east England, whereas *H. mantegazzianum* is widely scattered with occasional extensive areas, as is *I. glandulifera*; the latter is concentrated in the south-east and north-central midland areas of England. The overall frequencies in the U.K. based upon one occurrence per 10 × 10 km map square, are from the most frequent to the least frequent:

I. glandulifera > *F. japonica* > *H. mantegazzianum*

However, there are some regional differences with twice as much *H. mantegazzianum* in N. Ireland as in England, whereas Scotland has about a quarter as much *F. japonica* or *I. glandulifera* as England and Wales (Table 1). Extensive stands occurred on average at 13% of sites.

Table 1. The percentage occurrence of *F. japonica*, *H. mantegazzianum*, and *I. glandulifera* in countries of the U.K.

Country/Species	*H. mantegazzianum*	*F. japonica*	*I. glandulifera*
England & Wales	3.7–4.3	8.5–8.9	14.0–15.1
Scotland	3.8	2.3	3.5
N. Ireland	7	6.9	15
U.K. average	4	6.6	11

The characteristic physical habitat of bankside sites invaded by these three invasive species is predominantly (a) streams and rivers of any size but more frequently medium sized streams up to small rivers; (b) at any altitude but most frequently in lowlands less than 170 m (90%); (c) at any bed-slope but frequently shallow to 10 m per km; (d) a bad or poor water quality for *I. glandulifera* and F. japonica which are associated with urban areas; but, the habitat is independent of other factors such as the distance from source of the watercourse, the bank-top width, (Figure 2). The sites invaded were often in the less (37%) rather than in the extensively modified watercourses and, also, surprisingly, 17.6% of the 'semi-natural' sites had all three species.

Each species has a characteristic habitat despite their general wide ranging occurrences, thus, *F. japonica* is more commonly associated with the banksides of higher energy streams at higher altitudes and with larger bed substrates eg gravels or cobbles, artificial sections or more modified sections of streams (Figure 3). *I. glandulifera* and *H. mantegazzianum* are more often associated with rivers of lesser stream power, lower altitudes and with finer channel and bank

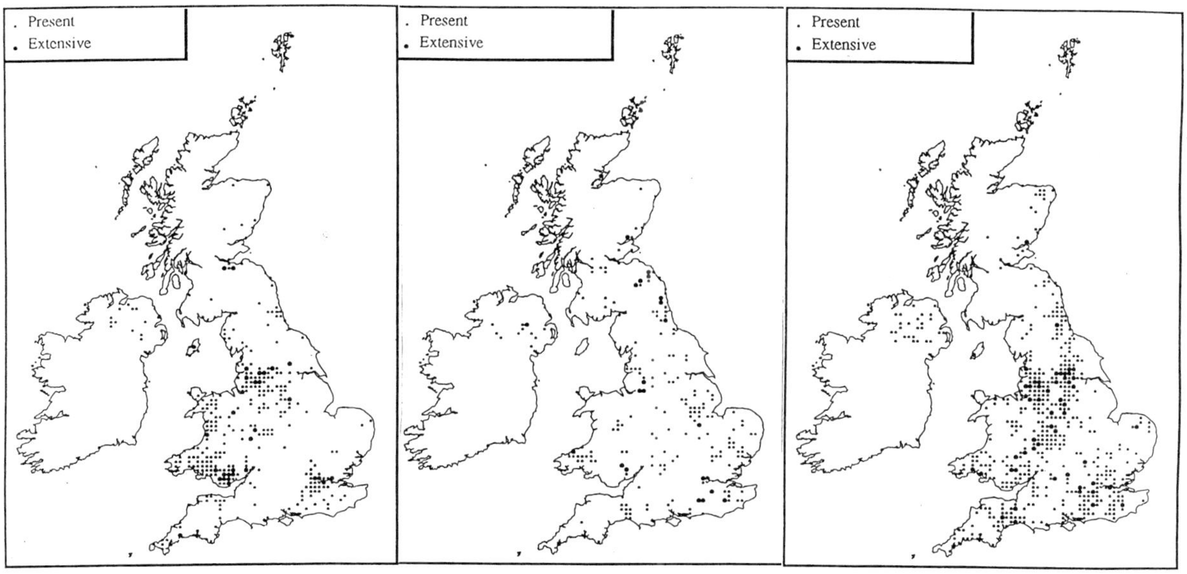

Figure 1. The occurrence of (a) *Fallopia japonica*, (b) *Heracleum mantegazzianum* and (c) *Impatiens glandulifera* in the U.K. for 1995–96 data (key: • = present; ● = extensive; data are centred on the 10 × 10 km squares).

materials, but all are found existing over a wide range of conditions presumably indicating their ability to colonise most habitats if they can initially gain access. The ability to predict the potential occurrence of these species was tested using several combinations of the RHS physical habitat data, such as bank substrate and flow, for various scales of predominance and frequency of data, but little specific predictive conditions were found despite the use of site-specific data from the regular transects. The data on habitat requirements for the plants were either too general or the plants have a very broad range of tolerance in their habitat requirements; this indicates that there is likely to be a great potential for further spread of the plants if conditions for dispersal are met.

The comparison of the frequency of stands of nuisance plants of similar habit to the invasive species, showed that there were several species which were common but only one, *Urtica dioica* L, as common as that of the invasive species. These other bankside species appeared to occur in other sites and broadly, in other areas of the country to that of the three invasives (Figure 4). The proportions of species found varied significantly between country and between years (1995 and 1996, Table 3). Few sites (2%) had more than one of these other species recorded, but did when they occurred most commonly as a combination of *U. dioica* and *P. aquilinum*, or occasionally, *R. fruticosa agg.* and *P. spinosa*.

Assessment of the problems caused by the presence of alien species

The main adverse effect is considered to be the dense and lank growth of 1–5m in height which shades out the shorter native plants that give stability to bank structure. Short vegetation especially grass, is generally preferred by river engineers (Hemphill & Bramley, 1989) as this is considered to minimise erosion of the river bank, because dense short stems and leaves become flattened in floods and protect the bank material. Taller plants especially these three invasive species, shade out short vegetation and in autumn die back, often leaving exposed bank material. Channel capacity for flood flows may also be reduced by their presence or by the effects of shoals of trapped sediment; both add to the cost of bank maintenance. In designated conservation sites, these invasive plants may outcompete native plants of defined value by reducing or destroying the valued habitat and the landscape value. Access to the river bank may be impeded. *H. mantegazzianum* is a health hazard because of its photo-sensitising sap which causes irritation, swelling and recurrent blistering (NRA, 1994).

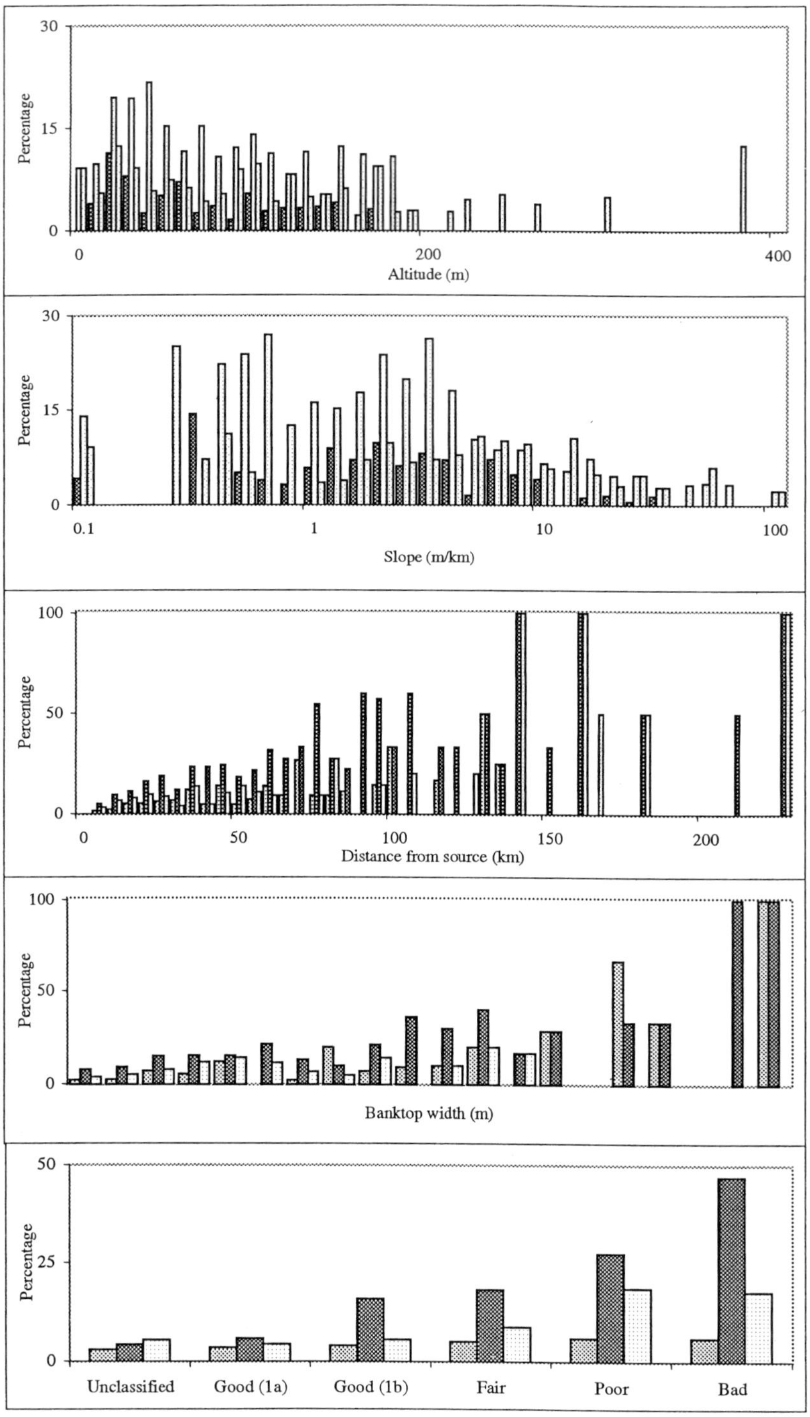

Figure 2. The percentage occurence of *Heracleum mantegazzianum, Impatiens gladulifera* and *Fallopia japonica,* in each RHS map-based dataset of (a) altitude, (b) slope of stream-bed (c) distance from source of watercourse, (d) bank-top width of watercourse in bands of 5 m to 100 m and (e) water quality class (1985), in the U.K. [key: *Heracleum mantegazzianum* (mid-stipple), *Impatiens gladulifera* (dark stipple) and *Fallopia japonica* (light stipple)].

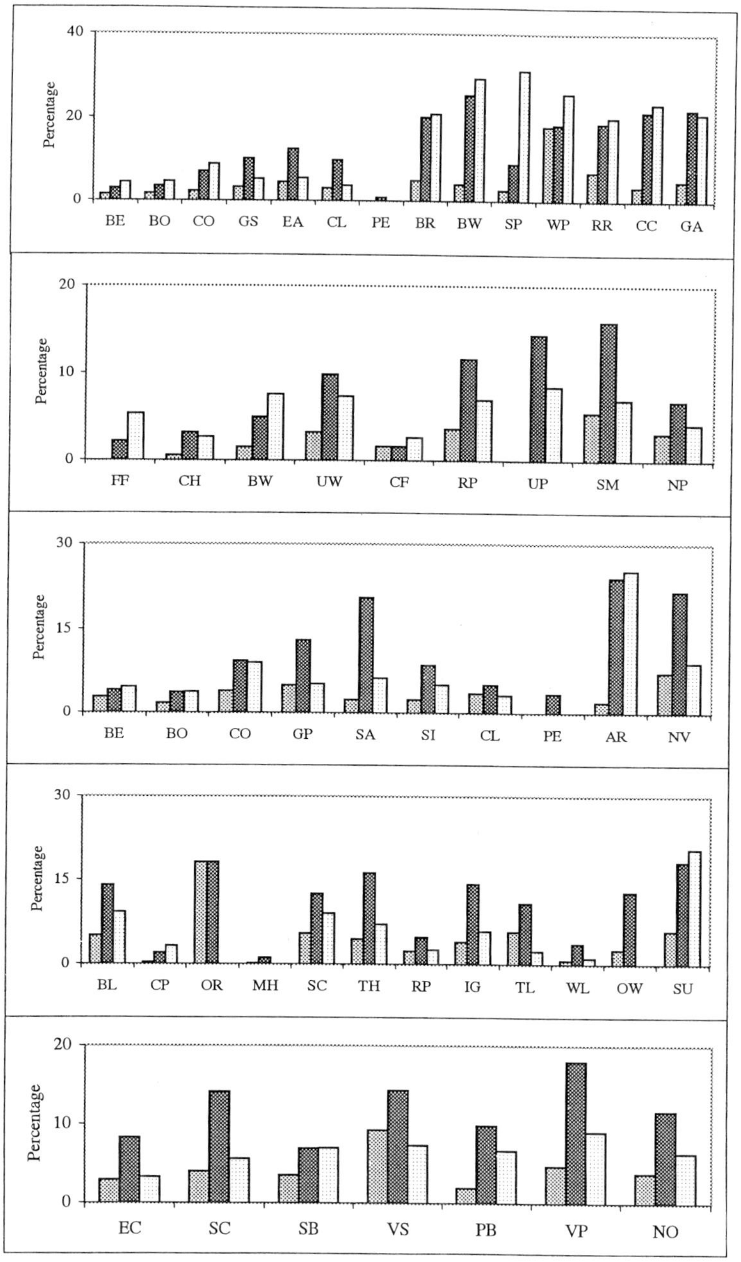

Figure 3. The percentage occurrence of *Heracleum mantegazzianum*,*Impatiens glandulifera* and *Fallopia japonica*, in each RHS type of predominant (a) bank material, (b) flow type (c) channel substrate, (d) bankside land class and (e) bank feature in the U.K. [key: *Heracleum mantegazzianum* (mid-stipple), *Impatiens glandulifera* (dark stipple) and *Fallopia japonica* (light stipple); **bank material** – bedrock (BE), boulders (BO), cobbles (CO), gravel-sand (GS), earth (EA), clay (CL), peat (PE), and artificial – brick or laid stone (BR), builders waste (BW), sheet piling (SP), wood piling (WP), rip-rap (RR), concrete (CC) and gabion (GA); **flow types** – free-fall (FF), chute-flow (CH), broken standing-waves (BW), unbroken standing waves (UW), chaotic (CF), rippled (RP), upwelling (UP), smooth (SM), and 'no perceptible flow' (NP); **channel substrate** – bedrock (BE), boulders (BO), cobbles (CO), gravel-pebble (GP), sand (SA), silt (SI), clay (CL), peat (PE), artificial (AR) and 'not visible' (NV); **bankside land class** – broadleaf or mixed woodland (BL), coniferous plantation (CP) orchard (OR), moorland or heath (MH), scrub (SC), tall herbs or rank vegetation (TH), rough or improved pasture (RP), improved or semi-improved grassland (IG), tilled land (TL), wetland (WT), open water (OW) and suburbans or urban development (SU); **bank feature** – eroding cliff (EC), stable cliff (SC), unvegetated (<50%) side bar (SB), vegetated (>50%) side bar (VS), unvegetated point bar (PB), vegetated point bar (VP) and no obvious feature (NO)].

Table 2. The percentage occurrence by year of large stands of nuisance and other species of similar habit to the three invasive plants, recorded in some bankside habitats in U.K.

Specific and common names of plant	Occurrence at RHS sites				
	E & W		Scotland	N. Ireland	U.K.
	1995	1996	1995/6	1995–96	
Urtica dioica L., Nettle	11.2	27.6	2		16.8
Pteridium aquilinum Gled. -Scop., Bracken	2.0	2.6	20		2.9
**Rhododendron ponticum* L., Rhododendron	0.6	1.4	9		1.3
Aegopodium podograria L., Ground Elder	0.2	0.2	9		0.6
Rubus fruticosa agg., Bramble,		1.2			0.5
**Symphoricarpos albus* (L.)Blake, Snowberry		0.2		2.8	0.3
**Mimulus guttatus* DC., Monkey flower		0.1	6		0.2
Prunus spinosa L., Thornbushes, Blackthorn		0.7			0.2
Petasites spp., Butterbur		0.1			0.1
**Pseudosasa* spp., Bamboo		0.1			0.1
Prunus lusitanica. Laurel,		0.2		0.4	0.1
Cirsium spp., Thistle,		0.2			0.1
Other occasional species	2.2	0.1			2.2
Numbers of sample sites	899	1064	100	246	2063

Table 3. Functions of Environment Agency areas with an interest in the problems of invasive weeds in the area

Interested parties	**Reasons**
Flood defence	Routine maintenance
	Heavy maintenance
	Capital works
	Erosion after winter die-back
	3rd party applications for development and consents
	Control of invasives at Agency installations
Hydrology	Interference with flows from excessive or decaying vegetation
	Control of invasives at Agency installations
Emergency work unit	Implications for all work on the ground.
	Control programmes
	Control of invasives at Agency installations
Environment protection	Herbicides to water – WQ (inc spray irrigation)
Water resources	Herbicides to water – WQ (inc spray irrigation)
	Impoundments (capital works)
Waste regulations	Consent to carry
	Application to landfill – management of site to prevent further spread
Customer service centre	Conditions to be attached to planning application and consent approvals.
All functions	Site visits = Awareness of problems for the Agency
Ecology	Spreading distribution and invasion of designated sites

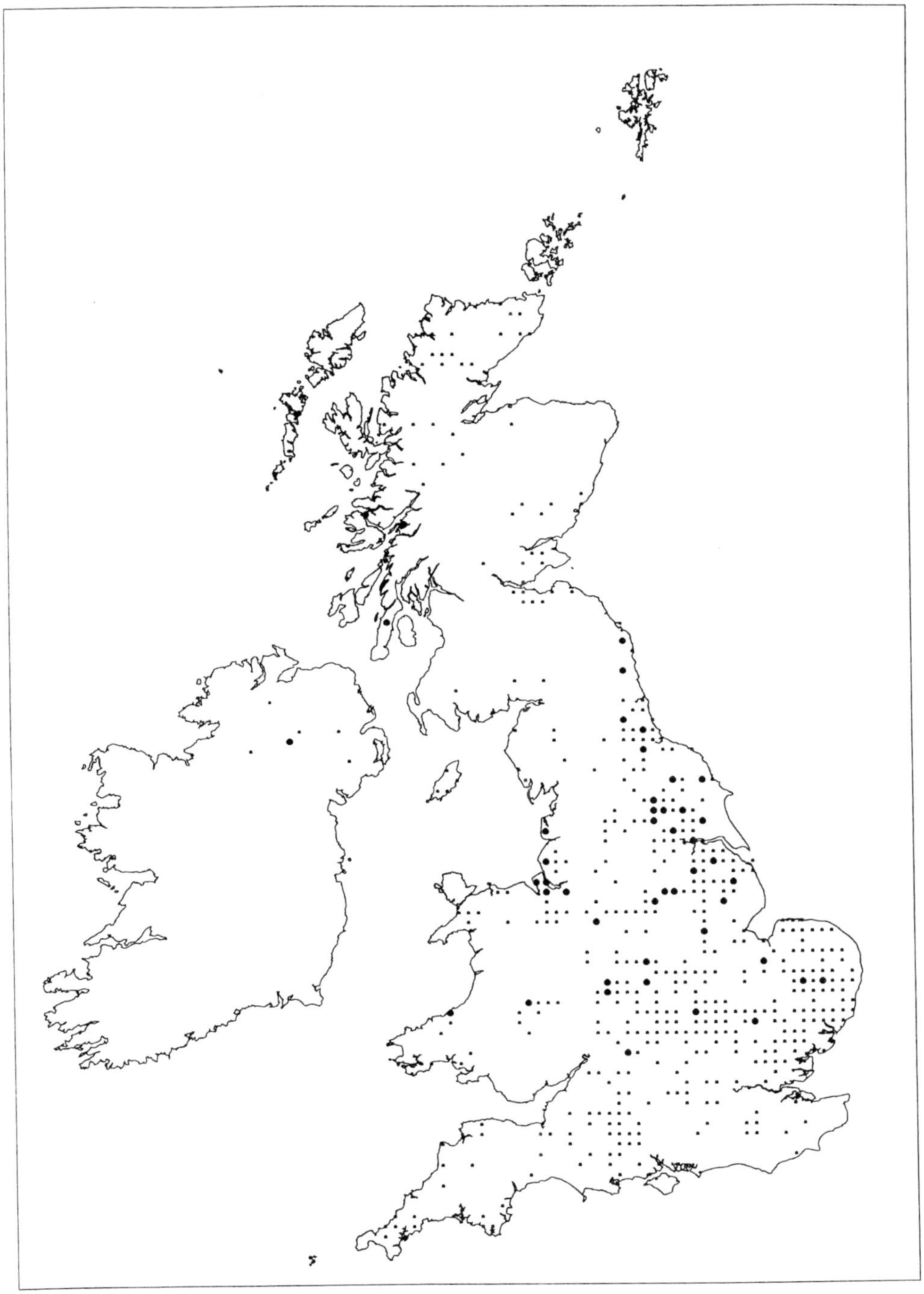

Figure 4. The occurrence of large stands of nuisance and other plants of similar habit to the three invasive plants recorded in some bankside habitats in U.K. (key: • = present; ● = extensive; data are centred on the 10 × 10 km squares).

A number of different dispersal mechanisms have been observed but not quantified, during the course, and as a result, of normal river maintenance in the Mersey catchment of north-west England. Dispersal mechanisms include:

1. by water: seeds (*I. glandulifera* and *H. mantegazzianum*) and fragments (*F. japonica*) are carried downstream, thus reducing dispersal in the upper catchment is critical for effective control;
2. by vegetation management of watercourse banks: routine maintenance of bankside vegetation by hand and machinery has a high potential for spreading at seed time (*H. mantegazzianum, I. glandulifera*); *F. japonica* fragments are spread by flailing;
3. by transfer of material and associated actions: by loss from footware to footpaths and roads; by vehicles, especially their wheels when travelling between on- and off-road sites; by movement of agricultural machinery from field edges to mid-field sites; by seeds during mowing, ploughing etc.;
4. by movement of animals especially cattle;
5. by recreational activities eg. golfing including footware, trolleys and vehicle wheels around sites and between sites;
6. by clearance for site management or development; on-site spoil movements; removal of affected spoil and incidental spread. In the U.K., the Wildlife and Countryside Act (1981) prohibits transfer to other wild sites; thus for capital works and removal of potentially contaminated spoil from sites, conditions should always attached to the 'consents' granted to applications for development as part of the planning process (Table 3);
7. natural dispersal by the plants includes aerial projection of seeds and wind dispersal. Seeds of *I. glandulifera* and *H. mantegazzianum* can be carried or thrown for 5 metres when enhanced by air turbulence.

The study of the control programme of the southern area of the North-west region of the Environment Agency which started in 1995 found that there were two principal problem areas for the control of *H. mantegazzianum,* the River Bollin (South of Manchester) and the River Croal (North of Bolton) through to Salford on the River Irwell, although small populations were being controlled on several other watercourses. Currently the funding for this control of c. 250 km is £30 000 per annum from the Flood Defence revenue and this rate has been projected for the three years (1999–2002) for the continuing herbicide (glyphosate) spraying programme on all river populations. A considerable amount of effort and time has also been expended in liasing with land owners, local authorities and various agencies who manage land on which this plant was growing at a distance from the river system. So to achieve meaningful success on the rivers, the off-river colonies need to be controlled with the same vigour as is being applied to bankside colonies.

Discussion and recommendations

This survey of the U.K. confirms and extends the previous data on distribution and gives detail on the habitat requirements of these three plants. Although site-specific details about colonies were not necessarily coincident, the general nature and variability of short sections of watercourse (500 m) were available for comparison. Colonies were, however, found on a wide range of habitats and although some differences between species were revealed, use of these data in predictions produced little useful data on the susceptibility of sites. The conclusion, which is pessimistic, is that almost all U.K. river banksides are susceptible to invasion by one or more of these species. This emphasises the importance of the need for having measures in place for early and appropriate control to reduce the colonisation, particularly at upstream sites and from adjacent land, by these undesirable species of plant and reveals the degree of vulnerability of river systems to wider infestation.

Large bankside stands of other nuisance species were never very common and appeared to lie in other areas of the country to those of the invasives. *Urtica dioica* was much more common than any other species being recorded by many surveyors, but its presence on banksides may relate more to adjacent land uses especially resulting from nitrogen-enrichment activities. Stands of these plants are present in the general countryside, as is *Fallopia japonica*, but do not seem to use water courses for dispersal or even as a preferred habitat.

The fundamental policy of control upon finding a new colony is for immediate and appropriate action. Firstly, following identification of a new colony, a mechanism must be in place to initiate immediate control because it is easiest and cheapest to control colonies at an early stage of their development. Secondly, the general policy should be: to contain plant material and treat on site; to start immediately; to treat from upstream to downstream; to ensure that a long-term management policy has been formulated and which includes survey; to fully control isolated colonies; and never to consider partial or incomplete control measures.

Although it is recognised that it may be impractical to control all colonies immediately, it is important that a progressive series of measures are planned and progressed at a sufficiently rapid rate. Each species has different priorities for control, thus for:

1. *Heracleum mantegazzianum*. The main objective is to prevent flowering before taking control of the colony. Larger areas or infrequently visited sites are best treated chemically but physical control may be more environmentally-sound for small populations but only if frequent attention can be given through the year.
2. *Impatiens glandulifera*. The main objective is to prevent flowering. Chemical control is effective but hand pulling or hoeing at frequent intervals from early in the season can also be used.
3. *Fallopia japonica*. Chemical control is best in most circumstances because of the nature of the extensive root system which can readily regenerate new plant growth after physical disturbance and even after chemical application.

Specific details of control are given elsewhere (e.g. NRA, 1994).

Acknowledgements

We are grateful to the RHS surveyors from the Environment Agency and IFE who undertook the field work including Paul Henville, Michael Gravelle and Peter Scarlett, who also undertook other tasks particularly assembly of the RHS database.

References

Beerling, D. J. & J. M. Perrins, 1993. Biological flora of British Isles: *Impatiens glandulifera* Royle (*Impatiens roylei* Walp). J. Ecol. 81: 367–382.

Bunce, R. G. H., C. J. Barr, M. K. Gillespie, D. C. Howard, W. A. Scott, S. M. Smart, H. M. Van der Poll & J. W. Watkins, 1999. Vegetation of the British Countryside – Countryside Vegetation System (Ecofact Vol. I), Department of Environment, Transport & the Regions, Bristol, U.K., 224 pp.

Environment Agency, 1997. River Habitat Survey: 1997 Field Survey Guidance Manual including SERCON. Environment Agency, Bristol, England: 100 pp (circulated).

Hemphill, R. W. & M. E. Bramley, 1989. Protection of river and canal banks. Construction Industry Research and Information Association. Butterworths, London: 200 pp.

National Rivers Authority, 1994. Guidance for the control of invasive plants near watercourses: Japanese Knotweed, Giant Hogweed & Himalayan Balsam. National Rivers Authority, U.K., 19 pp.

Raven, P. J., P. J. A. Fox, M. Everard, N. T. H. Holmes & F. H. Dawson, 1997. River Habitat Survey: a new system to classify rivers according to their habitat quality. In Boon, P. J. (ed.), Freshwater Quality: Defining the Indefinable. HMSO, London, England: 215–234.

Pysek, P. & K. Prach, 1994. How important are rivers for supporting plant invasions? In de Waal, L. C., L. E. Child, P. M. Wade & J. H. Brock (eds.), Ecology and Management of Invasive Riverside Plants. Wiley & Sons, Chichester: 19–26.

Raven, P. J., N. T. H. Holmes, F. H. Dawson, P. J. A. Fox, M. Everard, I. Fozzard & K. J. Rouen, 1998. River Habitat Quality. The physical character of rivers and stream in the U.K. and Isle of Man. River Habitat Survey Report No. 2. Environment Agency: 100 pp.

Roblin, L., 1994. Alien invasive weeds – an example of National Rivers Authority sponsored research. In de Waal, L. C., L. E. Child, P. M. Wade & J. H. Brock (eds.), Ecology and Management of Invasive Riverside Plants. Wiley & Sons, Chichester: 189–194.

Tiley, G. E. D., F. S. Dodd & P. M. Wade, 1996. Biological flora of British Isles: *Heracleum mantegazzianum* Sommier & Levier. J. Ecol.81: 297–319.

de Waal, L. C., L. E. Child & M. Wade, 1994b. The management of three alien invasive riparian plants *Impatiens glandulifera* (Himalayan balsam), *Heracleum mantegazzianum* (giant hogweed) and *Fallopia japonica* (Japanese knotweed). In Harper, D. M. & A. J. D. Furgeson (eds), The Ecological Basis for River Management. Wiley & Sons, Chichester: 315–322.

de Waal, L. C., L. E. Child, M. Wade & J. H. Brock (eds), 1994a. Ecology and Management of Invasive Riverside Plants. John Wiley & Sons, Chichester: 217 pp.

Hydrobiologia **415:** 203–206, 1999.
J. M. Caffrey, P. R. F. Barrett, M. T. Ferreira, I.S. Moreira, K. J. Murphy & P. M. Wade (eds),
Biology, Ecology and Management of Aquatic Plants.
© 1999 *Kluwer Academic Publishers. Printed in the Netherlands.*

Azolla filiculoides Lamarck (Pteridophyta: Azollaceae), its status in South Africa and control

M. P. Hill & C. J. Cilliers
Weeds Division, Plant Protection Research Institute, Agricultural Research Council, Private Bag X 134, Pretoria, 0001, South Africa
Tel: [+ 27] 12 329 3276. Fax: [+ 27] 12 329 3278. E-mail: RIETMH@PLANT2.AGRIC.ZA

Key words: Red water fern, biological control, mechanical control, herbicide control, utilization

Abstract

Azolla filiculoides Lamarck (Pteridophyta: Azollaceae) (red water fern) is native to South America and was first recorded in South Africa in 1948. The lack of natural enemies and the presence of enriched waters have contributed to its establishment throughout the country where dense infestations cause severe degradation of aquatic ecosystems and limit their utilisation. The biology of *A. filiculoides* precludes it from being controlled manually or mechanically and herbicidal control in the aquatic environment is undesirable. Biological control is suggested as the only sustainable control method for this weed and a natural enemy, *Stenopelmus rufinasus* Gyllenhal (Coleptera: Curculionidae), was released for this purpose in late 1997. The weevil has established well and has caused a significant decline in the *Azolla* mat at a number of sites, suggesting that the weevil will be an efficient tool in the management of *A. filiculoides* in South Africa.

Introduction

The genus *Azolla* grows in a symbiotic association with the heterocystous cyanobacterium, *Anabaena azollae* Strasburger, that occurs within the dorsal leaf lobes of the plant. The alga can fix atmospheric nitrogen and is able to fulfil the nitrogen requirements of the fern, making it successful in nitrogen-deficient waters (Ashton, 1992). As a result, *Azolla* is an economically important fern, having been used in South-east Asia as a green manure associated with wetland rice cultivation for at least 200 years. There is a large volume of literature devoted to this topic; this is excellently reviewed by Lumpkin & Plucknett (1982). Red water fern, *Azolla filiculoides* Lamarck, is a small aquatic annual or perennial heterosporous fern, rarely larger than 25 mm. Although native to South America (Lumpkin & Plucknett, 1982), it is widely distributed throughout the world and is often regarded as a weed (Ashton, 1992). Each plant consists of a short, branched rhizome; bearing small, alternate, overlapping leaves and roots which hang into the water. The plant turns a red/magenta color, especially in winter; this is due to the production of anthocyanins (red plant pigment) induced by a combination of direct sunlight and low temperatures.

Azolla filiculoides is able to undergo rapid vegetative reproduction throughout the year through elongation and fragmentation of small fronds. Under ideal conditions the daily rate of increase may exceed 15%, and doubling time of the fern may be 4–5 days (Lumpkin & Plucknett, 1982). In addition, the fern can reproduce sexually by the production of spores during spring and summer. These spores overwinter and are resistant to extreme desiccation, allowing the fern to re-establish after drought.

Azolla filiculoides was first recorded in the Oorlogspoort River near Colesburg (Eastern Cape Province (30° 52′ 16″ S/25° 19′ 22″ E)) in 1948 (Oosthuizen & Walters, 1961). There are no details regarding the introduction of this plant into South Africa, but it was probably as an aquarium plant, which eventually found its way into a river. For many years the fern was confined to small streams and farm dams in the Colesburg area in the centre of the country. However, the lack of natural enemies (Hill, 1998a) and the presence of enriched waters has lead to its inevitable spread by man, waterfowl and floods to many sites around the country (Ashton, 1992). In June 1996, the Southern African Plant Invaders Atlas database (SAPIA) had records of *A. filiculoides* in 65 quarter-degree squares (15-minute squares) in South Africa. By May 1998, the plant had been recorded in 152 quarter-degree squares. Although it is tempting to cite this increase as an explosion of the weed, these rates of expansion should be treated with caution as they may reflect increased sampling rather than an actual increase in weed abundance.

Impact of Azolla filiculoides *in South Africa*

The increasing abundance of *A. filiculoides* in conservation, agricultural, recreational and suburban areas over the last ten years is cause for great concern. There are major consequences of the dense mats (5–30 cm thick) of the weed on still and slow-moving water bodies in South Africa. These are: reduced quality of drinking water, such as bad odour, colour and turbidity; an increase in waterborne, water-based and water-related diseases; increased siltation of rivers and dams; reduced water surface for recreation (fishing, swimming and water-skiing) and water transport; reduction in aquatic biodiversity; clogging of irrigation pumps; drowning of livestock; and reduced water flow in irrigation canals.

Control options

Mechanical and herbicide control options have been suggested for red water fern. However, mechanical control is very labour-intensive. Small infestations of the weed can be removed manually using fine meshed nets. However, the rate of increase of the plant is such that this method is impractical for larger infestations. In addition, if eradication is achieved, re-establishment of the weed from spores would be inevitable. The herbicidal control of *A. filiculoides* using glyphostate (Steÿn et al., 1979; Ashton, 1992), paraquat and diquat (Axelsen & Julien, 1988) and kerosene mixed with a surfactant (Diatloff & Lee, 1979) has been implemented. Herbicide control can result in rapid and effective control of the weed, but it requires an extensive follow-up programme to control the plants continually germinating from spores and there is a danger of spray drift on to other vegetation. In view of this, biological control was seen as a sustainable option for the control of this increasingly invasive weed.

Two insect species were identified as possible candidates for the biological control of this weed (Hill, 1997). Both are frond-feeding beetles, *Pseudolampsis guttata* (Leconte) (Chrysomelidae) and *Stenopelmus rufinasus* Gyllenhal (Curculionidae) recorded from *A. filiculoides* in Argentina (S. Neser, pers. comm.) and *A. caroliniana* in Florida (Richerson & Grigarick, 1967; Buckingham & Buckingham, 1981). Both species do extensive damage to the plants in the country of origin, and it is expected that without their natural enemies they might control the weed in South Africa. The weevil *Stenopelmus rufinasus* was imported from Florida, USA in late 1995.

The adult female inserts an egg into the tip of the frond, which hatches after about four days. There are three larval instars, all of which feed voraciously on the fronds. The first instar larvae mine the upper leaf lobe of the fronds, while the second and third instars feed externally. Pupation occurs in an ovoid black chamber in the upper fronds of the plant. Duration of immature stages from egg to adult eclosion ranged from 16 to 23 days (Hill, 1998b). The adults are small, about 1.7 mm in length, and are a grey-black colour, covered with red, black and white scales in a variable pattern. The adults are long-lived (55–60 days) and the females can produce between 128 and 474 offspring. This insect has a high rate of increase and, as the larvae and adults are voracious feeders, it is capable of decimating mats of *A. filiculoides*.

The host specificity of *S. rufinasus* was determined on 31 plant species in 19 families. Although some feeding and development was recorded on other species of *Azolla*, it was far inferior to that on *A. filiculoides*, indicating that they would not support field populations of the weevil (Hill, 1998b). *Stenopelmus rufinasus* was cleared for release as a natural enemy of *A. filiculoides* in South Africa in December 1997.

Between December 1997 and December 1998, the weevil has been released at 30 sites of *A. filiculoides* throughout South Africa. To date establishment has been recorded at 16 of the release sites while at 14 of these sites there has been a significant decline in the mat of the weed. The weevil has established well in all climatic regions of the country, including the high elevation areas which are characterised by cold winters.

Utilization of Azolla

Azolla filiculoides has primarily been utilized as a source of nitrogen for wetland rice cultivation (Lumpkin & Plucknett, 1982). However, it may also be used as fodder for pigs, poultry, cattle and fish, but should be supplemented with other feeds, as it cannot be used as the only protein source. The advantages of its use are that it has a high nutrient content, it grows quickly on natural water bodies, it is available throughout the year and requires no processing. *Azolla* can also be composted or used decoratively on ponds and fish tanks. In India it is eaten in fried dishes. It may also be used in mosquito control because a complete mat disrupts larval development. However, an incomplete mat has quite the opposite effect because it affords the larvae protection from predators (Lumpkin & Plucknett, 1982). The utilization of *A. filiculoides* as a nitrogen source for Africa has also been suggested (Wagner, 1997; van Hove et al., 1994).

Potential for the utilization of *A. filiculoides* in South Africa is limited. Commercially produced, inorganic nitrogen-based fertilizers are cheap and freely available in South Africa as they are produced as a byproduct of oil extraction from coal. Considering the limited potential of *A. filiculoides* in South Africa and its severe environmental impact, there is no conflict of interest in the control of this weed.

Discussion

Despite its usefulness in other parts of the world (Lumpkin & Plucknett, 1982), *A. filiculoides* is a damaging weed in South Africa. The runoff of nutrients, particularly phosphates, from fertilizers used in agriculture and the slowing of rivers and streams through the building of dams and weirs has created ideal conditions for the weed to flourish. Biological control appears to be an effective tool in the management of this weed in South Africa. The decline in weed mats at a number of weevil release sites in a short period of time (less than one year) suggests that the interaction between *S. rufinasus* and *A. filiculoides* is stochastic. High female fecundity combined with high feeding rates cause rapid and extensive damage to the mats, causing them to sink, leaving no residual *A. filiculoides* population. The weevil population then undergoes massive larval mortality with the sinking mat. In Florida, USA, the same behaviour has been reported (Center, pers. comm.) where the adult weevils then disperse from the sinking mat to colonize another mat elsewhere. However, the weed is likely to return to the "controlled" site in the following summer due to germination of spores and re-colonization by the insects will be crucial for sustained control.

The preliminary results indicate that *S. rufinasus* will be an efficient biological control agent for *A. filiculoides* in South Africa. Therefore, at this stage, other potential natural enemies need not be considered.

Acknowledgements

The SAPIA data was provided by Ms. L. Henderson (Plant Protection Research Institute). We thank the Water Research Commission of South Africa for funding the research on the control of *Azolla filiculoides* and for supporting the attendance of C.J.C. at the 10th EWRS Symposium on Aquatic Weeds in Lisbon, 1998 where this paper was presented. The financial support of the Agricultural Research Council of South Africa is gratefully acknowledged.

References

Ashton, P. J., 1992. *Azolla* infestations in South Africa: history of the introduction, scope of the problem and prospects for management. Water Qual. Information Sheet, 7 pp.

Axelsen, S. & C. Julien, 1988. Weed control in small dams. Part II. Control of salvinia, azolla and water hyacinth. Queensland Agricultural Journal, September–October 1988: 291–298.

Buckingham, G. R. & M. Buckingham, 1981. A laboratory biology of *Pseudolampsis guttata* (Leconte) (Coleoptera: Chrysomelidae) on waterfern, *Azolla caroliniana* Willd. (Pteridophyta: Azollaceae). Coleopt. Bull. 35: 181–188.

Diatloff, G. & A. N. Lee, 1979. A new approach for control of *Azolla filiculoides*. Proc. 7th Asian-Pacific Weed Sci. Soc. Conf. 1979: 253–255.

Hill, M. P., 1997. The potential for the biological control of the floating aquatic fern *Azolla filiculoides* Lamarck (red water fern/rooivaring) in South Africa. Report to the Water Research Commission by the Plant Protection Research Institute Agricultural Research Council. WRC Report No. KV 100/97, 31 pp.

Hill, M. P., 1998a. Herbivorous insect fauna associated with *Azolla* species (Pteridophyta: Azollaceae) in southern Africa. Afric. Ent. 6: 370–372.

Hill, M. P., 1998b. Life history and laboratory host range of *Stenopelmus rufinasus*, a natural enemy for *Azolla filiculoides* in South Africa. BioControl 43: 215–224.

Lumpkin, T. A. & D. L. Plucknett, 1982. *Azolla* as a green manure: Use and Management in crop production. Westview Tropical Agriculture Series No. 5. Westview Press, Boulder, Colorado, 230 pp.

Oosthuizen, G. J. & M. M. Walters, 1961. Control of water fern with diesoline. Farming in South Africa 37: 35–37.

Steÿn, D. J., W. E. Scott, P. J. Ashton & F. S. Vivier, 1979. Guide to the use of herbicides on aquatic plants. Technical Report TR 95, Department of Water Affairs, South Africa, pp. 1–29.

Richerson, P. J. & A. A. Grigarick, 1967. The life history of *Stenopelmus rufinasus* (Coleoptera: Curculionidae). Ann. ent. Soc. Am. 60: 351–354.

Wagner, G. M., 1997. *Azolla*: A review of its biology and utilization. Bot. Rev. 63: 1–26.

Van Hove, C., C. Colot & L. Ghesquiere, 1994. From the research station to the field: the missing link. In M. Sidiki & A. Hilali (eds), Recent developments in biological nitrogen fixation research in Africa. Proceedings of the 5th International Conference of the African Association for Biological Nitrogen Fixation (AABNF), held at the Hassan II Institute of Agronomy and Veterinary Medicine (IVA), Rabat, Morocco, 14–19 September 1994: 542–547.

Hydrobiologia **415**: 207–211, 1999.
J. M. Caffrey, P. R. F. Barrett, M. T. Ferreira, I.S. Moreira, K. J. Murphy & P. M. Wade (eds), Biology, Ecology and Management of Aquatic Plants.
© 1999 *Kluwer Academic Publishers. Printed in the Netherlands.*

Loss of native aquatic plant species in a community dominated by Eurasian watermilfoil

Charles W. Boylen[1,*], Lawrence W. Eichler[1] & John D. Madsen[2]
[1] *Darrin Fresh Water Institute, Rensselaer Polytechnic Institute, Troy, NY 12181, U.S.A.*
[2] *U.S. Army Engineer Waterways Experiment Station, Vicksburg, MS 39180, U.S.A.*

Key words: milfoil, *Myriophyllum spicatum*, aquatic plant communities, exotic species, invasive species, submersed aquatic macrophytes

Abstract

Ecological evaluation of the impact of an exotic species upon native plant species is frequently a combination of historical data prior to introduction and after full establishment with little observation in between. The introduction of *Myriophyllum spicatum* L. (Eurasian watermilfoil) into Lake George, New York, U.S.A. was first noted in 1985. In 1987, a few newly established plants were allowed to grow unimpeded by human management to document the rate of colonization of this species into a new habitat with its ultimate dominance over, and systematic elimination of, native species. This changing community has been closely monitored over the past decade. Initially a 6 m^2 grid system composed of 144 0.5 m^2 plots with four radiating transect lines was laid out with the isolated *M. spicatum* plants as the center. The site was revisited between 1987 and 1997 to mark the extent of the developing *M. spicatum* bed and its effect on the composition of the native plant community. Since 1987, the area of dense growth has expanded in all directions, impeded only where some physical barrier, such as upper or lower water depth limits or sediment type curtailed its growth. Concurrently, with this expansion, a decline in species richness and abundance of native species was observed.

Introduction

Myriophyllum spicatum L. (Eurasian watermilfoil), an alien species to North America, was first noticed in nuisance proportions in the Chesapeake Bay in the United States in the 1940s (Couch & Nelson, 1985). In the last five decades, its spread throughout much of North America has created nuisance level growth while drastically altering native plant communities (Smith, et al., 1966; Coffey & McNabb, 1974; Newroth, 1985). At the present time, it is found from Florida to Quebec in the east, and California to British Columbia in the west (Smith & Barko, 1990).

The proliferation of this species has serious impacts on aquatic ecosystems. From a human perspective, dense growth interferes with water-based recreation, disrupts use of waterways for waterflow, irrigation and drinking purposes, as well as contributing to flooding (Newroth, 1985). Dense canopies shade out native vegetation, alter the community composition of aquatic macroinvertebrates and may impair the ability of some fish species to spawn in the littoral zone (Aiken et al., 1979; Newroth, 1985). The high rates of plant sloughing and leaf turnover, as well as the decomposition of high biomass at the end of the growing season, can significantly increase the internal loading of phosphorus and nitrogen to the water column (Nichols & Shaw, 1986).

Domination of littoral zone vegetation by *M. spicatum*, and suppression of native plants has been noted in the literature, but generally not studied while plant community changes were actually occurring. For instance, Coffey & McNabb (1974) noted that *Vallisneria americana* appeared to survive under a canopy of *M. spicatum*, but other native species were already absent from dense stands of the exotic. Studies in British Columbia noted that 2–3 years were required for *M. spicatum* to dominate a given area of littoral zone (Aiken et al., 1979; Newroth, 1985). The present

* Author for correspondence

study is a continuation of observations made during 1987–1989 (Madsen et al., 1991a).

Description of study site

Lake George is a large (114 km^2) and deep (18 m average depth, 58 m maximum) lake located in northeastern New York State, U.S.A. *M. spicatum* was first observed in the lake in 1985 when three sites were discovered. Lakewide surveys have found approximately 10–15 new sites yearly. By 1997, *M. spicatum* was established at 123 different sites in the lake. Currently most of these sites are managed in some way. In 1986, 10 isolated plants were found at the Northwest Bay site. In 1987, a small bed had formed. Because of the small size of the population and physical characteristics of the site (e.g. no shoreline development, depth range, silty sediments and diverse native plant community), the site was selected for long-term study.

Materials and methods

In 1987, a permanently-fixed grid (6 m × 6 m) was centered over the newly established bed in Northwest Bay. The grid was subdivided every 0.5 m, creating a grid-work of 144 contiguous quadrats, each with an area of 0.25 m^2. Four radiating transects were set up from the central grid system, deployed along the four cardinal axes of the compass. The transects were 25 m long with a marker every meter.

The sediment type of the observation area was sandy silt. The bed covered a depth range from 2–4 m. Percent cover was estimated using the Daubenmire scale (Daubenmire, 1968). Within the grid, percent cover was estimated for each species within each quadrat of 0.25 m^2. For each transect, a portable 0. 1 m^2 quadrat was used to estimate percent cover. Daubenmire values were converted to the centroid of each cover range for calculations of (a) average percent cover of species (where total cover could exceed 100%), (b) percent of community – percent cover relative to total coverage of all plants (relative cover of all species adds up to 100% for each quadrat), and (c) percent frequency (percent of quadrats containing a given species). The expanding bed was mapped by divers using compass bearings and measuring marked intervals along the bed perimeter.

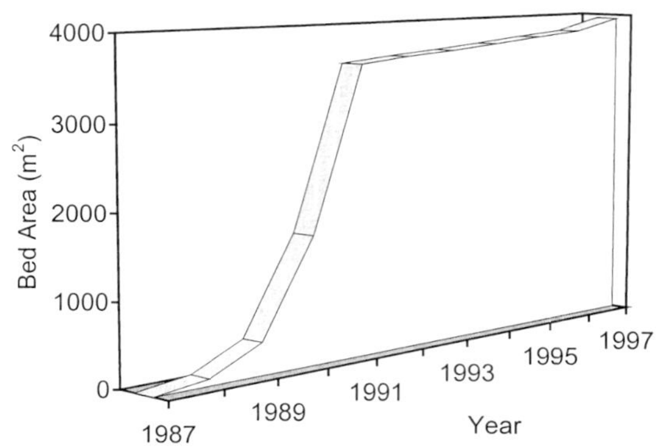

Figure 1. Areal expansion of the bed of *Myriophyllum spicatum* in Northwest Bay, Lake George, New York.

Results and discussion

The areal extent of the bed expanded rapidly between 1987 and 1991, but has leveled off in recent years (Figure 1). In 1997, the maximum area attained was approximately 4000 m^2, representing an expansion consistent within allowable environmental constraints (upper or lower depth limits and discontinuity in sediment type). In Lake George, the increase in bed density and area has been primarily through the growth of root crowns and stolons.

Within the quadrats of the central grid and along the four radiating transects (N, S, E and W), *M. spicatum* showed consistent increases in both abundance and presence (as measured by frequency and percent cover). The initial bed, where percent cover was greater than 50%, began in the grid center and spread outward. In 1987, almost all quadrats in the grid had some *M. spicatum*, with percent cover less than 30%. In 1988, percent cover in the grid was almost 80% and approached 100% by 1989. Since 1989, dense growth of *M. spicatum* has completely filled the grid and has spread outward along the four radiating transects. The east and south transects contained *M. spicatum* in only 10% of the quadrats in 1987 but over 80% had *M. spicatum* in 1989. The spread along the north and west transects has been due in part to the physical bathymetry of the site. Percent cover (Figure 2) increased from almost zero in 1987, to 30–40% in the east, north and west transects, and over 60% in the south transect.

Concomitantly, the percent frequency of *M. spicatum* found in the quadrats has increased (Figure 3). *M. spicatum* represented only 15% of the community within the grid in 1987 (Figure 3) to be-

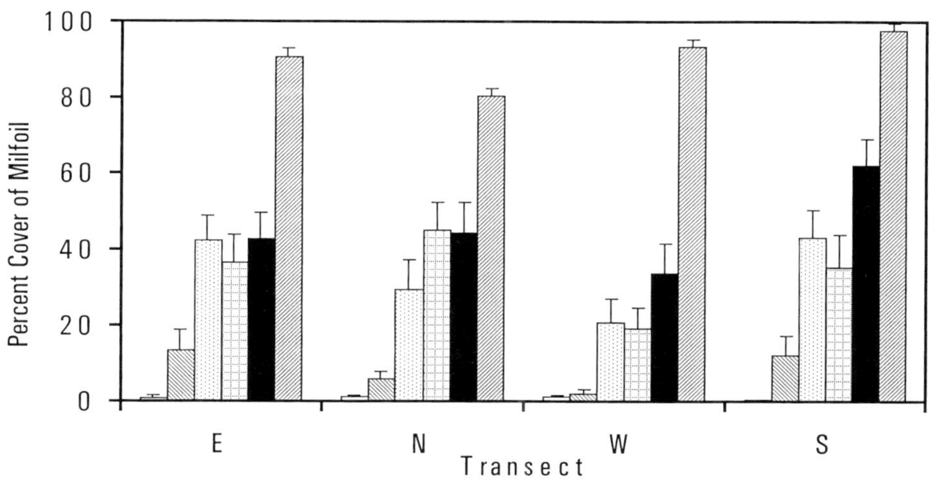

Figure 2. The percent cover of milfoil (*Myriophyllum spicatum*) in each of the transects radiating from the central grid. Bars for each transect represent data from years 1987, 1988, 1989, 1990, 1991 and 1997. Error bars equal the standard error of the mean, n=25.

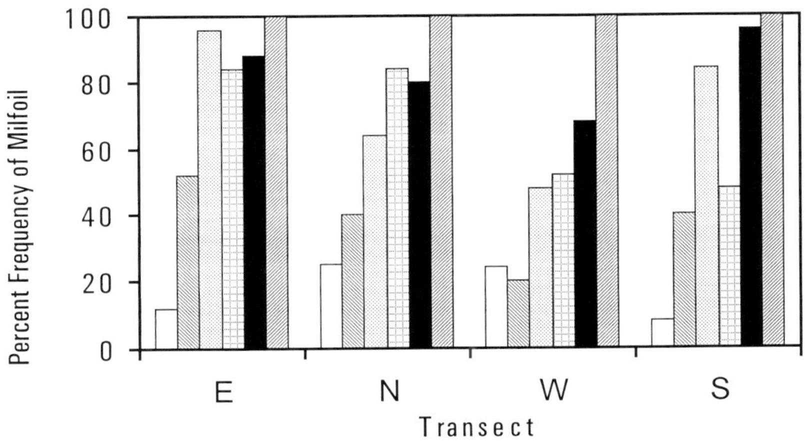

Figure 3. The percent frequency of milfoil (*Myriophyllum spicatum*) in each of the transects radiating from the central grid. Bars for each transect represent data from years 1987, 1988, 1989, 1990, 1991 and 1997.

come the overwhelming dominant species in 1989 in excess of 95%. Within three years (1989), the formerly native communities were reduced to only a few stems of native plants under a dense canopy of *M. spicatum*. Species richness declined from an average of 5.5 species per quadrat in 1987 to slightly over two in 1989 (of which one species was always *M. spicatum*). During 1990–1997, *M. spicatum* continued to suppress the native vegetation in this locality with species richness less than two species /quadrat for all years. The total number of species found in the grid (36 m^2) has also decreased linearly over time, from 20 in 1987 to 14 in 1988 and to nine in 1989. Since 1990, the number of species in the grid has leveled out at seven.

Not only did *M. spicatum* dominate along the radiating transects beyond the central grid work, the number of species per 0.1 m^2 quadrat measured at 1 m intervals along each transect also decreased (Figure 4). In 1997 the north transect had an average of three species per 0.1 m^2 quadrat, whereas the east, west and south transects had an average of 1.5 species per quadrat. Along the transects, the presence of native species which were abundant before *M. spicatum* became established has diminished considerably. Affected species include *Bidens beckii* Torr., an *Eleocharis* sp.,

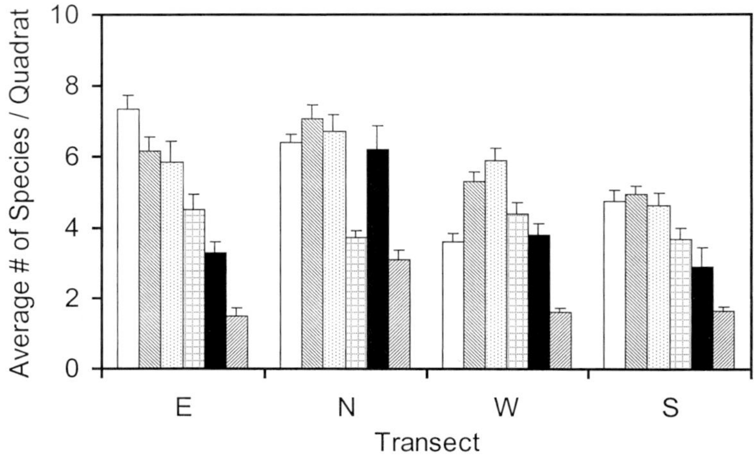

Figure 4. The average number of species found in each of the transects radiating from the central grid. Bars for each transect represent data from years 1987, 1988, 1989, 1990, 1991 and 1997. Error bars equal the standard error of the mean, n=25.

Elodea canadensis (Michx) Planchon., *Heteranthera dubia* (Jacq.) MacM., *Juncus pelocarpus* Meyer, *Myriophyllum tenellum* Bigel., *Najas flexilis* (Willd) Rostk. & Schmidt., *Nuphar lutea* (Durand) Beal, *Potamogeton amplifolius* Tuckerm., *P. gramineus* L., *P. praelongus* Wulfen, *P. robbinsii* Oakes, *P. spirillus* Tuckerm., *P. zosteriformis* Fern., *Sagittaria graminea* Michx., a *Sparganium* sp., *Utricularia vulgaris* L. and *Vallisneria americana* Michx. Species which were infrequent within the native plant community in 1987 (*Myriophyllum alterniflorum* DC, *Najas guadalupensis* (Spreng.) Morong., *Potamogeton perfoliatus* L., *P. pusillus* L., *P. vaseyi* Robbins, *Ranunculus longirostris* Godr. and *Utricularia resupinata* B.D.Greene.) had disappeared altogether by 1997.

The mechanism by which *M. spicatum* dominates other species has been the topic of some discussion. The allocation patterns of plant biomass and growth form allow *M. spicatum* to create a dense upper canopy that contributes substantially to reduced light intensity available to plants under the canopy (Madsen et al., 1991b). Two Lake George species, *V. americana* and *P. robbinsii*, have continued to survive at reduced abundance under the dense *M. spicatum* canopy. Similar tolerance of low light levels under a *M. spicatum* canopy has been reported for these two species in other lakes (Coffey & McNabb, 1974). In addition, the dense root mass formed by *M. spicatum* may competitively exclude some species. Prolific formation of stem fragments and an ability to exploit disturbed habitats also may be key elements to its success.

Conclusions

Invasion and subsequent explosive growth of *Myriophyllum spicatum* severely impacts the density and diversity of native aquatic plant communities. Over an 11-year period, 13 of 20 native plant species were eliminated and the abundance of the remaining species was substantially reduced under a dense canopy of *M. spicatum*. Species richness declined from an average of 5.5 species per quadrat (0.25 m^2) in 1986 to two species per quadrat in 1997, one of which was always *M. spicatum*. Loss of species richness and habitat complexity as a result of plant community dominance by *M. spicatum* may be extended to other components of the food web.

Acknowledgments

The authors greatly acknowledge the support of the FUND for Lake George, New York Department of Environmental Conservation and the Endowment of the Darrin Fresh Water Institute. Numerous DFWI student interns and technicians provided assistance.

References

Aiken, S. G., P. R. Newroth & I.Wile, 1979. The biology of Canadian weeds. 34. *Myriophyllum spicatum* L. Can. J. Plant Sci. 59: 201–215.

Coffey, B. T. & C. D. McNabb, 1974. Eurasian watermilfoil in Michigan. Mich. Bot. 13: 159–165. Couch, R. & E. Nelson,

1985. *Myriophyllum spicatum* in North America. In Proc. First Int. Symp. on watermilfoil (*Myriophyllum spicatum*) and related Haloragaceae species. Vancouver, BC, Canada. Aquat. Plant Mgmt Soc. 8–18.

Daubenmire, R., 1968. Plant Communities: A Textbook of Synecology. Harper & Row, NY: 300 pp.

Madsen, J. D., C. F. Hartleb & C. W. Boylen, 1991a. Photosynthetic characteristics of *Myriophyllum spicatum* and six submersed aquatic macrophyte species native to Lake George, New York. Freshwat. Biol. 26: 233–240.

Madsen, J. D., J. W. Sutherland, J. A. Bloomfield, L. W. Eichler & C. W. Boylen, 1991b. The decline of native vegetation under dense Eurasian watermilfoil canopies. J. Aquat. Plant Mgmt 29: 94–99.

Newroth, P. R., 1985. A review of Eurasian Watermilfoil impacts and management in British Columbia. In Proc. First Int. Symp. on watermilfoil (*Myriophyllum spicatum*) and related Haloragaceae species, Vancouver, BC, Canada. Aquat. Plant Mgmt Soc. 139–153.

Nichols, S. A. & B. H. Shaw, 1986. Ecological life histories of the three aquatic nuisance plants, *Myriophyllum spicatum, Potamogeton crispus*, and *Elodea canadensis*. Hydrobiologia 131: 3–21.

Smith, C. S. & J. W. Barko, 1990. Ecology of Eurasian watermilfoil. J. Aquat. Plant Mgmt 28: 55–64.

Smith, G. E., T. F. Hall & R. A. Stanley, 1966. Eurasian watermilfoil in the Tennessee Valley. Weed: 15: 95–98.

Hydrobiologia **415**: 213–222, 1999.
J. M. Caffrey, P. R. F. Barrett, M. T. Ferreira, I.S. Moreira, K. J. Murphy & P. M. Wade (eds),
Biology, Ecology and Management of Aquatic Plants.
© 1999 Kluwer Academic Publishers. Printed in the Netherlands.

Spartina patens as a weed in Galician saltmarshes (NW Iberian Peninsula)

D. G. SanLeón[1,*], J. Izco[1] & J. M. Sánchez[2]
[1]*Laboratorio de Botánica, Facultade de Farmacia - Universidade de Santiago de Compostela, E-15706 Spain*
E-mail: bvdgslce@uscmail.usc.es
[2]*E.U. de Ingeniería Técnica Forestal, Universidad de Vigo, E-36005 Pontevedra, Spain*

Key words: biological invasion, biodiversity, *Spartina versicolor*, Spain

Abstract

Spartina patens (Gramineae) is an American coastal grass which grows in a wide range of coastal habitats in its native area. After it was discovered in some Galician wetlands two years ago, we undertook a regional survey whose preliminary results reveal that this grass is actually a rather common species in Southwestern Galicia, being present in several coastal habitats also. Nevertheless, it is in saltmarshes where it seems to compete more successfully, invading rush communities that have been traditionally harvested for cattle bedding. In many estuaries, it tends to form dense monospecific stands and could become a serious threat to high marsh plant diversity. In order to evaluate the impact of *S. patens* as an invasive weed in saltmarshes, we have performed a transect study, which seems to indicate that *S. patens* establishes preferentially in the upper marsh (but reaching higher coverages in the uppermost part of its altitudinal rank), where it has a significant negative effect on species diversity as well as on total cover of other species.

Introduction

Spartina patens (Ait.) Muhl. (Gramineae) is a perennial rhizomatous grass that grows in a wide range of coastal habitats. It has long been known to be present along Western Mediterranean coasts, but had not been recorded on the northern Iberian coast up to the present (Bueno, 1997; Campos & Herrera, 1997). On the other hand, former records exist from other parts of the Iberian Peninsula between the Mediterranean Spanish–French and Atlantic Spanish–Northern Portuguese borderlines.

There is no agreement on the origin of this taxon on the Atlantic coasts of Europe, since it has traditionally been reported as *S. versicolor* Fabre and considered as a native Mediterranean species (Fabre, 1849; Tutin, 1980; Van der Maarel & Van der Maarel, 1996). It has also been known as *S. patens* (Aiton) Muhlenberg, previously described as original from the Atlantic coasts of North America. Nevertheless, since the early days, many authors have considered them to be the same single species (Cosson & Durieu, 1867; Fernald, 1929; Mobberley, 1956; Hultén, 1958, and other people later), introduced from America to the Mediterranean Sea (see Hultén, 1958: 287, Figure 268, for distribution map). While the North American populations are very variable (Silander & Antonovics, 1979), the European ones always show a very reduced flowering (Fabre, 1849; Pignatti, 1982; our own data) and their morphology resembles that of the American variety *S. patens* var. *monogyna* (Curtis) Fernald (Hultén, 1958). Nevertheless, in his review of the genus, Mobberley (1956) regarded all the infraspecific taxa as synonyms of *S. patens*, not recognizing any consistent differences among the North American populations or between them and the European ones.

It is probable that the plant was introduced first to various ports around the Mediterranean, as it was used as packing material in ships boxes and crates. So, it might have been established locally from this material and eventually spread out along the coast (Silander, *in litt.*). Saint-Yves (1932) also supports this theory, since the Gulf Stream itself would not have been enough to transport the plants so far.

Being a strong rhizomatous invasive species (Daehler & Strong, 1996), the presence of *S. patens*

* Author for correspondence

in NW Iberian saltmarshes could represent a considerable impact on the native plant communities. Looking to these precedents, this work intends to present initial data concerning the distribution of *S. patens* in Galicia, its frequency and its preferred habitats. The preliminary results of this regional survey lead us to select a representative site on the SW Galician coasts to undertake a more detailed study on the frequency of the weed in this part of Galicia. As for its ecological characterization and impact assessment, we try to delimit its altitudinal position with respect to plant communities of NW Iberian saltmarshes and to evaluate the effect of its intrusion on plant diversity.

Material and methods

The plant

In contrast to the huge amount of studies on the ecology of *S. patens* in the U.S.A., this plant has commonly been neglected in Europe, probably because its populations, wich rarely flower, are easy to confuse with other common coastal grasses (Pignatti, 1982).

Based on the existing studies carried out in North America, *S. patens* does not tolerate waterlogging and the growth and size of the plant is controlled largely by drainage: in poorly drained areas it tends to be eliminated and replaced by other species (Chapman, 1974). On the other hand, according to Silander (*in litt.*), *S. patens* is not very drought tolerant, but there is ecotypic variation between the American populations, and some clones are better adapted to certain locations than others (however, this variation depends largely on the competing species present). It grows well along a gradient of soil moisture conditions from sand dunes to marshes, and the ability of *S. patens* to develop aerenchyma in response to flooding may be important in maintaining flood tolerance and competitive ability in habitats at various positions on a soil waterlogging gradient (Burdick, 1989). Also, Hester et al. (1996) proved that *S. patens* displays a highly significant intraspecific variation in lethal salinity level, as well as in plant morphological variables (leaf rolling, leaf expansion rates; aboveground, belowground and total biomass, and aboveground/belowground ratio). Nevertheless, none of these variables was associated with the variation in salt tolerance.

In America, sterile clones have been found mainly in the tropics, and less frequently elsewhere (Silander, *in litt.*); in contrast, but in common with other places of Europe, this grass rarely if ever flowers in Galicia (only one spike has been found to date). Looking at its wide distribution it seems to spread easily by rhizome fragments carried by currents and tides (Saint-Yves, 1932). Apparently, the Galician plants grow mainly in saltmarshes – where they reach the highest densities, coverages and heights. Similar results were obtained by Silander & Antonovics (1979) in America. However, this plant can also be found in other places in Galicia, those with tidal influence or wet saline substrate, like beaches, rocky shores, wet dunes and dune slacks.

Distribution in Galicia

In order to assess the current status of the *S. patens* invasion in Galicia, we undertook a broad survey along the coastline, searching for the presence of the species in every 10×10 km quadrat of the UTM grid reference system but recording sites with a 1×1 km accuracy. In order to determine its distribution pattern on a smaller scale we carried out a more detailed study within one of the 10×10 km quadrats (UTM 29TMH91) at the Corrubedo Natural Park, where *S. patens* stands were mapped on a 1:5000 scale which allows us to locate them with a margin of error of less than 10 m.

Altitudinal range and effects on plant diversity

In that site, in order to check for any changes in composition of plant communities invaded by this grass, we sampled locally along three belt transects (T1, T2, T3) across saltmarsh stands of *S. patens* located in the landward zone of a coastal lagoon (see Figure 2). Here, vegetation was analyzed in 50×50 cm quadrats along the entire length of each transect (220 m in total), following the methodology of Sánchez et al. (1996). For each species in each quadrat, the cover was estimated visually on a modified Braun-Blanquet scale (1=<10%, 2=10–25%, 3=25–50%, 4=50–75%, 5=>75%). The altitude of each quadrat was determined at the centre of each side between consecutive quadrats, with the aid of a Fuji-Koh FL25M engineer's level and level rods, at the nearest cm. Altitudes were related to the highest tidal level recorded for each transect during the second semester of 1998. Vegetation was characterized based on relevés representative of different plant communities, following Izco et al. (1992) and Izco & Sánchez (1996).

The following plant communities were visually identified along the transects: *Limonio serotini - Juncetum maritimi* subass. *typicum* (henceforth **LJ**), *Li-*

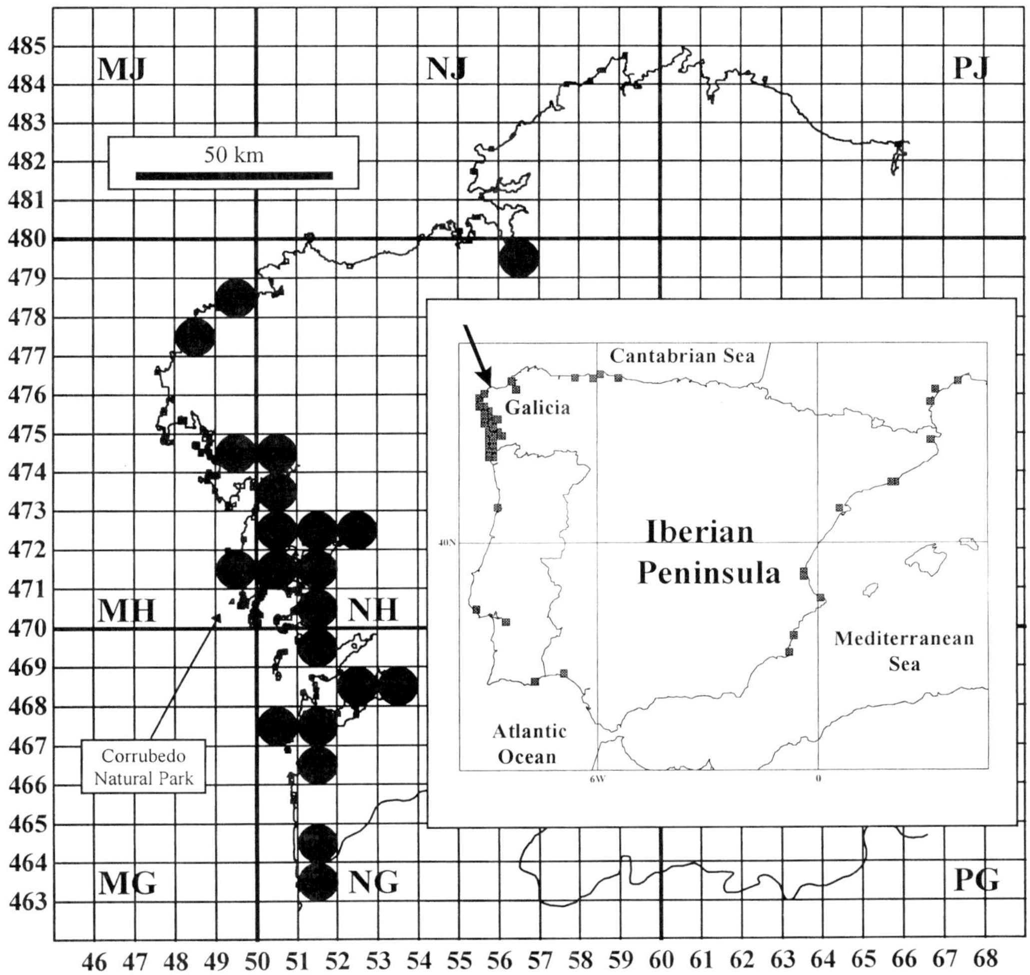

Figure 1. Distribution of *Spartina patens* in Galician saltmarshes in a grid of 10×10 km quadrats (UTM zone 29T) and localization within the Iberian Peninsula.

monio serotini - Juncetum maritimi subass. *halimionetosum portulacoidis* (**LJh**), *Limonio serotini - Juncetum maritimi* subass. *paspaletosum vaginati* (**LJp**), *Agrostio stoloniferae - Juncetum maritimi* subass. *typicum* (**AJ**), *Agrostio stoloniferae - Juncetum maritimi* subass. *stenotaphretosum secundati* (**AJs**), *Bostrychio scorpioidis - Halimionetum portulacoidis* (**BH**), *Puccinellio maritimae - Arthrocnemetum perennis* (**PS**) and patches dominated by *Arthrocnemum perennis* subsp. *alpini* (**Sa**) and by *Spartina patens* (**Sv**). Nomenclature follows Izco et al.. (1992) and Izco & Sánchez (1996).

Data analysis

In order to know the relationship between altitudinal range and abundance of *S. patens* as well as their influence on plant diversity, only the 55 quadrats with a presence of *S. patens* were used in the statistical analysis:

1. Altitudinal range of *S. patens*: mean altitude (non-weighted and weighted by abundances) was calculated, together with Spearman's correlation between altitude and coverage index.
2. Impact of *S. patens* on plant diversity (number of species) and total cover of other plants was tested with two regression analyses, considering

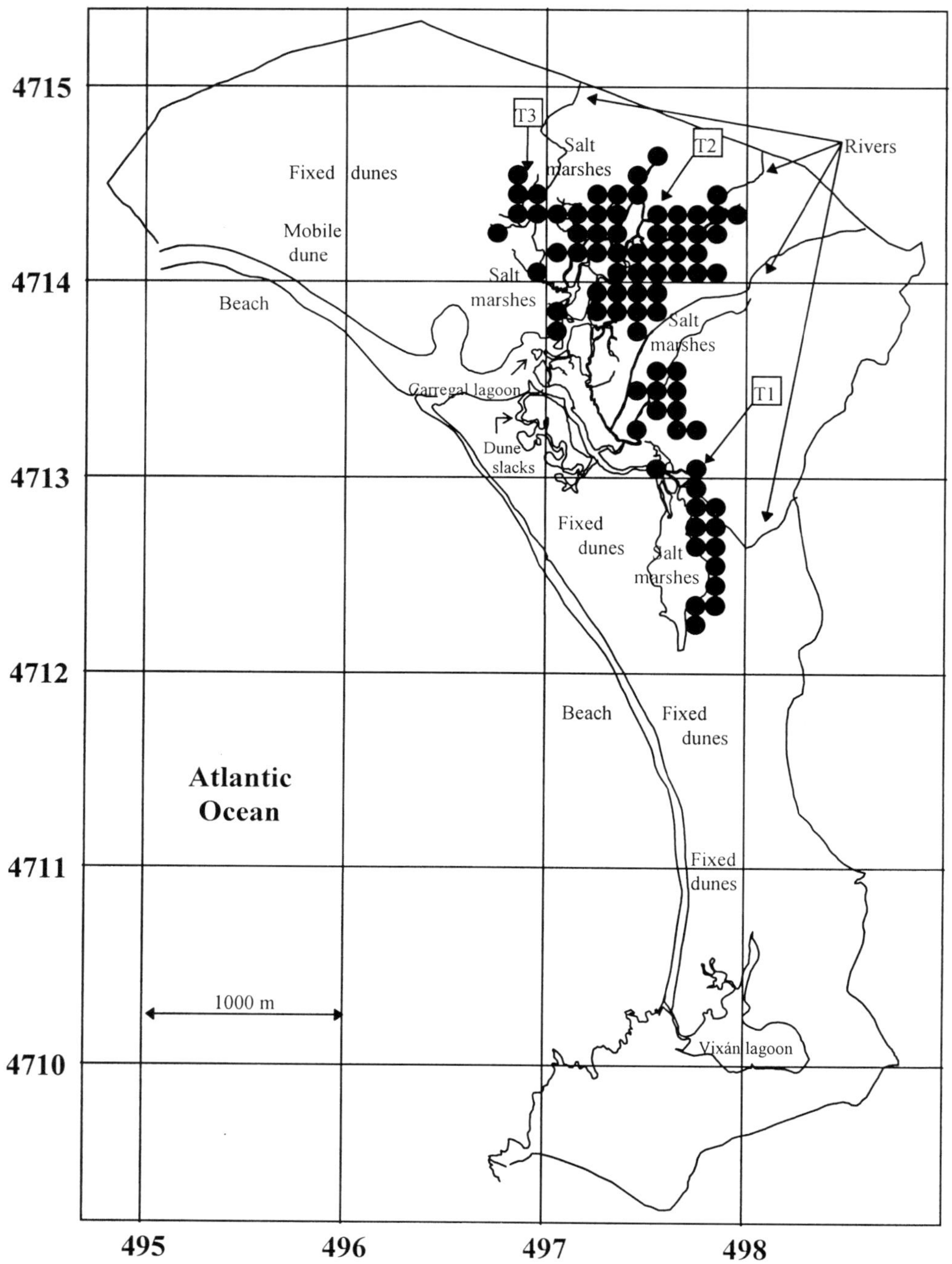

Figure 2. Distribution of saltmarsh *Spartina patens* patches in the studied area (Corrubedo Natural Park) in a grid of 100×100 m quadrats (UTM zone 29TMH).

'*S. patens* cover' and 'altitude' as independent variables. All data were arcsin transformed prior to analysis in order to improve normality.

Results

Distribution in Galicia

The species is already present in 45.83% of the 10×10 km coastal quadrats of the UTM grid (see Figure 1, where only saltmarsh records are represented). However, it becomes much more frequent in the southern estuaries (the proportion rising to 85.29% between cape Fisterra and Portugal) than in the northern ones (only 7.89%). In total, it was found in about 100 sites (1×1 km UTM grid), but only five of them were north of cape Fisterra (29 TMH74). Most of them are in saltmarsh habitats associated with estuaries and bay inlets, and also with coastal lagoons on the western-most coasts of Galicia (29TMH 100×100 km quadrat).

Our detailed study of Corrubedo N.P. saltmarshes (see Figure 2, where a 100×100 m grid was used for better visualization) shows that many little patches of *S.patens* (usually less than 10 m across) occur in the middle of the high marsh rush communities (see transect 3, Figure 5). It also forms linear patches along the elevated margins of saltmarsh creeks (like in quadrats 29–45 of transect 2, Figure 4) and colonizes drift lines in the upper high marsh boundary in contact with terrestrial communities (see transect 1 in Figure 3, and quadrats 175–180 of transect 2, Figure 4).

Altitudinal range

In the studied area, *S. patens* grows in high levels of the marsh, between −40 and 20 cm with respect to the highest tidal level recorded for each transect during the second semester of 1998 (see Figures 3–5); although this is enough to get a reference level and relativize the altitudes of different plant communities, further work is necessary to get more accurate data in order to calculate the inundation time at different levels. The mean altitude of the 55 sampling quadrats with *S. patens* presence is −33.53 cm respect to high tide level. However, the mean altitude rises to −29.48 cm if we weigh the altitude of the quadrat multiplying it by the *S.patens* coverage index, so giving more influence on the mean to those quadrats with higher abundance of *S. patens*.

Furthermore, for the same subsample of quadrats, there is a positive correlation (Spearman's r_S=0.46, $p<0.001$) between altitude and coverage of *Spartina patens*.

Table 1. Results of the two multiple regression analyses to assess the weight of 'altitude' and '*S. patens* cover' on the number and total cover of native species (n.s. = non significant; * = significant, $p<0.05$)

	Total cover of other species	Species number
Independent variables	Standardized Coefficient	Standardized Coefficient
altitude	0.09 (n.s.)	0.28 (n.s.)
S.patens cover	−0.69*	−0.49*
R^2	0.42	0.17

Effects on plant diversity

Looking to the transects represented in the Figures 3–5, the most affected community seems to be the *Agrostio stoloniferae – Juncetum maritimi* association, which shows a very strong reduction in the accumulated cover indexes for all the species other than *S. patens*. This community is the one that usually occupies the highest zones of the saltmarsh. Nevertheless, *S. patens* can also replace the *Limonio serotini – Juncetum maritimi* communities when these are located at a certain altitude. These rush communities are traditionally harvested for cattle bedding in many Galician saltmarshes. Sometimes, the dominance of *S. patens* is so high that the original plant community is no longer recognizable (see Figure 3).

Keeping in mind Figures 3–5, it seems clear that the species number tends to increase in the higher marsh levels, also according to Sánchez et al. (1996). Nevertheless, when *S. patens* is present with high cover indexes there is a remarkable decrease in the species number. This is even clearer if we look at the accumulated cover indexes of all the species other than *S. patens* (of course, bare soil patches have the same effect, this being the reason to represent them in the transects). A multiple regression analysis of data from the quadrats with a presence of *S. patens* produced the results shown in Table 1.

Discussion

Looking at the results of the regional survey, *S. patens* seems to be a very widespread species in southwest-

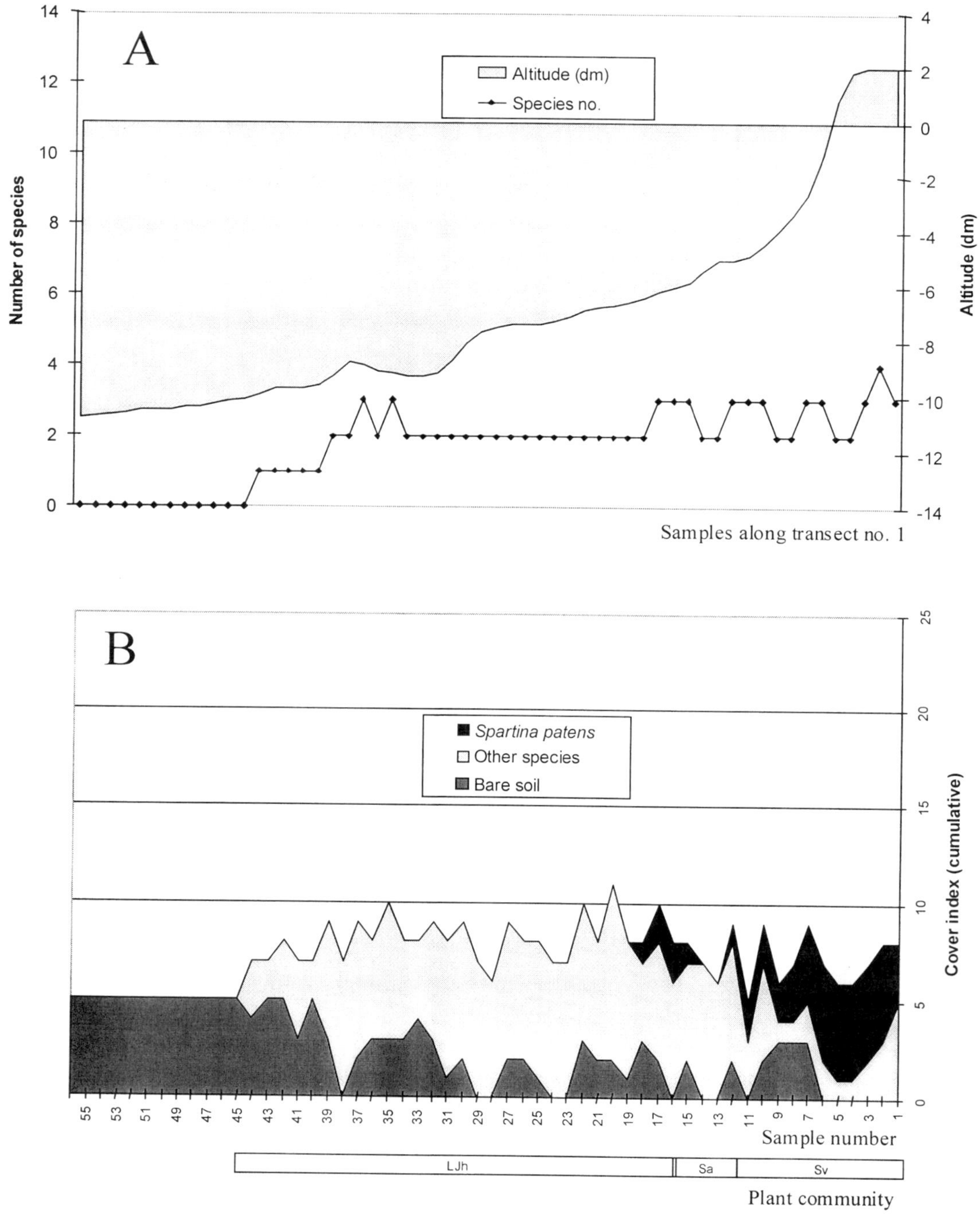

Figure 3. Vertical arrangement of samples along transect 1: (A) altitudes and number of species per sample; (B) Cumulative coverage indexes for *S. patens*, other species and bare soil; abbreviations correspond to names of plant communities cited in the text.

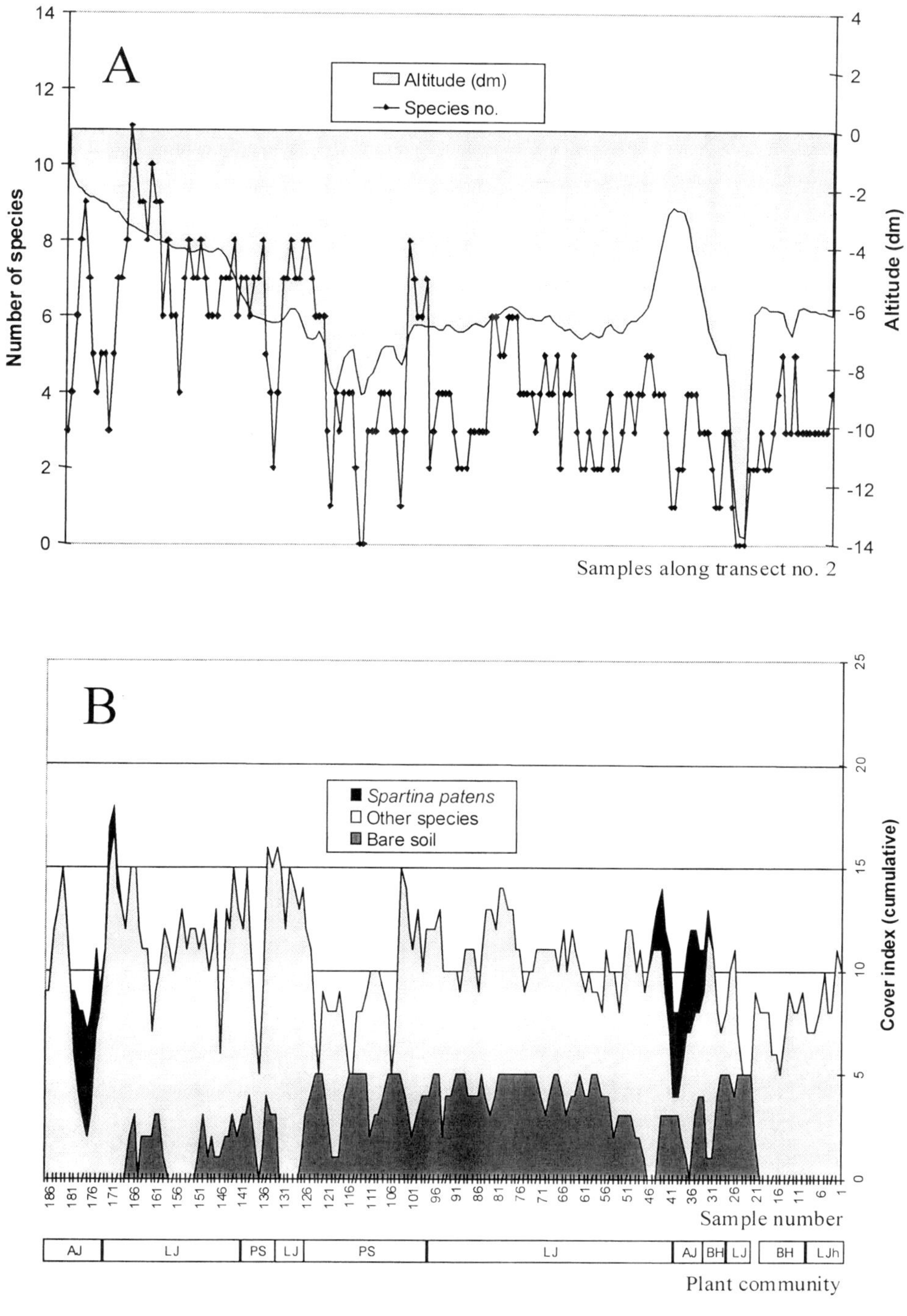

Figure 4. Vertical arrangement of samples along transect 2: (A) altitudes and number of species per sample; (B) Cumulative coverage indexes for *S. patens*, other species and bare soil; abbreviations correspond to names of plant communities cited in the text.

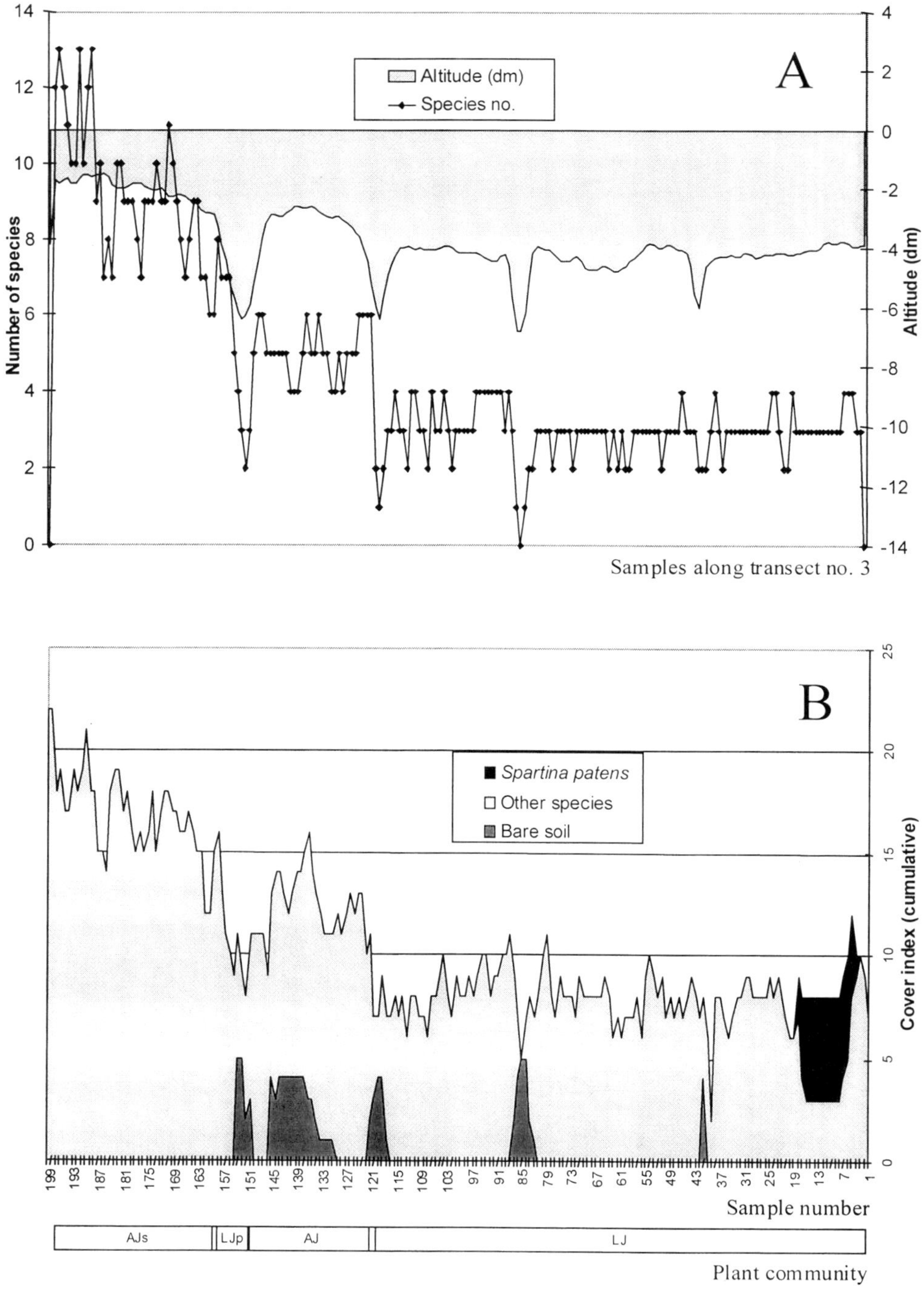

Figure 5. Vertical arrangement of samples along transect 3: (A) altitudes and number of species per sample; (B) Cumulative coverage indexes for *S. patens*, other species and bare soil; abbreviations correspond to names of plant communities cited in the text.

ern Galician estuaries. This is rather surprising, taking in account that the present work constitutes the first published record of this species in Galicia.

Moreover, the results of the local survey carried out in Corrubedo N.P. on a more detailed scale resemble the distribution pattern shown by this plant throughout the southwestern Galician coasts: it becomes apparent that – despite being a neophyte – it is extremely frequent there, dotting the coastline with numerous populations quite variable in size (but locally reaching a considerable extension and density in certain estuaries where big pure stands of *S. patens* dominate the saltmarsh physiognomy).

This suggests that it must be quite an old introduction, not detected until the present date due to its lack of flowering. It is possible that the coastline orientation with respect to dominant winds (SW and NE) could have an influence on the apparently easy establishment of the plant in this area, in contrast to its relative scarcity in Northern Galicia. Another important factor could be that the invasion may have been migrating northward along the Iberian Atlantic coastline, so having had more time to colonize the southern coasts of Galicia.

As for its position within the altitudinal zonation of saltmarsh plant communities, our results clearly show the preference of *S. patens* for the higher levels of the marsh, probably due to a limited ability to oxygenate its rhizosphere in anoxic soils (Bertness, 1991b); nonetheless, the plant only seems to exclude the other main dominant perennial – *Juncus maritimus* Lam. – in the landward zones of the upper high marsh, in contrast to the results of Bertness (1991a) and Bertness & Shumway (1993). On the other hand, the study of sample quadrats with a presence of *S. patens* (both the non-weighted and weighted mean altitudes as well as the correlation between altitude and *S.patens* coverage index) suggests that the plant grows better in the highest part of its altitudinal range, but it can spread downwards and grow new vegetative shoots connected by rhizomes to the clone it grew from (clonal integration *sensu* Hester et al., 1993; see also Shumway, 1995). These shoots would not achieve a very high density in lower levels because the plant is not very waterlogging tolerant, according to the aforementioned results of Burdick (1989), and Burdick & Mendelssohn (1987, 1990). Shumway (1995) suggests that physiological integration of ramets colonizing disturbance-generated bare patches may explain the predominance of vegetative invasion over sexual recruitment in marsh succession. Despite the fact that *S. patens* in our territory seems to invade well established communities rather than disturbance-generated gaps, physiological integration could be also valid to explain the way this plant spreads downwards from upper high marsh boundaries and creek elevated margins to colonize lower marsh levels.

With regard to the influence on plant diversity, and despite the fact that it increases with altitude (see Figures 3–5 and Sánchez et al., 1996), our results seem to indicate that this may not be true where *S. patens* has penetrated, since altitude is not a significant factor in our regression analysis (considering both its effect on species number or on total plant cover).

So, we can conclude from this work that *Spartina patens* is a new alien species for the Galician flora, although probably introduced a long time ago, being specially frequent – and locally very abundant – on the southwestern Galician coasts. Its preferred habitat is the upper limit of saltmarshes, becoming invasive and with a noxious effect on plant diversity of native communities when it forms very dense stands (often pure patches with 100% cover of this grass). Keeping in mind cases of *S. patens* introductions elsewhere (Frenkel, 1987; Frenkel & Boss, 1988; Mumford et al., 1991; Daehler & Strong, 1996), we should consider it as a threat to the biological diversity of Galician saltmarshes and take action towards eradicating the grass or at least curbing its expansion.

Following Daehler & Strong (1996), *Spartina patens* occurs over predictable tidal ranges within its native (e.g. Bertness & Ellison, 1987; Adam, 1990) and non-native (Frenkel & Boss, 1988) habitat range; within a vulnerable site and at a fine scale, the mean tidal range can be used to predict the extent of spatial spread of *Spartina patens* after colonization. Taking in account the rather low presence of this weed in northern Galician coasts, these prediction techniques might be used to identify and prioritize sites for protection against future invasions.

Acknowledgements

This research was supported by the Galician Government (XUGA20309B96) and by grant from the FPU Program of the Spanish Government, awarded to D. G. SanLeón. We thank Minia Bongiorno García for translation of the original document.

References

Adam, P., 1990. Saltmarsh Ecology. Cambridge University Press, Cambridge

Bertness, M. D., 1991a. Interspecific interactions among high marsh perennials in a New England salt marsh. Ecology 72: 125–137.

Bertness, M. D., 1991b. Zonation of *Spartina patens* and *Spartina alterniflora* in a New England salt marsh. Ecology 72: 138–148.

Bertness, M. D. & S. W. Shumway, 1993. Competition and facilitation in marsh plants. Am. Nat. 142: 718–724.

Bertness, M. D. & A. M. Ellison, 1987. Determinants of pattern in a New England salt marsh plant community. Ecol. Monogr. 57: 129–147.

Bueno, A., 1997. Flora y vegetación de los estuarios asturianos. Cuadernos de Medio Ambiente NATURALEZA 3. Principado de Asturias. Consejería de Agricultura.

Burdick, D. M., 1989. Root aerenchyma development in *Spartina patens* in response to flooding. Am. J. Bot. 76: 777–780.

Burdick, D. M. & I. A. Mendelssohn, 1987. Waterlogging responses in dune, swale and marsh populations of *Spartina patens* under field conditions. Oecologia 74: 321–329.

Burdick, D. M. & I. A. Mendelssohn, 1990. Relationship between anatomical and metabolic responses to soil waterlogging in the coastal grass *Spartina patens*. J. exp. Bot. 41: 223–228.

Campos, J. A. & M. Herrera, 1997. La flora introducida en el País Vasco. Itinera Geobotanica 10: 235–255.

Chapman, V. J., 1974. Salt marshes and salt deserts of the World. J.Cramer, Lehre (2nd, supplemented reprint edition).

Cosson, E. & Durieu de Maisonneuve, 1867. Introduction á la flore d'Algérie. Phanérogamie. Groupe des Glumacées (seu descriptio glumacearum in Algeria nascentium). Exploration Scientifique de l'Algérie, publiée par ordre du Gouvernement. Sciences Naturelles. Botanique. Imprimerie Impériale. Paris.

Daehler, C. C. & D. R. Strong, 1996. Status, prediction and prevention of introduced cordgrass *Spartina* spp. invasions in Pacific estuaries, U.S.A. Biol. Conserv. 78: 51–58.

Fabre, M. E., 1849.Description d'une nouvelle espèce de *Spartina*, abondante sur une portion du littoral méditerranéen. Ann. Sci. Nat. Bot. Paris 3: 122–125.

Fernald, M. L., 1929. Some relationships of the floras of the northern hemisphere. Proc. Int. Congr. Pl. Sci. 2: 1487–1507.

Frenkel, R. E., 1987. Introduction and spread of cordgrass (*Spartina*) into the Pacific Northwest. Northwest Envir. J. 3: 152–154.

Frenkel, R. E. & T. R. Boss, 1988. Introduction, establishment and spread of *Spartina patens* on Cox Island, Siuslaw Estuary, Oregon. Wetlands 8: 33–49.

Hester, M. W., I. A. Mendelssohn & K. McKee, 1996. Intraspecific variation in salt tolerance and morphology in the coastal grass *Spartina patens* (Poaceae). Am. J. Bot. 83: 1521–1527.

Hester, M. W., K. L. McKee, D. M. Burdick, M. S. Koch, K. M. Flynn, S. Patterson & I. A. Mendelssohn, 1993. Clonal integration in *Spartina patens* across a nitrogen and salinity gradiente. Can. J. Bot. 72: 767–770.

Hultén, E., 1958. The amphi-atlantic plants and their phytogeographical connections. Almquist & Wiksell. Stockholm.

Izco, J. & J. M. Sánchez, 1996. Los medios halófilos de la ría de Ortigueira (A Coruña, España). Vegetación de dunas y marismas. Thalassas 12: 63–100.

Izco, J., P. Guitián & J. M. Sánchez, 1992. La marisma superior cántabro-atlántica meridional: estudio de las comunidades de *Juncus maritimus* y de *Elymus pycnanthus*. Lazaroa 13: 149–169.

Mobberley, D. G., 1956.Taxonomy and distribution of the genus *Spartina*. Iowa State College Journal of Sciences 30 471–574.

Mumford, T. F. Jr., P. Peyton, J. R. Sayce & S. Harbell (eds), 1991. *Spartina* Workshop Record: Seattle, Washington, November 14–15, 1990. Washington Sea Grant Program. College of Ocean & Fishery Sciences. University of Washington.

Pignatti, S., 1982. Flora d'Italia 3. Edagricole. Bologna, Italia.

Saint-Yves, A., 1932. Monographia Spartinarum. Candollea 5: 19–100.

Sánchez, J. M, J. Izco & M. Medrano, 1996. Relationships between vegetation zonation and altitude in a salt-marsh system in northwest Spain. J. Veget. Sci. 7: 695–702.

Shumway, S. W., 1995. Physiological integration among clonal ramets during invasion of disturbance patches in a New England salt marsh. Ann. Bot. 76: 225–233.

Silander, J. A. & J. Antonovics, 1979. The genetic basis of the ecological amplitude of *Spartina patens*. I. Morphometric and physiological traits. Evolution 33: 1114–1127.

Tutin, T. G., 1980. *Spartina* Schreber. In Tutin, T. G. et al.: Flora Europaea, 5: 259–260. Cambridge University Press. Cambridge, U.K.

Van der Maarel, E. & M. Van der Maarel-Versluys, 1996. Distribution and conservation status of littoral vascular plant species along the European coasts. J. Coast. Conserv. 2: 73–92.

Hydrobiologia **415:** 223–228, 1999.
J. M. Caffrey, P. R. F. Barrett, M. T. Ferreira, I.S. Moreira, K. J. Murphy & P. M. Wade (eds), Biology, Ecology and Management of Aquatic Plants.
© 1999 *Kluwer Academic Publishers. Printed in the Netherlands.*

Phenology and long-term control of *Heracleum mantegazzianum*

J. M. Caffrey
Central Fisheries Board, Mobhi Boreen, Glasnevin, Dublin, 9, Ireland

Key words: Giant Hogweed, invasive plants, riparian zones, angling

Abstract

The growth characteristics and reproductive capacity are described for *Heracleum mantegazzianum*, an invasive perennial plant that has significantly increased its geographical range in Ireland in recent years. The effect of cutting on growth and seed production is described. Based on research findings, a detailed Giant Hogweed control programme is presented which could result in the elimination of the plant. A Giant Hogweed eradication programme on the Mulkear River catchment (c. 670 km^2), using this protocol, commenced in 1998 and will continue until 2002.

Introduction

Heracleum mantegazzianum Sommier and Levier is an alien species that was introduced to Ireland in the late 19th Century as an ornamental plant. It has since become naturalised in a number of locations. Up to the late 1960s, the plant formed only localised dominant stands although, since that time, it has noticeably expanded its range and in 1997 was recorded from 30 different 10 × 10 km squares (Wade et al., 1997).

In Ireland, *H. mantegazzianum* most commonly occurs along the banks of rivers and streams, reflecting its dependence on flowing water for long-distance seed dispersal. Following effective colonisation of long stretches of river channel throughout Ireland, a secondary spread onto roadways and adjacent habitats is occurring, mediated by wind and anthropogenous factors.

The uncontrolled spread of *H. mantegazzianum* in Ireland poses a serious threat to human health and to the ecology of infested habitats (Williamson & Forbes, 1982; Tiley et al., 1996; Wade et al., 1997), and threatens to destabilise the delicate equilibrium present in salmonid rivers. Dense stands of this tall, leafy plant can suppress and ultimately exclude indigenous herbaceous plant species, which play an important role in river bank stabilisation. This can result in erosion during winter floods (Tiley & Philp, 1994; Dodd et al., 1994), with large quantities of soil being washed into the river. The relatively fine particulate suspension settles in slow flowing sections of channel, where it will alter substrate characteristics and, as a consequence, make conditions more favourable for abundant aquatic plant growth (Caffrey, 1990). Fine soil and silt particles will clog gravel interstices and render them unsuitable for salmonid spawning. Research in Ireland (Gargan & Caffrey, 1991; Caffrey, 1992) and elsewhere (E.I.F.A.C., 1974; Reiser & White, 1988) has shown that salmon and trout eggs buried in gravel can only develop satisfactorily when a current of water is passing through the gravel. An oxygen deficit to the developing ova results when silt causes a reduction in intragravel percolation. Thus, by altering the substrate and clogging the spawning gravels with silt, *H. mantegazzianum* indirectly has the potential for serious impact on productive salmonid fisheries (Wade et al., 1997).

Why *H. mantegazzianum* has successfully established where other alien plants have failed, or why it has only begun to spread geometrically within the country in the past two decades, is unclear. It could be due to:

(a) the paucity of natural pests and diseases which could limit its spread;

(b) the plant becoming adapted to local conditions through natural selection;

(c) the plant generally growing in areas which are little affected by human activities; or

(d) the plant being highly prolific.

The object of this paper is to describe the phenology of *H. mantegazzianum* in Ireland and to present a control/eradication strategy for this aggressive invasive plant.

Materials and methods

Seven sites where *H. mantegazzianum* has established large dominant stands, often greater than 100 m^2, were studied during 1993. The phenology of seedling, immature and flowering plants was examined. At each site, 10 plants were marked soon after emergence in spring, using different coloured threads to distinguish between seedling, immature and mature plants. Due to vandalism and unseasonal flooding, it was only possible to monitor the growth of marked, immature plants at two sites, on the Mulkear and Shannon Rivers. Sufficient mature, marked plants were available at all but two sites (the Tolka and Dargle Rivers) to provide useful growth data.

At Portmarnock, where spoil had been deposited over tarmac to a depth of $\leqslant$ 15 cm, the effect of substrate depth on growth and seed production was monitored. This represented the 'shallow' site in the study, from which 10 mature plants were selected to compare growth and reproductive capacity with plants from an adjacent site with no soil depth restriction.

At each site, a total count of seed numbers was recorded from 10 mature plants, selected to include a range of plant heights. The relationship between plant height and seed number was tested using a Pearson Correlation coefficient. Seedling density was measured at Portmarnock and Mulkear at 5×1 m^2 permanent quadrats on seven occasions between March and August 1993. Similar quadrats were established in late September at 2 m intervals from the edge of the parent plant stand to determine the extent of seed dispersal.

The effects that cutting had on growth and seed production were examined. In order to determine the effect that the timing of a cut had on seed production, plants in *circa* 100 m^2 plots at Portmarnock and Mulkear were cut to ground level in late March (early cut) and mid-May (late cut). Total seed counts were made from 10 mature plants, to include the range of plant heights in each of the plots, and an ANOVA was used to test statistical differences. The length and width of 50 seeds from the centre of the terminal umbel in each treatment at Portmarnock were measured to examine the possibility that smaller (and possibly less viable) seeds might result after a late cut.

Results and discussion

Seedling density and growth

The fraction of seeds that germinate from the soil bank and establish seedlings in a growing season is small, varying between 0.001 and 10% (Mortimer, 1990). The first seedlings emerged in February, two to three weeks after the first leaves of previous years' plants had emerged. Peak seedling density was recorded at both sites (Portmarnock and Mulkear) in late April, after which numbers declined (Figure 1). By the end of August, only 1.2% of the seedlings recorded in April remained at Portmarnock while at Mulkear, 13.7% of the April population survived. This compares to survival rates of between 11 and 23% recorded on the lower River Shannon by Warde (1985). The difference in survival rates between the two sites probably reflects the smaller peak density of seedlings at Mulkear, although the impact that regular disturbance by walkers at Portmarnock had on survival should also be considered. In Scotland, spring germination produced 400 first leaf stage seedlings m^{-2}, which fell to 33 m^{-2} with full canopy cover of adjacent adult plants in mid-summer (Tiley et al., 1996).

In autumn 1992, large numbers of seedlings that presumably were suppressed from the spring were observed. Seedlings appeared in early September and grew until the end of October. The leaves then died back and the plants overwintered as small taproots. Similar observations were recorded by Warde (1985), Tiley & Philp (1997) and Tiley et al. (1996).

Phenology

Based on studies conducted on the life cycle of *H. mantegazzianum* in Ireland between 1990 and 1993, it appears that most plants flower in their third or fourth year. This agrees with the findings of Stewart & Grace (1984). In the plant's second year, growth from elongated taproots generally begins in early February. Initial growth is relatively slow and maximum stem and leaf enlargement was recorded between mid-May and mid-June. While some of the leaves achieved a width in excess of 1 m, few plants grew taller than 1 m. The majority of plants died back in late August.

Heracleum mantegazzianum plants that were to flower in the current growing season generally

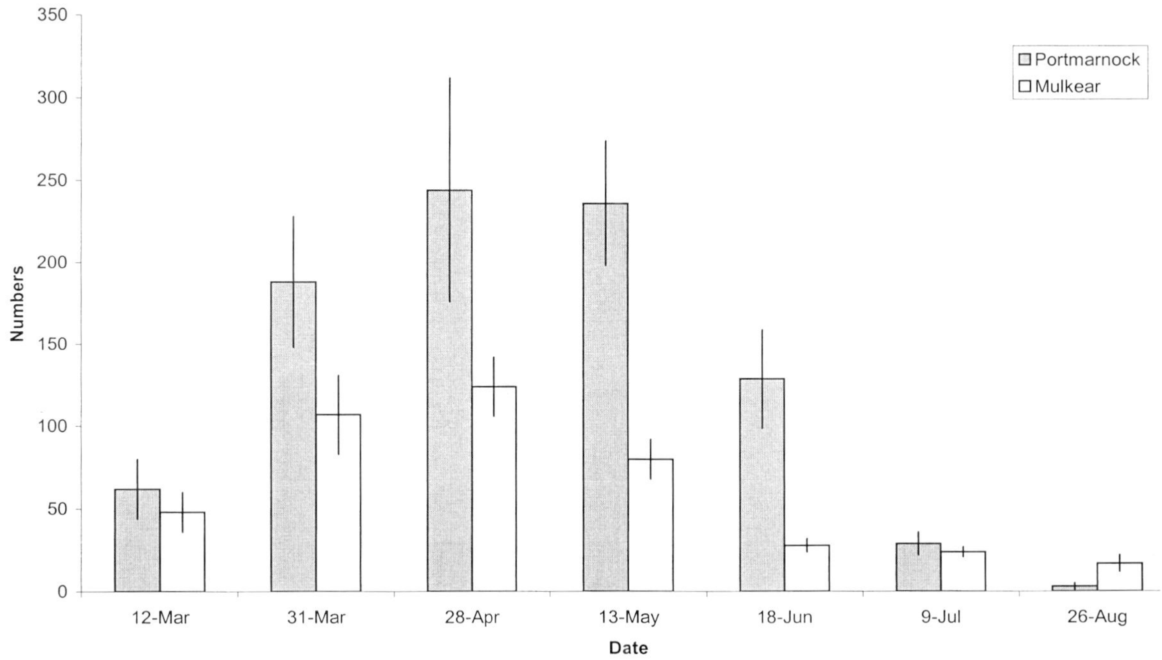

Figure 1. Mean density of *H. mantegazzianum* seedlings m^{-2} recorded in five 1 m^2 quadrats at Portmarnock and Mulkear.

emerged in early February. In the first few weeks of growth, the plant produced large leafy rosettes which shaded light from potential competitor species. Vegetative growth continued until early May, by which time leafy canopies up to *circa* 1 m tall were produced (Table 1).

From May until the end of the growing season (late July), the dramatic increments in stem height resulted from flower stalk elongation. Peak flowering occurred in late June and most seeds had ripened and dispersed by the end of August. The differences in the ultimate plant heights achieved (Table 2) reflect local environmental and substrate conditions.

Effect of substrate depth

The effect of substrate depth on plant height and seed production is shown for plants at Portmarnock in Tables 1 and 2. There was a significant difference ($P < 0.001$) between the maximum height of plants recorded on normal and on shallow substrates. This agrees with Tiley et al. (1996) who observed that plants are shorter and smaller where rooting is restricted, and stressed the need for a sufficient depth of soil to allow the development of the tap-root. Likewise, seed numbers were significantly greater ($P < 0.001$) on normal compared with shallow substrates (Table 2). Plants growing on shallow soil had fewer satellite and axillary umbels. It is noteworthy that the plants growing on shallow soils did not possess true tap-roots but instead adapted to the unfavourable conditions by developing a highly-branched, laterally-spreading root system which provided good anchorage and sufficient food reserves for growth.

Seed production

Williamson & Forbes (1982) state that mature *H. mantegazzianum* plants shed "up to 5000 or more seeds". Warde (1985) recorded between 14 000 and 29 000 seeds and Brondegaard (1990) reported 27 000 seeds per plant in Ireland and Germany, respectively. Based on estimated counts on single large (2.74 m) flowering plants, Tiley & Philp (1997) recorded 52 800 seeds. Mean seed number per plant for the seven sites examined ranged between 13 884 at Portmarnock and 68 748 on the Dargle. The highest number of seeds recorded from a single plant was 107 984 from a 3.3 m tall specimen on the River Tolka. This plant supported one large apical umbel, nine lateral umbels and 14 tertiary umbels.

Table 1. Mean height (cm) of marked, mature *H. mantegazzianum* plants (n=10) at selected river sites between February and July 1993

Date	8/2		15/3		6/4		3/5		3/6		27/7	
	cm	S.E.	cm	S.E.	cm	S.E.	cm	S.E.	cm	S.E.	cm	S.E.
Portmarnock - normal	3	1.1	43	11	55	21	94	24	146	30	268	55
- shallow	2	0.5	23	4	29	6	43	10	54	12	137	31
Shanganagh	5	2.3	28	9	34	14	65	9	97	26	220	48
Newport	2	0.7	47	13	55	17	64	16	102	19	284	61
Mulkear	4	1.1	48	19	68	18	107	23	147	26	300	54
Shannon	2	0.5	26	8	39	7	62	11	111	30	291	72

Table 2. Plant height and seed numbers in mature *H. mantegazzianum plants* (n=10) from selected river sites. Figures are presented as mean, standard error (S.E.), minimum and maximum values

	Plant height (cm)				Seed number			
	Mean	S.E.	Min.	Max.	Mean	S.E.	Min.	Max.
Portmarnock - normal	278	19	201	369	42 068	5170	17 790	64 254
- shallow	149	13	82	195	13 884	4117	1516	21 857
Tolka	291	7	259	329	44 264	9632	17 688	107 984
Shannon	295	10	226	320	42 820	7144	5154	72 420
Mulkear	306	13	244	366	46 386	6452	13 676	76 992
Newport	299	9	262	366	52 920	4125	42 704	81 410
Dargle	311	6	290	351	68 748	4804	39 084	92 702
Shanganagh	205	10	253	347	18 528	1564	12 478	26 068

A Pearson correlation showed a significant correlation ($P < 0.05$) between plant height and seed number for the 80 plants examined. This finding is supported by Tiley et al. (1996) who report that the number, size and weight of the inflorescence rays correlate with each other, with umbel size of the same order and with overall plant size.

Seed dispersal

Seed dispersal is passive and normally mediated by flowing water or wind. The seeds are large and flat and, even in conditions of strong wind, are unlikely to travel more than a few metres from the parent plant (Clegg & Grace, 1975). Results from quadrat counts conducted in the vicinity of dense plant stands at Portmarnock and Mulkear during the present study showed that in excess of 95% of the seeds fell within 10 m of the colony. Only isolated seeds were observed up to 50 m from the parent stands. Sheppard (1991) observed that few *H. sphondylium* seeds, which are similar to those of *H. mantegazzianum*, travelled more than 50 m from the parent plant.

Long distance seed dispersal, where not mediated by flowing water, commonly results from human activities (Tiley et al., 1996). The presence of isolated, mature stands of *H. mantegazzianum* on major roadways throughout the country, long distances from known areas of infestation, are suggestive of inadvertent human contribution to the dispersal of the species.

Effect of timing of cut on H. mantegazzianum

No mortalities were recorded among plants which were cut to ground level in either early March or late May. Results from these trials are presented in Tables 3 and 4. Significant differences ($P < 0.05$) between all treatments in respect of plant height and seed numbers were recorded.

Seeds from early-cut plants were significantly ($P < 0.05$) longer than those from uncut plants, although no statistical difference (at the 95% level) was recorded for seed width. The seeds produced by late-cut plants were shorter ($P < 0.05$) and narrower (NS) than those recorded from uncut or early-cut plants (Table 4).

Table 3. Effect of cutting in early March and late May on plant height (cm) and seed number

Type of cut	No. plants	Plant height (cm)	S.E.	Significance	Seed no.	S.E.	Significance
No cut	30	244	15	$P<0.05$	34 113	3997	$P<0.05$
Early cut (March)	15	181	13	$P<0.05$	16 880	3596	$P<0.05$
Late cut (May)	30	127	5	$P<0.05$	4281	523	$P<0.05$

Table 4. Effect of cutting in early March and late May on seed size of *H. mantegazzianum* plants at Portmarnock in 1993

Type of cut	No. seeds	Seed length	S.E.	Significance	Seed width	S.E.	Significance
No cut	50	12.8	0.18		8.14	0.14	
Early cut	50	13.7	0.1	$P<0.05$	8.46	0.09	NS
Late cut	50	11.5	0.22	$P<0.05$	7.27	0.15	NS

Tiley & Philp (1997) demonstrated that cutting in late June significantly reduced total inflorescence yields and the seeds produced were viable.

Control strategy for *H. mantegazzianum*

Studies on the distribution of *H. mantegazzianum* in Ireland and the U.K. show that the plant is still in an expansive mode and that further expansion into new ecologically and socially sensitive areas is likely. These studies have further revealed that there are still a number of large catchments in Ireland which are presently free of the plant (Wade et al., 1997). Every effort must be made, therefore, to safeguard these catchments and to eliminate or effectively curtail the spread of the plant in already infested regions.

Prior to 1998, no serious attempt had been made to limit the expansion of *H. mantegazzianum* in Ireland. Long-term control and ultimate eradication can only be achieved through the implementation of a national, co-ordinated and comprehensive management strategy that involves all interested groups (Wade et al., 1997). Failure to embark on such a strategy will lead to even greater problems that will prove more costly to resolve.

H. mantegazzianum populations can only be perpetuated via seeds. Control measures applied before flowering and seed set will limit recruitment to subsequent generations and, if applied systematically over a number of years, will ultimately deplete the seed bank reserve. The longevity of seeds in the soil is unknown, although there are indications that the vast majority of viable seeds germinate within one year (Tiley et al., 1996).

Results from extensive trials have demonstrated the susceptibility of *H. mantegazzianum* to glyphosate (Williamson & Forbes, 1982; Powell, 1988; Caffrey, 1994; Tiley & Philp, 1997). Presented below is a step-by-step protocol for the long-term control of *H. mantegazzianum*, using glyphosate. If strictly adhered to, and the areas are not reinfested from external sources, it could lead to the successful eradication of the plant from any treated catchment.

1. Compile maps which accurately detail the distribution of *H. mantegazzianum* in each affected catchment throughout the country.
2. A catchment approach to treatment with glyphosate must be taken and spraying must commence at the farthest upstream site from which the plant is recorded.
3. Spraying should commence in March or early April, when the plants have expanded their leaves and reached a height of >15 cm. Experience in Ireland has shown that young seedlings and small plants (<15 cm) are less susceptible to glyphosate than their larger counterparts.
4. A dose rate of 5 l ha^{-1} is sufficient to kill treated vegetation.
5. Sections treated in March/April should be retreated in May or later, if required.
6. The whole catchment must again be surveyed in July and any plants that have flowered, or are likely to flower, should be deheaded or chopped down before seeds are produced. The cut umbels must

be removed from the area and destroyed. Any regrowth should be sprayed with glyphosate as the plant will again attempt to flower and set seed.

7. A further glyphosate treatment in September throughout the catchment will kill late-developing plants or seedlings that survive earlier treatments.

If the above protocol is rigorously applied over a four year period, which is the normal lifespan of *H. mantegazzianum*, the seed reserve will be depleted and the risk of serious reinfestation from within the catchment will be minimal.

Eradication programme on Mulkear Catchment

Large localised populations of *H. mantegazzianum* were reported in the Mulkear catchment by Praeger (1939). The level of infestation has increased dramatically in the past two decades and this has affected practically all beneficial usage of the watercourse (Caffrey, 1994). In 1997, the Office of Public Works compiled maps which detailed the distribution of the plant in the catchment. Planimeter readings indicated that an area of *circa* 35 km^2 was overgrown by *H. mantegazzianum* (A. Casey, pers. comm.). A contractor was commissioned to undertake a four-year control/eradication programme, with follow-up monitoring, on the entire catchment using the above protocol. This detailed programme commenced in March 1998 and will be closely monitored until its completion.

Acknowledgements

The author wishes to thank Ms C. Monahan and Ms M. Fallon for their considerable assistance in the field and laboratory. Thanks are also extended to Mr Gordon Tiley for his constructive comments.

References

Brondegaard, V. J., 1990. Massenausbreitung des Barenklaus. Naturw. enschaftliche Rundschau: 43: 438–439.

Caffrey, J. M., 1990. The classification, ecology and dynamics of aquatic plant communities in some Irish rivers. Ph.D. Thesis, University College, Dublin: 254 pp.

Caffrey, J. M., 1992. Salmon ova planting experiments in the River Deel catchment, North Mayo. North Western Regional Fisheries Board commissioned report, Cent. Fish. Bd, Dublin: 13 pp.

Caffrey, J. M., 1994. The spread and management of Giant Hogweed *Heracleum mantegazzianum* (Somm. and Levier) along Irish river corridors. In de Waal, L. C. et al. (eds), Ecology and Management of Invasive Riverside Plants. John Wiley & Sons: 67–76.

Clegg, L. M. & J. Grace, 1975. The distribution of *Heracleum mantegazzianum* (Somm. and Levier) near Edinburgh. Trans. Bot. Soc. Edinburgh 42: 223–229.

Dodd, F. S., L. C. de Waal, P. M. Wade & G. E. D. Tiley, 1994. Control and management of *Heracleum mantegazzianum* (Giant Hogweed). In de Waal, L. C. et al. (eds), Ecology and Management of Invasive Riverside Plants. John Wiley & Sons: 111–126.

EIFAC, 1974. Water quality criteria for European freshwater fish. Report on Finely Divided solids and Inland Fisheries. EIFAC Technical Paper (1): 21 pp.

Gargan, P. & J. M. Caffrey, 1991. Results of a preliminary survey of the Rivers Deel and Owenmore, Co. Mayo, with particular reference to peat siltation. North Western Regional Fisheries Board commissioned report, Centr. Fish. Bd, Dublin: 41 pp.

Mortimer, C., 1990. Weeds. In Hance, R. J. & K. Holly (eds), Weed Control Handbook: Principles. Blackwell Scientific Publications, Oxford.

Powell, F., 1988. Giant Hogweed control in north-east Scotland. Plant Press Nr. 4: 4.

Praeger, R. L., 1939. A further contribution to the flora of Ireland. Roy. Irish Acad. 45B: 231–254.

Reiser, D. W. & R. G. White, 1988. Effects of two sediment size-class on survival of Steelhead and Chinook salmon eggs. North am. J. Fish. Mgmt 8: 432–437.

Sheppard, A. W., 1991. *Heracleum sphondylium* L. J. Ecol. 79: 235–258.

Stewart, F. & J. Grace, 1984. An experimental study of hybridization between *Heracleum mantegazzianum* Somm. & Lev. and *H. sphondylium* L. ssp. *Sphondylium* (Umbelliferae). Watsonia 15: 73–83.

Tiley, G. E. D. & B. Philp, 1994. Giant Hogweed and its control in Scotland. In de Waal, L. C. et al. (eds), Ecology and Management of Invasive Riverside Plants. John Wiley & Sons: 101–110.

Tiley, G. E. D., F. S. Dodd & P. M. Wade, 1996. *Heracleum mantegazzianum* Sommier & Levier. Biological Flora of the British Isles. J. Ecol. 84: 297–319.

Tiley, G. E. D. & B. Philp, 1997. Observations on the flowering and seed production of *Heracleum mantegazzianum* in relation to control. In Brock, J. H. et al. (eds), Plant invasions. Backhuys Publishers, Leiden: 123–138.

Wade, M., E. J. Darby, A. D. Courtney & J. M. Caffrey, 1997. *Heracleum mantegazzianum*: a problem for river managers in the Republic of Ireland and the United Kingdom. In Brock, J. H. et al. (eds), Plant Invasions. Backhuys Publishers, Leiden: 139–152.

Warde, C., 1985. Investigation of the biology and ecology of the Giant Hogweed *Heracleum mantegazzianum* along the lower Shannon. B.Sc. Thesis, Thomond College, Limerick: 70 pp.

Williamson, J. A. & J. C. Forbes, 1982. Giant Hogweed *Heracleum mantegazzianum*: its spread and control with glyphosate in amenity areas. Proc. 1982 Brit. Crop Protection Conf. – Weeds: 967–972.

Hydrobiologia **415:** 229–234, 1999.
J. M. Caffrey, P. R. F. Barrett, M. T. Ferreira, I.S. Moreira, K. J. Murphy & P. M. Wade (eds), Biology, Ecology and Management of Aquatic Plants.
© 1999 *Kluwer Academic Publishers. Printed in the Netherlands.*

Aquatic weeds and their management in Portugal: insights and the international context

I. Moreira[1,*], T. Ferreira[1], A. Monteiro[1], L. Catarino[2] & T. Vasconcelos[1]
[1]*Instituto Superior de Agronomia, Tapada da Ajuda, 1349-017 Lisboa, Portugal*
E-mail: ilidimor@esoterica.pt
[2]*Centro de Botânica, Instituto Investigação Científica Tropical, tv. Conde da Ribeira 7, 1349-142 Lisboa, Portugal*

Key words: Aquatic Weed Symposium, *Eichhornia crassipes*, *Myriophyllum aquaticum*, *Typha* spp., *Phragmites australis*

Abstract

A historical insight into the connections between Portuguese researchers working on aquatic weeds and the EWRS Symposia (1964–1998) is given. A review is provided of the major aquatic weeds in Portuguese waters, including fluvial systems, reservoirs, drainage and irrigation canals, and rice fields along with the types of control and aquatic plant management practised in Portugal. Aquatic weed control in Portuguese waters is performed mostly by mechanical means, in spite of several experimental attempts at chemical control and biological control with fish and insects. Aquatic weed studies in other Mediterranean countries are also referred to.

The 10th Aquatic Weed Symposium in Lisbon

The first International Aquatic Weed Symposium took place in France (1964), with subsequent symposia in Germany (1967), the United Kingdom (1971) and Austria (1974), and the 5th in the Netherlands (1978). The following meetings were held in Loughborough (U.K.) in 1986, and in Uppsala (Sweden) in 1990. In 1994, the event moved west, to Dublin (Ireland).

These previous symposia gathered together a large number of people working in this field of research, and were seen as important milestones in the development of aquatic plant management. Caffrey & Wade (1996) reviewed the contents of these International Symposia on Aquatic Weeds sponsored by the European Weed Research Society (EWRS), the main objectives of the EWRS Working Group on Aquatic Weeds and the activities of other groups with an interest in this topic.

The 10th Aquatic Weed Symposium in Lisbon, the last of the series in the 20th century, was organised on the theme *"towards an integrated aquatic plant management"*. Aquatic plant assemblages are important biological features of freshwater ecosystems. However, human activities have now changed and disturbed natural ecosystems to a point of no return, and biological and limnological processes must be invoked as driving concepts for engineering practices. As pieces of the ecosystems puzzle, aquatic plants and their weedy behaviours are now being viewed in a wider context of watershed management.

The 122 participants in the 10th Symposium came from 29 different countries from all continents and contributed 108 papers to the Proceedings (Monteiro et al., 1998b), the largest number of contributions an Aquatic Weed Symposium has ever had. Besides 21 contributions from Portugal, the organising country, the United Kingdom and the Netherlands increased the number of their contributions, to 18 and 11, respectively, and maintained their traditional interest in aquatic plant management, as in previous symposia (Caffrey & Wade, 1996). There were six contributions from Argentina, and five each from France and Hungary. The remaining contributions came from Australia, Austria, Belgium, Brazil, the Czech Republic, Egypt, Finland, Germany, India, Ireland, Japan, New Zealand, Poland, Romania, Senegal, Slovenia, South Africa, Spain, Sweden, Switzerland, Taiwan, the U.S.A. and Yugoslavia.

* Author for correspondence

In total, there were 48 oral and 60 poster presentations, organised into five subject areas: biology and distribution of aquatic plants (33 papers); functional ecology and ecosystem approach (21); invasive and weedy plants (18); control of aquatic weeds and its side effects (29); and the utilisation of aquatic plants (7). In comparison with previous symposia, the interest in biological and physical control seemed to decrease, with only three papers on each topic. However, there were 11 papers on chemical control and its side effects and 10 on the integration of different control methods. The control of algae using straw was the subject of two papers, though it was considered a subject of expanding interest in the last symposia (Caffrey & Wade, 1996).

In Portugal, the number of people involved in aquatic plant management is very small. Firstly, river rehabilitation programmes, generally including dredging and resectioning of the channel and plant cutting, are still performed as a strict engineering activity. It is only recently that ecologically-based directives have started to be implemented during these procedures, and even then at a slow pace. Secondly, Iberian watercourses are generally poorer in the number of species and in aquatic plant cover than other regions with more regular and permanent river flow. Most weed problems are related to the excessive growth of emergent plants by the banks of lowland rivers and man-made canals. These are easily dealt with by the farmers using mechanical harvesting methods.

The massive development of two South American aquatic plants in the mid 1970s – water hyacinth *Eichhornia crassipes* (C.F.P. Mart.) Solms-Laub. and parrotfeather *Myriophyllum aquaticum* (Vell. Conc.) Verdc. – has substantially changed this situation, and experimental forms for their control began to raise researchers interest, as well as the factors that triggered their weedy behaviours. The EWRS and the Aquatic Weed Symposia soon became the fundamental way to present and discuss results and to keep up-to-date with new developments in this research area. After 20 years, it was the task of the Portuguese working group on aquatic weeds to organise the 10th Symposium on Aquatic Weeds.

Aquatic weeds in Portugal

Vasconcellos' book (1970) on the taxonomy of aquatic and emergent plants was the first to address these biological communities in scientific literature. More than 300 *taxa* are listed in this book. Haslam (1987) considered that there are only about a hundred common river macrophytes in the rivers of Western Europe and listed *c*. 70 as the most common species. The majority of these are referred to by Vasconcellos (1970). Later, Franco & Rocha-Afonso (1980) presented the distribution of the most troublesome aquatic plants. Both works were essentially descriptive and did not approach more ecological issues such as the type of water bodies where nuisance growth occurred and the environmental conditions of the occurrence.

To establish the exact dimension of weed problems in Portuguese freshwater ecosystems, a prospective enquiry was recently undertaken covering all organisations involved in water resource management, utilisation and planning, namely Municipalities, Regional Authorities of Agriculture, Forestry and Environment and Farmers' Associations (Aguiar et al., 1996). The results show that aquatic weed problems are locally important whilst not geographically widespread.

Algal mats were the primary cause of concern and quoted as a particular problem in irrigation canals and river systems associated with crop areas, with special emphasis on *Cladophora glomerata*. Artificial irrigation canals provide an aquatic environment that is extremely favourable for the growth of periphytic macroalgae and the swift flow probably enhances primary production. Filamentous macroalgae stands can attain more than 1 m in length and 20 cm in thickness (Ferreira et al., 1999c), although only a small number of species appear to dominate.

Emergent plants such as *Juncus* spp. and *Scirpus* spp. (rushes) were in second place, followed by the common reed (*Phragmites australis* (Cav.) Steudel), the cane (*Arundo donax* L.), galingales (or sedges) (*Cyperus* spp.) and reedmaces (*Typha* spp.). Emergents are considered troublesome in many river systems, preventing recreation and angling activities. This situation is similar to other Mediterranean countries as indicated by Haslam (1987), Arsenovic & Konstantinovic (1990) and Hulina (1990). The latter author found 26 semi-aquatic weeds. Stands are frequently monospecific or species-poor, composed of such species as *Typha* sp. and *Phragmites australis*. These two *taxa* are frequently referred to as invasive weeds that take over the native flora, as occurred in 'Albufera de Valencia', Spain (Jambrino & Fernández-Anero, 1997). The same observation was made by Costa et al. (1998a) in Portuguese coastal lagoons and some local effort has been targeted at preventing the disappearance of these waterbodies by dredging the lagoon bed and harvesting these plants.

Parrotfeather and water hyacinth have a South-American origin and Central Portugal corresponds to the highest northern latitude of their ecological range. Both species have become weeds in many lowland aquatic systems, namely in the Ribatejo-Oeste and Beira Litoral. Water hyacinth is not a widespread problem in southern Europe, only in Portugal (Pieterse & Murphy, 1990). The biology and behaviour of water hyacinth and parrotfeather have been studied and summarised in Moreira et al. (1989) and Moreira et al. (1999a, b). Parrotfeather is also nearly unique to Portugal. It is mentioned in *Flora Europaea* (Tutin et al., 1968) as frequently casual in other European countries and locally naturalized in S.W. France.

Earth drainage channels in Portugal are frequently dominated by these two species, although other species can also proliferate. These include charophytes, *Potamogeton* spp., *Polygonum* spp., *Lemna* spp., *Azolla filiculoides* Lam., *Sparganium erectum* L., *Zannichellia palustris* L., *Apium nodiflorum* (L.) Lag., *Alisma* spp., grasses such as *Echinochloa crus-galli* (L.) Beauv., *Paspalum paspalodes* (Michx.) Scribner and *Phragmites australis* and graminoids such as *Typha latifolia* L. (Ferreira & Moreira, 1990; Ferreira et al., 1998a; Moreira et al., 1998a).

Many small wetland areas and more or less permanent pools closely connected with superficial aquifers are considered important biodiversity reserves and particularly valued for their bird populations (e.g. Lagoas de Santa Margarida, Paul da Tornada, Paul de Madriz, Paul de Boquilobo and Alverca da Golegã). Many of these wetland areas have also become heavily infested with parrotfeather and water hyacinth.

A bloom of *Azolla filiculoides* occurred on one or two occasions in the River Guadiana (Carrapiço et al., 1998). These infestations are probably related to a combination of early, mild Mediterranean springs, the enrichment of the water due to untreated sewage inflows, and abnormally low flow conditions related to dry years and river regulation. In fact, disturbed or man-made habitats are frequently associated with plant invasions, such as *Eichhornia crassipes* and *Myriophyllum aquaticum* in drainage canals (Ferreira et al., 1998) or the exotic *Eryngium pandanifolium* Cham. & Schldl. in the canalised River Mondego (Morais et al., 1994). Lowland rivers and their valley floors have always experienced many types of human disturbance. Due to the lack of historical vegetation records, invasive movements are difficult to assess. However, in many segments of the River Tagus, native aquatic species, such as *Potamogeton crispus* L., *Ceratophyllum demersum* L. and *Myriophyllum spicatum* L., are currently being displaced by *Myriophyllum aquaticum* (Ferreira & Moreira, 1995).

Some studies have also addressed rice field weeds (Rosa & Espírito-Santo, 1984; Vasconcelos & Luzes, 1990). In addition to rice weeds that are common to other Mediterranean countries (such as *Cyperus difformis* L., *Cyperus eragrostis* L., *Leersia oryzoides* (L.) Swartz., *Paspalum paspalodes, Alisma* spp. *Ammania coccinea* Rottb., *Baldellia ranunculoides* (L.) Parl., *Cotula coronopifolia* L., *Eclipta prostrata* (L.) L. and *Lindernia dubia* (L.) Pennell), the South American *Heteranthera limosa* (S.W.) Willd. and *Heteranthera reniformis* Ruiz & Pavón were recently reported in Portuguese rice weed flora (Vasconcelos et al., 1999).

Management of aquatic plants in Portugal

Mechanical control

Mechanical control is the only weed control method extensively used in Portugal, in spite of its high costs. In drainage canals, hydraulic excavators are the traditional way of harvesting water hyacinth, parrotfeather and other floating and submerged weeds. They are also used to remove common reeds and reedmaces from the banks. Manual harvesting has been used in circumstances where access is difficult, such as near bridges. Weed development following harvesting and the time taken to recolonise were studied by Ferreira & Moreira (1990).

Floating harvesters have also been used. One was recently imported to remove water hyacinth from artificial lagoons at Golegã and an important wetland nature reserve in the same region, Boquilobo. Army boats were also used, supplied with a frontal grid that pushes plant masses to the margin where they can be removed.

In Lezíria Grande, a partial control of water hyacinth was achieved by increasing water depth and canal flushing rates. Net barriers are also placed at the entrance of small canals, which prevents reinfestation and helps to accumulate the weed in places where it can be harvested (Figueiredo et al., 1984).

Chemical control

The first attempts to use chemical control of aquatic weeds in Portugal date from the mid-1970s (Fernandes et al., 1978), but it was not until later that Monteiro & Moreira (1990) summarised the results of chemical control trials on *Myriophyllum aquaticum*. The active ingredients 2,4-D and dichlobenil showed some efficacy for this species. In Portugal, 2,4-D is not a trademark and there are severe restrictions against the application of dichlobenil in water which might be used for irrigation (Rocha, 1998). In those trials (Monteiro & Moreira, 1990), glyphosate did not satisfactorily control parrotfeather, though Machado & Rocha (1998) recently sprayed the plant after previously drying out the channel and obtained good results. Glyphosate is an interesting herbicide from the ecological point of view because its safety interval is very short. After chemical treatment, however, large masses of dying plants can have an ecological impact, particularly on water quality, as studied in Lezíria channels (Monteiro et al., 1988; Moreira et al., 1989).

Several trials against *Eichhornia crassipes* showed the efficacy of 2,4-D and diquat (Fernandes et al., 1978; Moreira et al., 1989) but the latter active ingredient has the inconvenience of a relatively long safety interval (10 days, Rocha, 1998), which does not occur with glyphosate. When tested in extensive applications, glyphosate was found to be very effective (Fernandes & Moreira, 1987).

Chemical control trials against *Phragmites australis* are summarised in Monteiro et al. (1999a) and Moreira et al. (1999c). They include the effect of a previous cutting of common reed on the efficacy of the herbicides glyphosate and gluphosinate, and the results of different doses, times of application and types of sprayer. There are very few studies concerning other weed species. Calha et al. (1999), for example, described *Alisma* spp. resistance to sulfonylurea herbicides.

Biological control

The first attempt at biological control of parrotfeather was performed in the early 1980s in the drainage channels of Lezíria Grande, by increasing the stock of common carp (*Cyprinus carpio* L. f. *specularis* Lac.) so that plant development would be inhibited by the increase in turbidity as a result of bottom feeding activities (Agusti et al., 1983). Plant growth effectively stopped but the initial cover of the weed remained the same as that recorded at the beginning of the trial.

More recently, Catarino et al. (1997) studied the feeding preferences of the grass carp (*Ctenopharyngodon idella* Val.) from the age 0+ to 3+, for the aquatic weeds occurring in Lezíria Grande drainage channels. The experiments were set in tank and pond conditions using combinations of macrophytes occurring as typical weed assemblages, including *Myriophyllum aquaticum* and *Eichhornia crassipes* (Ferreira et al., 1998a). Using Ivlev's index, the floating soft-tissue species, *Azolla filiculoides* and *Lemna* sp., were clearly shown to be preferred by grass carp, whilst other species such as *Potamogeton pectinatus* L.and *Myriophyllum spicatum* were moderately eaten. Exotic species were avoided up to the age of +2 and then only moderately consumed if no preferred species were present. Selectivity decreased as the grass carp's age increased.

These results showed a modest possibility for using grass carp in drainage channels for the control of the two major weeds – *Eichhornia crassipes* and*Myriophyllum aquaticum*. They also showed that there is some possibility that grass carp might control macroalgae in artificial channels, though further experimentation is needed to confirm the compatibility of grass carp with this type of environment and to determine the optimum stocking rates.

The control of water hyacinth by *Neochetina* spp. insects, imported from the U.S.A. in 1995, was experimentally attempted in greenhouses before the insects were released as larvae on water hyacinth plants in small ponds. They did not survive, presumably because they were at an inadequate stage of development and because of the cold winter that followed, and further experimentation is needed.

Final comments

A review of the weed control methods used in Portugal and of the possibilities of combining physical, chemical and biological methods against the most important vascular weed species (*Eichhornia crassipes, Myriophyllum aquaticum* and *Phragmites australis*) has previously been undertaken by Moreira et al. (1989), and more recently Moreira et al. (1998). Weed problems in Portugal are quite particular when compared to other Mediterranean areas because of the uniqueness of the two major weeds, *Myriophyllum aquaticum* and *Eichhornia crassipes*.

Other particular problems arise from the macroalgae weed occurring in artificial irrigation channels, similar to those in other irrigation systems, such as in California and South Africa and other Mediterranean regions. The experiments being conducted at present include assessment of the environmental factors that are involved in algae development (Ferreira et al., 1998a) and several types of control such as allelopathic inhibition by barley straw, shade and anti-fowling paints.

On a European scale, weed problems in Portugal thus present a certain degree of uniqueness in terms of major weed species and of channel related weed features, and are locally quite important. In fact, much can still be done, studied and discussed on the subject of aquatic weed management in Portugal. Surely the same comment can be applied to other regions and we hope the 10th Symposium on Aquatic Weeds represented a useful and meaningful contribution for the understanding and management of aquatic plants and the aquatic ecosystem.

References

Aguiar, F., I. Moreira & M. T. Ferreira, 1996. A percepção da vegetação infestante em ecossistemas dulciaquícolas pelas entidades gestoras dos recursos hídricos. Rev. Ciênc. Agrárias 19: 35–56.

Agusti, S., C. Duarte & T. Guimarães, 1983. Influência da carpa espelhada (*Cyprinus carpio* L., f. *specularis* Lac.) no desenvolvimento das infestantes aquáticas em valas de drenagem. II Seminário de Aquacultura: 378–402.

Arsenovic, M. & B. Konstantinovic, 1990. Aquatic weed control with herbicides in drainage system in Yugoslavia. Proc. EWRS 8th Int. Symp. on Aquatic Weeds: 13–17.

Caffrey, J. M. & P. M. Wade, 1996. The European Weed Research Society and the Management and Ecology of Freshwater Plants. Hydrobiologia 340: ix–xii.

Calha, I. M., C. Machado & F. Rocha, 1999. Resistance of *Alisma plantago aquatica* to sulfonylurea herbicides in Portuguese rice fields. Hydrobiologia 415 (Dev. Hydrobiol. 147): 289–293.

Carrapiço, F., M. L. Costa, G. Teixeira & V. Baioa, 1998. A new evaluation of some *Azolla* blooms: the Guadiana cases. Proc. 10th EWRS Symp. on Aquatic Weeds: 259–262.

Catarino, L. F., M. T. Ferreira & I. S. Moreira, 1997. Preferences of grass carp for macrophytes in Iberian drainage channels. J. Aquat Plant Mgmt 36: 79–83.

Costa, J. C., J. Capelo, M. D. Espírito Santo, M. Lousã, A. Monteiro, S. Mesquita, M. T. Vasconcelos & I. Moreira, 1999. Plant communitites of the lagoons of the Portuguese Coastal Superdistrict – a multivariate approach. Hydrobiologia 415 (Dev. Hydrobiol. 147): 67–75.

Fernandes, J. D., A. R. Guerreiro, T. Vasconcelos & I. Moreira, 1978. Essais de lutte contre les plantes aquatiques au Portugal. Proc. EWRS 5th Int. Symp. Aquatic Weeds: 189–194.

Fernandes, E. & I. Moreira, 1987. Aplicação de glifosato no combate ao jacinto aquático *(Eichornia crassipes (Mart.) Solm.* Centro Nac. Protecção da Produção Agrícola. PPA (HF): 23/87.

Ferreira, T. & I. Moreira, 1990. Weed evolution and ecology in drainage canals of central Portugal. Proc. 8th EWRS Symp. Aquatic Weeds: 97–102.

Ferreira, T. & I. Moreira, 1995. The invasive component of the macroflora from river systems within agricultural Mediterranean areas. In Pysek, P., K. Prach, M. Rejmanek & P. M. Wade (eds), Plant Invasions – General Aspects and Special Problems. SPB Academic Publishing, Amsterdam: 117–27.

Ferreira, T., L. Catarino & I. Moreira, 1998. Aquatic weed assemblage from Iberian drainage channels and related environment factors. Weed Res. 38: 291–300.

Ferreira, M. T., A. Franco, L. Catarino, I. Moreira & P. Sousa, 1999. Environmental factors related to the establishment of algal mats in concrete irrigation channels. Hydrobiologia 415 (Dev. Hydrobiol. 147): 163–168.

Figueiredo, J., C. Duarte, I. Moreira & S. Agusti, 1984. As infestantes aquáticas nos sistemas de irrigação e drenagem do Ribatejo. Recursos Hídricos 5: 5–14.

Franco, J. A. & M. L. Rocha-Afonso, 1980. Distribuição em Portugal das principais infestantes. I Cong. Port. Fitiatria e Fitofarmacologia 3: 1–22.

Haslam, S. M., 1987. River plants of Western Europe. Cambridge University Press: 512 pp.

Hulina, N., 1990. Aquatic weeds in open canals in the upper Sava river valley (Croatia-Yugoslavia). Proc. EWRS 8th Int. Symp. on Aquatic Weeds: 123–129.

Jambrino, J. I. D. & F. J. Fernández-Anero, 1997. Resultados en la recuperación de la biodiversidade en el Racó de l'Olla (l'Albufera de Valéncia) tras la aplicación selectiva de um herbicida de baja peligrosidad. Bol. San. Veg. Plagas 23: 17–37.

Machado, C. & F. Rocha, 1998. Control of *Myriophyllum aquaticum* in drainage and irrigated channels of the Mondego river valley, Portugal. Proc. 10th EWRS Symp. on Aquatic Weeds: 373–376.

Monteiro, A., T. Ferreira & I. Moreira, 1988. Evolução das características fisico-químicas de uma vala infestada por *Myriophyllum aquaticcum* após a aplicação de herbicida. 1ª Conf. Nac. Qualidade do Ambiente: 198–205.

Monteiro, A. & I. Moreira, 1990. Chemical control of parrotfeather (*Myriophyllum aquaticum*). Proc. 8th EWRS Symp. Aquatic Weeds. Uppsala: 63–164.

Monteiro, A., I. Moreira & E. Sousa, 1999. Effect of prior common reed (*Phragmites australis*) cutting on herbicide efficacy. Hydrobiologia 415 (Dev. Hydrobiol. 147): 305–308.

Monteiro, A., T. Vasconcelos & L. Catarino (eds), 1998. Management and Ecology of Aquatic Plants. Proc. 10th EWRS Int. Symp. on Aquatic Weeds, Lisboa: 444 pp.

Morais, H. P., M. T. Almeida & J. P. Lima, 1994. Weeds found in agricultural irrigation canals in a corn/rice growing area Low Mondego region – Portugal. Proc. 5th EWRS Mediterranean Symp: 269–276.

Moreira, I., L. Catarino, A. Monteiro & T. Ferreira, 1998a. Gestão de Infestantes aquáticas. Estudos e Documentos de Trabalho 4. Instituto da Água: 60 pp.

Moreira, I., T. Ferreira & A. Monteiro, 1989. Aquatic weed bioecology and control in Portugal. A review. In Bianchi, A. (ed.), Portuguese-German Cooperation in Applied Agricultural Activities. University of Vila Real Publication, Vila Real: 71–106.

Moreira, I., A. Monteiro & T. Ferreira, 1999a. Biology and control of parrotfeather (*Myriophyllum aquaticum* (Vell) Verdc.) in Portugal. Ecology, Environment and Conservation (in press).

Moreira, I., A. Monteiro, T. Ferreira, L. Catarino, J. C. Franco & T. Rebelo, 1999b. Estudos sobre biologia e combate do jacinto aquático (*Eichhornia crassipes* (Mart.) Solms-Laub.) em Portugal. Garcia de Orta (série Botânica, 14(2). (in press).

Moreira, I., A. Monteiro & E. Sousa, 1999c. Chemical control of common reed (*Phragmites australis*) by foliar herbicides under different spray conditions. Hydrobiologia 415 (Dev. Hydrobiol. 147): 299–304.

Pieterse, A. & K. L. Murphy, 1990. Aquatic weeds. The ecology and management of nuisance aquatic vegetation. Oxford Science Publications: 593 pp.

Rocha, F., 1998. Chemical control of aquatic weeds in the countries of the EU. Proc. 10th EWRS Int. Symp. on Aquatic Weeds: 17–19.

Rosa, M. L. & M. D. Espírito-Santo, 1984. Évolution des mauvaises herbes des rizières portugaises. Proc. 3rd EWRS Symp. on Weed problems in the Mediterranean Area 2: 411–417.

Tutin, T. G., V. H. Heywood, N. A. Burges, D. M. Moore, D. H. Valentine, S. M. Walters & D. A. Webb, 1968. Flora Europaea.Vol.2 Rosaceae to Umbelliferae. Cambridge. Univeersity Press.

Vasconcellos, J. C., 1970. Plantas (Angiospérmicas) aquáticas, anfíbias e ribeirinhas.Estudos e Divulgação Técnica. Direcção-Geral dos Serviços Florestais e Aquícolas: 253 pp.

Vasconcelos, T. & D. Luzes, 1990. Ervas daninhas dos arrozais. Rhône-Poulenc: 79 pp.

Vasconcelos, T., M. Tavares & N. Gaspar, 1999. Aquatic plants in the rice fields of the Tagus valley, Portugal. Hydrobiologia 415 (Dev. Hydrobiol. 147): 59–65.

Hydrobiologia **415:** 235–241, 1999.
J. M. Caffrey, P. R. F. Barrett, M. T. Ferreira, I.S. Moreira, K. J. Murphy & P. M. Wade (eds),
Biology, Ecology and Management of Aquatic Plants.

© 1999 *Kluwer Academic Publishers. Printed in the Netherlands.*

Integrated submerged aquatic vegetation management in an urban New Zealand river

Anne-Maree Schwarz & Ton Snelder*
National Institute of Water and Atmospheric Research Ltd, P.O. Box 8602, Christchurch, New Zealand

Key words: resource management, harvesting, *Potamogeton, Nitella*

Abstract

A comprehensive overview of the waterway network in Christchurch City, New Zealand, classified the reaches of three rivers and numerous smaller waterways according to the values for which submerged aquatic vegetation should be managed. The priority values; drainage, recreation, fish habitat and native plants, were assigned according to Council's historical information on the river and with regard to comments they receive from the general public. The classification recognised that aquatic plants have positive and negative effects on the range of values that the waterway network supports, and that these values can be prioritised. In reaches selected on the basis of their value-based classification, field trials of modified management options were designed to establish seasonal growth patterns and the implications of a reduced level of maintenance. Results from trials that involved harvesting the central strip of deeper reaches of the Avon River, leaving submerged macrophyte habitat on either bank, and selective removal of *Potamogeton crispus* by hand raking in shallower areas, showed considerable seasonal and interannual variability in species dominance. This has implications for timing and frequency of harvesting. The long term aim is to establish management options that optimise for competing values, minimise maintenance costs and are based on a more comprehensive understanding of growth characteristics of the range of aquatic plant species in the river.

Introduction

Understanding the nature of a submerged aquatic weed problem and accounting for the needs of various users has been an ongoing challenge for waterway managers world-wide. The multipurpose nature of many waterbodies results in aquatic plants being considered as detrimental in relation to one aspect of the water body and beneficial in relation to another (Petr, 1993; Mitchell, 1996).

Various paths towards assessing aquatic weed problems, and producing and monitoring management plans have been suggested (e.g. Mitchell, 1979; Coffey, 1987; Nichols et al., 1988). To be effective, such plans must account for multiple uses of lakes and rivers. Integrated management programmes are, however, rarely formulated because of a lack of a mechanism to evaluate the relative beneficial and nuisance attributes of the plants in relation to the perceived values (Johnstone, 1986).

In 1991, New Zealand's previous environmental legislation was replaced by the Resource Management Act (RMA) and this has provided the impetus to critically examine current management practices. The RMA introduced the concept of 'sustainable management' of natural and physical resources and requires resource mangers to recognise and provide for amongst other things; "*the preservation of natural character, the protection of significant habitats of indigenous fauna and the maintenance and enhancement of public access*". In addition, the Act requires managers to have regard to: "*the efficient use and development of resources, the maintenance and enhancement of amenity values and the intrinsic value of ecosystems*". Under the RMA, local government, such as City Councils, must gain 'resource consents' from regional government to undertake activities in the beds of rivers such as aquatic vegetation clearance.

* Author for correspondence

Christchurch City has been built on drained wetlands and the maintenance of adequate drainage has long been a continuing priority. For more than 90 years, the Christchurch City Council has regularly controlled submerged aquatic vegetation, for the single purpose of maintaining efficient drainage, by hand raking and use of harvesting boats. Submerged aquatic plants which are considered to have the potential to impede drainage are the introduced *Potamogeton crispus* which extends from the banks into water approximately 1 m deep, and the native *P. ochreatus* which dominates in deeper water (up to 3.5 m). Most native species in the Christchurch City waterways, including *Nitella hookeri*, *Myriophyllum propinquum* and *P. cheesemanii* tend to form relatively low growing communities which are considered less problematic. Little attention has been paid to differentiating between species in the management regime. Consequently, the maximum removal of biomass on a regular basis, regardless of species, has been the approach to date.

The new requirements of the RMA meant that Council must now consider the wider implications of its weed control program. We therefore firstly recognised the value of efficient drainage and the need to undertake aquatic vegetation management to provide for this. However, a number of other aspects are also considered to be values associated with the waterways. These are habitat for a range of native and introduced fish and bird species and public access for recreational pursuits such and rowing, canoeing and fishing. Aquatic vegetation has important implications for all of these. The conflicting requirements means that all values cannot be provided in every location.

A comprehensive approach to managing conflicting environmental demands is well recognised (e.g. Born & Sonzogni, 1995) and was the basis of our approach to this study. Natural seasonal growth patterns play an important role in management decisions. However, a consistent harvesting regime in Christchurch's waterways over the last 90 years has resulted in a paucity of the necessary quantitative data, hampering the ability to refine management decisions. The aims of the studies reported here were; to prepare a comprehensive overview that assessed the values for which submerged aquatic vegetation is managed in Christchurch waterways, to assess the potential for reducing harvesting effort by the Council, thereby reducing their management costs and, to begin collecting quantitative growth information to aid in future decision making. The comprehensive overview was followed up by trials of alternative maintenance strategies based on the principle of selective removal of components of the community (Johnstone et al., 1983). We present the results of the comprehensive overview and some preliminary findings from subsequent monitoring.

Study sites

The Christchurch City Council manages approximately 390 km of waterways in a 400 km^2 area that includes three rivers and an extensive above ground drainage system. All waterways that are subject to the existing vegetation management regime were included in the comprehensive overview. The Avon River was the subject of alternative management trials described here. The river flows through the city and discharges into the Avon-Heathcote estuary. The waters are of generally high clarity and relatively high nutrient concentrations. This, coupled with a stable hydraulic regime and a temperate climate, provides ideal conditions for the proliferation of aquatic plants.

Methods

Classification of the drainage network for aquatic vegetation management

In developing the comprehensive overview for aquatic vegetation, we developed a classification system that recognised the hydrological, ecological and social values that were affected by plants, as well as the physical conditions that impose specific requirements for their management. The classification system was then applied to the entire waterway network to develop a management program.

The existing aquatic plant management regime divided the drainage network of the city into 500 individual 'reaches'. Our system classified each reach on the basis of a three level hierarchy. The first level was on the basis of morphology. The morphological categories that were recognised (Table 1) were defined with relevance only to the management of aquatic vegetation. These were: the lower reaches including those with tidal influence and those beyond tidal influence but deep enough to be navigated by weed cutter boats; the shallow upper reaches that are managed by hand clearing of vegetation with slow flows and predominantly silty substrate; and shallow swift flowing reaches where substrates include gravel that provide spawning habitat for trout.

Table 1. Reach classifications and the management regime. Weeds = Predominantly invasive weed species that are likely to form nuisance growths, Mix = Equal mix of invasive and desirable native species or no aquatic plants, Native = Predominantly desirable native species

Morphological category	Priority value category	Plant type category	Recommended management regime
Channels with tidal influence and deep channels beyond tidal influence	Drainage	Weeds	1
		Mix	2
		Native	3
	Recreation	Weeds	1
		Mix	2
		Native	3
	Fish habitat	Weeds	4
		Mix	4
		Native	5
	Native plants	Weeds	4
		Mix	5
		Native	5
Shallow with fine substrates	Drainage	Weeds	1
		Mix	2
		Native	3
	Recreation	Weeds	1
		Mix	2
		Native	3
	Fish habitat	Weeds	4
		Mix	4
		Native	5
	Native plants	Weeds	4
		Mix	5
		Native	5
Shallow with gravel substrates	Drainage	Weeds	1
		Mix	2
		Native	3
	Recreation	Weeds	1
		Mix	2
		Native	3
	Fish habitat	Weeds	1
		Mix	1
		Native	3
	Native plants	Weeds	4
		Mix	5
		Native	5

At the next hierarchical level, we further subdivided the reaches on the basis of how aquatic vegetation affected the priority values in each reach. Priority values were assigned according to Council's historical information on the river and with regard to comments they receive from the general public. Values were categorised as drainage, recreation, fish habitat and native plants (Table 1). The relative importance of each value in each reach was determined on the basis of the Council's hydraulic studies of the drainage system, a separate assessment of the relative importance of reaches for recreation activities and aesthetic values, and available information on dominant fish (Eldon & Kelly, 1992) and aquatic plant species (Robb et al., 1994). The initial premise was that the priority for aquatic vegetation management should be in the following order:

1. Maintaining an effective drainage system for both low flows and flood flows.
2. Maintaining recreation values in important recreational reaches.
3. Optimising the waterway habitat for fish and birds i.e. maintaining aquatic vegetation to provide food sources or cover and maintaining channels with gravel substrates clear of vegetation for trout spawning.
4. Providing for areas with vegetation that is dominated by native plant species with high intrinsic value.

The final hierarchical level assigned a category on the basis of the aquatic plant community in the reach (Table 1).

1. Predominantly invasive weed species that are likely to form nuisance growths.
2. Equal mix of invasive and desirable native species or no aquatic plants.
3. Predominantly desirable native species.

Alternative weed clearance regimes

Prior to 1996, the waterways were routinely (annually or twice yearly) cleared of all visible submerged aquatic vegetation by hand weeding or mechanical harvesting depending on water depth. During 1996 and 1997, we worked with the Christchurch City Council to test and monitor the effectiveness of clearance regimes in specific reaches that differed from this approach. Two reaches which were classified at the first hierarchical level (Table 1) as being different morphologically were targeted for trials. The basis for recommending a level of management intensity for each reach is described more fully in the results. The classification of both reaches, and the recommendation, was that no values were adversely affected by aquatic vegetation, and the vegetation, which was dominated by native species, enhanced fish habitat; management should be primarily on the intrinsic value of the plant community with the curtailing of the existing maximum plant removal being desirable.

Avon Reach 1

The first reach was in an area with a high public profile in the City's botanic gardens. At this point the river was approximately 16 m wide and less than 1 m deep with a gravel/sand bottom and patchy submerged vegetation dominated by *Potamogeton crispus, P. cheesemanii* and *Nitella hookeri.* The area had patches of silty sands and gravel, and had historically been routinely cleared of all submerged vegetation by hand weeding using rakes. There was a relatively high aquatic plant species diversity compared to deeper parts of the river downstream with a high proportion of native species (Robb et al., 1994), including *Nitella hookeri, P. cheesemanii, P. ochreatus, Myriophyllum propinquum* and a range of mosses. The most frequently occurring introduced species was *P. crispus.* With the aim of promoting the growth of native species in this reach, contractors familiar with these plants were prepared to attempt the selective weeding of *P. crispus,* using hand weeding with rakes, rather than removing all plants. Such an approach was feasible because of the tendency for a clear demarcation between patches comprised of different species. In March and October 1996, a selective clearing regime was implemented where *P. crispus* was identified and selectively removed. Native species were left untouched as much as possible. The species composition and percent cover were monitored at four sites which included a control (unharvested site) prior to the first clearance and subsequently at two monthly intervals for the period of a year. At each site, five transects, 4 m apart, were placed across the river. On each transect, the species composition, plant height and percent cover were noted at 1 m intervals across the river.

Avon Reach 2

The second reach was downstream of Avon Reach 1 where the river widens to 20 m with water depths up to 1.5 m. This area is utilised by canoeists and for fishing. Most of the river bed was colonised by submerged vegetation with demarcation between beds of *P. crispus, P. ochreatus, N. hookeri* and *Elodea canadensis.*

This reach was classified as an intermediate reach beyond the influence of saline water but still affected by tidal water backup. It is deep enough to be navigated by weed cutter boats and has historically been managed by a twice yearly full cut of all submerged vegetation. For this trial, a variation on the usual method of bank to bank harvesting was carried out which involved harvesting a strip in the middle of the river, retaining a 5 m strip of aquatic plants on either bank. We were asked to assess the effects of this method after it had been carried out in March 1996. In the absence of a pre-harvest survey, two strip harvested sites were surveyed to obtain information on seasonal growth patterns under this management regime. Comparisons were made with a control site which had been harvested in the normal bank to bank manner.

The sites were monitored for 8 months following the first harvesting event and before and after a second harvesting event in September 1997. Water depth, aquatic plant species, plant height and percent cover were recorded at 1 m intervals across the river. Measurements were made at high tide to obtain the maximal measure of plant height and to provide consistency between samplings.

Results

Classification of the drainage network for aquatic vegetation management

A management regime was recommended for each reach (Table 1) on the basis of the categories which were assigned according to priority values.

Five management regimes were developed and are summarised as:

1. Aquatic plants have the potential to adversely affect drainage or recreation values; continuation of maximum plant removal on a regular basis is desirable.
2. Aquatic plants have the potential to adversely affect drainage or recreation values and comprise a mix of introduced (weed) and native species; reduction in clearance effort may be possible in some sites to promote conservation of the native plants.
3. Conflict between the values (drainage, recreation or trout spawning areas) and the intrinsic value of the native plant community; potential for active management procedures to optimise the identified values of the reach.
4. No values are adversely affected by aquatic vegetation, which comprises a mix of introduced (weed) and native species; management should be primarily on the intrinsic value of the plant community; lower priority area for control and means of enhancing growth of native species could be explored.
5. No values are adversely affected by aquatic vegetation, which is dominated by native species and enhances fish habitat; management requirements are minimal.

The classifications and management regimes for each of the 500 reaches were compiled in a spreadsheet and the summarised output was a map of the cities' waterways with each management regime represented by a different colour (Snelder & Schwarz, 1996). The results indicated that the existing management regime was necessary in only 30% of the waterway network and that the remaining 70% of the network would benefit from reduced clearance effort or a modified regime. In general, these changes would reduce the intensity and frequency of vegetation clearance, implying reduced maintenance costs for the Council and increased value of the waterway network.

Avon Reach 1

Over the course of one year, similar seasonal changes in average percent cover were evident at both the hand weeded monitoring sites and the control site. There was an increase in the occurrence of native species (primarily attributable to *Nitella*) in spring and a marked reduction in percent cover of all species in mid summer. Visual observations suggested that the hand removal of *P. crisp*us was very successful. However, from data collected at points along the transects there was no difference between the percent frequency of *P. crispus* before hand weeding, one month after hand weeding and at the end of the years monitoring at the hand weeded sites. This anomaly was probably due to the retention of root mat that ensured there was enough plant material remaining to keep recorded percent cover values at a similar level. Thus, there was no evidence of a repressive effect on the *P. crispus* community using selective hand weeding in this reach of river.

Avon Reach 2

After mechanical harvesting, some plants remained rooted to the bottom in all harvested areas. Thus 68% of the width of the river bed at the strip harvesting site and 29% at the total harvest site remained colonised by submerged vegetation immediately after the harvesting event. Harvesting the entire width of the river removed most of the *P. crispus*, which tends to grow from the banks into 1 m water depth (approx), and thus immediately following harvesting the vegetation comprised *E. canadensis* and *N. hookeri* at less

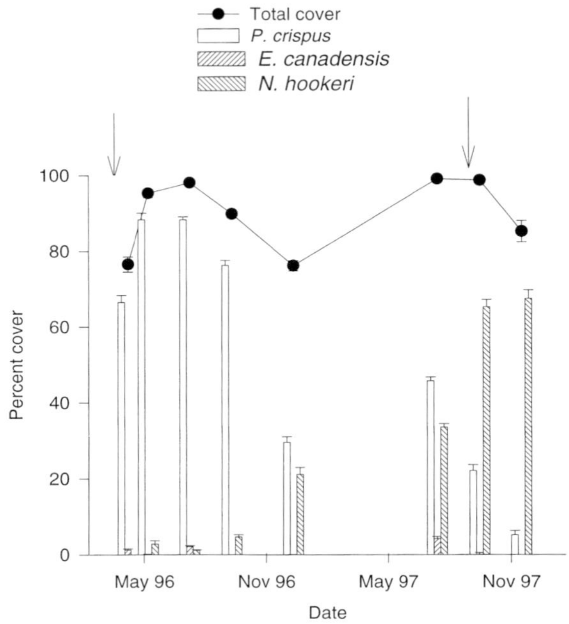

Figure 1. Percent cover of three dominant species (bars) at one strip harvested site in the Avon River between April 1996 and November 1997. Values are mean and standard error of 28 records across the width of the river. The total cover of submerged vegetation, which includes other species not illustrated by bars is shown by the line. Timing of strip harvesting is shown by the arrows.

that 20% cover each. At all sites, a seasonal pattern emerged with a tendency for total % cover to decline in summer, apparently unrelated to harvesting events (Figure 1). During 1996, % cover at the strip harvested site was dominated by *P. crispus* at 60–90% with *E. canadensis* and *N. hookeri* at less than 20% (Figure 1). However, there was a shift during 1997 with a tendency for *P. crispus* to decrease after this time with a concurrent increase in the percent cover of the low growing native *N. hookeri*. During the three months following strip harvesting there was a steady increase in plant height on both the strip-harvested and total harvest sites but the natural river flow ensured the central channel remained clear of surface weed growth under both regimes.

Discussion

The comprehensive overview assessed priority values for each of the 500 reaches in the Christchurch City area. Recommended management regimes suggested that there was potential for 70% of the waterways to benefit from a reduced or modified management regime. Trials of some alternative regimes that met the requirements of the priority value in each reach, suggested that there was considerable scope for a reduction in historical harvesting intensity. Selective hand weeding in shallow areas did not result in a marked decrease in the 'weed' species *P. crispus* which has been described as disturbance tolerant and therefore enhanced by pressures imposed by aquatic vegetation control (Sabbatini & Murphy, 1996). All species showed periodic shifts in dominance on variable time scales and this is likely to depend on season, degree of disturbance and environmental conditions. A review by Sand-Jensen (1997) suggested that the likelihood of survival of patches of individual species will increase, as they grow in size and we, therefore, expect the retention of an inoculum of native species in the Avon River to enable more effective competition with *P. crispus*. However, to date there is insufficient evidence to show whether selective hand weeding enhances this process and no control at all may be just as effective an option.

Strip harvesting which resulted in strips of weedbeds left on the river edges did not lead to more prolific growths than complete harvesting in the long term and retained a clear central channel for flood management and recreational use. Retention of marginal vegetation maintains a variety of niches for macro-invertebrates (Lillie & Budd, 1992), as well as habitat for fish and an inoculum of a range of favourable aquatic plant species. There were indications of species changes from year to year which were independent of the timing of harvesting. Surveying of these sites is ongoing and causal relationships will be able to be established with time. It is of interest that there appears to be a tendency for a decline in biomass in mid – late summer in the Avon River. Such observations will have implications for future timing of harvesting. The possibility that enhancement of the occurrence of the native *N. hookeri* by the removal of the taller growing *Potamogeton* species is a desirable long term outcome. However, as shown in the clear waters of Lake Aratiatia (Howard-Williams et al., 1996), this is likely to be a temporary effect prior to the regrowth of the taller species. Alternative methods to annual or biannual maximum biomass removal at all sites have shown that there are likely to be ways of reducing maintenance costs while maintaining the values of the system such as drainage, access and fish habitat.

Further trials in other sections of the river as well as focused research on the long-term effects of different harvesting regimes are required. To date it appears that the system that has been devised for deciding on management methods for specific reaches goes some way to providing a tool for managers of this system on which they can base decisions to reduce or modify harvesting effort in this particular urban environment.

Acknowledgements

We acknowledge the Christchurch City Council for permission to publish the results of this study. The manuscript was improved by journal referees' comments.

References

Born, S. M. & W. C. Sonzogni, 1995. Integrated environmental management: Strengthening the conceptualisation. Envir. Mgmt 19: 167–181.

Coffey, B. T., 1987. Submerged macrophytes: survey, conservation and control. Lake Managers Handbook. Water and Soil Misc. Publ. 103: 92–103.

Eldon, G. A & G. R. Kelly, 1992. Fisheries survey of the Avon River 1991–1992. Fish. Envir. Rep. 137: 33 pp.

Howard-Williams, C., A. -M. Schwarz & V. Reid, 1996. Patterns of aquatic weed regrowth following mechanical harvesting in New Zealand hydro-lakes. Hydrobiologia 340 (Dev. Hydrobiol. 120): 229–234.

Johnstone, I. M., 1986. Macrophyte management: an integrated perspective. New Zealand J. Mar. Freshwat. Res. 20: 599–614

Johnstone, D. L., D L. Sutton, V. V. Vandiver, Jr. & K .A. Langeland, 1983. Replacement of Hydrilla by other aquatic plants in a pond with emphasis on growth of American Lotus. J. Aquat. Plant Mgmt 21: 41–43.

Lillie, R. A. & J. Budd, 1992. Habitat architecture of *Myriophyllum spicatum* L. as an index to habitat quality for fish and macroinvertebrates. J. Freshwat. Ecol. 7: 113–125.

Mitchell, D. S., 1979. Assessment of aquatic weed problems. J. Aquat. Plant Mgmt 17: 19–21.

Mitchell, D. S., 1996. Interactions between national and local strategies for the management of aquatic weeds. Hydrobiologia 340 (Dev. Hydrobiol. 120): 153–156.

Nichols, S. A., S. Engel & T. McNabb, 1988. Developing a plan to manage lake vegetation. Aquatics 10: 10–19.

Petr, T., 1993. Aquatic weeds and fisheries production in developing regions of the world. J. Aquat. Plant Mgmt 31: 5–13.

Robb, J. A., M. J. Manning, A. E. Marshall & A. McGil, 1994. A botanical survey of the Avon, Heathcote and Styx Rivers and their tributaries and the City Outfall Drain 1993–94. A report prepared for the Christchurch City Council: 143 pp.

Sabbatini, M. R. & K. J. Murphy, 1996. Response of Callitriche and *Potamogeton* to cutting, dredging and shade in English drainage channels. J. Aquat. Plant Mgmt 34: 8–12.

Sand-Jensen, K., 1997. Macrophytes as biological engineers in the ecology of Danish streams. In Sand-Jensen, K. & O. Pedersen (eds), Freshwater Biology Priorities and Development in Danish Research: G.E.C. Gad Publishers Ltd., Copenhagen: 74–101.

Snelder, T. & A. Schwarz, 1996. Aquatic vegetation management for Christchurch. NIWA Wat. & Atmos. 4: 22–24.

Hydrobiologia **415**: 243–247, 1999.
J. M. Caffrey, P. R. F. Barrett, M. T. Ferreira, I.S. Moreira, K. J. Murphy & P. M. Wade (eds),
Biology, Ecology and Management of Aquatic Plants.
© 1999 Kluwer Academic Publishers. Printed in the Netherlands.

Succession of *Egeria densa* in a drinking water reservoir in Morbihan (France)

A. Dutartre[1,*], J. Haury[2] & A. Jigorel[3]
[1]*Cemagref, Unité Qualité des Eaux, 50, Avenue de Verdun, F-33612 Cestas Cedex, France*
[2]*ENSA-INRA, 65, Route de Saint Brieuc, F-35042 Rennes Cedex, France*
[3]*INSA, 20, Avenue des Buttes de Coesmes, CS 14315, F-35043 Rennes Cedex, France*

Key words: Egeria densa, succession, water quality, management, reservoir, France

Abstract

Located in western France, the Pen Mur dam of Muzillac (Morbihan) is mainly used for the production of drinking water and for fishing. Over the past few years, the upstream part of this eutrophic waterbody has been invaded by *Egeria densa*. To define suitable methods for the management of this plant, a 1996 study focused on the morphometric characteristics of the reservoir, the quality of its water and sediments, the diversity of its vegetation and various aspects of colonisation by *E. densa*. Analyses of usage and pollution were also made, as well as a review of the techniques available to control the development of this species. Based on the results of this study, a management plan was put forward. Studies in 1997 and 1998 showed a significant decline in the level of colonisation. This was probably due to the substantial floods which uprooted submerged plants in winter, followed by massive ice formation during winter and warm temperatures in spring leading to a development of cyanobacteria which are detrimental to macrophytes. Monitoring the competition between *E. densa* and the cyanobacteria could lead to determination of the causes of the predominance of one or the other.

Introduction

Excessive growth of imported species is becoming an increasing problem in France (Dutartre et al., 1996). Among these, *Egeria densa* Planchon (Hydrocharitaceae) is extremely widespread in tropical and subtropical parts of the world (Cook & Ürmi-König, 1984). This underwater plant with its dense foliage and ramified stems can reach a length of three metres. It can grow on various substrates but prefers organic silt.

It was first observed in France in 1960 : Feuillade (1961) reported it in a reservoir in the Manche (Western Normandy). Since then, the species has spread along the entire Atlantic coast.

The Pen Mur Reservoir in Muzillac belongs to the Morbihan County Council. It is mainly used for the production of drinking water, for fishing and leisure activities.

Its waters receive large inputs of nutrients from intensive agriculture in the catchment area. Thus, eutrophication causes the development of cyanobacteria which damage the water quality during the summer. Over the past few years, the reservoir has also been invaded by *E. densa*.

The purpose of the study was 1. to identify the pattern of its colonization in Pen Mur reservoir, 2. to define the methods of managing this plant in the basin, and 3. to use this example to assess the necessity of integrated management of aquatic plants in other reservoirs and lakes.

Site, material and methods

The Pen Mur reservoir is approximately 3 km long and generally less than 120 m wide. The maximum water depth is approximately 3 m, and the mean volume is 900 000 m^3.

The bathymetry and distribution of the plants were monitored in 1996 at contact points distributed over

* Author for correspondence

Table 1. Sucession of *E. densa* on some transects (frequency in %)

Transect	1996	1997	1998
Upstream part of dam (no 4)	100	5	0
Larger part of dam (no 7′)	90	57	11
Immediately downstream of larger part of dam (no 8)	90	55	0
Downstream part of dam (no 13)	9	0	0

17 cross sections. In 1997 and 1998, four reference transects were studied in different parts of the dam. Observations were also made over the whole area of the reservoir to specify the biological diversity, check the distribution of *E. densa* and assess the negative effects which it brings about.

Physical measurements and chemical samples were taken in the water during each site survey, including analyses of nutrients (nitrogen and phosphorus). Cores samples made it possible to determine the depth of the silt and to take samples for chemical and particle size distribution analyses. Chemical analyses of the plants (organic matter, phosphorus, nitrates, Kjeldahl nitrogen) were also performed.

Results

The study confirmed the ecological value of the site: over 100 Spermatophyta taxons were observed, including species of special interest such as *Luronium natans* (L.) Rafin, which is now protected throughout Europe, *Potentilla palustris* (L.) Scop., *Littorella uniflora* (L.) Ackerson, *Eleocharis multicaulis* (Sm.) Desv., *E. acicularis* (L.) Roemer & Schultes, *Trapa natans* L. and *Ludwigia palustris* (L.) Elliott (Dutartre et al., 1997).

Colonization of the reservoir by *E. densa* had already been quite substantial before 1996. In 1996, the species was found at 49% of 456 contact points (Figure 1a). Nevertheless, in relation to the depth (Figure 1b), this high frequency varied considerably between the upstream and downstream parts of the reservoir (Figure 1c). The first upstream section had a very small population, which rapidly increased to exceed 80% over approximately one quarter of the length of the reservoir. It then dropped considerably, remaining generally below 20% in the middle part, and decreasing to 2–3% in the downstream part. The maximum depth of occurrence of *E. densa* is 2.5 m, which could enable it to colonise practically the entire waterbody. In 1997 and 1998, *E. densa* had decreased considerably in the reservoir, probably due to high floods and severe frost (15 cm of ice for several weeks) in winter 1997. Comparisons of results in some transects on about 170 contact points show this important decrease (Table 1). The mean frequency of *E. densa* on the dam decreased from 49% in 1996, 35% in 1997 to 6% in 1998.

The decrease was much faster in the upstream and narrowed part of the dam under direct influence of winter floods. In the larger part, *E. densa* remained only on old root crowns and small stems. In the downstream and deeper part of the dam, *E. densa* had already disappeared in 1997.

Based on literature (Cook & Ürmi-König, 1984; Getsinger & Dillon, 1984) and from analysing data obtained in the current study (frequency and abondance of *E. densa* on points contacts in the reservoir), estimates of the biomass of *E. densa* indicated 50 tons of dry weight for 1996, just over 13 tons for 1997 and below 1 ton in 1998.

The thickness of the reservoir bottom increased moving from upstream (20–40 cm) to downstream (about 1.5 m). The total volume of deposits was estimated to be 170 000 m^3. The particle size distribution of sediments indicated a distinct upstream/downstream evolution. The predominantly fine sandy fraction upstream was progressively replaced by a fine clay and silt fraction downstream. Observations under the scanning electron microscope made it possible to differentiate the detrital fraction, due to erosion of the catchment area, from the biogenic fraction, produced by the reservoir's aquatic flora (Jigorel & Bertru, 1993).

The mineral detrital fraction decreased from the upstream to the downstream part. The vertical variations in the deposits indicated that the biogenic fraction had increased over the years, resulting in a recent rise in the trophic level of the reservoir, related to an increase in the input of nutrients. The phosphorus content of the sediments was higher in the upstream part of the reservoir, with heavy colonization of *E. densa* (Figure 1f). The submerged plant community trapped a considerable amount of fine sediments, which was confirmed by the fact that over the last few years the fluctuations in the suspended solids content at the water treatment plant had been noticeably smaller (Compagnie Générale des Eaux, pers. comm.).

The nitrate content (approximately 30 mg/l) and phosphate content (up to nearly 0.06 mg/l) of the water

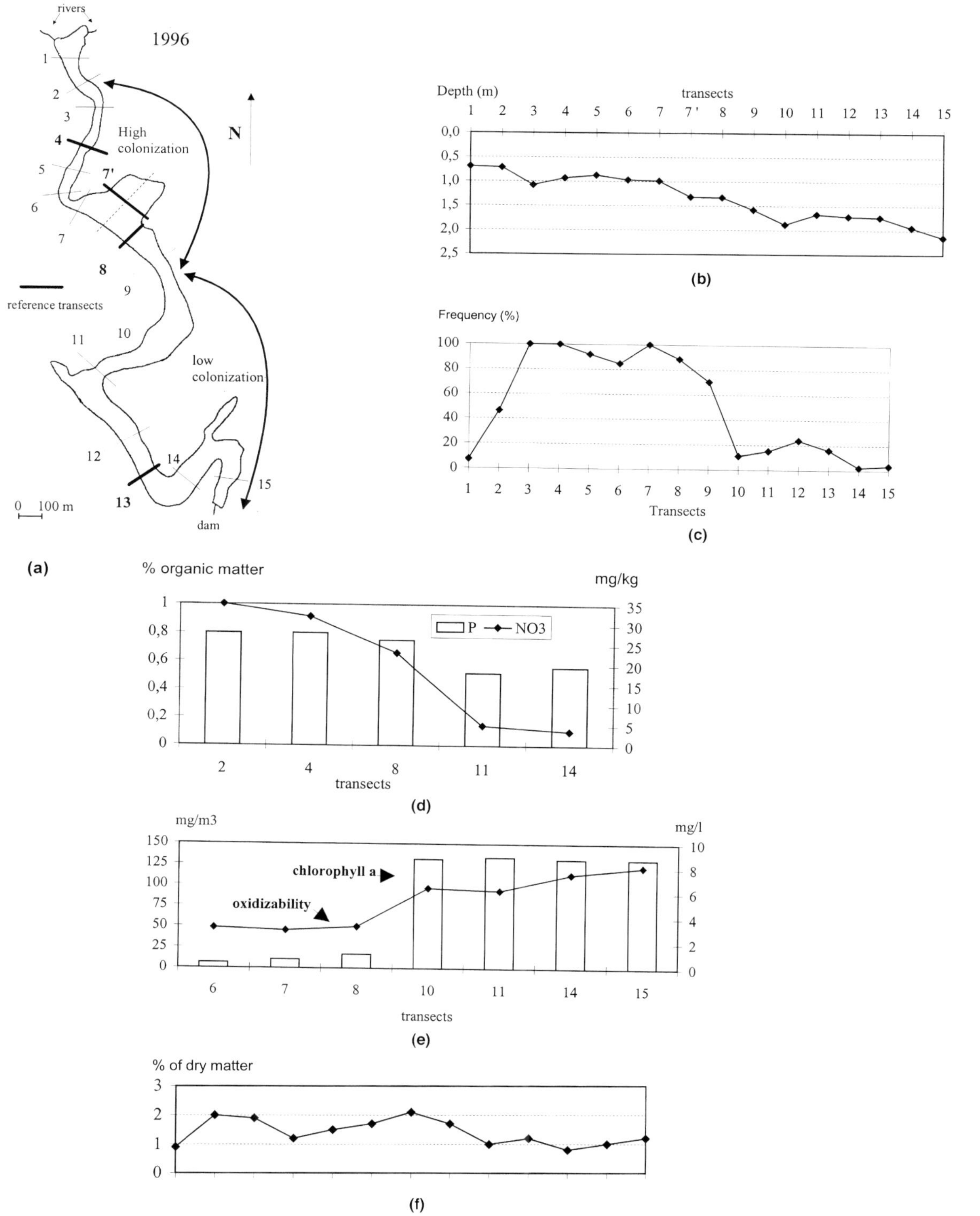

Figures 1 a–f. Environmental parameters in transects and characteristics of colonization and nutrients contents of *Egeria densa* (1996). (a) Location of transects and characteristics of *E. densa* colonization, (b) mean depth (m) of transects, (c) frequency of *E. densa* on differents transects (%), (d) chemical composition of *E. densa* (P in% of dry matter NO^3 in mg of wet plants /kg), (e) chlorophyll a (mg/m^3)and oxidizability (mg/l) f - phosphorus in sediment (% of dry matter). (d) chemical composition of *E. densa* (P in% of dry matter NO^3 in mg of wet plants /kg), (e) chlorophyll a (mg/m^3)and oxidizability (mg/l) (f) phosphorus in sediment (% of dry matter).

were relatively high and explained the summer development of cyanobacteria. The predominant species were *Microcystis flos-aquae* (Wittr.) Kirchner and *M. aeruginosa* Kütz. There were high concentrations of chlorophyll *a* which exceeded 120 mg/m^3 in the summer. They resulted in a major drop in the transparency of the water, which decreased from over 2 m in the spring to less than 0.4 m in summer. During the survey in September 1996, the cyanobacteria were developing immediately downstream of the dense communities of *E. densa* with high concentrations of chlorophyll *a* and increase oxidizability (dissolved organic matter) (Figure 1e). Over a distance of a few hundred metres, the chlorophyll a content in the water increased from 5–15 mg/m^3 (content where dense plant communities of *E. densa* occurred) to 120–130 mg/m^3 in the part of the reservoir where *E. densa* was scarce.

Analyses of the plants (Figure 1d) indicated the close relationships between the nutrient content of the species and of the water and sediments of the site. A sharp decrease in the phosphorus and nitrate contents downstream indicated the purifying capacity of *Egeria*. The phosphorous concentration of the plant decreased from 0.8 upstream to 0.56 downstream (in percentage of dry matter), and the concentration in nitrates dropped from 35 to less than 4 (in mg/kg) (Figure 1d).

Discussion

It is obvious that the dynamics of the colonization of the reservoir by *E. densa* were strongly influenced by winter climatic events – rise in the waters of water level in the tributaries and freezing of the open stretch of the watercourse.

In 1996 and 1997, the reservoir presented a significant upstream–downstream gradient which concerned the colonization of *E. densa*, the chemical composition and particle size distribution of the sediments, the phosphorus and nitrate content of the macrophytes, and the summer development of cyanobacteria.

Observations made in 1996 and 1997 seemed to indicate that competition between the macrophyte and the planktonic populations resulted in a predominance of *E. densa* in the upstream part of the reservoir, whereas the planktonic algae predominated downstream. This longitudinal distribution of vegetal populations was the result of the input of nutrients at the upstream end of the reservoir and of the elongated morphology of the basin.

The observations made in 1998 showed a very clear decrease of *E. densa* colonization. Only small parts of plants remained in the larger part of the dam under weak influence of winter floods.

Greater depth and low transparency values were major limiting factors of *E. densa* distribution on the dam. The organic matter content in superficial layer of sediment were relatively high, up to 15–25%. In the upstream part of the reservoir, though the mineral detrital fraction is upper than in downstream, the organic matter content (about 15–20%) were not a limiting factor for *E. densa* growing.

Winter floods and ice cover in 1997 seemed to have also injured *E. densa* populations.

In 1998, early development of cyanobacteria due to high water temperature in spring had probably caused additional damage to *E. densa*.

After the observations of 1996, management proposals were prepared. At that time, eventual colonization of the entire reservoir by *E. densa* seemed possible and would have created problems for anglers and biodiversity. However, no intervention was proposed at that time because the annual fluctuations were unknown. Regular surveillance of colonization by anglers according to a clearly defined protocol was proposed. For better *in situ* application of this protocol, this surveillance was undertaken with help from anglers during the 1997 and 1998 summers.

Harvesting and dredging of *E. densa* would have been possible. Nevertheless, in such a eutrophic river basin with high nutrients concentrations in the water, it appeared to be preferable to use *E. densa* beds for water cleaning in upper part of the dam, particularly for trapping fine sediment particles, in particular phosphorus and for restricting cyanobacterial blooms by competition for nutrients. Consequently, it was probably advantageous to keep it as a natural filter in the upstream part of the reservoir and to restrict any intervention to a local level, by mechanical means, according to the requirements of anglers, over the middle and downstream parts. At some later date, partial dredging of the upstream part of the pond would make it possible to remove large quantities of phosphorus contained in the sediments and in the plants themselves. Regular monitoring is recommended to implement possible interventions at the right time.

Observations of the fast *E. densa* decrease in 1997 and 1998 have changed our problem analysis, although the surveillance is going on. Could this *E. densa* regression lead to total disappearance or is it only a lower stage of a biological cycle as described by Dr Jacoby (pers. comm.)?

The survival of this species is possible under ice cover (Cook & Ürmi-König, 1984), but the optimal temperature of growth is between 15 and 25 °C (e. g. Haramoto & Ikusima, 1988). This species is also adapted to various water qualities and can grow in nutrient enriched waters (Reddy et al., 1987). In laboratory conditions, its nutrient uptake was relatively high. Its light requirements are also relatively weak (Clayton J., National Institute of Water and Atmospheric Research, Hamilton, New Zealand. pers. comm.). In laboratory experiments, Tanner et al. (1993) showed that *E. densa* is likely growing up to 25–35 g/m^{3}of suspended solids. In New Zealand, this species is frequent and often dominant in meso and eutrophic water bodies (Howard-Williams, 1993).

It seems that ice cover and winter flooding in 1997 could not totally explain the fast decrease of *E. densa* in the Pen Mur dam. In this ecosystem, many others ecological conditions, such sediment and water quality could be considered as influencing factors of this decrease.

The main phenomenon in this evolution was probably the competition between cyanobacteria and the macrophyte. This competition vary between years and the successive damage to *E. densa* in 1997 (ice cover, flooding, high temperature in spring) may have allowed cyanobacteria to become dominante. On the other hand, allopathic substances from cyanobacteria could have negative effects on *E. densa* growth. Alternatively or additionally, the biological cycle of *E. densa* is perhaps also responsible for its decrease.

The future management proposals will depend on the relative nuisance of *E. densa* and cyanobacteria in the reservoir. For drinking water production from the reservoir, a water treatment station could take out cyanobacteria. However, the great variations of cyanobacteria concentrations and high concentrations of suspended solids create some difficulties in treatment. According to the treatment manager, the winter concentrations of suspended matter were relatively stable during the years of the macrophyte invasion. The macrophyte decrease could allow greater variations of these suspended matter and subsequently increase treatment difficulties.

On the other hand, harvesting of *E. densa* when macrophyte beds were very dense and dredging reservoir bottom could reduce amounts of nutrient in the dam, particularly phosphorus. The actual decrease of *E. densa* prevented this operation.

In the future, management of this reservoir could integrate the cyanobacteria/macrophyte competition and regular surveillance will be continued to verify this functional hypothesis.

Acknowledgements

This study, commissioned by the Morbihan County Council, was granted by the Morbihan County Water Board, the Brittany Regional Council and the Loire-Brittany Water Board. Gilles Boulande, from the local Anglers Association, Mr Dayon, a local farmer, Myriam Jaffré, Aline Moreau, Gaëlle Royer, Christophe Laplace, Paul Simon and Olivier Touzot all participated in the site surveys.

References

Cook, C. D. K. & K. Urmi-König, 1984. A revision of the genus *Egeria* (Hydrocharitaceae). Aquat. Bot. 19: 73–96.

Dutartre, A., J. Haury, A. Jigorel & C. Laplace, 1997. Possibilités de gestion de l'invasion de la retenue de Pen Mur (Muzillac, Morbihan) par une plante aquatique exotique: *Egeria densa*. Cemagref, ENSA/INRA, Bordeaux Rapport pour le Conseil Général du Morbihan: 142 pp.

Dutartre, A., J. Haury & A. M. Planty-Tabbachi, 1997. Macrophytes aquatiques et riverains introduits en France. Bull. Fr. Pêche Piscic. 344–345 (1–2): 407–426.

Feuillade, J., 1961. Une plante aquatique nouvelle pour la France *Elodea densa* (Planch.) Casp. Bull. Soc. Linnéenne Normandie 10 (2): 47–51.

Getsinger, K. D. & C. R. Dillon, 1984. Quiescence, growth and senescence of *Egeria densa* in lake Marion. Aquat. Bot. 20: 329–338.

Haramoto, T. & I. Ikusima, 1988. Life cycle of *Egeria densa* Planch., an aquatic plant naturalized in Japan. Aquat. Bot. 30: 389–403.

Howard-Williams, C., 1993. Processus of aquatic weed invasions: the New Zealand example. J. Aquat. Plant Mgmt 31: 17–23.

Jigorel, A. & G. Bertru, 1993. Endogenic development of sediments in a eutrophic lake. Hydrobiologia 268: 45–55.

Reddy, K. R., J. C. Tucker & W. F. Debusk, 1987. The role of *Egeria densa* in removing nitrogen and phosphorus from nutrient enriched waters. J. Aquat. Plant Mgmt. 25: 14–19.

Tanner, C. C., J. S. Clayton & R. D. S. Wells, 1993. Effects of suspended solids on the establishment on growth of *Egeria densa*. Aquat. Bot. 45: 299–310.

Hydrobiologia **415**: 249–255, 1999.
J. M. Caffrey, P. R. F. Barrett, M. T. Ferreira, I.S. Moreira, K. J. Murphy & P. M. Wade (eds),
Biology, Ecology and Management of Aquatic Plants.
© 1999 Kluwer Academic Publishers. Printed in the Netherlands.

Typha control efficiency of a weed-cutting boat in the Lac de Guiers in Senegal: a preliminary study on mowing speed and re-growth capacity

Seppo Hellsten[1,*], Charles Dieme[2], Marthe Mbengue[2], Georg A. Janauer[3], Nick den Hollander[4] & Arnold H. Pieterse[4]

[1]*VTT, Communities and Infrastructure, Water Engineering and Ecotechnology, P.O. Box 19042, 90571 Oulu, Finland*
[2]*Ministère de l'Environnement, Direction des Parcs Nationaux Senegal, Dakar-Fann, Senegal*
[3]*University of Vienna, Institute for Plant Physiology, Althanstrasse 14, 1090 Vienna, Austria*
[4]*Royal Tropical Institute (KIT), Department of Agriculture and Enterprise Development, Mauritskade 63, 1092 Amsterdam, The Netherlands*

Key words: Typha australis, Senegal River, Lac de Guiers, mowing boat, aquatic weed, control

Abstract

Prolific growth of *Typha australis* in the lower part of the Senegal River and the Lac de Guiers resulted from changed ecological conditions following the construction of two high dams in the Senegal River. Fluctuation of the water level has decreased markedly and the water has changed from brackish to fresh as the inflow of salt water from the ocean is prevented. The efficiency of a hydraulic weed cutting boat (model Conver 480 H) has been tested in various plots in the Lac de Guiers. This lake, connected with the Senegal River, is economically very important as it is the city of Dakar's major drinking water supply and also provides irrigation water for vast agricultural areas. Cutting, performed at 20 cm and 50 cm below the water surface, was very effective in removing *Typha* stands. However, due to the excessively thick stems of the *Typha* plants, cutting took much more time compared to similar operations in Europe. It was estimated that approximately 35 h were required to clear one ha of dense *Typha* stands. Directly after cutting re-growth was observed during a relatively brief period (approximately three months), however, subsequently all newly formed growth tips in the plots had suddenly disappeared. It was concluded that this sudden collapse was linked to a destruction of the root zone due to anoxic conditions. Eventual re-growth was relatively slow as after one year the plots were still largely free from *Typha* plants. Efficiency of mowing was not related significantly to depth of cutting.

Introduction

The hydrology of the Senegal River system has changed markedly in recent years. As a result of the construction of the Diama dam (27 km upstream of St. Louis near the mouth of the river, completed in 1985), and the Manantali dam (situated in Mali, approximately 1200 km upstream from St. Louis, completed in 1988), the river has become navigable throughout the year. In addition, water stored between the dams makes it possible to irrigate about 255 000 ha of farmland. This has led to a significant change in the area's ecological conditions. Water level fluctuations have decreased significantly. Consequently, annual floods no longer occur and the vast floodplains remain dry throughout the year. As salt intrusion is prevented (before the construction of the Diama dam tides reached around 250 km upstream), water in the river has turned from brackish to fresh (Cogels et al., 1993). On the other hand, large areas have become permanently flooded and new limnic ecosystems have been formed.

These changes in ecological conditions have resulted in an explosive growth of aquatic weeds (Thiam et al., 1993). This applies in particular to an approximately 30 km long stretch just upstream of the Diama dam, where, due to the construction of embankments,

* Author for correspondence

the Senegal River has become approximately 6 km wide and the Lac de Guiers, a lake with a surface area of approximately 240 km^2 (Cogels et al., 1993), which is connected to the river. These waterbodies have become overgrown mainly by *Typha australis* Schum. et Thonn. The Lac de Guiers is economically very important as it constitutes the city of Dakar's major water supply providing irrigation water for large agricultural areas. *Typha* fields in the Senegal River, preventing access of local villagers and fishermen to open water, are burnt regularly, causing large-scale smoke development and the deposit of scorched plant debris over long distances.

In 1996, an EU-financed project was started, carried out by the Ministry of Environment of Senegal in collaboration with the Royal Tropical Institute (KIT), The Netherlands, VTT Communities and Infrastructure, Finland and the University of Vienna, Austria. The major objective of this project was to develop an ecologically sound, effective and integrated control programme for aquatic weeds in the Lower Senegal River Basin in both Senegal and in Mauritania. The current study, conducted within the framework of this project, is concerned with the efficiency and costs of mechanical control of *Typha* by means of a hydraulic mowing boat commonly used for aquatic weed control in Europe. The major objective was to estimate the time required for removing *Typha* plants with this type of equipment and to obtain information on the re-growth capacity of *Typha* after cutting.

Materials and methods

The Lac de Guiers is connected with the Senegal River and is situated approximately 50 km from the coast (14° 09′ N, 16° 08′ E, see Figure 1). This shallow lake (maximum depth 3 m) was brackish before dam construction as a result of salt water intrusion via the Senegal River. It is to a large extent situated below sea level. Originally, the water level fluctuated up to 3 m annually (John, 1986). In 1947, a dam was built in Richard-Toll near the outlet of the lake into the Senegal River, to prevent salt water intrusion. Following the construction of the Diama dam in the mouth of the Senegal River and the Manantali dam upstream in Mali, fluctuations have decreased to fewer than 1 m.

A hydraulic mowing boat (model Conver 480 H) was tested equipped with a 50 hp diesel engine and a T-cutter bar (2 m horizontal, 1.2 m vertical). T-cutters could be replaced by a 1.5–3 m wide hydraulically operated weedrake.

In February 1997, mowing experiments were conducted in the Lac de Guiers in the vicinity of the Knit pumping station near the western shore. The experimental area included four mowing plots (each plot 30 m × 60 m), two of which (plot 1 and plot 2) were cut 20 cm below the water surface and two (plot 3 and plot 4) 50 cm below the water surface (Figure 1). These mowing plots were situated on the edge of the vegetation along open water. Development of vegetation and water quality were measured weekly at five measuring sites (2 m × 2 m) in each plot until April 11, 1997 and every second week between April 11 and June 26, 1997. Re-growth was measured by means of ropes, tied to sticks at various heights placed around these plots, i.e. on the surface (0 cm) and 15 cm, 30 cm, 50 cm, 100 cm and 150 cm above water level. All plant tips exceeding the rope were counted and a special re-growth index (R_i) was calculated as follows:

$$R_i^2 = \sum_{k=l}^{6} (n_j * k),$$

where n_j = (n_0, n_{15}, n_{30}, n_{50}, n_{100}, n_{150}) number of plant tips reaching the specific height. The theoretically highest re-growth value would be 24.5, calculated by using a mean observed stem density of 25 individuals per square metre, which are all reaching the level of 150 cm. All individuals were marked by a ribbon and the number of new plant tips reaching the surface was also used to determine a regeneration index (G_i). The re-growth index is related to the total plant biomass, whereas the regeneration index indicates the number of new plant tips.

Water quality analysis was based on measurements on the water surface in each plot. It included water temperature (°C), conductivity (μS/cm) and dissolved oxygen (mg O_2/l) (Figure 1). Conductivity was measured by an ATI orion 130 and dissolved oxygen by a YSI 59 model. Weekly measurements of water temperature, conductivity and dissolved oxygen at the intake of the Knit pumping station (SDE) were used as reference data. The reference point was situated at approximately 100 m from the experimental area and it should be taken into consideration that measurements on this site could only be indicative for open water areas in the close vicinity of *Typha* stands. During the study, the water level in the lake decreased from 1.85 m (February, 3) to 1.58 m (June, 23) above sea level.

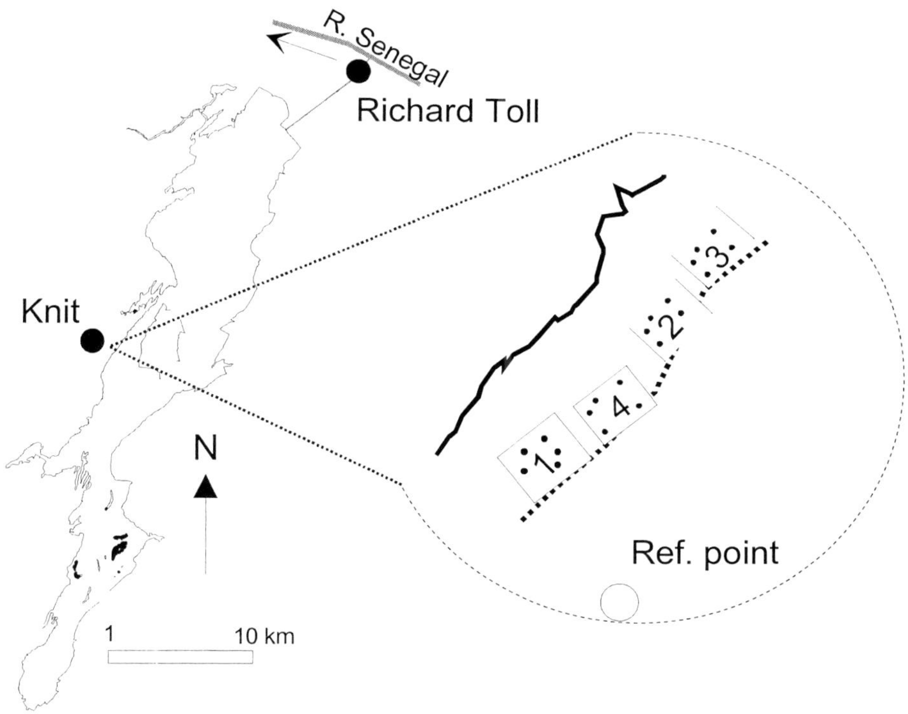

Figure 1. Location of the experimental mowing area in the Lac de Guiers. Mowing plots are delineated by a rectangle and measuring sites are marked by filled dots. Reference point (SDE) is marked by open circle. The outer edge of *Typha* has been marked by a dotted line.

Between 27 May and 25 June 1997 mowing efficiency studies were conducted in 13 different areas in the vicinity of the experimental mowing area. The period required for mowing a certain area and the quantity of diesel used, were measured. To estimate the amount of removed *Typha* biomass, three different plots (1 m^2) at depths of 0.1 m, 0.7 m and 1.5 m were harvested manually and total wet biomass was determined by weighing the plants.

Results

Re-growth of *Typha* started in plot 1 only one week and in plot 2 three weeks after cutting (Figure 2a-b). Initially the re-growth index was relatively high, especially in plot 1 and the regeneration index indicated that there were still new plant tips reaching the surface at the end of March. At the end of May, however, almost all *Typha* plants had disappeared and *Typha*-stands had been partly replaced by submerged plants (*Potamogeton schweinfurthii* A. Benn. and *Potamogeton octandrus* Poir.). At the end of June there were no *Typha* stems in these plots.

Results of plot 3, where stems were cut 50 cm below water level, did not differ significantly from the first two plots (Figure 2c). Re-growth and regeneration started one week after cutting, but came to a standstill in the middle of April. On the other hand, during the experimental period there was hardly any re-growth in plot 4 (Figure 2d). Some *Nymphaea*-species as well as the submerged plants *P. schweinfurthii* and *P. octandrus* became quite common in both plots. Even after one year, in February 1998, the plots were largely free from *Typha*.

Water quality did not show any significant changes, with the exception of a season related increase in water temperature during the study (Figures 3a–b). Conductivity and temperature values of the mowing plots and the reference point largely corresponded. However, dissolved oxygen showed a marked fluctuation, probably due to a breakdown of dead plant material (Figure 3c). In plot 1 and plot 2, where *Typha* was cut 20 cm below the water surface, nearly anoxic conditions occurred at the beginning of May. On the other hand, in plot 4, where *Typha* was cut 50 cm below the water surface and re-growth was very slow, there were two periods in which the oxygen level was low during the

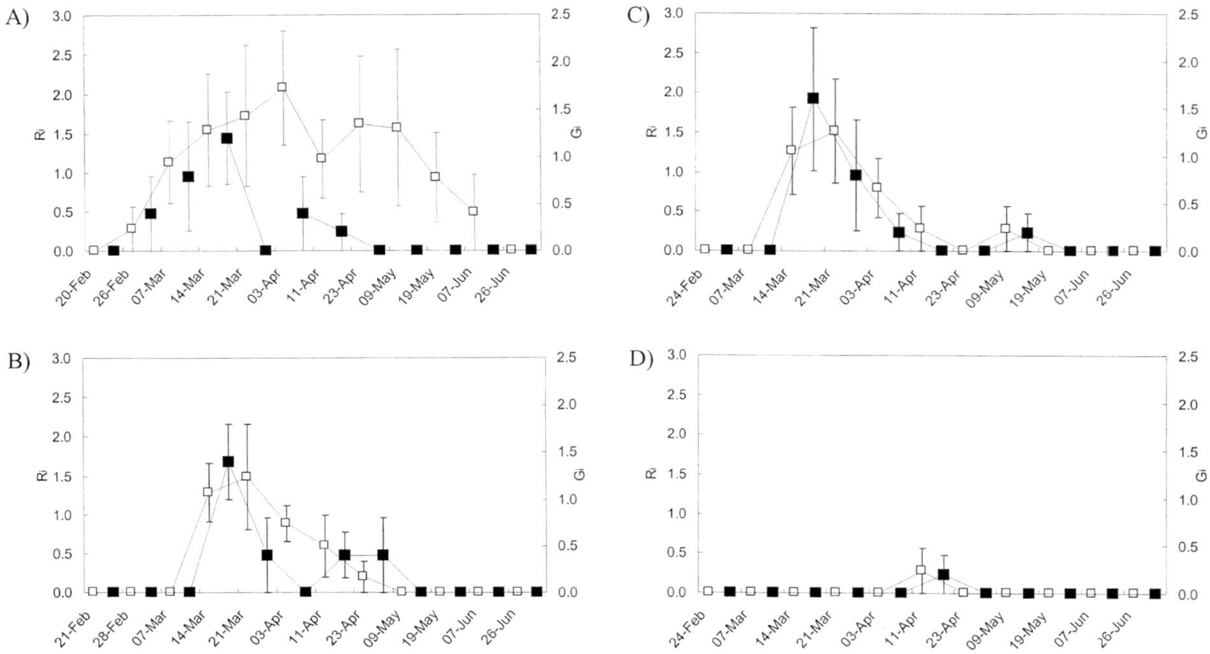

Figure 2. Results of mowing experiments near Knit in the Lac de Guiers. Each square represents the mean of five samples with S.E. for each sampling time. White squares = re-growth index (R_i), black squares = re-generation index (G_i). (A) Plot 1 cut at a depth of 0.2 m, total water depth of 1.29 m, (B) Plot 2 cut at a depth of 0.2 m, total water depth of 1.06 m, (C) Plot 3 cut at a depth of 0.5 m, total water depth of 1.08 m, (D) Plot 4 cut at a depth of 0.5 m, total water depth of 1.19 m.

experiment, i.e. in March and in May (Figure 3c). In plot 3, where *Typha* was also cut 50 cm below the water surface, dissolved oxygen during the research period corresponded with the reference level.

Mowing efficiency was not related significantly to cutting depth (Table 1), but varied markedly, depending on overheating of the engine, caused by blocking of the propeller by *Typha* roots, as well as by *Typha* seeds blown into the radiator, preventing normal cooling. Directly after cutting, floating *Typha* plant material was collected by means of the weed rake. This procedure took twice as long as cutting. Only 0.04 ha was cleared in 1 h. In addition to the relatively long time for collecting plant material, installation of the weed rake took 50 min and it also took some time to re-install the cutting bar. This markedly delayed the speed of mowing.

Discussion

The experiments have shown that dense stands of *Typha* in the Lac de Guiers can be effectively reduced by means of a hydraulic mowing boat. Although a marked, temporal re-growth was observed during the first three months after cutting in plots 1, 2 and 3, all plots were still more or less free from *Typha* after one year.

Low oxygen levels in the water in plots 1, 2 and 4 in May are probably connected with dying of the plants following anoxic conditions in the rhizomes. *Typha* -species, as also many other emergent aquatic plants, have well developed internal aerenchymatous tissue, which acts as an efficient diffusion pathway for gaseous interchange between organs growing under anaerobic conditions and the atmosphere (Armstrong, 1978). According to Sale & Wetzel (1983), who worked in Michigan in the United States, cutting stems of *Typha latifolia* L. and *Typha angustifolia* L. below the water surface will prevent oxygen transportation between stems and rhizomes, which results in anaerobic respiration in the rhizomes and roots. Under these conditions the authors measured the production of ethanol in the rhizomes, which eventually led to a decay of plant material below the water surface. Mowing trials conducted by Sale & Wetzel (1983) showed that three cuts during the growing season below the water surface were sufficient to kill almost all underwater biomass. On the other hand, after cutting above the

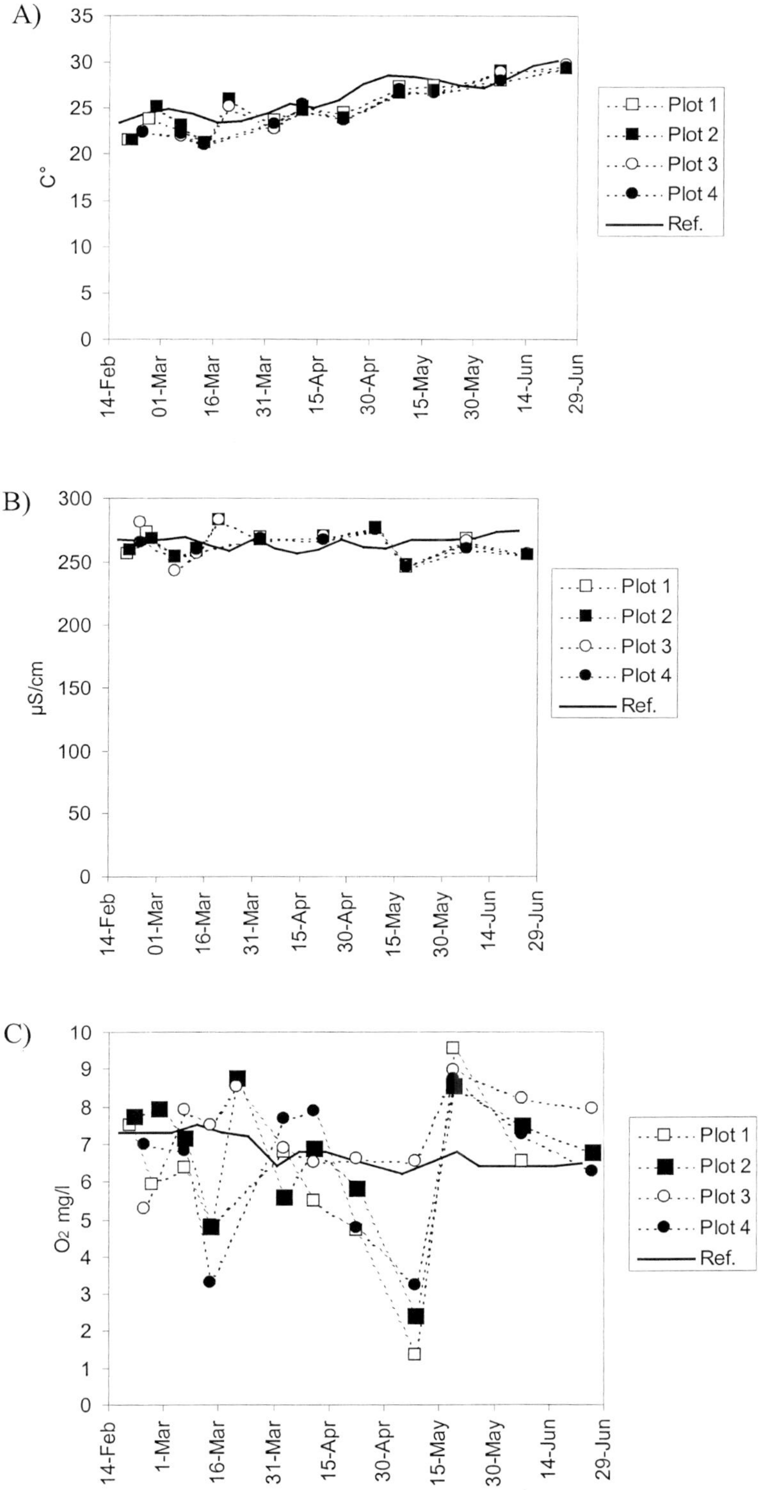

Figure 3. Water quality parameters of surface water (0–10 cm) in mowing plots and in a reference point (SDE) in the Lac de Guiers. Each mowing plot point is calculated as a mean of five samples, reference point is sampled weekly. (A) Temperature (C°) (B) Conductivity (μS cm^{-1}), (C) Dissolved oxygen (mg O_2 l^{-1}),. For other details, ref. Figure 2.

Table 1. Mowing efficiency tests in the Lac de Guiers (May 27–June 25 1997). Total wet weight calculated by using a value of 6.6 kg m^{-1}

Cutting depth (m)		0.5	0.2
Number of experiments		8	5
Area mowed/hour (ha/h)	Mean (min–max)	0.1 (0.03–0.21)	0.11 (0.1–0.13)
Total wet weight cut/hour (kg/h)		6600	7260
Diesel (l/h)		6	6

water surface, much of the underwater biomass remained healthy and capable of regenerating. In this context it should be noted that three cuts during one growing season of *Typha angustata* Bory & Chaub in India, 5 cm above the water level, did not kill the plants either, although above ground as well as below ground biomass decreased rapidly (Sharma & Kushwaha, 1990).

Taking into consideration that the effectiveness of cutting *Typha* plants below the water surface in the Lac de Guiers is in accordance with the conclusions of Sale & Wetzel (1983), it may be that also the time of the year has been advantageous. During the experimental period *Typha* plants flowered profusely. As shown by Husák (1978) with *T. angustifolia* in Central Europe and Singh et al. (1976) with *T. angustata* in India, a single cut is most effective during the time the flowering spike is growing rapidly. Under these circumstances the rhizome's carbohydrate reserves are used for the growth of aerial plant parts.

Changes in oxygen concentration of the water seem to be related to a breakdown of death plant material, particularly marked in plots in which mowing was carried out at a depth of 20 cm. However, these changes can also be partly related to increased oxygen demand of bottom sediment promoted by decreased oxygen transportion into sediment (Sale & Wetzel, 1983). In this context it is difficult to draw conclusions on differences between plots 3 and 4 as far as changes in the oxygen content of the water are concerned. Apparently, less decaying plant material has been present in plot 3 or water exchange with open water areas has been more efficient. In connection with the higher re-growth rate in plot 3 it is also possible that the availability of oxygen in the sediment was higher compared to plot 4.

As far as the final results are concerned differences between the depth of cutting were unexpectedly small. Although the water level decreased 27 cm during the research period, plots cut 20 cm below the surface did not show any higher *Typha* survival compared to plots cut 50 cm below the surface. Evidently the higher cutting level did not provide better conditions for stems to reach the surface.

Although mowing was quite effective, the speed at which plants were removed was quite low. In addition, taking into consideration the subsequent removal of plant material after cutting, it was calculated that clearing one hectare would take at least 35 h.

Consequently, from an economic point of view this type of boat, commonly used for clearing waterways in Western Europe, is less appropriate for controlling *Typha* under prevailing conditions in Senegal. In this context, it should also be taken into consideration that stems of *Typha* plants in Senegal are quite thick compared to *Typha* plants in Western Europe and that the motor regularly became overheated. Therefore, a more powerful mowing boat (see Wade, 1990) is required, whereas this type of mowing boat would only be suitable for opening small waterways for local villagers.

Acknowledgements

This research was partly financed within the framework of the EU (DG XII INCO/DC) project no. IC18-CT96-0080. The substantial contribution of Hanna Ahonen, Indega Bindia, Sandra Kloff, Jaana Rantakokko and Mandiaye Sall regarding field work and data collection is especially acknowledged. Staff of the Knit water pumping station (SDE) are acknowledged for providing reference data on water quality. Special thanks are also due to Mr Sara Diouf, Director of the Biological Station in the Djoudj National Park, for his advice.

References

Armstrong, W., 1978. Root aeration in wetland conditions. In Hook, D. D. & R. M. M. Crawford (eds), Plant Life in Aerobic Environments. Ann Arbor Science publishers, Ann Arbor: 269–297.

Cogels, F. X., A. Thiam & J. Y. Gac, 1993. Premier effets des barrages du fleuve Sénégal sur le lac de Guiers. Rev. Hydrobiol. trop. 26: 105–117.

Husák, J., 1978. Control of reed and reed mace stands by cutting. In Dykyjova, D. & J. Kvet (eds), Pond Littoral Ecosystems: Ecological Studies 28. Springer Verlag, Berlin: 404–408.

John, D. M., 1986. The inland waters of tropical West Africa. An introduction and botanical review. Arch. Hydrobiol. Beih. 23: 104–219.

Sale, P. J. M. & R. G. Wetzel, 1983. Growth and metabolism of *Typha* species in relation to cutting treatments. Aquat. Bot. 15: 335–348.

Sharma, K. P. & S. P. S. Kushwaha, 1990. Effect of cutting of above-ground organs of *Typha angustata* Bory & Chaub on its growth and total chlorophyll content. Aquat. Bot. 36: 293–296.

Singh, S. P., S. S. Pahuja & M. K. Moolani, 1976. Cultural control of *Typha angustata* at different stages of growth. In Varshney, C. K. & J. Rzóska (eds), Aquatic Weeds in S.E. Asia. Proceedings of a Regional Seminar on Noxious Aquatic Vegetation. New Delhi. Dr W. Junk Publishers, The Hague: 245–247.

Thiam, A., R. Ndiaye & M. Ouattara, 1993. Macrophytes aquatiques et zooplancton du lac de Guiers. Institut des Sciences de l'Environnement (ISE). Projet 2: ISE/FUL, Report: 55 pp.

Wade, M., 1990. Physical control of aquatic weeds. In Pieterse, A. H. & K. J. Murphy (eds), Aquatic Weeds. Oxford University Press, Oxford: 93–135.

Hydrobiologia **415:** 257–263, 1999.
J. M. Caffrey, P. R. F. Barrett, M. T. Ferreira, I.S. Moreira, K. J. Murphy & P. M. Wade (eds),
Biology, Ecology and Management of Aquatic Plants.
© 1999 *Kluwer Academic Publishers. Printed in the Netherlands.*

Aquatic plant management: ecological effects in two streams of the Swiss Plateau

Barbara R. Kaenel* & Urs Uehlinger
Swiss Federal Institute for Environmental Science and Technology (EAWAG), CH-8600 Duebendorf, Switzerland

Key words: plant management, stream, disturbance, macrophytes, invertebrates, habitat conditions, oxygen concentrations

Abstract

Effects of plant removal on habitat conditions, stream metabolism and benthic invertebrates were studied in two macrophyte-rich streams (Chriesbach, Mühlibach) of the Swiss Plateau. We monitored a control reach (no treatment) and two impact reaches (removal of plants by cutting or dredging) in each stream. Sampling was conducted during a 2–4 month period before and a 9 month period after the removal of 84–94% of the plant biomass. Oxygen concentrations were continuously recorded for 3–4 months. Plant removal decreased water depth and increased current velocity. The total number of invertebrates decreased by about 65%. Plant cutting mainly affected taxa that used macrophytes as habitat. Highly mobile taxa and taxa living on or within the bed sediment were less affected. Invertebrate densities recovered within 4–6 months. The removal of plants resulted only in a moderate increase in nocturnal oxygen concentrations. In the stream where plants were cut in spring, macrophytes recovered within the same growing season. In the other stream, where plant growth started later, plants were cut in summer and no recovery of plants occurred until the following spring.

Introduction

Aquatic macrophytes are a common feature of many nutrient-rich unshaded lowland streams with stable flow regimes (Pieterse & Murphy, 1990; Dawson et al., 1991). Macrophytes change habitat conditions, influence the oxygen balance and modify the structure and spatial distribution of benthic invertebrates in streams (Edwards & Owens, 1962; Marshall & Westlake, 1990; Ward, 1992). Excessive plant growth may increase the risk of bank-overtopping and result in large diel oxygen fluctuations due to the photosynthetic and respiratory activity of macrophytes and associated epiphytic algae (Dawson, 1978; Hearne & Armitage, 1993). However, plant removal is traditionally related to the improvement of the hydraulic conditions.

Plant removal is a disturbance that affects the structure of habitats and communities. Studies on the impacts of plant removal on streams have usually focused on single aspects such as invertebrates, plants or metabolism (e.g. Simonsen & Harremoës, 1978; Ham et al., 1982; Caffrey, 1990; Dawson et al., 1991). In this paper, we summarise the effects of plant management techniques (cutting and dredging) on habitat conditions, oxygen concentrations and invertebrates in two streams of the Swiss Plateau. We discuss the applied management regimes in relation to their ecological impacts and their efficiency in reducing interference with human demands. Based on these results, we make suggestions for future plant management practices.

Methods

Study sites

The study was performed in two nutrient-rich streams of the Swiss Plateau, Chriesbach and Mühlibach, that were similar with respect to their chemical and physical characteristics (Table 1). The wetted channels of both streams were sparsely shaded. Macrophyte communities were dominated by *Ranunculus fluitans* Lam. in the Chriesbach and by *Ranunculus trichophyllus*

* Author for correspondence

Table 1. Morphometric and chemical properties of the Chriesbach and Mühlibach in 1995. Mean annual values and standard deviation are based on continuous measurements of depth, discharge and temperature. Mean annual values and standard deviation of nutrient concentrations were calculated from monthly samples

	Chriesbach	Mühlibach
Channel width at base flow (m)	4	3.2
Slope (‰)	1.3	1.9
Elevation (m a.s.l.)	430	418
Mean annual discharge (m^3/s)	0.51±0.21	0.38±0.28
Mean annual depth (m)	0.75±0.15	0.25±0.05
Mean annual temperature (°C)	11.20±2.70	10.30±2.70
NH_4–N (μg/l)	20.9±9.5	14.4±10.0
NO_2–N (μg/l)	22.7±7.0	10.7±10.5
NO_3–N (mg/l)	7.00±0.81	8.56±0.57
Soluble reactive phosphorus (μg/l)	50.8±13.4	80.8±79.3

Chaix in the Mühlibach. In summer, macrophytes cover ca. 100% of the streambed. Macrophytes are removed twice a year by water management authorities to prevent bank overtopping (Chriesbach) or water logging of adjacent agricultural fields (Mühlibach). A detailed description of both streams is given by Kaenel et al. (1998).

Experimental design

Macrophyte biomass, current velocity and water depth were monitored in three stream reaches subject to different management regimes: 1. no management = control, 2. cutting and 3. dredging. Benthic invertebrates were sampled in the control and in the cut reaches. Oxygen concentration and water temperature were measured at the up- and downstream ends of the control and impact sections. Control reaches were located upstream of the impact reaches. Additional replication was not possible due to the changing characteristics of the macrophytic vegetation upstream and downstream of the study reach. Samples were taken using a stratified random design (after Green, 1979; length of strata = 50 m) during a 2–4 month period before and a 9 month period after the removal of macrophytes. The four strata within each reach correspond to a spatial interspersion of replicates that has been described as clumped segregation (Hurlbert, 1984). Macrophytes were removed on 22–23 May 1995 in the Chriesbach and on 24–26 July in the Mühlibach (for more details see Kaenel et al., 1998; Kaenel & Uehlinger, 1998).

Methods

Invertebrates and plant biomass were simultaneously collected with a Surber sampler modified to allow sampling in water up to 1.2 m in depth (surface area = 0.09 m^2, mesh size = 250 μm, Mensch et al., 1997). Macroinvertebrates were separated from plants and stored at –20 °C. Invertebrates were identified to order level (except Mollusca to class, Crustacea to species and Diptera to family level) and counted under a stereomicroscope. Plants were sorted to species and dried to constant weight at 85 °C. Water depth and current velocity were measured within the sampling area. Estimates of mean channel depth were based on continuous water level recordings and four staff gauges that were read at weekly – biweekly intervals. Mean current velocities were determined by monitoring the downstream movement of an uranin cloud with an *in situ* pulse light fluorometer at different discharges. Oxygen concentrations were continuously monitored at the ends of the control and impact reaches. Measurements were performed with YSI equipment (probe 5739 with stirrer 5795A and O_2-meter 54 or 58, YSI Inc., Yellowsprings, Ohio, U.S.A.). Signals were averaged for 3 – 10 min and stored in 10 min intervals (Tattletale Model IV single card datalogger, Onset Corporation, N. Falmouth, Massachusetts, U.S.A). Macrophyte biomass, invertebrates and habitat conditions on control and impact reaches (dredged and cut) were compared with the two-sided Mann-Whitney U-test. Differences were assumed to be significant if p = 0.05.

Results

Plant biomass

In both streams, plant biomass was similar in all reaches before plant removal. Macrophyte removal reduced plant biomass by 87% in the cut and by 94% in the dredged reach in the Chriesbach and by 84% in the cut and by 90% in the dredged reach in the Mühlibach (Figure 1). Plants exhibited different recovery trajectories in the two streams. In the Chriesbach, where *R. fluitans* had been removed before flowering, macrophytes recovered within the same growing season and recovery was faster in the cut than in the dredged reach. In the Mühlibach, where plants had been removed after flowering of *R. trichophyllus*, plants did not recover until the following spring.

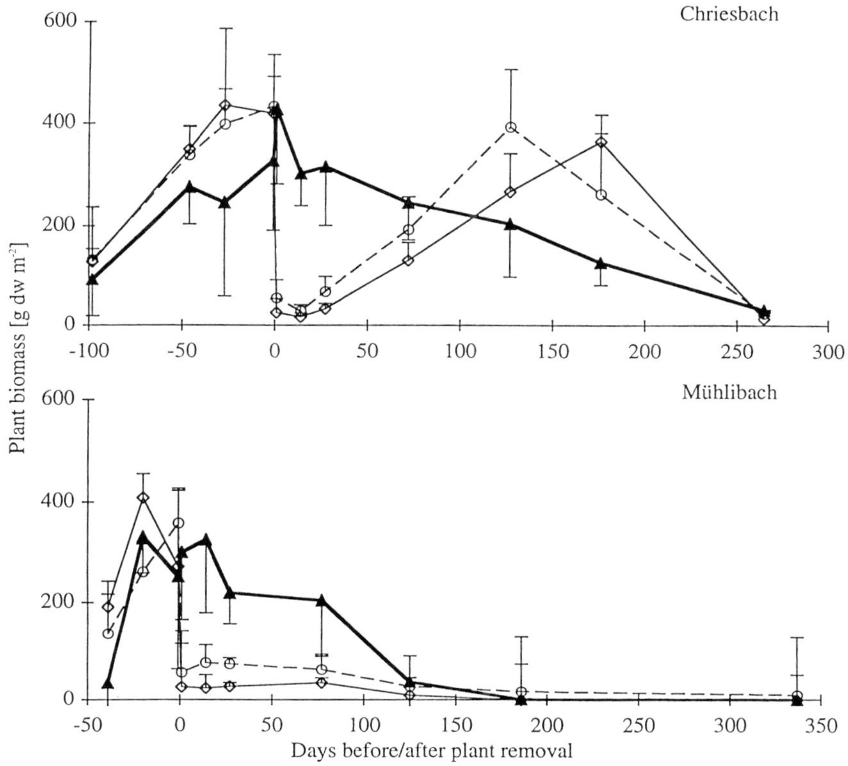

Figure 1. Macrophyte biomass (g dry weight m^{-2}) in the control (solid line), cut (dashed line) and dredged reaches (fine line) of both streams. Error bars represent standard errors.

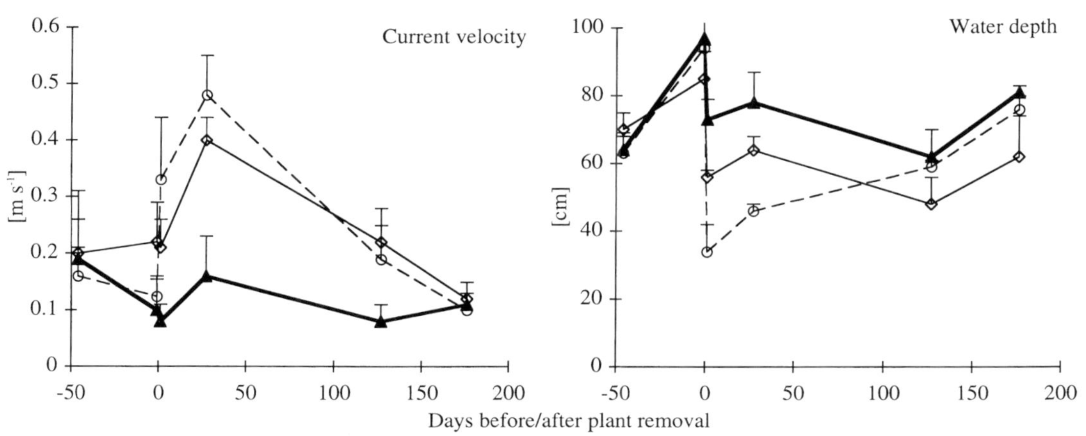

Figure 2. Current velocity and water depth in the control (solid line), cut (dashed line) and dredged reaches (fine line) in the Chriesbach. Error bars represent standard errors.

Habitat conditions

Apart from the dredged reach in the Mühlibach, depth and current velocity in impact and control reaches were similar before plant removal. Plant removal reduced mean depth and increased mean current velocity in both streams (Figure 2). Plant dredging did not have a significantly greater effect on current velocity than cutting. In the Chriesbach, water depth and current velocity were not significantly different between control and impact reaches 127 d (depth) and 176 d (velocity) after plant removal. In the Mühlibach, both parameters were similar 125 d after plant management.

Oxygen concentrations

Before plant removal, diel O_2 fluctuations were higher at the end of the impact reaches than at the end of the control reaches in both streams (Figure 3). The gradual increase of daily O_2 maxima until the spate of 26 May in the Chriesbach continued only in the control reach after plant removal, in the impact reach nocturnal O_2 concentrations increased from 7 to 8 mg/l. In the Mühlibach, nocturnal O_2 concentrations increased from 6 to 7 mg/l after plant removal.

Invertebrates

In both streams, plant removal decreased the total number of invertebrates by about 65% (Figure 4). Plant cutting affected mainly taxa that used macrophytes as habitat (e.g. Chironomidae; Figure 4), whereas highly mobile taxa (e.g. Ephemeroptera) and taxa living on or within the bed sediments (e.g. Bivalvia, Trichoptera) were less affected. Although recovery of macrophytes differed between streams, taxa affected by plant removal recovered within 4–6 months in both streams. In the Mühlibach, plant removal affected benthic invertebrates less distinctly than in the Chriesbach (Kaenel et al., 1998).

Discussion

Ecological effects of plant management

Macrophytes modify habitat conditions in streams. As expected, mean plant biomass after plant removal was lower in dredged than in cut reaches, but these differences were rarely significant. Cutting and dredging neither differed in their effect on depth and current velocity. The dredging method used in this study might explain the only small differences between the two management procedures. In the dredged reaches of our study, plants were cut immediately above the stream bottom and parts of the below ground plant material were eliminated by raking. There is some evidence that the periodic removal of macrophytes results in a plant community that is dominated by disturbance tolerant species (Kaenel, 1998).

Plant removal decreases community respiration and enhances the oxygen exchange between water and atmosphere, and thus, reduces the risk of low oxygen concentrations at night (Simonsen & Harremoës, 1978; Kaenel, 1998) which may threaten fish and benthic invertebrates (Gammeter, 1988; Alabaster & Lloyd, 1998). For example Brooker et al. (1977) reported massive fish kill when water temperatures varied between 21 and 28 °C and diel oxygen concentrations between 0.5 and 5 mg/l. However, oxygen conditions in Chriesbach and Mühlibach were far from such extremes, and thus, do not justify plant removal in this respect.

Plant removal is a disturbance with a distinct short-term effect on benthic invertebrates, mainly taxa that use macrophytes as habitat. Kaenel & Uehlinger (1998) demonstrated that the spatial distribution of plants after plant removal was more heterogeneous in the Mühlibach than in the Chriesbach. This suggests that maintaining a high spatial heterogeneity of the remaining plants can moderate the impact of plant removal on invertebrates (Kaenel et al., 1998). However, recovery of invertebrates did not coincide with plant recovery, indicating that factors other than plant biomass, current velocity and depth influence rate of recolonization. Such factors may include individual life history, the presence of undisturbed reaches upstream and the timing of plant removal (Mackay, 1992; Ward, 1992). Plant removal in spring has presumably a more severe impact on stream invertebrates than removal in summer (Pearson & Jones, 1978; Kaenel et al., 1998).

Efficiency of plant management

Plant removal immediately reduced the risk of bank-overtopping and agricultural fields becoming waterlogged because water depth decreased and current velocity increased. However, the long-term benefits of plant removal with respect to improved channel hydraulics were limited by the macrophyte's ability to recover. Dredging retarded the recovery of plant biomass compared to cutting, and thus was more effective

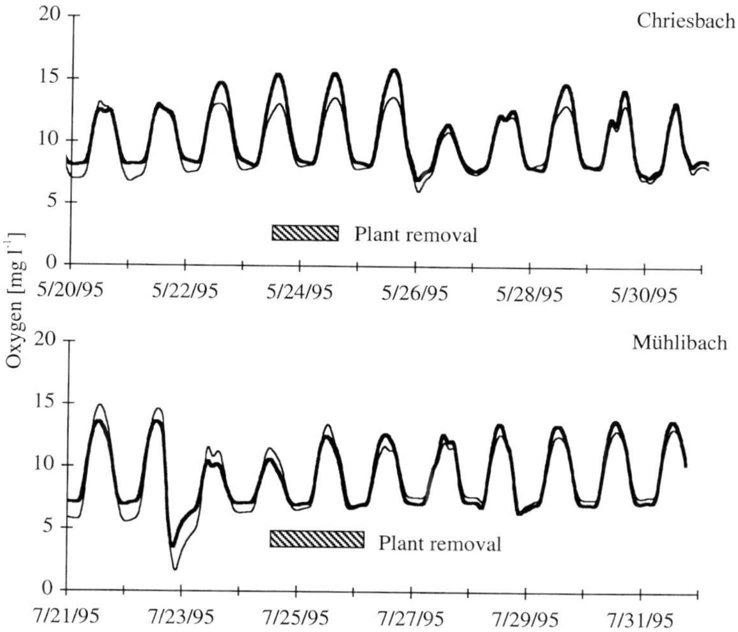

Figure 3. Oxygen concentrations at the end of the control (solid line) and cut (dashed line) reaches of both streams immediately before and after plant removal.

in maintaining the drainage capacity of a channel. Plant recovery can probably further be delayed if the dredging procedure normally used by management authorities (Haslam & Wolseley, 1987; Wade, 1992) is applied instead of raking, the method used in this study. However, the negative impact on the stream ecosystem (e.g. benthic invertebrates, fish) of the traditional method would be more severe. Plant recovery also depends on the timing of plant removal. Fast recovery of macrophytes after plant removal in spring is apparently a common phenomenon (e.g. Murphy et al., 1987; Caffrey, 1990; this study). Macrophytes are stimulated to regrow and flower if plants are cut prior to flowering (Dawson, 1978). If plants are left uncut, decay starts soon after flowering (Ham et al., 1982; this study). Plant recovery may further be delayed if plants are removed along a whole stream and not only from relatively small reaches; i.e. untreated stream reaches provide plant propagules and also may buffer the impact of spates.

Suggestions for further plant management

We recommend removing plants after flowering in summer or autumn. Then, plant removal is most effective with respect to plant recovery and thus, reduces the cost of stream maintenance. Plant removal in summer or autumn is also preferable for ecological reasons because the impact on benthic invertebrates appears to be less severe than in spring. We would like to emphasise that macrophytes considerably improve environmental conditions for invertebrates and fish in channelized streams with low habitat heterogeneity. If plant management is inevitable, negative impacts on benthic invertebrate communities should be reduced by maintaining a minimum degree of spatial heterogeneity of the habitat. This can be obtained by leaving some macrophyte beds untreated.

However, the need for plant management should be critically evaluated. Plant management is typically based on a traditional approach, which perceives the main function of a river as the efficient drainage of a catchment (Wade, 1992). Plant management is often done because of tradition and not necessarily because it is an effective management strategy (Haslam & Wolseley, 1987). Plant removal techniques do not result in a long-term reduction of plant biomass, as they do not reduce growth conditions for aquatic plants. Shading a stream by riparian vegetation (Krause, 1977; Dawson & Kern-Hansen, 1979; Can-

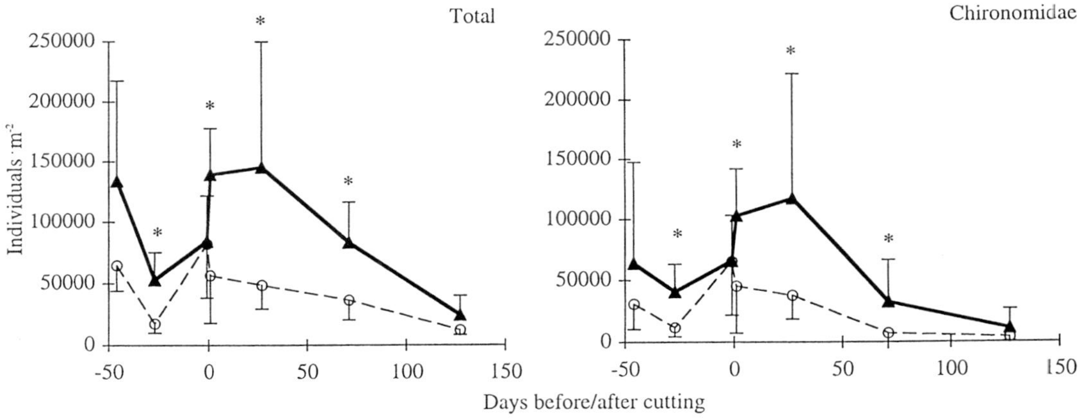

Figure 4. Abundance of all invertebrates and Chironomidae in the control (solid line) and cut (dashed line) reaches of Chriesbach. Error bars represent standard errors. * indicates significant differences between reaches ($p = 0.05$).

field & Hoyer, 1988) or leaving streams unmanaged after several years of regular plant removal (Dawson, 1976) are two management methods that can result in a sustainable long-term reduction of plant biomass; and thus are interesting from an ecological and economic point of view.

Acknowledgements

The Water Protection Authority of the Kanton Zurich (AGW) funded this project. We thank T. Boesch, U. Holliger and R. Mensch for sample processing. We are also grateful to P. Burgherr, C. Robinson, R. Zah, Prof. J. V. Ward and two anonymous reviewers for critical comments on the manuscript.

References

Alabaster, J. S. & R. Lloyd, 1980. Water quality criteria for freshwater fish. Butterworths, London: 297 pp.

Brooker, M. P., D. L. Morris & R. J. Hemsworth, 1977. Mass mortality of adult salmon, *Salmo Salar*, in the R. Wye 1976. J. appl. Ecol. 14: 409–417.

Caffrey, J. M., 1990. Problems relating to the management of *Potamogeton pectinatus* L. in Irish rivers. In Proc. 8th Symp. Aquatic Weeds. Uppsala: 61–68.

Canfield, D. E. & M. V. Hoyer, 1988. Influence of nutrient enrichment and light availability on the abundance of aquatic macrophytes in Florida streams. Can. J. Fish. aquat. Sci. 45: 1467–1472.

Dawson, F. H., 1976. The annual production of the aquatic macrophyte *Ranunculus penicillatus* Var. *calcareus* (R. W. Butcher) C. D. K. Cook. Aquat. Bot. 2: 51–73.

Dawson, F. H., 1978. The seasonal effects of aquatic plant growth on the flow of water in a stream. In Proc. 5th Symp. Aquatic Weeds. Amsterdam: 71–78.

Dawson, F. H. & U. Kern-Hansen, 1979. The effect of natural and artificial shade on the macrophytes of lowland streams and the use of shade as a management technique. Int. Rev. ges. Hydrobiol. 64: 437–455.

Dawson, F. H., E. M. F. Clinton & M. Ladle, 1991. Invertebrates on cut weed removed during weed-cutting operations along an English river, the River Frome. Aquacult. Fish. Mgmt 22: 113–121.

Edwards, R. W. & M. Owens, 1962. The effects of plants on river conditions IV. The oxygen balance of a chalk stream. J. Ecol. 50: 207–220.

Gammeter, S., 1988. Kurzzeiteffekte anthropogener Belastungsparameter auf Fliesswasser-Makroinvertebraten. Diploma Thesis, Swiss Federal Institute of Technology, Zurich.

Green, R. H., 1979. Sampling design and statistical methods for environmental biologists. John Wiley & Sons Inc, New York: 257 pp.

Ham, S. F., J. F. Wright & A. D. Berrie, 1982. The effect of cutting on the growth and recession of the freshwater macrophyte *Ranunculus penicillatus* (Dumort.) Bab. var. *calcareus* (R. W. Butcher) C. D. K. Cook. J. Envir. Mgmt 15: 263–271.

Haslam, S. M. & P. A. Wolseley, 1987. River plants of Western Europe: the macrophytic vegetation of watercourses of the European Economic Community. Cambridge University Press, Cambridge: 512 pp.

Hearne, J. W. & P. D. Armitage, 1993. Implications of the annual macrophyte growth cycle on habitat in rivers. Regul. Riv.: Res. Mgmt 8: 313–322.

Hurlbert, S. H., 1984. Pseudoreplication and design of ecological field experiments. Ecol. Monogr. 54: 187–211.

Kaenel, B. R., 1998. Effects of aquatic plant removal on lotic ecosystems. PhD Thesis, Nr. 12651, Swiss Federal Institute of Technology, Zurich.

Kaenel, B. R. & U. Uehlinger, 1998. Effects of plant cutting and dredging on habitat conditions in streams. Arch. Hydrobiol. 143: 257–273.

Kaenel, B. R., C. D. Matthaei & U. Uehlinger, 1998. Disturbance by aquatic plant management in streams: effects on benthic invertebrates. Regul. Riv.: Res. Mgmt 14: 341–356.

Krause, A., 1977. On the effect of marginal tree rows with respect to the management of small lowland streams. Aquat. Bot. 3: 185–192.

Mackay, R. J., 1992. Colonisation by lotic macroinvertebrates: a review of processes and patterns. Can. J. Fish. aquat. Sci. 49: 617–628.

Marshall, E. J. P. & D. F. Westlake, 1990. Water velocities around water plants in chalk streams. Folia geobot. phytotax. 25: 279–289.

Mensch, R., B. Kaenel & U. Uehlinger, 1997. Kurzfristige Auswirkungen einer Entkrautung auf einen Mittellandbach (Chriesbach bei Dübendorf, ZH). Vjsschr. Natf. Ges. Zurich 142: 23–31.

Murphy, K. J., A. M. Fox & G. Hanbury, 1987. A multivariate assessment of plant management impacts on macrophyte communities in a Scottish canal. J. appl. Ecol. 24: 1063–1079.

Pearson, R. G. & N. V. Jones, 1978. The effects of weed-cutting on the macro-invertebrate fauna of a canalised section of the river Hull, a northern English chalk stream. Envir. Mgmt 7: 91–97.

Pieterse, A. H. & K. J. Murphy, 1990. Aquatic Weeds: the ecology and management of nuisance aquatic vegetation, Oxford University Press, Oxford: 593 pp.

Simonsen, J. F. & P. Harremoës, 1978. Oxygen and pH fluctuations in rivers. Wat. Res. 12: 477–489.

Wade, P. M., 1992. Management of macrophytic vegetation. In Calow, P. & G. E. Petts (eds), The Rivers Handbook: Hydrological and Ecological Principles. Blackwell Scientific Publications, Oxford: 363–385.

Ward, J. V., 1992. Aquatic insect ecology, John Wiley & Sons Ltd, New York: 438 pp.

Hydrobiologia **415:** 265–269, 1999.
J. M. Caffrey, P. R. F. Barrett, M. T. Ferreira, I.S. Moreira, K. J. Murphy & P. M. Wade (eds),
Biology, Ecology and Management of Aquatic Plants.
© 1999 *Kluwer Academic Publishers. Printed in the Netherlands.*

Submerged macrophyte control with herbivorous fish in irrigation channels of semiarid Argentina

A. A. Dall Armellina*, C. R. Bezic & O. A. Gajardo
Universidad Nacional del Comahue, Centro Regional Zona Atlántica, Ayacucho y Esandi (8500) Viedma, Río Negro, Argentina

Key words: grass carp, submerged weeds, biological control, channels, stocking rate

Abstract

Two experiments were conducted in Río Negro province (Argentina) to study macrophyte response to different stocking rates of the grass carp (*Ctenopharyngodon idella* Val.). The first of these was performed in nine polyethylene-lined ponds, 12 m^2 each, planted with Illinois pondweed (*Potamogeton illinoensis* Morong) at two fish stocking rates, 100 and 360 kg ha^{-1}. The second experiment was set up under natural conditions in a channel full of *Potamogeton pectinatus* L., *Myriophyllum aquaticum* (Vell.) Verd. and *Chara contraria* A. Braun ex Kurtz, at 100 and 200 kg ha^{-1}. In both cases, the initial individual fish weight was between 22 and 40 g. A complete randomised block design was used in both experiments. Five samples of plant biomass were taken per experimental unit. Data were subjected to ANOVA and LSD was used as a comparative test. In both cases, the experimental period was 2 months because of the herbivory under the high stocking rate completely depleted both the leaves of *P. illinoensis* under pond conditions and the aboveground plant biomass in the channel. The results showed a high potential of biocontrol under tested conditions and suggest a stocking rate lower than 100 kg ha^{-1} as enough for local conditions and for a prolonged fish presence in the area.

Introduction

Most of the artificial or natural water bodies around the world are covered with aquatic plants (El Gharably et al., 1978; Pieterse, 1979; Sabbatini, 1988). In many cases, this vegetation is useful to the ecosystem because it provides food and protection to many organisms, it stabilises the sediment, reduces the water velocity and increases the oxygen in this environment. Oppositely, in irrigation channels of semiarid regions, when its growth exceeds the limit imposed by human beings, the vegetation becomes a weed (Sabbatini, 1988), severely interfering with the channel operation.

In the province of Río Negro (38–42° S), aquatic weeds constitute one of the main problems under both irrigation and drainage channels. Organisations spend more than $1.5 million annually to control weeds in different irrigation systems (Dall Armellina et al., 1996). Chaining is the method for channel maintenance (mechanical) which has a low efficiency because of the rapid regrowth of perennial weeds (Dall Armellina et al., 1996). A large number of treatments are needed throughout the season, which increases the operational costs significantly. Chemical control is considered dangerous since water is used for humans as well as for animals. Biological control was never used in the area and no research has been conducted.

One of the most widespread agents of biocontrol is the cyprinid grass carp (*Ctenopharyngodon idella* Val.), which was used successfully in many parts of the world. In The Netherlands, 190 kg ha^{-1} were recommended for a rapid control and 75–120 kg ha^{-1} as an optimum for long term maintenance (van der Zweerde, 1993). In Egypt, optimal results were obtained with 200 kg ha^{-1} in irrigation and drainage channels (El Gharably et al., 1978; Pieterse, 1979). By contrast, 180 kg ha^{-1} were not enough to control macrophytes in Egyptian channel systems near Port Said (Dubber et al., 1980).

For these reasons, local experience is essential to determine the possibilities of the method. Environmental conditions and the macrophyte communities

* Author for correspondence

may differ strongly from place to place and invalidate any recommendation taken from other situations. In the irrigation channels of the Lower Valley of the Rio Negro, 90% of the macrophyte biomass comprises the rhizomatous perennial *Potamogeton illinoensis* Morong (Bezic et al., 1996). Other weeds develop strongly when the above plant is absent (Dall Armellina et al., 1996). These two conditions normally appear in the south-east area of the province, the first one in the Lower Valley main irrigation channel and the second one in small channels far from the previous cited.

The objectives of this study were 1 to test the possibilities of controlling submerged weeds with the grass carp under artificial and natural conditions, and 2 to search for the appropriate stocking rate under two conditions: (a) a *Potamogeton illinoensis* pure population (from monoculture) and (b) a natural submerged weeds community where *P. illinoensis* is not the dominant species.

Materials and methods

Pond experiment

The research was carried out in nine ponds, each 12 m^2 and 1.5 m deep, built below the soil surface and lined with black 200 μ polyethylene. All the ponds were provided with a water recirculation system, which takes freshwater from an irrigation channel. The ponds were at an intermediate level and connected with both the irrigation and the drainage channel and water freely flowed by gravity among them. Each pond had a 25-cm layer of soil. Each pond was planted in the spring 1995 with 48 plants of *P. illinoensis* (4 pl m^{-2}) which were produced from rhizomes collected in main channel of the Lower Valley (Bezic et al., 1996; Dall Armellina et al., 1996).

Juvenile diploid grass carp, weighing between 22–40 g, were stocked on February 12th, 1996, when the plant biomass was considered sufficient for the experiment (Table 1) at stocking rates of 100 (L = Low) and 360 kg ha^{-1} (H = High). Plant biomass (DW-72 h, 105 °C)) for leaves, shoots and rhizomes of *P. illinoensis* was recorded three times in the experiment on target plants grown in 50 l plastic bags placed into the ponds. The experiment was concluded in April 18th, 1996, because of the total depletion of the leaf biomass for the H treatment.

Table 1. Control of *P. illinoensis* growing in pond monoculture by juvenile diploid white amur at rates of 100 k ha^{-1} (Low) and 360 k ha^{-1} (High). Effects on different plant parts over a period of 63 days from February to April, 1996. Data are average ± standard deviation of dry weight per plant. Values followed by the same letter have no statistical differences at the level of 5% (p<0.05), comparisons on the row sense

		Control	Low	High
February 12				
Leaves + Ligules	NS	6.2 ± 1.6	6.2 ± 0.5	6.8 ± 1.1
Shoots	NS	1.8 ± 0.4	2.1 ± 0.1	1.9 ± 0.2
Rhizomes	NS	2.0 ± 0.5	2.2 ± 0.2	1.7 ± 0.3
Total Aboveground	NS	8.0 ± 2.1	8.4 ± 0.6	8.8 ± 1.4
Total (whole plant)	NS	10.0 ± 2.6	10.6 ± 0.6	10.6 ± 1.6
March 27				
Leaves + Ligules	S	9.5 ± 1.7 a	6.4 ± 2.0 b	2.5 ± 0.3 c
Shoots	NS	3.4 ± 0.4	4.7 ± 1.2	3.1 ± 0.6
Rhizomes	NS	3.9 ± 0.7	3.3 ± 0.7	3.1 ± 0.5
Total Aboveground	HS	12.8 ± 2.4 a	11.5 ± 0.9 a	5.8 ± 1.1 b
Total (whole plant)	HS	16.9 ± 2.8 a	14.8 ± 0.5 a	8.9 ± 1.6 b
April 18				
Leaves + Ligules	HS	7.4 ± 0.8 a	3.5 ± 0.9 b	1.5 ± 0.4 c
Shoots	NS	2.3 ± 0.9	3.1 ± 0.9	2.7 ± 0.3
Rhizomes	NS	3.9 ± 0.5	4.7 ± 0.3	3.7 ± 0.2
Total Aboveground	NS	9.7 ± 1.4	6.3 ± 1.8	6.2 ± 0.8
Total (whole plant)	NS	13.7 ± 2.0	11.0 ± 2.1	7.9 ± 1.0

Statistical differences according LSD test: (NS) no significant, (S) significant, (HS) highly significant.

Channel experiment

This experiment was performed in a medium size channel (regular trapezoidal section, 4 m at the top and 2 m at the bottom) with a discharge of 0.5 m^3 s^{-1} and water velocity of 0.3 m s^{-1}. The site was characterised by a natural community dominated by *Potamogeton pectinatus* L., *Myriophyllum aquaticum* (Vell.) Verd. and *Chara contraria* A. Braun ex Kurtz. The experimental area was divided into nine 30-m long plots separated by iron barriers with plastic nets. Juvenile diploid grass carp were stocked on November 11th, 1996, at 100 (L) and 200 kg ha^{-1} (H). The aboveground plant biomass on a DW basis (72 h, 105 °C) was recorded on four sampling occasions. Five samples of 0.09 m^2 (square) per experimental unit were taken. The experiment was concluded in February 5th, 1997, because of the total eradication of the aboveground cover in the H plots.

In both experiments, a complete randomised block design was used with three blocks, three treatments of

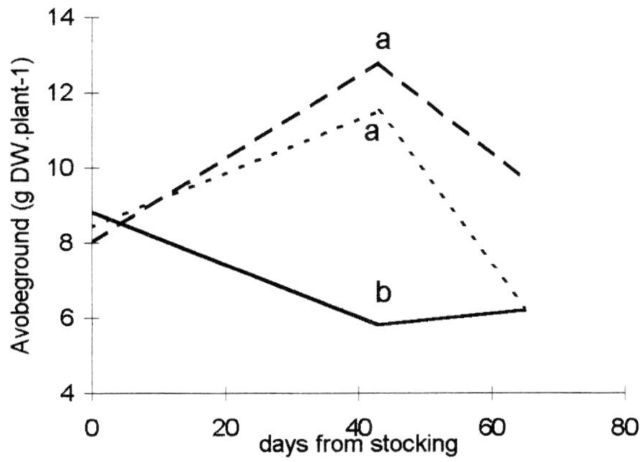

Figure 1. Juvenile diploid grass carp feeding under pond conditions in Río Negro, Argentina. Effects on the total aboveground biomass of *P. illinoensis* in monoculture. The dotted line is C: control without fish, the pointed line is L: low fish biomass (100 kg ha^{-1}), and the full line is H: high fish biomass (360 kg ha^{-1}). Different letters above the lines indicates significant differences ($p<0.05$) among treatments in a specific moment (days from stocking).

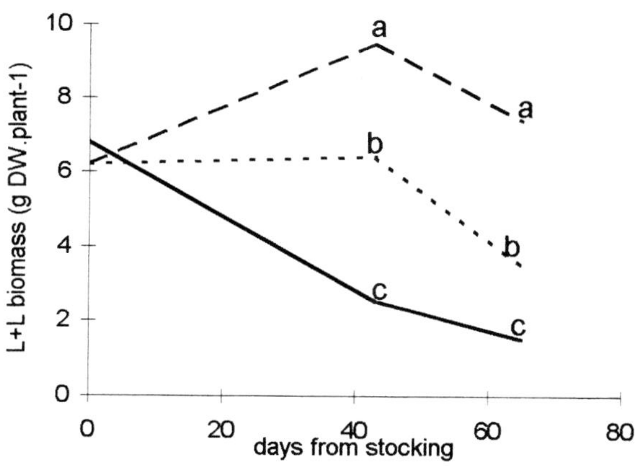

Figure 2. Juvenile diploid grass carp feeding under pond conditions in Río Negro, Argentina. Effects on the leaves plus ligules biomass of *P.illinoensis*. The dotted line is C: control without fish, the pointed line is L: low fish biomass (100 kg ha^{-1}), and the full line is H: high fish biomass (360 kg ha^{-1}). Different letters above the lines indicates significant differences ($p<0.05$) among treatments in a specific moment (days from stocking).

fish biomass (H for high, L for low and C for control, free of fish) and five samples per experimental unit. Data were subjected to ANOVA and means were compared with an LSD test.

Results and discussion

Pond experiment

At the stocking time in February 12th, no differences among plots were detected for any plant part of *P. illinoensis* (Figures 1 and 2). By March 27th, 43 days after the fish introduction, significant differences were detected ($p<0.05$) for H<L,C in the total aboveground biomass. This means a decrease of *circa* 50% relatively to the original value (Figure 1). The C and L plots increased their aboveground biomass in an equivalent proportion reaching *circa* 12 g DW plant^{-1} (Figure 1). The leaves plus ligules biomass (Figure 2) exhibited the same behaviour, proportional to fish biomass (H<L<C, $p<0.05$). No differences were detected in shoot and rhizome production (Table 1). Highly significant differences ($p<0.01$) were found in the total plant biomass (H<L,C where H=8.9; L=14.8; C=16.9), probably influenced by the consumption of leaves plus ligules (Table 1).

In the third evaluation, which was conducted on April 18th, highly significant differences were detected in leaves plus ligules biomass (Figure 2), no differences among any other variable could be found (Table 1).

The overall behaviour of both the aboveground biomass (Figure 1) and the leaves plus ligules biomass (Figure 2) clearly showed the effects of fish herbivory. The question is why in the third evaluation, no differences were found among treatments by the aboveground biomass. The answer may be related to the fact that: (a) a natural decline in the aboveground biomass is a characteristic of these macrophytes (Bezic et al., 1996) as temperatures decline and plant material, mainly shaded parts, decomposes – this behaviour is clearly observed in the control plot (Figure 1) – and (b) the main source of food for the small fishes used in this experiment was the leaves. Rate of herbivory and preference (plant species and tissue composition) are jointly related to animal size (Eaton et al., 1992; Van der Zweerde, 1993; Cooke et al., 1993). Because of the absence of leaves to eat, between the second and third evaluation, the aboveground biomass was the same in the H treatment (Figures 1 and 3). In this interval, the L treatment lowered the aboveground biomass because of both the natural decline of plant biomass and the leaves biomass decrease resulting from herbivory (Figures 1 and 3).

The consumption of leaves is clearly shown in the leaf/shoot ratio (Figure 3) where the ratio of both H and L treatments declined, based on the decrease in the leaves.

In the leaves plus ligules variable (Figure 2), the L treatment was able to maintain the same biomass, at least for the initial 43 days. This means that the rate

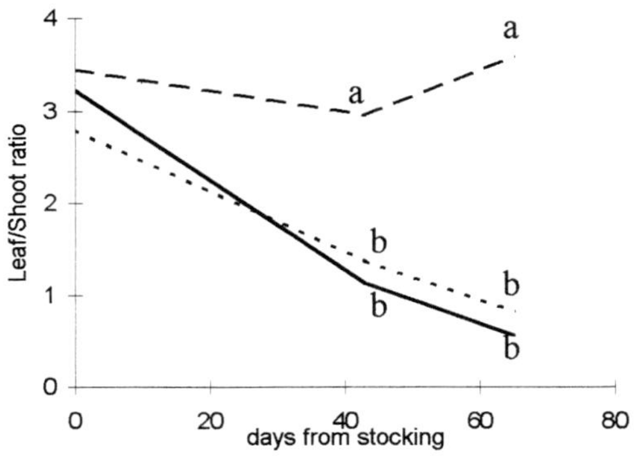

Figure 3. Juvenile diploid grass carp feeding under pond conditions in Río Negro, Argentina. Effects on the *P. illinoensis* leaf/shoot ratio. The dotted line is C: control without fish, the pointed line is L: low fish biomass (100 kg ha^{-1}), and the full line is H: high fish biomass (360 kg ha^{-1}). Different letters above the lines indicates significant differences ($p<0.05$) among treatments in a specific moment (days from stocking).

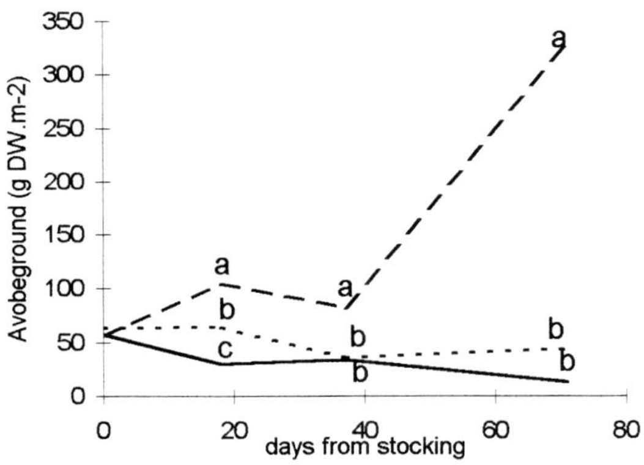

Figure 4. Juvenile diploid grass carp feeding under channel conditions in Río Negro, Argentina. Effects on the total aboveground biomass of a channel natural community dominated by *P.pectinatus*, *M. aquaticum* and *Chara* sp. The dotted line is C: control without fish, the pointed line is L: low fish biomass (100 kg ha^{-1}), and the full line is H: high fish biomass (200 kg ha^{-1}). Different letters above the lines indicates significant differences ($p<0.05$) among treatments in a specific moment (days from stocking).

of herbivory equals the leaf growth rate during this period.

Channel experiment

The first evaluation, at the date of stocking (November 28th , 1996), showed an initial avobeground biomass of approximately 50 g DW m^{-2} and no differences among plots were detected (Figure 4). A significant difference among treatments (H<L<C) was recorded in the second evaluation (December 16th, 1996). After 18 days, the plant biomass in the L treatment remained unchanged while, at the same time, that biomass in the C grew, in H declined. This indicates a balance in L treatment between herbivory and growth rates (Figure 4). For the third evaluation (January 2nd, 1997) no differences were detected between fish plots at the time that C was significant higher (C>L,H).

Finally, at the fourth evaluation (February 5th, 1997), it was possible to see a significant depressor effect exerted by fish after 71 days, compared with C (Figure 4). No significant differences were detected between fish treatments (L=H).

The results indicate that a positive effect on macrophyte control by juvenile diploid grass carp, weighing less than 50 g each, was demonstrated (objective 1). It is necessary to differentiate between two conditions: (a) sites with *P. illinoensis* in which small fish consumed the tender leaves, and (b) sites without *P. illinoensis* where the control was more efficient due to the presence of more palatable plants. Under both conditions, the H treatments were excessive, only to be recommended as an impact density for a rapid control. The L biomass (100 kg ha^{-1}) was sufficient for both conditions, providing an effective control (objective 2). Two months was the maximum time allowed for the presence of fish under the L stocking rate. This is a short period for channel operation. When a period longer than 2 months is required, the fish biomass must be lower than 100 kg ha^{-1}. To increase the efficacy in the control of *P. illinoensis* due to shoot consumption, larger fish must be used.

Acknowledgements

This work was supported by the Secretaría de Investigación de la Universidad Nacional del Comahue, Departamento Provincial de Aguas (province of Río Negro) and Instituto de Desarrollo del Valle Inferior del Río Negro (IDEVI).

References

Bezic, C. R., A. A. Dall Armellina & O. A. Gajardo, 1996. Studies on vegetative production of *Potamogeton illinoensis* Morong in southern Argentina. Hydrobiologia 340: 7–10.

Cooke, G. D., E. B. Welch, S. A. Peterson & P. R. Newroth, 1993. Rest. Mgmt of Lakes and Reservoirs. 2nd edn: 368–399.

Dall Armellina, A. A., O. A. Gajardo, C. R. Bezic, E. C. Luna, A. Britto & V. Dall Armellina, 1996. Mechanical aquatic weed management in the lower valley of the Rio Negro, Argentina. Hydrobiologia 340: 225–228.

Dubber, F. A. A., S. Ghoneim, M. E. Siemelink, Z. El Gharably, A. H. Pieterse & J. E. Blom, 1981. Aquatic weed control in irrigation drainage canals in Egypt by mean of grass carp (*Ctenopharyngodon idella)*. Proc. 5th Int. Symp. on Biological Control of Weeds. Brisbane 1980: 261–271.

Eaton, J. W., M. A. Best, J. A. Staples & K. O'Hara, 1992. Grass carp for aquatic weed control. A users manual. National Rivers Authority, Bristol, U.K: 91 pp.

El Gharably, Z., A. A. Tolba, A. H. Pieterse & A. H. Druijff, 1978. Preliminary experiments with grass carp for control of aquatic weeds in Egypt. Proc. EWRS 5th Symp. on Aquatic Weeds 1978: 369–373.

Pieterse, A. H., 1979. Aquatic weed control in tropical and subtropical regions. Proc. of a symposium. Berichte Fagebeit Herbologie, Universitat Hokenheim 1: 130–136.

Sabbatini, M. R., 1988. Biología y Manejo de *Chara contraria* A. Braun ex Kurtz. Thesis MSc. Prod. Veg. Univ. Nac. del Sur, Bahía Blanca, Argentina: 101.

Van der Zweerde, W., 1993. Biological control of aquatic weeds by means of phytophagous fish. In Pieterse, A. H. & K. J. Murphy (eds), Aquatic Weeds. The Ecology and Management of Nuisance Aquatic Vegetation. Oxford University Press, Oxford, U.K: 201–221.

Hydrobiologia **415**: 271–276, 1999.
J. M. Caffrey, P. R. F. Barrett, M. T. Ferreira, I.S. Moreira, K. J. Murphy & P. M. Wade (eds), Biology, Ecology and Management of Aquatic Plants.
© 1999 *Kluwer Academic Publishers. Printed in the Netherlands.*

Lysathia n.sp. (Coleoptera: Chrysomelidae), a host-specific beetle for the control of the aquatic weed *Myriophyllum aquaticum* (Haloragaceae) in South Africa

C.J. Cilliers
ARC, Plant Protection Research Institute, Private Bag X134, PRETORIA 0001, South Africa
Tel: [+27] 12 3293276. Fax: [+27] 12 3293278. E-mail: rietcjc@plant2.agric.za

Key words: Myriophyllum aquaticum, Lysathia n. sp., Chrysomelidae, biological control

Abstract

The plant, *Myriophyllum aquaticum* (Vell.) Verdc. (parrot's feather), is one of five alien water weeds causing problems in South Africa. *M. aquaticum* is presently considered a minor weed problem, but it is expected that it will become more important as the more troublesome water weeds are brought under control. For this reason, research was initiated on the possibility of biological control of *M. aquaticum*. Exploratory surveys on *M. aquaticum* in South America, the continent of origin, yielded several promising insect natural enemies. A leaf-feeding beetle, *Lysathia* n. sp. (Chrysomelidae), was the first biological control candidate imported. Its biology was studied and under quarantine conditions it was found that *Lysathia* n. sp. was host-specific, feeding and reproducing only on *M. aquaticum*. Permission for release was sought and given by the National Department of Agriculture, Directorate of Plant and Quality Control. This insect was released at five sites throughout the country in December 1995. Post-release studies on this beetle have been in progress for three years. An account follows of its field performance at one of these sites: Culmpine, Vaalwater, Northern Province on the Mokolo River. The results obtained at the Culmpine site indicate that *Lysathia* n. sp. shows promise as a biological control agent. A second agent may however be required at a later stage.

Introduction

There are two species of the genus *Myriophyllum* (Family: Haloragaceae) that occur in South Africa (Southern African Plants W. 9/1980). *Myriophyllum aquaticum* (Vell.) (parrot's feather) is an emergent species, whereas *M. spicatum* L. (spiked water-milfoil, Eurasian water-milfoil) is a submerged plant with the inflorescence only emergent for a short period when flowering and cannot be confused with *M. aquaticum*. By law all *Myriophyllum* spp. are declared weeds and should be controlled (Conservation of Agricultural Resources of 1983: Act no. 43 of 1983). The only other plants in the family Haloragaceae in South Africa are two *Laurembergia* spp. and these are not listed as weeds or troublesome plants (Henderson et al., 1987; Wells et al., 1987).

M. aquaticum can be identified by its light green emergent shoots with pinnate leaves arranged in whorls. The emergent shoots are 0.2–0.5 m above the water surface and stems can be 1.5 (–3) m long. Many long roots arise from the nodes (Henderson, 1995). Dense masses of shoots and roots grow above and below the water surface to a depth of 1.5 m or deeper. Only female plants occur in South Africa and they propagate vegetatively. Small pieces which break off, or are left behind when the plant is removed from the water, can reinfest new areas. Only about 30% of the plant emerges above the water in summer. It grows throughout the year in subtropical areas, but the emergent parts are damaged or killed by frost in winter at higher elevations. The dead top growth forms a protective cover for the very short shoots arising at the nodes. Dense stands of this plant cause problems by reducing stream flow, blocking water pumps, and interfering with recreation, e.g. swimming, angling, skiing and boating. This weed not only disrupts indigenous aquatic ecosystems, but also interferes with irrigation activities. For example, in the Vaalwater area along the Mokolo River, farmers found that tobacco was tinted red when irrigation water was withdrawn from the root zone of this weed. The discolouring of the tobacco results in a 50% reduction in price (P. Vermaas, Agricultural Extension Officer, Nylstroom, 1995). The plant is also capable of increasing evapotranspiration – water loss is doubled when the water surface is covered with *M. aquaticum* (H. H. Bosman, Department of Water Affairs and Forestry, Pretoria, 1996, pers. comm.).

Present distribution of *Myriophyllum aquaticum*

The weed is found primarily in the Eastern and Western Cape regions (e.g. the Berg, Breë, Liesbeek, Eerste and Buffels Rivers), on the Vaal and Crocodile (West) Rivers in the North West Province and the Crocodile (East) River in the Mpumulanga Province. It also occurs on Hessie-se-water and the Mokolo River in the Northern Province, in KwaZulu-Natal on the Umhlatuzan and Mgeni Rivers and throughout South Africa in smaller farms and larger impoundments like Shongweni, Hazelmere, Hartebeespoort, Vaal Dam, etc. (Stirton, 1978; Van der Merwe, 1992; Henderson, 1995).

Five alien floating macrophytes have attained problem status in South Africa. These are *Eichhornia crassipes* (Mart.) Solms-Laub. (water hyacinth), *Salvinia molesta* D. S. Mitchell (salvinia), *Pistia stratiotes* L. (water lettuce), *M. aquaticum* (parrot's feather) and *Azolla filiculoides* Lam. (red water fern). All five of these plants are of South American origin and biological control programmes against these weeds have already been initiated in South Africa.

Biological control of salvinia and water lettuce was achieved with only one host-specific beetle on each weed (Cilliers, 1991a; Cilliers et al., 1996). The biological control programme on water hyacinth is in progress and that of red water fern was initiated in 1996, with the first agent released in December 1997 (Cilliers, 1991b; M. P. Hill, Plant Protection Research Institute, Pretoria, 1997, pers. comm.). It was predicted that parrot's feather would present an increasing problem as water hyacinth was brought under control. According to the Plant Protection Reseach Institute, parrot's feather was regarded as more important than water hyacinth in the 1960s (unpubl. data). At present there are no herbicides registered for use against *M. aquaticum* in South Africa (Vermeulen et al., 1996). Therefore it became essential to investigate the possibility of biological control before the problem escalates. In addition, biological control of other aquatic weeds can be adversely affected by indiscriminate use of herbicides.

For these reasons, *M. aquaticum* was surveyed for natural enemies in its country of origin, South America. Some insects were already known to be associated with *M. aquaticum*, but more were found (Habeck, & Wilkerson, 1980; Cordo, & DeLoach, 1982; S. Neser, Plant Protection Research Institute, Pretoria, 1992, pers. comm.). One of the insects found to be associated with *M. aquaticum* was the leaf-feeding beetle *Lysathia* n. sp. (Coleoptera: Chrysomelidae). This beetle was collected in the vicinity of Penedo (10°.17′S 36°36′W), Rio de Janeiro Province, Brazil (S. Neser, Plant Protection Research Institute, Pretoria, 1992 pers. comm., Cilliers, 1995, Unpublished report on file, Plant Protection Research Institute, Pretoria). *Lysathia* n. sp. is, as yet, an undescribed species. The beetle was identified by Dr D. G. Furth, (current address, Smithsonian Institute, Washington), and differs from the known *L. flavipes* (Boheman) and *L. ludoviciana* (Fall) previously found on *M. aquaticum* (E. Grobbelaar, Plant Protection Research Institute, Pretoria, 1994, pers. comm.). Voucher specimens have been deposited at the Plant Protection Research Institute, Biosystematics Division, Pretoria. The beetles used in the host specificity tests and those eventually released are thus *Lysathia* n. sp. A summary of the biology, the host range of *Lysathia* n. sp. and its effect on *M. aquaticum* are reported here.

Methods

The beetles were reared on *M. aquaticum* in a quarantine glasshouse with temperatures fluctuating between 25 and 32 °C. Humidity was not controlled as *M. aquaticum* was grown in aquaria in which the beetles could also be contained. Test plants were exposed to the beetles according to the centrifugal host specificity testing system (Wapshere, 1989). A series of economic plants and plants growing in or near the aquatic environment were also exposed to *Lysathia* n. sp. to confirm host specificity. The life history of *Lysathia* n. sp. was studied and a summary is given in this paper. Fifty adult *Lysathia* n. sp. were released on *M. aquaticum* in December 1994 and 70 in January 1995. Field establishment followed these releases. Sampling was started in January 1995 and continued at four- to six-week intervals, when possible, until January 1998 where *Lysathia* n. sp. was closely monitored to determine its effect on the plant. A square (0.1 m^2) was thrown at random in the area where the beetles were released; ten of these squares gave a sample of 1 m^2. The number of shoots per square and those showing feeding damage by *Lysathia* sp. were counted, the increasing and decreasing number of damaged shoots giving an indication of insect activity. During 1997/98 the number of different insect stages was also noted. The sample size used was the same as described by Cilliers (1987).

Results

Biology of Lysathia *n. sp. (Coleoptera:Chrysomelidae)*

At temperatures fluctuating between 25 and 32°C, the eggs hatched in 3–6 days. The four larval instars were completed in 14–18 days and the pupal stage lasted 3–6 days. The preoviposition period of the adult females was 4–6 days and a generation was completed in 24–36 days. Females deposited their eggs on the growth tip between or on the leaflets, either grouped or singly. The yellow eggs were covered in a layer of fine white waxy powder and were approximately 0.7 mm long. The newly-hatched larvae were yellowish-brown with dark brown patches and changed to dark brown to black after the first moult. Larvae measured 1–8 mm long. Under glasshouse conditions, the larvae pupated on the plants or against the wall of the rearing containers. When in the proximity of water hyacinth, both in the laboratory and the field, larvae bored into and pupated inside the petioles. This phenomenon did not occur on any other plants. The adults are metallic green, on average 3.8 mm in length. The light yellow legs are characteristic of this species, with the tibiae and tarsi dark brown to black. Both the larvae and adults fed on the shoots and leaflets of *M. aquaticum*, retarding the unlimited growth of the weed.

Host specificity testing

From Table 1 it can be seen that *Lysathia* n. sp. fed and developed only on *M. aquaticum.*

Field evaluation

At Culmpine, the mean maximum summer temperature ranges between 27 and 36 °C, and the water temperature between 22 and 27 ° C. During the winter months, the mean maximum temperature was between 22 and 25 °C and the water temperature between 11 and 14 ° C. The results presented in Figure 1 show the total number of emergent shoots and those showing feeding damage by *Lysathia* n. sp./m^2.

It is evident that *Lysathia* n. sp. populations increased rapidly in the field after the initial release, as indicated by the number of shoots damaged by *Lysathia* n. sp.. In April 1995, when 70 adults and 300 larvae were counted in the sample, all emergent shoots examined showed slight to medium *Lysathia* n. sp. feeding damage. Sampling stopped in June 1995 and was resumed in August 1995, when insect activity was still low after winter. Of the 616 emergent shoots sampled in September 1995, only 37 showed slight feeding damage. More importantly, we observed that *Lysathia* n. sp. could survive a winter in South Africa. By September, the feeding damage was mainly restricted to the leaves on the shoot tip but was heavier than before the winter. Sampling was not possible from December 1995 to March 1996 because of floods causing very high water levels and fast-flowing water. During this time, most of the emergent shoots were submerged except for one to two cm and the sampling sites could not be reached. By March 1996, when the water had subsided, sampling was resumed and it was clear that *Lysathia* n. sp. had survived the floods. There were also fewer shoots present in March 1996 than in November 1995 because of the effect of the floods. Rapid plant growth ensued, but also an increase in shoots damaged by *Lysathia* n. sp. Towards the end of summer, in April, 521 of 723 shoots had some feeding damage, with more than 55% showing medium to heavy damage. After winter in August 1996, the majority of shoots again exhibited feeding damage (908 of 958). Only slight feeding damage was seen in September 1996, that is 43 shoots damaged out of 563/m^2; this was because mainly eggs, the predominant stage at this time of year, were present. Relatively few shoots with feeding damage were counted from December 1996 to February 1997, because of high water levels following heavy rains; those shoots above the water were few and short. By April, 422 of 488 shoots showed damage. Feeding damage was now heavy compared to the previous years, as 85% of the leaves on the emergent shoots were eaten. In September 1997 heavy feeding damage was recorded on at least 30% of the 651 damaged shoots (total/m^2 = 801). By November all 424 shoots/m^2 examined were completely defoliated. Most of the shoots were still

Table 1. Plants exposed to *Lysathia* sp. in host specificity testing

Plant species	Family	Feeding damage/development
Myriophyllum aquaticum (Vell.) Verdc	Haloragaceae	Feeding and full development
M. spicatum L.	Haloragaceae	None
Laurembergia repens Berg subsp. *brachypoda*	Haloragaceae	None
Gunnera perpensa L.	Haloragaceae	None
Ludwigia stolonifera (Guill. & Perr.) Raven	Onagraceae	None
Oenothera grandiflora Ait.	Onagraceae	None
Trapa natans L. var. *pumila* Nakano ex Verdc.	Trapaceae	None
Tested for additional information:		
Polygonum acuminatum H.B. & K. var. *capense*	Polygonaceae	None
Limnophyton obtusifolium (L.) Miq.	Alismataceae	None
Alisma plantago-aquatica L.	Alismataceae	None
Nymphaea capensis Burm. f. var. *caerulea* (Sav.) Verdc.	Nymphaeceae	None
Pistia stratiotes L.	Araceae	None
Zanthedeschia aethiopica (L.) Spreng.	Araceae	None
Salvinia molesta D.S. Mitchell	Salviniaceae	None
Eichhornia crassipes (Mart.) Solms-Laub.	Pontederiaceae	None
Cyperus sexangularis Nees	Cyperaceae	None
Cyperus laevigatus L.	Cyperaceae	None
Lactuca sativa L. var. *capitata*	Asteraceae	None
Zea mays L.	Poaceae	None
Cucurbita pepo L.	Cucurbitaceae	None
Lycopersicon lycopersicum (L.)	Solanaceae	None
Solanum melogena L. var. *sativus*	Solanaceae	None
Nicotania tabacum L.	Solanaceae	None
Daucus carota L. var. *sativus*	Apiaceae	None
Beta vulgaris L. var. *cicla*	Chenopodiaceae	None
Raphanus sativus L.	Brassicaceae	None
Brassica oleracea L.	Brassicaceae	None
Pisum sativum L.	Fabaceae	None
Chlorophytum comosum (Thunb.) Jacq.	Asphodelaceae	None
Commelina L. var. *africana*	Commelinaceae	None
Berula erecta (Hudson) Cov. subsp. *thunbergii* (DC) B.L. Burtt	Apiaceae	None
Aponogeton distachyos L.	Aponogetonaceae	None

(Scientific plant names as in Arnold & de Wet, 1993; Wells et al., 1986).

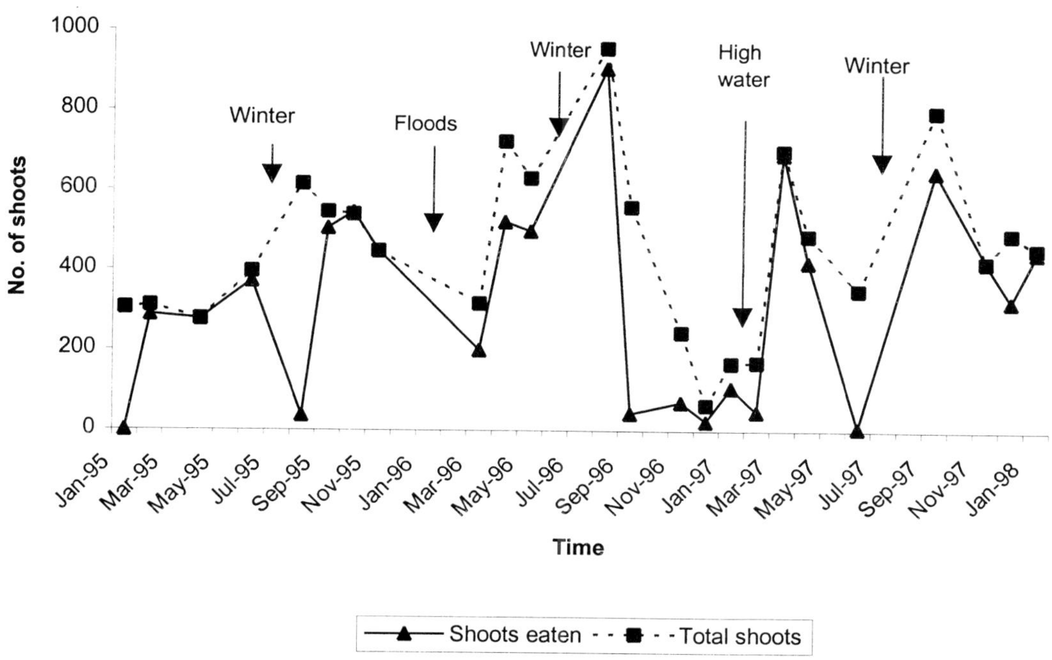

Figure 1. Culmpine. *Myriophyllum aquaticum*. Emergent shoots damaged by *Lysathia* n. sp. and total number of emergent shoots.

bare in December 1997, when insect feeding and damage to the plant was high. New growth (1–2 cm long) of *M. aquaticum* had already started. Bare shoots had sprouted one to three shoots, 1-2 cm long, at some of the nodes. It was observed that *Lysathia* n. sp. migrated to *M. aquaticum* where new growth on the bare shoots was more advanced. In January 1998, 448 of 456 shoots still showed heavy damage and had very few leaves.

The shoots were brown and denuded of leaves. Two samples were taken in February 1998; one in an area where *M. aquaticum* was not damaged by *Lysathia* n. sp., and one in an area severely damaged by *Lysathia* n. sp.. The number of shoots/m^2, undamaged by *Lysathia* n. sp., was 1281 and in a sub sample of 50 of these shoots, the mean shoot length was 19.09 cm (S. D. 5.03) and nine side shoots were counted. In contrast, the number of shoots/m^2 damaged by *Lysathia* n. sp. was 558 and, in the sub-sample of 50 shoots, the mean length was 9.98 cm (S. D. 3.97) and 96 side shoots were counted. The dry weight of the m^2 undamaged plants was 186.5g and that of the damaged plants was 62.9 g. The statistical difference was highly significant ($p = 0.00$) for all parameters measured and counted (Student's t-Test).

During the sampling period from January 1995 to January 1997, several indigenous South African natural enemies were observed attacking *Lysathia* n. sp.. In the 1996 and 1997 summer samples, two predatory bugs, *Dorycoris pavinus* Westwood and ?*Dorycoris* sp. (Hemiptera: Pentatomidae), were found attacking *Lysathia* n. sp. adults and larvae. During the period of heavy rain in 1996 and 1997, the fungus *Beauveria bassiana* (Bals.) Vuill. was also was found infecting larvae in the field. The diseased larvae adhered to the emergent plant shoots. The effect of these natural enemies on the population of *Lysathia* n. sp. was not examined in this study. A pathovar of *Xanthomonas campestris* (Pammel) Dowson 1993 was also regularly found on *M. aquaticum* in South Africa. This bacterial pathovar was studied by Ferreira (1994) and not included here.

Discussion and conclusions

The biology of *Lysathia* n. sp. was studied in the laboratory and was found to be host-specific to *M. aquaticum*. The relatively short period of 23–36 days required to complete a generation allowed rapid build-up of populations during summer. An unusual phenomenon occurred, in that the larvae of *Lysathia* n. sp. usually pupated externally on the host plant, but when in the proximity of *E. crassipes* some larvae bored into the leaf petioles and pupated inside. This did not occur on any other plant growing in the proximity of *M. aquaticum*.

Lysathia n. sp. survived winters and periods of high water and flooding in South Africa. During flooding the plant and insect population were low (December 1995 to March 1996, and again December 1996 to February 1997; Figure 1). When there was no flooding and resultant high water from September 1997 to January 1998, the damage inflicted by *Lysathia* n. sp. on *M. aquaticum* increased dramatically (Figure 1). The field studies conducted during 1995–1998 showed that *Lysathia* n.sp. retarded the growth of *M. aquaticum*. As populations increased, feeding on the plant increased and it took longer for the plants to recover. In the 1997/98 summer samples, the emergent shoots were often totally bare of leaves and some growing tips had died back. Shoots were also thinner and smaller. Plants that were heavily attacked showed new growth after three weeks. Full recovery took approximately six weeks. *Lysathia* n. sp. tended to migrate from heavily damaged plants to areas where food was more plentiful (C.J. Cilliers, unpubl. obs.). The number of shoots/m^2 increased because each shoot, when damaged, usually sprouted 1–4 new shoots. In the study area at Culmpine, the total water surface covered by *M. aquaticum* declined slowly from 50 to 20% over three years (C. J. Cilliers, unpubl. obs.). Because *Lysathia* n. sp. severely affected and retarded the growth of *M. aquaticum*, it shows promise as a biological control agent. However, the widely fluctuating plant and insect populations indicate that a second biological control agent might be required to reduce *M. aquaticum* to an acceptable level. Further field monitoring of *Lysathia* n.sp. may be desirable to ascertain the long-term effect beyond the first three years after field releases. A full description of *Lysathia* n. sp. would be desirable; however neither Dr Furth nor any other specialist taxonomist is currently working on this genus (Furth, D. G. Pers. comm., 1993).

Acknowledgements

Ms M. C. Bolton, Directorate of Plant and Quality Control, handled the submission for permission to release *Lysathia* n. sp.. All my colleagues in Weeds Division and in the Biosystematics Division, PPRI are thanked for assistance in the field, identifications and comments on this paper. The Department of Water Affairs and Forestry and the ARC, PPRI are thanked for financial support.

References

Arnold, T. H. & B. C. De Wet, (eds), 1993. Plants of southern Africa: names and distribution. Mem. Bot. Surv. S.A. No. 62. National Bot. Inst., Pretoria. 825 pp.

Cilliers, C. J., 1987. First attempt at and early results on the biological control of *Pistia stratiotes* L. in South Africa. Koedoe 30: 35–40.

Cilliers, C. J., 1991a. Biological control of water fern, *Salvinia molesta* (Salviniaceae), in South Africa . Agric. Ecosyst. Envir. 37: 219–224.

Cilliers, C. J., 1991b. Biological control of water hyacinth, *Eichhornia crassipes* (Pontederiaceae), in South Africa. Agric. Ecosyst. Envir. 37: 207–215.

Cilliers, C. J., D. Zeller & G. Strydom, 1996. Short- and long-term control of water lettuce (*Pistia stratiotes*) on seasonal water bodies and on a river system in the Kruger National Park, South Africa. Hydrobiologia 340: 173–179.

Cordo, H. A. & DeLoach, C. J. 1982. The flea beetle, *Lysathia flavipes*, that attacks *Ludwigia* (water primrose) and *Myriophyllum* (parrot feather) in Argentina. The Coleopterists Bull. 36: 298–301.

Ferreira, W., 1994. Studies on the ethiology of a bacterial wilt disease of the water weed *Myriophyllum aquaticum*. Unpub. M.Sc. Thesis. University of Stellenbosch, Stellenbosch.

Habeck, D. H. & R. Wilkerson, 1980. The life cycle of *Lysathia ludoviciana* (Fall) (Coleoptera: Chrysomelidae) on parrot feather, *Myriophyllum aquaticum* (Velloso) Verdc. The Coleopterists Bull. 34: 167–170.

Henderson, L., 1995.Plant Invaders of Southern Africa. Plant Protection Research Institute, Private Bag X134, Pretoria 0001. Handbook no. 5. 177 pp.

Henderson, M., D. M. C. Fourie, M. J. Wells & L. Henderson, 1987. Declared weeds and alien invader plants in South Africa. Dept. Agric. and Water Suppl., Dir. Info., Pretoria. Bull. 413. 167 pp.

Southern African Plants. Farming in South Africa. 1980. Water plants. Parrot's feather (*Myriophyllum aquaticum* (Vell) Verdc.) Familie Haloragaceae. Bot. Res. Inst., Pretoria. 2 pp.

Stirton, C. H. (ed.), 1978. Invader Plants. Beautiful, but dangerous. Dept. Nat. Envir. Conserv., Cape Prov. Admin., Cape Town. 175 pp.

Van der Merwe, B. C., 1992. Distribution of five aquatic weeds in South Africa. Dept. of Water Aff. For., Hydr. Res. Inst. Unp. Rep. 50 pp.

Vermeulen, J. B., M. Dreyer, H. Grobler & K. Van Zyl, 1996. A guide to the use of herbicides. Nat. Dept. Agric., Rep. S. A., Dir. Agric. Prod. Inp. 15th edn. 153 pp.

Wapshere, A. J., 1989. Testing a sequence for reducing rejection of potential biological control agents for weeds. Ann. appl. Biol. 114: 515–526.

Wells, M. J., A. A. Balsinhas, H. Joffe, V. M. Engelbrecht, G. Harding & C. H. Stirton, 1986. A catalogue of problem plants in southern Africa. Mem. Bot. Surv. S.A. Bot. Res. Inst., Pretoria. 658 pp.

Hydrobiologia **415**: 277–282, 1999.
J. M. Caffrey, P. R. F. Barrett, M. T. Ferreira, I.S. Moreira, K. J. Murphy & P. M. Wade (eds),
Biology, Ecology and Management of Aquatic Plants.
© 1999 Kluwer Academic Publishers. Printed in the Netherlands.

Fate of dichlobenil in the River Ivel. 1: Determination of the residues of dichlobenil in sediment and water

P. R. F. Barrett[1,*] & R. J. Makepeace[2]
[1]*Consultant in Aquatic Plant Management, 8 Sunderland Avenue, Oxford OX2 8DX, U.K.*
[2]*Oxford Agricultural Consultants Ltd, 3 Lower Farms Barn, Bainton Road, Bucknell, Nr Bicester, Oxon OX6 9LT, U.K.*

Key words: dichlobenil, residues, water, sediment

Abstract

A section of river was treated with the slow release formulation of dichlobenil, Casoron GSR, at a dose rate equivalent to just over 1 mg/l. Water and sediment samples were taken at various points over a 1.75 km section downstream of the treated section for up to 84 d to determine residue concentrations. Residues in the water reached a peak 3 h after application of just over 5000 ng/l and then declined but were still detectable after 84 d. Sediment residues were higher than the water residues but had declined close to the limit of detection after 84 d. The residue data derived from this experiment were used by Makepeace & Glaisher (1999) to show that lettuces, grown hydroponically and irrigated with water containing residues of dichlobenil similar to those found in the river, were damaged or killed if exposed to flowing, but not in static, water.

Introduction

Various formulations of the herbicide dichlobenil are approved in the U.K. by the Pesticides Safety Directorate for total and selective weed control on land and for use in water for the control of submerged and floating aquatic weeds (Whitehead, 1995). A review of this chemical by Pieters & de Boer (1971) suggested that its use in the aquatic environment was justified on safety grounds and that it did not affect established crops (including lettuce) when they were irrigated with water containing dichlobenil residues of 1 mg/l. For aquatic weed control, the recommended treatment concentration in water is 1.0 mg/l (1 part per million) of active ingredient and water treated with this herbicide can be used for agricultural irrigation after two weeks or when the concentration in the water has fallen below 0.3 mg/l (300 000 ng/l) (MAFF, 1995).

Dichlobenil controls a wide range of submerged and floating-leaved aquatic species as well as some emergent weeds. It was first approved for aquatic use in Britain in 1971, and has been used since then by water authorities and private owners as an effective and safe aquatic herbicide (Spencer-Jones, 1974). The preferred formulation for aquatic use (Casoron GSR) consists of a granular product containing 20% weight to weight (w/w) of the active ingredient dichlobenil (Miracle Garden Care Ltd – formerly Zeneca and ICI – Product Manuals). When applied to water, the granules sink rapidly and penetrate into soft, silty sediments. The active ingredient gradually dissolves into the water surrounding the granules but is adsorbed onto the sediment from where it is taken up by plant roots. The herbicide is translocated within the plant to the growing points where it inhibits cellulose biosynthesis leading to the death of the plants (Zeneca Product Manual, 1994). Casoron GSR can be used to treat entire water bodies or to control aquatic weeds in localised areas within a water body to reduce both the cost and the overall impact on the ecosystem resulting from the loss of plant habitats. In static water, the boundary between controlled and unaffected plants can be very precise with weeds being unaffected within 1–2 m of the site of application. In slow flowing water, some spread of the granules can occur when the current carries them downstream as they sink. In faster flowing water, the granules may be carried away from

* Author for correspondence

the intended area of treatment and, for this reason, the manufacturers recommend that the chemical should not be used in water flowing at more than 90 m/h.

Even though Casoron GSR is a slow release formulation, some dichlobenil is likely to dissolve into the water as the granules sink or lie on the sediment surface before they become buried. Other granules may be intercepted by the leaves and stems of submerged plants and held so that the active ingredient is released directly into the water. Nevertheless, extensive use of Casoron GSR has shown that, in many situations, the amount retained within the treated area is sufficient to control the target weeds and that losses of dichlobenil from the treated area are usually insufficient to affect adversely aquatic plants outside the target area.

Casoron GSR was used to control submerged weeds in the River Ivel in Bedfordshire, England in 1993 and 1994. A 525 m section of the river was heavily infested with *Hippuris vulgaris* (Mare's Tail) and, to a lesser extent, with *Callitriche* spp. (Starworts) which reduced channel capacity and created a potential flood risk. In order to minimise environmental impact, only one third of the width of the river was treated each year with the final treatment being planned for 1995. Treatment was applied to the centre section of the channel in 1993 and to the east side in 1994.

Following the second of the planned treatments in April 1994, a farmer noticed that lettuces in a glasshouse, 1.25 km downstream of the treated section of river, were showing symptoms of damage or disease. The lettuces were grown semi-hydroponically in peat blocks immersed and irrigated by water pumped from the river. Symptoms appeared first on 23 April 1994, less than 24 h after the application had been made. The lettuces died and subsequent analysis showed traces of dichlobenil present in the plant tissues and it was suggested that the herbicide treatment in the river could have been responsible for the contamination of the glasshouse.

In 1995, the weed control programme of the Internal Drainage Board called for the final application of Casoron GSR to the remaining third of the channel. This provided the opportunity to investigate the concentrations of dichlobenil which occurred in the water and sediment following treatment of the river. The purpose of this experiment was to replicate as closely as possible the treatment of Casoron GSR applied to the River Ivel in 1994, and to determine the concentrations of dichlobenil in water and sediment found at different times and distances downstream of the treated section as far as and beyond the point where water was abstracted from the river to irrigate the glasshouse.

Materials and methods

The section of the River Ivel used for this experiment was near the village of Stotfold in Bedfordshire from New Bridge (National Grid Ref: TL 2242 3635) to Stotfold Mill (TL 2228 3677). This reach of the river was impounded by a wier at Stotfold Mill which increased the depth of the water and reduced the velocity compared with upstream and downstream reaches of the river. The reduced velocity had allowed a sediment of soft silt to be deposited which had been colonised by dense growths of *H. vulgaris* and *Callitriche* spp. Much of this weed growth had been controlled in the sections treated in 1993 and 1994, but it remained dense along the west (untreated) side. Filamentous algae were also present growing amongst and entangled with the vascular plants. Various emergent weeds, mainly *Glyceria maxima* (Reed Sweet-grass) and *Sparganium erectum* (Bur-reed), were growing along the margins.

The treated length of river was 525 m and the width of the river in this section was between 10 and 13 m with a mean width of about 12 m. The depth and velocity of the river were measured in a transect approximately 40 m upstream of Stotfold Mill on 29 May 1995. The mean depth of the river at this point was found to be 0.66 m and the water velocity (measured using a Braystoke Mk2 Current Meter at 1.0 m intervals and at 2/3 water depth) was 0.0314 m/s (113.04 m/h). Using these data, the discharge of the river was calculated to be 0.261 m^3/s. Within the 4 m wide target section of the river, the mean depth was 0.7 m and the mean velocity was 0.036 m/s.

The sediment within the treated section of river was soft silt. The depth of the silt was not measured but attempts to wade at this point suggested that the silt was more than 0.5 m deep.

On 30 May 1995, a treatment was applied using the same dose of Casoron GSR (8.4 kg) which had been used in 1994. The granules were applied to a 4 m wide section of the river along the west (Stotfold) bank. This dose was equivalent to a theoretical concentration in the water of 1.14 mg/l (assuming no dilution from the flowing water and that all the active ingredient would be released immediately on application).

Application was made using motorised granule air blower (Fischer Granule Applicator) from a boat trav-

Table 1. Location of sampling sites

Sample site	Description, map reference and distance upstream (U/S) or downstream (D/S) of Sample Site A
U	40 m U/S of treated section, 565 m upstream of A (TL 2244 3633)
A	Downstream end of treated section at Stotfold Mill (TL 2227 3677)
B	Bend in river. 350 m D/S from A (TL 2222 3694)
C	Ford Bridge, Manor House Lane. 600 m D/S from A (TL 2223 3732)
D	Downstream end of playing field. 875 m D/S from A (TL 2233 3754)
E	Abstraction Point for glasshouse. 1250 m D/S from A (TL 2224 3787)
F	Taylors Mill. 1725 m D/S from A (TL 2211 3823)

elling upstream at a constant speed. Prior to use, the applicator was calibrated and quantities of Casoron GSR were weighed out into eight separate bags such that each bag contained sufficient material to treat a 60×4 m length of river. Markers were placed along the bank at 60 m intervals and a separate operator on the bank checked the speed and distance travelled by the boat to ensure accurate and uniform application in each 60 m length. The remaining 45 m at the upstream end of the plot were treated similarly. The treatments were applied between 10.10 a.m. and 11.20 a.m. on 30 May 1995.

Samples of both water and sediment were taken at various points downstream of the treated section for residue analysis. The location of the sampling points is shown in Table 1.

Because of the turbulance and mixing of the water in the wier at Stotfold Mill and in the channel downstream, it was considered that single samples of water and sediment taken from the centre of the channel at each sampling point would represent a sufficiently accurate measurement of the concentration of dichlobenil at each sampling time. In order to ensure that the water samples taken at the different points along the river in the first 3 h after application were exactly comparable, the time taken for water to flow from the treated section to all the downstream sampling points was determined using a florescein dye marker. A small quantity of florescein solution was applied to the water at Stotfold Mill and the time taken to each of the predetermined sample points was measured. Whenever the dye was found to be outside the main current of the river, an additional injection of dye was added into the fastest point of current opposite the existing marker. Thus, the time taken to reach each sampling point represented the maximum possible speed to travel of the chemical in solution. The time to each sampling point was added to the sampling time at Stotfold Mill as shown in Table 2. The time taken for water flowing through Stotfold Mill to reach the abstraction point (Point E) was 77 min.

The first samples (T1) were taken as soon as treatment started at Stotfold Mill. The second samples (T2) were taken 70 min later when treatment had been completed and all subsequent samples were timed from when the treatment had been completed. Sediment residues were collected according to the times shown in Table 3.

Water samples were collected from the centre of the river at approximately the mid-depth of the water column. New glass bottles were used for all water samples. Sediment samples were collected in wide necked plastic sample bottles which were dipped into the sediment close to the centre of the channel. The sealed bottles were transferred to Mountainheath Environmental Analysis and Consultancy, Unit 7, Pond Close, Walkern Road, Old Stevenage, Herts, SG1 3QP, U.K., who carried out the sample analyses.

Rate of sink of Casoron GSR granules

In order to determine if granules of Casoron GSR were likely to be carried over long distances downstream by the current in the River Ivel, a laboratory trial was undertaken to determine the rate at which the granules sank. The mean rate of sink was found to be 6.4 cm/s. In the treated section of the river, the mean depth was 0.7 m and the mean velocity was 0.036 m/s. Thus, the average time that the granules took to sink to the bottom was calculated to be 10.94 s and the mean distance downstream, which they would be carried in the current as they sank, was 0.39 m. Although the velocity exceeded the manufacturer's recommended upper limit (0.025 m/s), it was considered to be acceptible for the purposes of the experiment.

Weed control

Visual assessments of the effects of the treatment on weed growth were made on T7 and T8 (28 and 56 d after treatment). Visual estimates of the degree of weed control were made within the treated section and

Table 2. Water sampling regime. T1=Start of dichlobenil application (10.10 a.m.), T2=Application completed (11.20 a.m.). Sampling times (mins) shown against Sample Sites B to F at T1, T2 and T3 were added to allow for travel time of river water from Site A to downstream sampling sites

Sample time	Site						
	U	A	B	C	D	E	F
T1 1st possible time of arrival	*	0	*	+40	*	+77	*
T2 Post application	*	0	+30	+40	+55	+77	+115
T3 Application+3 h	*	0	+30	+40	+55	+77	+115
T4 Application+24 h	*	✓	✓	✓	✓	✓	*
T6 Application+7 d	✓	✓	✓	✓	✓	✓	✓
T7 Application+28 d	✓	✓	✓	✓	✓	✓	✓
T8 Application+56 d	✓	✓	✓	✓	✓	✓	✓
T9 Application+84 d	✓	✓	✓	✓	✓	✓	✓

*=Samples not taken.
✓=Samples taken working in a downstream direction over a period of approximately 1 h (travelling time to reach all sample points).

Table 3. Sediment sampling regime

Sample time	Site							
	U	A	B	C	D	E	F	
T4 Application+24 h	*	✓	✓	✓	✓	✓	✓	✓
T5 Application+48 h	*	✓	✓	✓	✓	✓	✓	*
T6 Application+7 d	*	✓	✓	✓	✓	✓	✓	✓
T7 Application+28 d	✓	✓	✓	✓	✓	✓	✓	✓
T8 Application+56 d	✓	✓	✓	✓	✓	✓	✓	*
T9 Application+84 d	✓	✓	✓	✓	✓	✓	✓	✓

*=Samples not taken.
✓=Samples taken immediately subsequent to and at the same point that water samples were taken.

on weeds growing downstream of the treated section between Stotfold Mill and Taylors Mill.

Results

Water residues

Water samples taken at T1 showed that dichlobenil had only reached the first sampling site by this time (Table 4). Subsequent samples showed that a peak concentration was reached 3 h after application by which time the highest residue levels were at sampling sites C and D. After this, the residue levels declined although there were considerable but unexplained fluctuations in individual values. After 84 d, dichlobenil residues were still detectable in the water. The mean concentrations at each sample site over the 84 d sampling period showed that the highest levels were recorded at sample sites C and D and not at points A and B closest to the treated section which might have been expected.

Sediment residues

Concentrations of dichlobenil were expectedly higher in the sediment (Table 5) than in comparable water samples as dichlobenil accumulates preferentially in sediments. Unfortunately, sediment samples were not taken at the site upstream of the treated section (Sample site U) until T7 (28 d after treatment) but the presence of dichlobenil residues in this and the T8 sample suggests that dichlobenil was entering the river from other sources upstream of the treatment. The sediment residue concentration at Sample site U at T8 (56 d) was similar to three of the residue levels found downstream of the treated section.

Weed control

On both of the assessment dates (T7 and T8), weed growth had been reduced in the treated 4 m wide section of the river. The main target weed *H. vulgaris* had been reduced by between 50 and 60%. However, *Callitriche* spp., the other submerged weed in the treated section of channel, had been controlled by only 20%. No effect on the emergent weeds growing along the banks was observed.

Downstream of the treated section, *H. vulgaris* was absent at the time of treatment and the dominant submerged weed species were *Callitriche* spp. and *Ranunculus penecillatus*. Various species of emergent weed, including *Sparganium erectum, Glyceria maxima* and *Carex* spp., grew along the margins. No control of any of these weeds was observed between

Table 4. Residues of dichlobenil found in water samples (ng/l)

Sample time	Site							Mean[1] (Sites A–E)
	U	A	B	C	D	E	F	
T1 First possible time of arrival	*	0.3	*	<0.1	*	<0.1	*	*
T2 Immediately after application	*	90	28	100	12	5.9	*	47.2
T3 Application+3 h	*	480	3150	4860	5060	2170	*	31440.0
T4 Application+24 h	*	790	23	1610	920	320	250	732.6
T5 Application+48 h	120	230	650	120	1140	270	*	482.0
T6 Application+7 d	0.1	34	380	65	50	615	870	228.8
T7 Application+28 d	0.2	1110	960	520	290	440	930	664.0
T8 Application+56 d	<0.1	299	278	326	276	358	264	307.4
T9 Application+84 d	<0.1	40	31	25	5	17	23	23.5
Mean concentrations at each sampling site[2] (excluding T1)		384.1	687.5	953.3	969.1	587.4		

*=Samples not taken.
[1]The mean concentration at each sampling time (excluding T1) between the treated section of river and the abstraction site for the glasshouse.
[2]The mean concentration at each sampling site over 84 d after application (excluding point U).

Table 5. Residues of dichlobenil found in sediment samples (ng/l)

Sample time	Site							Mean
	U	A	B	C	D	E	F	
T4 (24 h)	*	36 000	300	4100	5800	8400	6600	10 200
T5 (48 h)	*	5300	27 000	3200	3300	6500	*	9100
T6 (7 d)	*	7900	15 000	16 000	7300	12 000	4400	10 400
T7 (28 d)	300	900	1500	300	800	2100	2500	1350
T8 (56 d)	1600	1600	24 000	1600	1600	11 000	*	6900
T9 (84 d)	<1000	<1000	3000	<1000	<1000	<1000	<1000	<1300
Mean concentrations at each sampling site		<8800	11 800	<4400	<3300	<6800		

*Sample not taken.

Stotfold Mill and Taylors Mill. On 27 June, a section of the river at Astwick Mill, approximately 2 km downstream of the treated section, was inspected. This area contained a dense growth of *H. vulgaris* which appeared to be healthy and was showing no symptoms of being affected by any dichlobenil residues which might have reached that point.

Discussion

All the submerged weed species (except filamentous algae) within and downstream of the treated section of the River Ivel are susceptible to Casoron GSR (Zeneca Product Manual, 1994). However, the only area where weed control was observed was within the treated section of the river between New Bridge and Stotfold Mill. This confirmed the observations made by Terry et al. (1981) that, when dichlobenil was applied to localised areas of a water body, much of the herbicide remained within the treated area and only those plants growing within the treated area were controlled.

Even within the treated area, the level of control was only moderate. A higher level of control had been achieved in the two previous treatments in this river and elsewhere. A number of factors, such as water temperature and velocity, may have contributed to the lower than predicted level of control but two factors were identified as likely to of major importance in reducing the efficiency of this treatment. First, due to various delays, treatment was not applied until 30 May which is more than a month later than the two previous treatments in this river and approximately one month later than the recommended time of application which is in early spring when active growth has started but before plants reach the surface. After this, growth slows down and susceptibility decreases. By 30 May, much of the *H. vulgaris* and almost all of the *Callitriche* spp. had reached the surface and were in the stage of growth when the level of susceptibility is declining (Zeneca Product Manual, 1994). Second, because of the delayed treatment, the plants had produced dense vegetative growth which would have intercepted granules, preventing some of them sinking down to the sediment and being absorbed by the roots. Filamentous algae had also become entrained with the other vascular species and would have contributed further to the interception of granules. Therefore, because of the delay in applying the treatment in 1995, it is likely that more interception of granules would have occurred than in previous treatments and this could have resulted in a reduced level of weed control and in greater downstream losses of dichlobenil than had occurred in 1993 or 1994. Nevertheless, the observation that at least 50% of the *H. vulgaris* within the treated section was controlled indicates that sufficient dichlobenil was retained in contact with the plant roots to produce an effect.

The herbicide residue data showed a predictable pulse of dichlobenil moving downstream in the water soon after application. This pulse was probably released from the granules as they sank onto the sediment and from those granules retained by submerged plants. It reached a peak at T3 (3 h after application) with a mean value over all sampling sites of 3144 ng/l (0.00314 mg/l). This low concentration of dichlobenil represented less than 3% of the applied dose and demonstrated that the formulation had the slow release properties claimed by the manufacturer. Dichlobenil was still being released after 84 d when the concentration had declined to a mean level of 34 ng/l.

Sediment residues downstream of Stotfold Mill remained fairly constant at around 10 000 ng/kg for 7 d after treatment and then declined to a point close to the limits of detection after 84 d.

There is some doubt as to whether all the dichlobenil found downstream of the treated section originated from the known treatment applied to the river. Dichlobenil residues were found in water and sediment samples taken from upstream of the treated section. It would have been impossible for dichlobenil to move upstream to this sampling site and the presence of these residues indicates that dichlobenil was entering the river system from another source. Dichlobenil is a relatively commonly used herbicide on uncropped land and in water and the presence of dichlobenil in these samples suggests that contamination was occurring.

The residue data confirm that a low level of dichlobenil was present in the water flowing past the abstraction point (Site E) for at least 84 d after treatment. Because the treatment applied in 1995 repeated, as closely as possible, the previous treatments, it is likely that dichlobenil would also have been present in 1994 when lettuces were irrigated with water taken from the river. The concentrations of dichlobenil found in 1995 were well below the safe irrigation limit set by the Pesticide Safety Directorate and would have been unlikely to affect irrigated crops grown conventionally in soil. However, the lettuces in the glasshouse were grown semi-hydroponically and further trials were carried out by Makepeace & Glaisher (1999) to determine if these concentrations were high enough to affect lettuces. These trials showed that the test concentrations of dichlobenil had no effect on lettuce plants in static water but, when a dynamic system was used in which water flowed around the lettuce plants, the same concentrations caused damage or death to the lettuces. However, in these trials, visable symptoms did not appear until after several days of exposure so that the initial damage observed by the farmer less than 24 h after the treatment had been applied to the River Ivel could not have been caused by the dichlobenil applied to the river.

References

MAFF, 1995. Guidelines for the use of herbicides on weeds in or near watercourses and lakes. MAFF Publications. London SE99 7TP. 26–27.

Makepeace, R. J. & J. A. Glaisher, 1999. Fate of dichlobenil in the River Ivel. 2: Effects of dichlobenil residues on hydroponically grown lettuce. Hydrobiologia 415 (Dev. Hydrobiol. 147): 283–287.

Pieters, A. J. & F. G. de Boer, 1971. Dichlobenil in the aquatic environment. Proceedings EWRS 3rd international symposium on aquatic weeds. Oxford: 183–191.

Spencer-Jones, D. H., 1974. Some advances in the development of dichlobenil granules in the U.K. for the control of aquatic weeds. Proc. EWRS 4th international symp. on aquatic weeds. Vienna: 194–201.

Terry, P. J., T. O. Robson & S. Hanley, 1981. Localised control of aquatic weeds with dichlobenil. Proceedings of conference – Aquatic Weeds and their Control. Ass. Appl. Biologists: 165–176.

Whitehead, R., 1995. The U.K. Pesticide Guide. CAB International and BCPC. ISBN 0 85198 953 5: 295–296.

Zeneca Product Manual, 1994. What is Casoron G? Zeneca Product Manual. 6 pp.

Hydrobiologia **415:** 283–287, 1999.
J. M. Caffrey, P. R. F. Barrett, M. T. Ferreira, I.S. Moreira, K. J. Murphy & P. M. Wade (eds),
Biology, Ecology and Management of Aquatic Plants.
© 1999 *Kluwer Academic Publishers. Printed in the Netherlands.*

Fate of dichlobenil in the River Ivel. 2: Effects of dichlobenil residues on hydroponically grown lettuce

R. J. Makepeace* & J. A. Glaisher
Oxford Agricultural Consultants Ltd., 3 Lower Farm Barns, Bainton Road, Bucknell, Oxon, OX6 9LT, U.K.

Key words: phytotoxicity, flowrate, dynamic, freshweight

Abstract

Following claims that dichlobenil in river water had caused damage to lettuce (*Lactuca sativa*) in 1995, a series of trials were carried out to assess the effects of dichlobenil on lettuce grown using hydroponic techniques under glass. Plants grown in three inch pots of soil-based compost and watered with dichlobenil solution at concentrations up to 300 000 ng/l showed no differences in growth compared to the untreated controls. Plants grown in peat-based compost in a static nutrient solution showed a phytotoxic reaction when dichlobenil concentrations reached 100 000 ng/l or greater. Trials on plants grown using a dynamic nutrient film technique indicated that under glasshouse conditions, dichlobenil solutions as low as 10 ng/l could adversley affect the development of lettuce plants.

Introduction

In 1993, a programme of weed control was initiated to clear vegetation choking a section of slow moving river in southern England. 'Casoron GSR' granules containing 20% by weight of dichlobenil as the active ingredient (a.i.) was chosen as the most suitable product. When applied in water, it rapidly sinks to the river or lake bed. Plant roots then take up the dichlobenil from the sediment. To prevent deoxygenation of the river, it was decided to treat one third of the river each year over three years.

Some 1500 m down-stream from the treated area was an extraction point used by a grower to remove water from the river to irrigate crops. He was growing lettuces under glass using hydroponic techniques. Twenty-four hours after the second dichlobenil treatment up-stream, the farmer started to notice that his lettuces were showing signs of damage or disease. Further investigations found low levels of dichlobenil in his lettuce crop.

In 1995, studies were set up to determine the fate and breakdown of dichlobenil in the River Ivel (Barrett & Makepeace, 1998) and to investigate the effect of dichlobenil on lettuce grown under glass in order to establish the relationship between low levels of dichlobenil in river water and damage to lettuce plants.

* Author for correspondence

Materials and methods

Two varieties of lettuce, Salad Bowl and Lollo Bionda, were grown in cells containing peat based compost. When they reached an appropriate size, they were transferred to the trial area.

The first trial was designed as a range finding exercise. The lettuce were grown in plastic trays containing a nutrient solution with a range of dichlobenil concentrations. Each treatment had three trays (replicates) each containing five plants. The trays were covered with polythene sheeting, black underneath and white on top, to reduce evaporation. The plants were grown through 'x' shaped cuts made in the polythene.

Dichlobenil solutions were prepared by serial dilution from a 0.1 mg/ml standard of dichlobenil in methanol as follows: 1 000 000, 100 000, 10 000, 1000, 100 and 10 ng/l.

To these solutions, 1.0 ml/l of seaweed extract (Maxicrop Concentrate) and 0.5 ml/l (Omex) (containing 30% nitrogen as ammonium sulphate) were added. A control solution was prepared using the same nutrient concentration but no dichlobenil. Where low growth rates were observed, additional nutrients were

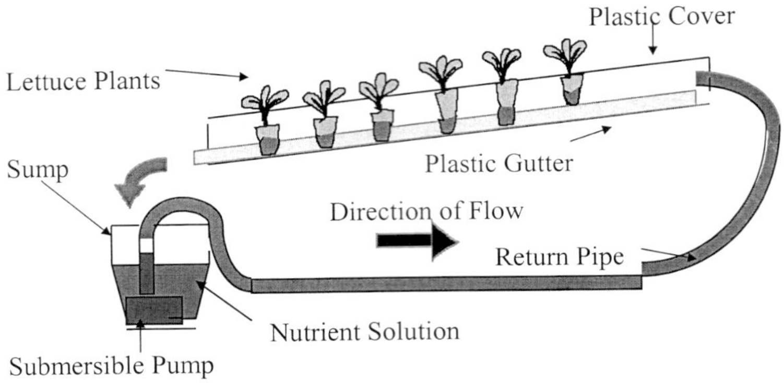

Figure 1.

added equally to all treatments at approximately 50% of the original concentration.

In the second trial, the plant plugs were grown in three inch pots of soil based compost and watered with dichlobenil solutions at rates of 10, 300, 3000 and 300 000 ng/l. The object of this trial was to assess the effects that a soil medium would have on lettuce grown conventionally in soil-based compost.

The third trial was designed to establish a hydroponic growing system based on similar methods to those used by the lettuce grower. Two identical hydroponic systems were constructed from 4 m lengths of plastic guttering covered with polythene, black underneath and white on top. In each gutter, 15 lettuce plant plugs were placed through 'X' shaped cuts in the polythene, with the shoots held above. The gutters were held at an incline of approximately 1:100. This was the incline used by the grower in his glasshouse unit. Nutrient solutions were pumped from sumps at the lower ends to the tops at a rate of approximately 4 l per minute, via plastic hoses and submersible pond pumps. The solution then flowed down the gutters, across the lettuce roots and back into the sumps, (Figure 1).

One hydroponic system was used as a control and contained nutrients at the same rates as the previous trial. The other system contained the same nutrient solution, but with dichlobenil added at 10.0 ng/l, prepared by dilution from a 0.1 mg/ml standard of dichlobenil in methanol.

The fourth trial used four gutters set up identically to the third trial, enabling three different concentrations (300, 3000 and 300 000 ng/l) of dichlobenil and a control to be run concurrently.

Table 1. Trial 1: Freshweight of Lettuce, (g) after 18 days in Static Dichlobenil Solution

Treatment No.	Replicate No. I	II	III	Mean of reps	Mean without[a]
1. Control	26.39	27.00	22.60	25.33	25.33
2. 0.01 ppb*	12.50[a]	29.70	22.70	21.63	25.67
3. 0.1 ppb*	20.19[a]	22.35	11.24[a]	17.93	22.34
4. 1.0 ppb*	24.50	14.76[a]	19.06	19.44	21.78
5. 10.0 ppb*	23.20	23.85	24.53	23.86	23.86
6. 100 ppb*	16.11	17.02	14.50	15.88	15.88
7. 1000 ppb*	6.55	6.06	4.70	5.77	5.77

Replicates I & II Salad Bowl, Replicate III Lollo Bionda.
[a] Indicates replicates in which one or more plants have died and become desiccated and therefore are not characteristic of the freshweights for that rep.
* 1.0 ppb/L=1000 ng/L.

The fifth trial was a refinement of the fourth. All conditions were identical with the exception of the incline of the gutter. This was increased to give an incline of 1:20, to increase the flow rate and potential uptake.

All trials were visually assessed on a regular basis using a vigour scale of 0–9, where 0 was dead and 9 was the same as the controls. The time, date and any symptoms/comments were recorded during each assessment. The end point of each trial was chosen prior to mortality of the treated plants, to prevent desiccation and ensure a representative freshweight. Trials ran between 18 and 31 days. At the end of each trial, the freshweight of the aerial parts of the plants were recorded.

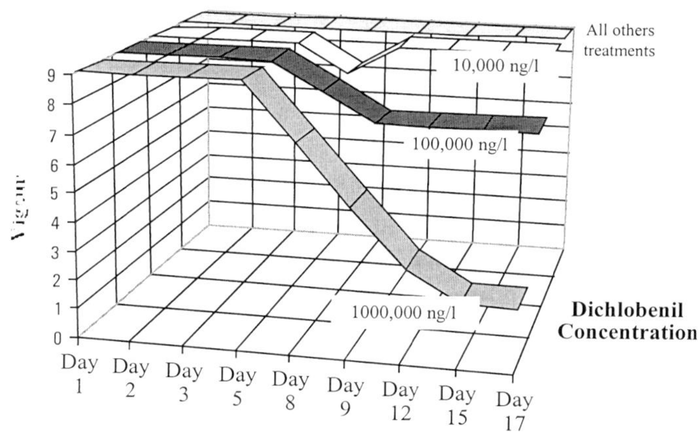

Figure 2.

Table 2. Trial 2: Mean Freshweight (g) of Lettuce exposed to different concentrations of dichlobenil (ng/l). Data are means±Standard Deviation (n=18)

Treatment	Freshweight (g)
Control	9.06±1.98
10.0	9.11±2.07
300	8.34±2.23
3000	8.94±2.40
300 000	8.89±2.44

Results

In the first trial, the lettuce plugs were placed in trays containing a range of dichlobenil solutions. No differences were apparent between the control and treated plants until the 7th day of the trial. At that stage, plants in 1 000 000 ng/l of dichlobenil started to die without showing any signs of growth. All other treatments grew, although those in 100 000 ng/l were noticeably stunted. The vigour of the lettuce plants was recorded (Figure 2) and the fresh weights recorded at the end of the trial (Table 1).

In the second trial, the lettuce plugs were planted in 3 inch pots of soil-based compost and watered with water or solutions of dichlobenil at 10, 300, 3000 or 300 000 ng/l. Throughout the 31 days over which the trial ran, no visible differences were noted between any of the treatments. There were no differences in freshweights, as shown in Table 2.

In trial three, the effects of using a flowing nutrient medium on the growth of lettuce were assessed.

Table 3. Trial 3: Freshweight of Lettuces after 24 days exposure to 10 ng/l. Data are means±Standard Deviation (n=15)

Treatment	Freshweight (g)
Control	5.54±S.D.
10.0	4.91±S.D.

Table 4. Trial 4: Freshweight of Lettuces after 25 days exposure to different concentrations of dichlobenil (ng/l). Data are means±Standard Deviation (n=15)

Treatment	Freshweight (g)
Control	3.68±S.D.
300 ng/l	2.18±S.D.
3000 ng/l	1.55±S.D.

Two 1:100 inclined re-circulating systems were used. In one system, nutrient solution alone was used and the other system a nutrient solution containing 10 ng/l of dichlobenil was used. After 24 days, the fresh weight of the aerial parts of the lettuce plants was recorded. The results showed a 10% reduction in fresh weight of lettuces to 10 ng/l of dichlobenil in the nutrient medium. The results are given in Table 3.

In trial four, two extra sets of equipment were set up and rates of 300, 3000 and 300 000 ng/l of dichlobenil were tested. In this trial, differences were noted between the different concentrations at 5 days after the start of the trial. By the 13th day, all plants in 300 000 ng/l concentration had died, and those in

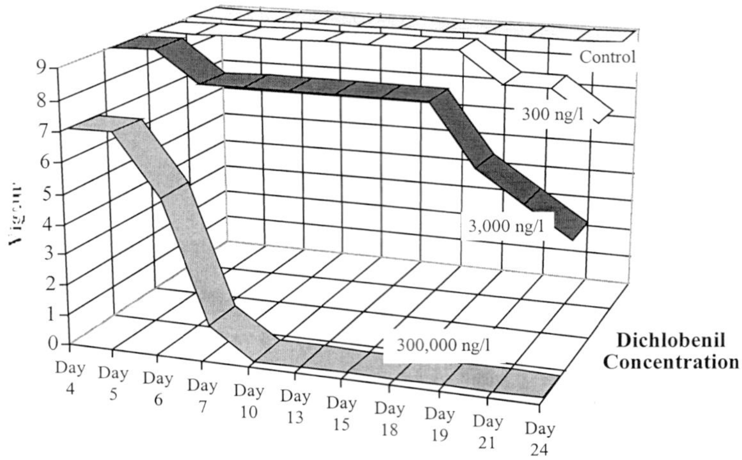

Figure 3.

Table 5. Trial 5: Freshweight (g) of two cultivars of lettuce after exposure to 10 ng/l dichlobenil for 22 days. Data are means±Standard Deviation

Treatment	Freshweight (g)
Lollo Bionda control n=8	4.58±S.D.
Lollo Bionda 10 ng/l	2.04±S.D.
Salad Bowl control n=7	2.81±S.D.
Salad Bowl 10 ng/l	1.44±S.D.

3000 ng/l concentration were visibly stunted. There was no visible difference between plants growing in 300 ng/l concentration and those in the untreated control. Four days of very hot weather starting on the 18th day caused the plants in the 300 ng/l solution to wilt more readily than the controls. At the end of the trial, examination of these plants showed that the roots were less numerous and less branched than the controls. The vigour of the lettuce plants throughout the trial is shown in Figure 3 and the freshweight results are shown in Table 4.

The fifth trial was set up to enable the test to flow more quickly across the lettuce roots in the plastic channels and increase uptake potential. This was achieved by increasing the incline to 1:20. Only one concentration of 10 ng/l was tested in this trial. Effects on the lettuce became noticeable after 10 days. The fresh weight of the aerial parts of the lettuce plants was recorded after 21 days. The results are given in Table 5. It was noticeable that hot weather in the last seven days of the trials caused treated plants to wilt compared to the controls.

Discussion

The use of dichlobenil granules in aquatic situations is well established for the control of certain aquatic weeds (MAFF, 1995). There are clear guidelines on the dose to be applied, the rate of flow of water to be treated and related aspects of treatment, such as local treatment and abstraction for irrigation. In the United Kingdom, treated water cannot be used for irrigation within two weeks of treatment or until the concentration of dichlobenil falls below 0.3 ppm (3000 ng/l). There is also a statutory requirement to get permission to abstract for irrigation from the local river authority.

In most cases of irrigation, the river water is used on field crops grown without any type of protection. The work described in this paper arose as a result of a grower abstracting water from the river adjacent to his glasshouse unit in which he was growing lettuce in hydroponic solution. The floor of the glasshouses were concreted with 3 cm deep channels moulded into the concrete. The floor had an incline of approximately 1:100 along the channels. Nutrient solution was made up from the river water, pumped to the higher end of the concrete floor and run down the channels by gravity to a collection point at the lower end of the floor sections. From this point, the nutrient is re-circulated after being checked for concentration of nutrient. Lettuce plants in peat blocks were stood in the channels and left to grow until ready for marketing.

In the case in question, dichlobenil as 'Casoron G-SR' had been correctly applied in both 1993 and 1994 to a section of the River Ivel, Bedfordshire, England, 1500 m upstream of the glasshouse unit. Damage occurred to the lettuce in 1994. As a result of independent investigations, traces of dichlobenil were found in lettuce in the glasshouse and was considered to be the cause of the damage observed. As there were no data on the relationship of dose of dichlobenil and phytotoxicity to lettuce, the studies described in this paper were instigated.

The results show that there is a dose related response to dichlobenil in hydroponically grown lettuce but not in lettuce grown conventionally in a soil-based compost. In the first trial, a static test with lettuce grown in trays of nutrient solution, lettuce showed symptoms of phytotoxicity after three days. Concentrations of 100 000 ng/l of dichlobenil and above caused phytotoxicity. Little damage occurred at 10 000 ng/l, but complete death occurred at 1 000 000 ng/l. In contrast to these results, virtually no effects were seen from watering lettuce plants grown in soil-based compost. The plants in this test were watered with dichlobenil solutions ranging from 10 ng/l to 300 000 ng/l. After 31 days of watering with dichlobenil solutions, there was no reduction in fresh weight from any concentration of dichlobenil.

Greater damage to the lettuce plants occurred in the dynamic tests increasing with increase in slope from 1:100 to 1:20. Using the same slope as in the grower's glasshouse, namely 1:100, the amount of damage caused by an 10 ng/l of dichlobenil solution was seen as an 11% reduction in fresh weight. Further tests showed that both vigour and fresh weight was reduced by a 300 ng/l solution. Serious damage was caused by 3000 ng/l and complete death by 300 000 ng/l. In the final trial, the slope of the plastic gutters was increased to 1:20. Only a concentration of 10 ng/l was tested. At this slope, the reduction in fresh weight increased markedly. Compared to the 11% reduction in fresh weight caused in the 1:100 slope the fresh weight of lettuce cultivar Salad Bowl at a slope of 1:20 was reduced by 51% and the fresh weight of lettuce cultivar Lolla Bionda by 44.5%. These results could be interpreted to show that exposure to dichlobenil is increased with increased flow of even a 10 ng/l concentration of dichlobenil.

These results show that lettuce grown in the type of hydroponic system described here are extremely sensitive to dichlobenil at concentrations as low as 10 ng/l and that the sensitivity increases with both the dose and the speed of circulation of the nutrient medium.

References

Barrett, P. R. F. & R. J. Makepeace, 1999. Fate of dichlobenil in the river Ivel. 1: Determination of the residues of dichlobenil in sediment and water. Hydrobiologia 415 (Dev. Hydrobiol. 147): 277–282.

MAFF, 1995. Guidelines for the use of herbicides on weeds in or near watercourses and lakes. MAFF Publications, London SE99 7TP: 26–27.

Hydrobiologia **415:** 289–293, 1999.
J. M. Caffrey, P. R. F. Barrett, M. T. Ferreira, I.S. Moreira, K. J. Murphy & P. M. Wade (eds),
Biology, Ecology and Management of Aquatic Plants.
© 1999 *Kluwer Academic Publishers. Printed in the Netherlands.*

Resistance of *Alisma plantago aquatica* to sulfonylurea herbicides in Portuguese rice fields

Isabel M. Calha*, Carlos Machado & Fátima Rocha
Direcção-Geral de Protecção das Culturas, Quinta do Marquês, 2780 Oeiras, Portugal

Key words: Alisma plantago-aquatica L., bensulfuron-methyl, sulfonylurea, resistance

Abstract

The first known case of herbicide resistance found in Portugal occurred in the aquatic environment: populations of *Alisma plantago-aquatica* have shown resistance to bensulfuron-methyl. A large survey was carried out in 1996, covering 81% of the river valleys of Sado and Caia, 41% of Tagus and Sorraia and 26% of Mondego. In this survey, 164 fields were selected for sampling (77, 55 and 32 for each region). The present paper presents the first results from single dose bioassays for resistance confirmation, using 98 samples (55 soil samples and 43 seed samples). Thirty six new resistant populations were confirmed. The Tagus and Sorraia valleys were the most affected areas with 23.8% of the total fields having *Alisma plantago-aquatica* populations resistant to bensulfuron-methyl, followed by the Sado and Caia, and Mondego valleys with 7.3% and 1.2%, respectively. A new set of bioassays with the remaining samples is in progress.

Introduction

Rice paddy fields in Portugal cover an area of about 30 000 ha (INE,1993), mainly in the Tagus and Sorraia valleys (10 000 ha), Mondego (8000 ha) and Sado and Caia valleys (7000 ha). The crop is treated annually with herbicides but, in spite of a high selection pressure, there was no occurrence of herbicide resistance until sulfonylurea herbicides started to be applied in Portugal in 1989. *Alisma* spp. are common weeds in rice culture and could become problematic weeds as herbicide resistance is a potential problem in this crop. In 1995, the first population of *Alisma plantago-aquatica* L. (Water-plantain) resistant (R) to bensulfuron-methyl was confirmed in a rice paddy field at Coruche (Sorraia valley), which had been treated with the herbicide for six years (Calha et al., 1995a). Bioassays confirmed a high level of resistance: plants survived at rates of 1000 g ha^{-1}; they also showed cross-resistance to other sulfonylureas. Two other resistant populations were identified in the south of Portugal, one from the Sado valley and the other from the Caia valley (Prates, 1996). The R population from the Sado valley survived at rates 200 times the susceptible one (Ribeiro,1997). As the extent of the resistance problem was unknown at that time, a survey in rice fields was started in 1997, comprising a questionnaire among rice growers and bioassays to confirm the occurrence of resistance.

* Author for correspondence

Material and Methods

Survey

A survey questionnaire was circulated through the networks of official regional services. Rice growers were selected at random, until 20% of the rice producing area was covered. A simple questionnaire checked the potential occurrence of resistance in the field. It was possible to find the herbicide history for each field and the more troublesome weeds. According to Gressel (1991), resistance to sulfonylurea herbicides can arise after three years of continuous spraying. Therefore, fields where bensulfuron-methyl was applied for two years or less were discarded and the remaining were classified into classes of selection pressure, according to their herbicide history: Low – sulfonylurea applied for three years only; Medium – sulfonylurea applied for more than three years but alternating with other herbicides and High – sulfonylurea applied for more than three consecutive years. Dividing the fields into selection pressure classes had two purposes: (a) collecting sample material (seed and soil) for the evaluation of resistance to bensulfuron-methyl in *Alisma plantago aquatica* in those fields where a high herbicide pressure was evident, and not in all the 335 fields of the survey, and (b) establishing risk areas which could be useful for predicting/identifying townships where the problem was more likely to occur.

Bioassays to confirm resistance

Sampling covered 112 out of 164 rice fields where sulfonylurea herbicides had been applied for three or more years. In 43 fields, seed samples of *A. plantago-aquatica* were collected from mature inflorescences of 40 plants, before rice harvest. In the remaining 69 fields, soil samples were collected: 25 subsamples, in each field, from the nodes of a net 2.5×2.5 m covering an area of 100 m^2. After a chilling treatment to break dormancy (one month at 2 °C in the dark), seeds were sown in shallow trays (20×26×5cm) filled with soil and continuously irrigated with a *drip-feed* system, to simulate the conditions in rice paddy fields. Seedlings were thinned to five plants per tray. At the 2–6 true-leaf stage, plants were sprayed with 300 g ha^{-1} bensulfuron-methyl herbicide solution (five fold rate). Irrigation was stopped on the day before spraying and trays were kept without water during 3–4 days. The experimental design was a randomised complete block with four replications, repeated twice. For soil samples, the procedure was identical, apart from the chilling treatment. For this bioassay, different species emerged from the soil seed bank, but were thinned at leaf-stages when it was possible to distinguish them from *Alisma* seedlings. The efficacy of the herbicide treatment was assessed weekly, following the principles adapted from Letouzé et al. (1997): 1. five fold rate of herbicide application to distinguish between susceptible (dead plant) and resistant individuals (living plant); 2. the percentage survival of emerged seedlings, assessed 31 days after treatment, represents the rate of resistant individuals (frequency of resistance) within each population; and 3. resistance is visible in the field, when the frequency of resistance is over 30% of the population (Gressel & Segel, 1982). According to this, a field has a Susceptible (S) population when the frequency of resistance was below 30%; above 30%, the field has a Resistant (R) population. Analysis of variance was used.

Townships at low, medium and high risk of resistance were identified, according to the number of rice fields belonging to the highest class of selection pressure. A low risk township was defined as one where less than 30% of the fields within the township were treated continuously for more than three years with sulfonylurea herbicides. Medium and high risk townships were defined as those where sulfonylurea herbicides were used on 30–50% and over 50% of the fields, respectively (Morrison & Bourgeois, 1995). Finally, the intention was to establish a correlation between the number of fields with the highest selection pressure (potencial risk of resistance) and the actual occurrence of resistance on those fields for each valley.

Results and discussion

Survey

The questionnaire survey covered an area of 11 828 ha (45% of the total rice area). From the 335 rice fields surveyed, 164 had been subjected to three or more years of bensulfuron-methyl application. An overview of the results is presented in Table 1. Sulfonylurea herbicides are used most in the Tagus and Sorraia areas, followed by Sado and Caia and least in the Mondego valley.

Usually, *Echinochloa* spp. are considered to be the most troublesome weeds in rice crop (Vasconcelos et al., 1998) and *Alisma plantago-aquatica* less significant as it is one of the most susceptible species to most herbicides, including bensulfuron-methyl. However, according to the results of the survey, *Alisma plantago-aquatica* causes more problems to rice farmers than grass weeds (*Echinochloa* spp.) and Cyperaceae (*Cyperus* spp., *Scirpus* spp.). This change in the farmersþ perceptions may indicate a lack of efficacy of bensulfuron-methyl and that, seven years after the introduction of the herbicide in Portugal, most of the growers already realize herbicide resistance in the field. In fact, *Alisma plantago-aquatica* is considered to be a more important weed than *Echinochloa* spp. in all regions except in the Mondego valley, the region where bensulfuron-methyl resistance is less relevant (Table 2).

Table 1. Survey results for each valley

Valley	Rice crop -area-	Questionnaire survey -area (n^{o} of fields)-	Bensulfuron-methyl spraying for 3 or more years -area (n^{o} of fields)-
Sado and Caia	7000 ha	5790 ha (176)	2287 ha (75)
Mondego	8000 ha	1927 ha (78)	942 ha (32)
Tagus and Sorraia	10 000 ha	4110 ha (80)	2149 ha (57)
Total	25 000 ha	11 828 ha (335)	5378 ha (164)

Bioassays to confirm resistance

From ninety eight samples (43 seed and 55 soil samples) studied, up to now it was possible to confirm 36 fields with R populations. Bioassays results follow the same pattern as the survey results: the Tagus and Sorraia valleys had more fields with R population (19) than the other areas (13 from the valleys of the Sado and Caia and 4 from the Mondego). Bensulfuron-methyl resistance of *Alisma plantago-aquatica* affects an area of 1764 ha, corresponding to 840.5 ha, 600.5 and 323 ha distributed among the Tagus and Sorraia, Sado and Caia, and Mondego valleys, respectively.

One of the aims of this study was to evaluate the possibility of using either seed or soil samples for confirmation of resistance in greenhouse bioassays. A comparison of results pointed out that the average frequency of resistance was always lower in seed bank bioassays (soil samples) than from seed bioassays. From seed bank bioassays, the average frequency of resistance ranges from 50%, 66.6% and 68.2% for the Mondego, Tagus and Sorraia and Sado, and Caia valleys, respectively; seed bioassays showed higher average frequency of resistance, at 90%, 98.1% and 95% for the same regions. As Van Oorschot & Straathof (1988) argue, the results of such bioassays probably underestimate the actual infestation with resistant biotypes and could only detect heavily infested fields. However, in some conditions 'seed bank samples' could be useful, as *Alisma plantago-aquatica* are aquatic weeds with seed dormancy and a slow emergence rate, because: (a) seeds in natural conditions in soil may germinate faster than seeds collected from plant inflorescence, kept for some time in dry storage conditions and then allowed to germinate in wet conditions; (b) 'seed bank samples' do not need time consuming treatment for breaking dormancy and (c) they are always available on the top soil layer even when there are no plants in the field; this could be seen as advantageous over seed samples.

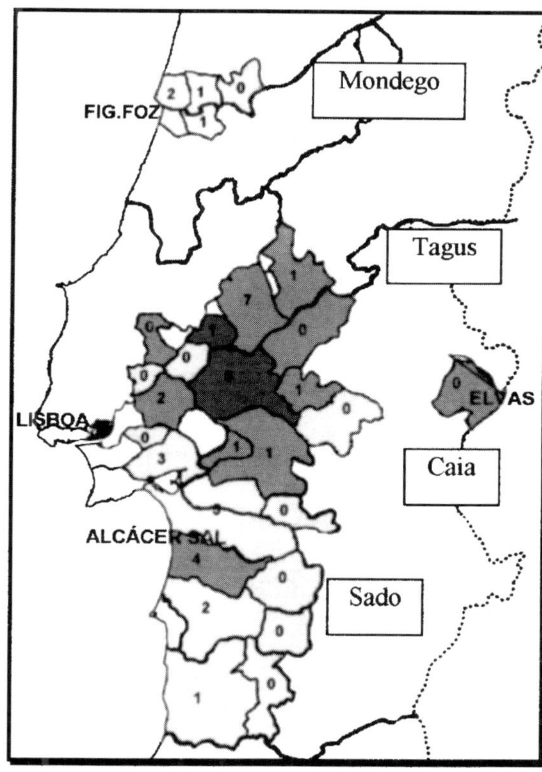

Figure 1. Townships of resistance risk ■ High (>50% rice fields within the township were treated continuously for more than 3 years with sulfonylurea herbicides) ▩ Medium (30–50% rice fields obey to the same condition) □ Low (< 30% rice fields obey to the same condition) and distribution of bensulfuron-methyl resistance (no of fields with R populations of *A. plantago-aquatica*) within those townships.

Fields in the class of high selection pressure, had the higher number of R populations (29) (Table 3). It was possible to define townships at risk according to

Table 2. Most troublesome weeds in rice according to farmers perception

Species	Mondego	Tagus and Sorraia	Sado and Caia	Total
Alisma plantago -aquatica	35	56	65	158
Echinochloa spp.	54	19	35	108
Cyperaceae	6	8	20	34
Others (*)	7	11	18	36

(*)*Leersia oryzoides* (L.) Swartz, *Glyceria declinata* Bréb., *Ammania xcoccinea* Rottb.

Table 3. Distribution of fields with resistant populations of *Alisma plantago-aquatica* L. for classes of low, medium and high selection pressure

Valley	Classes of selection pressure			
	Low	Medium	High	Total
Mondego	0	0	4	4
Tagus and Sorraia	0	4	15	19
Sado and Caia	1	2	10	13
Total fields with R pop.	1	6	29	36

the number of rice fields in this class (Figure 1). The actual occurrence of resistant weeds within these risk areas is as follows: 10% of fields from townships at low risk proved to have resistant populations. From townships at medium risk, 17% of rice fields had resistant populations while 32% of the fields from townships at high risk were resistant. 'High risk' townships were only present in the Tagus and Sorraia valleys, which have the highest number of fields with resistant populations. These valleys also contain medium and low risk townships. The Mondego rice producing area which represents only four fields with R populations is spread among four low risk townships. For the Sado and Caia valleys, 2/3 of the townships have a low level of risk of resistance and the remaining areas are of medium risk.

As Stallings et al. (1994) have indicated, although herbicide field history selects for resistance, the same herbicide history does not always lead to the same occurrence of resistance. As a result of this survey, it was highly probable that bensulfuron-methyl creates selection pressure for *Alisma plantago-aquatica* resistance in fields with an history of more than three years of consecutive spraying or which have had an alternation of one year with other herbicides. Thirty eight sulfonylurea-resistant weed species (29 dicotyledon and nine monocotyledon species) are known in a total of 11 countries (Heap, 1997). Among them, five are resistant to bensulfuron-methyl in rice. Resistance does not depends on a single factor so solutions to the problem need an holistic approach. For sulfonylureas, weed management strategies should include (a) restrictions in applications for more than three consecutive years (Calha et al., 1995b), (b) alternation with herbicides with different modes of action and (c) crop rotation.

Acknowledgments

This study was supported financially by the Research Project – PAMAF-IED-1017. The authors acknowledge the contributions of colleagues, Isabel Magalhães, António M. Lince and Conceição Canas Serra, from the Regional Services of Agriculture to the development of the Project. Thanks are also given to Alice Leitão and Filomena Figueiredo for statistical support; to A. P. Lavadinho for interesting discussions; to Cecília Marques, José Caetano, Aureliano Leitão and Manuel Barreto for field assistance; to Victor Oliveira and César Reis for computer support; to Victor Pereira of Dupont for providing samples of bensulfuron-methyl.

References

Calha, I. M., C. Machado & F.Rocha, 1995a. A survey of herbicide-resistant weeds in Portuguese fields. Proc. Int. Symp on weed and crop resistance to herbicides: Cordoba: 223–226.

Calha, I. M., C. Machado & F. Rocha, 1995b. A resistência da orelha-de-mula, colhereira ou colhereiro (*Alisma* spp.) ao herbicida bensulfurão-metilo em arrozais. Folheto de divulgação. IPPAA.CNPPA.

Gressel, J. & L. A., Segel, 1982. Interrelating factors controlling the rate of appearance of resistance: the outlook for the future. In Lebaron, H. M. & J. Gressel (eds), Herbicide Resistance in Plants. Wiley-Interscience Publication: 401.

Gressel, J., 1991. Why get resistance? It can be prevented or delayed. In: Caseley, J. C., G. W Cussans & R. K., Atkin (eds), Herbicide Resistance in Weeds and Crops. Butterworth-Heinemann, Ltd: 500.

Heap, I. M., 1997. The occurrence of herbicide-resistant weeds worldwide. Pestic. Sci. 51: 235–243.

INE 1993. Portugal Agrícola. 261.

Letouzé, A., D. Gasquez, D. Vaccara, D. Orlando, J. L. Leterrier., C. Roy & E. Bouvard-Derieux, 1997. Development of new reliable quick tests and state of grass weed herbicide resistance in France. Proc. Brighton Crop Protection Conference-Weeds: 325–331.

Morrison, I. N & L. Bourgeois, 1995. Approaches to managing ACCase inhibitor resistance to wild oat on the canadian prairies. Proc.Brighton Crop Protection Conference-Weeds: 567–576.

Prates, M. A. L., 1996. Impacte de herbicidas do grupo químico das sulfonilureias sobre *Alisma* spp. Trabalho de fim de curso. Universidade de Évora: 86.

Ribeiro, M. C. P., 1997. Determinação do nível de resistência de um biótipo de *Alisma plantago-aquatica*, proveniente de Alcácer do Sal. Relatório Global do Curso de Engenharia de Produção Agropecuária. Instituto Politécnico de Santarém: 64.

Stallings, G. P., D. C. Thill & C. A. Mallory-Smith, 1994. Sulfonylurearesistant russian thistle*(Salsola iberica)*. Survey in Washington state. Weed Tech. 8: 258–264.

Vasconcelos, T., M. Tavares & N. Gaspar, 1998. Aquatic plants in the rice fields of the Tagus valley, Portugal. Proc. 10th EWRS Symp. on Aquatic Weeds: 143–146.

Van Oorschot, J. L. P. & H. J. M. Straathof, 1988. On the occurrence and distribution of chloroplastic resistance of weeds to triazines in the Netherlands. VIIIé Coll. Intern. sur la Biologie l'Ecologie et la Systematique des Mauvaises Herbes. Ann. ANPP.03: 267–275.

Hydrobiologia **415:** 295–298, 1999.
J. M. Caffrey, P. R. F. Barrett, M. T. Ferreira, I.S. Moreira, K. J. Murphy & P. M. Wade (eds),
Biology, Ecology and Management of Aquatic Plants.
© 1999 *Kluwer Academic Publishers. Printed in the Netherlands.*

Ecology, distribution and chemical control of *Hydrocotyle ranunculoides* in the U.K.

Jonathan R. Newman[1]* & F. Hugh Dawson[2]
[1]*IACR-Centre for Aquatic Plant Management, Broadmoor Lane, Sonning, Reading RG4 6TH, U.K.*
[2]*NERC-Institute of Freshwater Ecology, Rivers Laboratory, East Stoke, Wareham BH20 6BB, U.K.*

Key words: herbicide, aquatic weed, alien species

Abstract

Hydrocotyle ranunculoides L.f. is reported in this paper to be present at only 29 sites in the south east of England and in South Wales, in a wide range of water body types, which are typically slow-flowing and eutrophic. The current infestation is likely to have originated from various introductions from a single clone available at aquatic garden centres and nurseries in the U.K.. Experiments on the optimum chemical control method have shown that the plant is most susceptible to the herbicide 2,4-D amine applied at 4.23 kg ha^{-1} active ingredient (a.i.). Initial observations suggest that the species is resistant to the herbicide glyphosate applied at 2.16 kg a.i. ha^{-1}. This resistance is postulated to be due to insufficient uptake of the herbicide through the leaf cuticle.

Introduction

Hydrocotyle ranunculoides L.f, the Floating Pennywort, is a native of North America, but is common in Central and South America (Preston & Croft, 1997) and also occurs in the Netherlands (Baas & Hoverda, 1996) and Western Australia (Ruiz-Avila & Klemm, 1996). It has recently become established in Britain and is rapidly becoming a major weed problem at 4 known sites. The primary source of the plant in the U.K. is from garden and aquatic centres open to the public, where it is often sold under the misnomer 'Marsh Pennywort', the common name of *H. vulgaris*, a native British species. It was first recorded as a naturalised alien in 1990 when discovered in the River Chelmer at Chelmsford. By 1992, it was present in the Chelmer and Blackwater Navigation up to 12 km downstream of Chelmsford (Payne, 1992). The second reported occurrence was on the River Wey (pers. comm. Environment Agency, 1994). By 1996, the growth of *H. ranunculoides* along the River Wey necessitated cutting to maintain navigation along the channel and to enable access to the water for anglers.

Effective control methods for aquatic plants are usually based on a sound knowledge of their ecology. During the brief time in which *H. ranunculoides* has been present in the U.K., little work on its growth or habitat requirements has been done, but successful control of this species using herbicides has been achieved in Western Australia (Klemm et al., 1993).

In Britain, reproduction is primarily vegetative, often resulting in extensive mats derived from single nodal fragments. Small greenish-white flowers have been produced in glasshouse conditions with natural daylight and in the Pevensey Levels where seeds have been observed. The plant possesses a number of characteristics typical of weedy species, including high growth rates, effective vegetative propagation, plasticity in growth response (e.g. over-wintering, aquatic and terrestrial growth forms) and high resistance to herbivory (McChesney, 1994). The ability to grow from single node fragments allows rapid recolonization of downstream areas following mechanical cutting. The fragments continue to gain weight and increase root biomass when they are floating, making successful establishment more likely when suitable substrates are encountered.

This paper reports the current distribution of the species in the British Isles and appropriate chemical methods of control.

* Author for correspondence

Table 1. Distribution of *H. ranunculoides* in Britain related to habitat occurrences

River	Stream	Ditch	Pond	Lake	Canal
13	1	6	7	1	1

Materials and methods

Data for distribution of the species were gained by collating information from Vice County Recorders of the Botanical Society of the British Isles. Samples were collected from the field and grown in culture in clay loam. Single node cuttings (3 cm) were taken and pinned down to the soil surface in 10 cm pots. The pots were placed in tap water (pH 8.2) to inundate the soil surface. The depth of water above the soil was 0.5 cm.

The Addlestone Bourne in SE England was chosen for chemical control experiments. This is a narrow flood relief channel connecting the River Wey to the River Thames. It was completely choked with *H. ranunculoides* for 150 m close to the River Wey (Appleby, 1997). Two trial plots were marked out within a 65 m section and subdivided into six treatment blocks. Two blocks were sprayed with 2,4-D amine, two with glyphosate and two were left untreated as control plots. Glyphosate was applied at a rate of 2.16 kg ha^{-1} active ingredient in 200 l water and 2,4-D amine was applied at a rate of 4.23 kg ha^{-1} active ingredient in 200 l water. Herbicides were applied on 10 June 1997 using a Cooper Pegler Falcon 10 ‘s’ sprayer with a long lance equipped with a red Zeneca polijet nozzle enabling a 2 m wide swath width to be applied from the bankside path. Before treatment, the wet weight of *H. ranunculoides* per square metre was measured in each plot using three random 25 cm^{2}quadrats per plot. All root and shoot material was cut and removed from the quadrat and weighed on site. The wet weight per unit area was measured each week for four weeks after herbicide application. Measurements of percentage cover were also made on each site visit until nine weeks after spraying.

Results

Observations made in 1996 and 1997 on the River Chelmer, the River Wey, the Basingstoke Canal, the Pevensey Levels and the Gwent Levels, show that *H. ranunculoides* roots in the shallow margins of slow-flowing eutrophic water bodies, particularly ditches, slow-flowing rivers, ponds and lakes (Table 1) and forms dense interwoven mats of vegetation which grow over the surface of the water. Particularly favoured habitats of the species included ponds, drainage ditches and wide, static or slow flowing channels (Table 1). Significant infestations were found in four locations, the River Chelmer (20 km), the River Wey (10 km), the Pevensey Levels (14 km) and the Gwent Levels (6 km), all of which are slow-flowing rivers or wide channels.

H. ranunculoides forms dense floating mats, rooted at the margins. These floating mats quickly cover the water surface, altering the ecology by excluding submerged aquatic species and interfering with amenity uses (angling and navigation) of the water body. The plant grows as long creeping rhizomatous stems, rooting from nodes at 40–120 mm intervals. The leaves are emergent, rising on stalks from the horizontally growing stem and have very waxy surfaces. Leaves are typically 25–40 mm in diameter, though they can attain maximum diameters of 180 mm. Leaves range from circular to kidney-shaped with deeply lobed margins. The plant grows up to 40 cm above the water surface, while the interwoven mat of roots and stems can reach down to a depth of 50 cm. It often occurs on soft, silty sediment types. Maximum biomass is attained from June onwards. Over-wintering probably occurs as a terrestrial growth form, characterised by prostrate growth and smaller leaf size.

Current distribution is restricted to sites in south east and central southern England and in south Wales. The infested river systems are unconnected and it is likely that the present distribution pattern reflects separate introductions to the wild from aquatic nurseries, garden centres and releases from private lakes and ponds. This pattern is probably due to the anthropogenic influence on distribution in the U.K.. Observations on the distribution of the plant at each site show that it spreads rapidly throughout river systems once established, displacing native vegetation and becoming dominant in less than two years.

Data from chemical control experiments are given in Table 2. Treatment with 2,4-D amine resulted in a significant (76%) decrease in biomass over the first month. Plants showed rapid die-back, the tissues wilting and shrivelling within 2 h of treatment and sinking and decomposing over the next month. Control was almost 100% effective, with only a few small leaves surviving in each treated plot by one month after treat-

Table 2. Mean biomass (wet weight) and percentage surface cover values of *H. ranunculoides* in untreated and herbicide-treated blocks. Data are means of six replicate samples ± standard deviation

Days after application	Untreated		Glyphosate		2,4-D amine	
	Biomass (kg m^{-2})	% Cover	Biomass (kg m^{-2})	% Cover	Biomass (kg m^{-2})	% Cover
0	68.0 ± 8.2	100	67.6 ± 6.7	100	68.6 ± 4.1	100
7	68.5 ± 6.3	100	61.7 ± 2.5	100	53.5 ± 6.6	100
14	67.0 ± 4.1	100	54.4 ± 4.6	100	47.1 ± 3.6	72.5
21	69.0 ± 3.6	100	53.5 ± 4.3	100	31.8 ± 1.5	62.5
28	66.0 ± 4.5	100	54.4 ± 4.1	72.5	16.8 ± 0.6	27.5
35	–	100	–	42.5	–	2.5
42	–	100	–	45.0	–	0.5
63	71.1 ± 3.6	100	62.3 ± 4.3	67.5	0.02 ± 0.00	0.5

ment. In contrast, treatment with glyphosate resulted in little biomass reduction (20%) after one month, with discoloration and wilting in only some of the uppermost leaves. These symptoms did not appear until three weeks after treatment. Many leaves and stems remained free of symptoms and significant regrowth was evident by nine weeks after treatment, when the percentage cover equalled that in the control. Untreated plots remained healthy with 100% cover throughout the experiment.

Discussion

The current distribution of *H. ranunculoides* in Britain shows rapid spread since the first recorded introduction in 1991. This is thought to be due to a combination of fast growth and vegetative reproduction (exacerbated by mechanical control), coupled with additional escapes from garden ponds, through sewage works and by deliberate introduction. Additional ecological advantage may be gained by the production of allelopathic anti-algal compounds by this species (Della Greca et al., 1994) .

Applications of glyphosate to dense infestations of *H. ranunculoides* appear to have little effect on the plant. The waxy leaf surface may restrict glyphosate uptake such that the amount available to reach new leaves growing from the submerged, interconnected mat of root material is too low to kill the plant. 2,4-D amine was considerably more effective, giving complete control within 6 weeks of treatment. Even so, because of the dense leaf canopy, repeated application of herbicide after two months was necessary to eradicate *H. ranunculoides*. Mechanical control is ineffective, creating fragments which disperse to recolonise downstream habitats with estimates of the cost of mechanical control up to £10 000 km^{-1}. 2,4-D amine offers the most effective chemical method of control, this is also supported by data in Newman & Greaves, 1997.

However, a subsequent treatment programme on the Pevensey Levels (a Site of Special Scientific Interest due to the presence of the Fen Raft Spider, *Dolomedes plantarius*) using 2,4-D amine in summer 1998 was not successful at eradicating the plant although biomass was reduced compared with previous years. The reduction in efficacy is thought to be due to the different growth patterns of the species at the two sites. At Addlestone, a 2 m wide watercourse, there was 100% surface cover at the time of spraying. In the Pevensey Levels, mats were treated when they were not fully established (from 5 to 20% surface cover). This left submerged shoot tips unaffected by the application of the herbicide. These shoot tips were able to grow and the result was often a green fringe surrounding a dead and decaying mat. This resulted in increased fragmentation of the mat and increased distribution of the plant within the drainage system, necessitating increased herbicide use. From these observations, it may not be possible to effectively control *H. ranunculoides* in the U.K. with the herbicides available and approved for use in water.

Future chemical control strategies for this highly invasive species will involve the use of broad-spectrum, fast-acting herbicides to limit the spread within existing sites and prevent spread to new sites.

The use of existing legislation and regulatory bodies for the prevention of sale, and further introduction, of alien invasive species to the U.K. and Europe are being actively investigated.

Acknowledgements

IACR-Centre for Aquatic Plant Management is funded by a consortium of subscribers. IACR receives grant aided support from the Biotechnology and Biological Sciences Research Council of the U.K. The CAPM acknowledges the support of the Environment Agency Thames and Southern Region for providing opportunities to experiment on control strategies.

References

Appleby, N. J., 1997. Survey report on the existence of floating pennywort *(Hydrocotyle ranunculoides)* along the River Wey and Wey Navigation. Centre for Aquatic Plant Management Report. Sonning, U.K.

Baas, W. J. & W. J. Holverda, 1996. *Hydrocotyle ranunculoides* L.f.: infiltrant in waterland? Gorteria 21: 193–198.

Della Greca, M., A. Fiorentino, P. Monaco & L. Previtera, 1994. Polyoxygenated oleanane triterpenes from *Hydrocotyle ranunculoides*. Phytochemistry 35: 201–204

Klemm, V. V., N. L. Seimon & R. J. Ruiz-Avila, 1993. *Hydrocotyle ranunculoides*: a control strategy for the Canning River National Park. Swan River Trust Report No. 6, Perth, Western Australia.

McChesney, C., 1994. Literature review of the genus *Hydrocotyle* L (Apiaceae), with particular reference on *Hydrocotyle ranunculoides* L.f. Swan River Trust Report No 18, Perth, Western Australia.

Newman, J. R. & M. P. Greaves, 1997. IACR-Centre for Aquatic Plant Management Progress Report 1996. IACR-Long Ashton Research Station: 46 pp.

Preston, C. D. & J. M. Croft, 1997. Aquatic Plants in Britain and Ireland. Harley Books, Colchester, Essex.

Payne, R. G., 1992. A new waterweed in Essex. The Essex Field Club Newsletter 9: 2–3

Ruiz-Avila, R. J. & V. V. Klemm, 1996. Management of *Hydrocotyle ranunculoides* L.f., an aquatic invasive weed of urban waterways in Western Australia. Hydrobiologia 340: 187–190.

Hydrobiologia **415:** 299–304, 1999.
J. M. Caffrey, P. R. F. Barrett, M. T. Ferreira, I.S. Moreira, K. J. Murphy & P. M. Wade (eds), Biology, Ecology and Management of Aquatic Plants.
© 1999 *Kluwer Academic Publishers. Printed in the Netherlands.*

Chemical control of common reed (*Phragmites australis*) by foliar herbicides under different spray conditions

I. Moreira*, A. Monteiro & E. Sousa
Departamento de Protecção das Plantas e de Fitoecologia, Instituto Superior de Agronomia, P-1349-017 Lisboa, Portugal
E-mail: ilidimor@esoterica.pt

Key words: cutting, glyphosate, glufosinate

Abstract

The efficacy of glyphosate and glufosinate applied with two sprayers, a motorised knapsack mistblower and a hydraulic sprayer, were evaluated for the control of common reed. *Phragmites australis* had been cut in the beginning of September 1994, 30 d before autumn application. The study sites were cleared of vegetation to within 10 cm of the surface using an hydraulic escavator equipped with a mowing bucket, in one operation it cuts and removes the weed. Herbicides were sprayed in October 1994 and April 1995 at different spray volumes and at different rates. Glyphosate rates were 0.81, 1.62, 3.24, 4.68 and 6.48 kg a.i./ha applied with a mistblower at spray volumes of 300; 800; 1100 l/ha and with a sprayer at 800; 1600 l/ha. Glufosinate rates were 0.3, 0.6, 1.2, 1.38, 1.8 and 2.4 kg a.i./ha applied with a mistblower at spray volumes of 800 l/ha and with a sprayer at 800; 1100 and 1600 l/ha. Glyphosate provided control over 2–3 yr for dense *P. australis* infestations in drainage channels when applied up to 1.62 kg/ha, independently of the type of sprayer, times of treatment or spray volume. Glufosinate produced rapid phytotoxicity in aerial parts; however the autumn treatment was more efficacious than the spring one. Regrowth, during 1995, was high and 1 yr after the first application the efficacy had become very low.

Introduction

A variety of herbicides have been reported to control common reed (*Phragmites australis* (Cav.) Steudel). Glyphosate, dichlobenil and diquat are the herbicides approved in Portugal for use in aquatic environments. Although glufosinate is registered in several countries to control common reed, in Portugal it is approved for use in non agricultural land but not in aquatic environments.

Combination of cutting and herbicide application has been considered to be a successful control measure for *Typha* sp. and common reed (Kliemand, 1974; Wilson, 1977; Sale & Wetzel, 1983; Buttler, 1992, Kay, 1995). According to Buttler (1992) this is probably due to the decrease of reducing, non-reducing and total sugar content of rhizomes and changes on the vegetation structure. Glyphosate trials have been carried out by several researchers using different rates, times of treatment and duration of control (Table 1). Some researchers found that good control was achieved with glyphosate rates up 1.8 kg/ha applied in autumn. However Fernandes et al. (1980) verified variable common reed control with that rate and Arsenovic & Konstantinovic (1990) observed only 70% efficacy 30 d after treatment and high regrowth next vegetation season.

Glufosinate applied in summer at 1.5 and 2 kg a.i./ha has been found to result in 90% control 30 d after treatment (DAT) but with 10–20% weed regrowth some months later (Arsenovic, 1986; Arsenovic & Konstantinovic, 1990).

The spray volumes reported in the literature showed considerable variation. A volume rate of 200–600 l/ha is recommended by glyphosate manufacturers for emergent weed control (Barrett, 1985; Fernandes & Moreira, 1985). Although many workers have reported higher volume rates (Fernandes et al., 1980; Stankovic et al., 1982). Differences in glyphosate phytotoxicity have been attributed to the quality of wa-

* Author for correspondence

Table 1. A review of glyphosate rates for control of common reed (*P. australis* and cattail (*Typha* spp.)

Species	Application rate (kg/ha)	Time of treatment	Duration of control (years)	Source
P. australis	3.6, 5.4	Spring	1	Nir (1976)
	2.7 –3	Spring	1	Evans (1978)
	5.4 – 10.8	Autumn	1–2	Fernandes et al. (1978)
	1.8 – 3.6	Autumn	1–2	Fernandes et al. (1980)
	1.8	Autumn	1–2	Seddon (1981)
	2.88 – 3.6	Autumn	1	Evans (1982)
	4.8	Summer	1	Stankovic et al. (1982)
	2.8, 3.2, 5.4	Summer	0–1	Fernandes & Moreira (1985)
	4.8	Summer	1	Arsenovic (1986)
	1.8, 2.7, 3.6	Summer	0–1	Arsenovic &Konstantinovic (1990)
	2.9	Autumn	1	Jambrino & Fernández-Anero (1997)
Typha spp.	2.7 – 3	Summer	1	Evans (1978)
T. angustifolia	5.4 – 7.2	Autumn	1–2	Fernandes et al. (1978)
	4.8	Summer	1	Arsenovic (1986)
	1.8, 2.7, 3.6	Summer	0–1	Arsenovic & Konstantinovic (1990)
T. latifolia	4.8	Summer	1	Arsenovic (1986)
	1.8, 2.7, 3.6	Summer	0–1	Aresenovic & Konstantinovic (1990)

ter used in the spray solution (Sandberg et al., 1978; Stahlman & Phillips, 1979; Jordan, 1981; Buhler & Burnside, 1983) and to differences in the dispersion pattern of the solution on leaf surfaces (Van Thai et al., 1986).The optimum timing for treatment using glyphosate is at flowering or just after (Evans, 1978). However good control has been achieved in both spring, summer and autumn treatments (Table 1). Early season treatment of common reed was reported by Evans (1978) with acceptable levels of control.

Efficacy of glyphosate for common reed through rope wick wipers is also described (Costa, 1981; Evans, 1982). The main advantages of this type of apparatus are the absence of drift, ease operation and increased herbicide efficiency, since much less of herbicide is inactivated in the soil (Costa, 1981). However, the wipe-on-herbicide application was found more recently to be not a practical method of *P. australis* control (Kay, 1995). Conventional sprayers, hydraulic and knapsack, have been currently used, giving effective control (Fernandes et al., 1980; Seddon, 1981; Arsenovic, 1986; Caffrey, 1996). Glyphosate (1.96 kg a.i./ha) was applied, using a microlight aircraft equipped with an atomiser micronaire at low volumes (20/ha) with controlled droplet application and it eradicated common reed under transmission powerlines in South Africa (van Rensburg, 1996). Studies carried out in Portugal with 2.8 and 5.4 kg a.i./ha glyphosate applied with the mistblower at 320 and 600 l/ha spray volume, respectively, pointed to a better and long term control of common reed than by sprayers (Fernandes & Moreira, 1985).

Cutting and sometimes burning of common reed are the methods used by drainage channel managers of Lezíria Grande de Vila Franca de Xira (LGVFX), in low Tagus river basin. Usually, cutting is done from August to October in order to facilitate removal of water hyacinth.

The aim of the experiments presented in this paper was to investigate the combination of cutting, with time of treatment and type of sprayer using two herbicides (glyphosate and glufosinate) at different application rates and spray volumes. From others experiments plant cutting was considered to contribute to glyphosate efficacy (Monteiro et al., 1999).

Material and methods

A dense infestation of common reed (50 shoots/m^2) growing in the largest drainage channel at LGVFX was used for the study. Common reed plants were cut on 5 September 1994. The study sites were cleared of vegetation to within 10 cm of the surface using an hydraulic escavator equipped with a mowing bucket,

Table 2. Herbicide efficacy for common reed applied in autumn (4 October 1994) and spring (19 April 1995) after cutting (5 September 1994) at different spray volumes using two types of sprayers – motorised knapsack mistblower and a hydraulic sprayer (wheelbarrow type)

Herbicide	Rate	Spray volume	Type of sprayer	Percentage weed control					
				Autumn application			Spring application		
	(kg a.i./ha)	(l/ha)		10 DAT	6 MAA	12 MAA	1 MAA	6 MAA	12 MAA
glyphosate	3.24	300	mistblower	0 b	100 a	85 a	50 b	90 a	85 a
glyphosate	3.24	800	mistblower	0 b	100 a	85 a	75 a	90 a	80 a
glyphosate	3.24	1100	sprayer	0 b	95 a	80 a	75 a	90 a	80 a
gluphosinate	1.38	800	mistblower	100 a	30 b	0 b	50 b	25 c	0 b
gluphosinate	1.38	1100	sprayer	100 a	40 b	0 b	75 a	50 b	0 b

Means within a column, for each application time, followed by the same letter are not significantly different at the 0.05 probablity level using Fisher's Protected LSD Test. DAT – days after treatment; MAA – months after application.

in one operation it cuts and removes the weed. A motorised knapsack mistblower and a hydraulic sprayer (a wheelbarrow type) were mounted in a boat from which applications were made. There were three sets of experiments.

Experiment 1

Plots, 15 × 4 m, were arranged end-to-end along the drainage channel in a split-split-split-plot design with three replications. The type of herbicide was the main factor, time of application the subplot factor, type of sprayer the sub-subplot factor and spray volumes the sub-sub-subplot factor. Herbicide rates, spray volumes and type of sprayer are given in Table 2. Herbicide treatments were done on 4 October 1994 and 19 April 1995, being the air temperature 25±2 °C and air humidity 60%.

Experiment 2

Plots, 30 × 4 m, were arranged end-to-end along the drainage channel bank in a split-plot design with three replications. Time of application was the main factor and glyphosate rates the subplot factor. Glyphosate rates are given in Table 3. The herbicide was sprayed with the hydraulic sprayer at a spray volume of 800 l/ha and 1600 l/ha; this higher volume was sprayed in two consecutive applications with half rates of the other. Autumn treatments of glyphosate were applied on 7 October 1994 and spring ones on 21 April 1995. The air temperature was 20±2 °C in Autumn and 25±2 °C in Spring and air humidity 60%.

Experiment 3

Plots, 30 × 4 m, were arranged end-to-end along the drainage channel bank in a split-plot design with three replications. Time of application was the main factor and glufosinate rates the subplot factor. Glufosinate rates are given in Table 4. Glyphosate at 3.24 kg a.i./ha was also included as herbicide reference. Herbicides were sprayed with the hydraulic sprayer at a spray volume of 800 l/ha and 1600 l/ha; this higher volume was sprayed in two consecutive applications with half rates of the other. The glufosinate and glyphosate treatments were applied on 10 October 1994 and on 26 April 1995. The air temperature was 23±2 °C and 24±2 °C, respectively and air humidity 60%.

Efficacy was measured by percentage cover in relation to control and was rated visually by two independent observers. It was based on a scale of 0 – 100 (0 = no reduction in weed biomass and 100 = no living weeds present), 10 DAT and 1, 6, 12 and at some cases 24 MAA (months after application).

Data were subjected to analysis of variance and means separated by Fisher's Protected LSD Test at the 5% level of significance. The efficacy (percentage weed control) was analysed using non transformed and transformed (arcsine of the square root) values. Preliminary analysis of variance of the combined data indicated significant treatment-by-time of application and treatment-by-date of evaluation interactions for all variables. Therefore, the data are presented separately for each time of application and date of efficacy evaluation.

Results and discussion

Experiment 1

Table 2 presents the efficacy obtained for the different treatments carried out with 3.24 kg/ha glyphosate and 1.38 kg/ha glufosinate applied, in autumn and

Table 3. Common reed control with glyphosate applied at different rates and spray volumes with a hydraulic sprayer on two dates, autumn (7 October 1994) and spring (21 April 1995), after cutting (5 September 1994)

Rate (kg a.i./ha)	Spray volume (l/ha)	Percentage weed control							
		Autumn application				Spring application			
		1 MAA	6 MAA	12 MAA	24 MAA	3 MAA	6 MAA	12 MAA	24 MAA
0.81	800	41 d	71 c	70 d	42 e	57 c	82 b	51 d	41 e
1.62	800	63 c	89 b	90 bc	77 c	93 a	97 ab	91 bc	75 c
1.62	1600	63 c	89 b	85 c	53 d	81 b	83 b	63 c	53 d
3.24	800	84 b	95 a	92 ab	86 b	97 a	97 b	94 abc	85 b
3.24	1600	80 b	95 a	90 bc	85 b	95 a	96 ab	92 bc	85 b
4.68	800	91 a	96 a	95 ab	90 ab	94 a	98 ab	94 abc	90 ab
6.48	800	91 a	96 a	95 ab	90 ab	97 a	98 ab	98 ab	95 a
6.48	1600	93 a	98 a	98 a	97 a	97 a	100 a	100 a	97 a

Means within a column, for each application time, followed by the same letter are not significantly different at the 0.05 probability level using Fisher's Protected LSD Test. MAA – months after application.

Table 4. Common reed control with glufosinate applied at different rates and spray volumes with a hydraulic sprayer on two dates, autumn (10 October 1994) and spring (26 April 1995) after cutting at 5 September 1994

Rate (kg a.i./ha)	Spray volume (l/ha)	Percentage weed control							
		Autumn application				Spring application			
		1 MAA	6 MAA	12 MAA	24 MAA	3 MAA	6 MAA	12 MAA	24 MAA
0.3	800	40 e	74 c	10 f	0 e	30 e	20 e	0 e	0 e
0.6	800	75 d	88 b	54 de	10 d	75 d	60 d	28 d	0 e
0.6	1600	80 d	83 b	50 e	10 d	75 d	50 d	25 d	0 e
1.2	800	85 c	95 a	60 d	15 c	85 c	70 c	45 c	12 d
1.2	1600	85 c	95 a	58 d	15 c	85 c	68 c	46 c	10 d
1.8	800	90 b	95 a	65 c	15 c	90 b	75 b	59 b	20 c
2.4	800	95 a	97 a	73 b	20 b	95 a	75 b	60 b	25 b
2.4	1,600	95 a	97 a	75 b	20 b	95 a	78 b	60 b	25 b
glyphosate 3.24	800	98 a	98 a	95 a	80 a	98 a	98 a	90 a	80 a

Means within a column, for each application time, followed by the same letter are not significantly different at the 0.05 probability level using Fisher's Protected LSD Test. MAA – months after application.

spring, by a mistblower and a sprayer using 800 and 1100 l/ha spray volumes respectively. The same rate of glyphosate was also delivered by a mistblower at 300 l/ha.

Six months after applications, there were no significant differences between autumn and spring herbicide treatments. The same was observed for active ingredient, spray volumes and type of sprayer.

Glufosinate (1.38 kg a.i./ha) resulted in higher efficacy one month after in both autumn application and spring application but common reed regrowth was rapid. Six months after application, the efficacy of the glufosinate spring application was slightly better however not significantly different than the autumn application when using the hydraulic sprayer but was nevertheless low.

Glyphosate (3.24 kg a.i./ha) applied in both autumn and spring, controlled common reed more effectivelly than glufosinate six months after spraying. Control ranged from 90 to 100% in both glyphosate time applications.

One year after herbicide application plant recovery in the glufosinate plots was 100% but on glyphosate plots the cover of common reed was only about 20%.

Experiment 2

The efficacy of glyphosate applied at different rates in autumn and spring, with a hydraulic sprayer (wheelbarrow type), is presented in Table 3. The analysis of variance for the efficacy observed 6-months after autumn and spring applications did not show significant differences.

The two lower rates of glyphosate application, 0.81 and 1.62 kg a.i./ha always resulted in low efficacies. Glyphosate at 3.24 kg/ha gave significantly better control than these lower rates. Although, between 1.62 and 3.24 kg/ha significant differences were observed mainly in autumn treatments. Higher rates produced only slightly better efficacy than 3.24 kg/ha, but was probably not economically worthwhile. For 3.24 kg/ha or higher rates, there was not a significant difference between the spray volumes tested but for 1.62 kg/ha the lower spray volume (800 l/ha) was significantly better than higher spray volume (1600 l/ha) in most observations. This observation agrees with the general assumption that glyphosate efficacy increases with the increase of spray concentration for the same rate. However, in situations of very dense stands glyphosate efficacy with low spray volumes could be inappropriate on account of poor dispersion of the herbicide (Fernandes & Moreira, 1985).

During the first 2 yr after application, the treated areas became totally covered by species including *Brassica nigra* (L.) Koch, *Picris echioides* L., *Ammi majus* L. , *Silybum marianum* (L.) Gaertner, *Lavatera trimestris* L. and *Convolvulus arvensis* L., which, apparently did not cause real inconvenience and stabilised the channel banks. After 3 years after the spring application some regrowth of common reed began.

Experiment 3

The results of common reed control with glufosinate are presented in Table 4. The analysis of variance indicated that herbicide application in autumn was significantly more efficacious than in spring.. A possible explanation for higher efficacy in autumn could be the depletion of rhizomes reserves due to cutting 30 d before (Buttler, 1992). Significant differences were also observed between herbicide rates. Higher rates of glufosinate (1.8 and 2.4 kg a.i./ha) resulted in similar efficacy to glyphosate (3.24 kg a.i./ha) 1, 3 and 6 months after application. Lower rates did not produce good control. Plant cover increased 1 yr after glufosinate applications; however the efficacy of glyphosate was still acceptable. No significant differences were obtained when the same rate of glufosinate was applied at 800 and 1600 l/ha. The results were in accordance with those referred by Arsenovic (1986) and Arsenovic & Konstantinovic (1990).

Conclusions

The results clearly reinforce the ability of glyphosate to provide long-term control of dense *Phragmites australis* infestations in drainage channels. The positive effect of plant cutting on herbicide efficacy is probably due to the elimination of dry aerial stems, the stimulation of many young leaves and depletion of rhizomes reserves. Glyphosate produced good control for at least 2 yr when applied at 3.24 kg/ha with a spray volume of 800 l/ha. In combination with plant cutting, in summer, autumn or spring glyphosate applications showed similar efficacies.

This range of experiments with glyphosate indicated that both the mistblower and the hydraulic sprayer were suitable for the treatments. The choice of equipment depends on field conditions; in many cases a knapsack mistblower is difficult to carry and it is preferable to use the hydraulic sprayer. However, the mistblower allows the use of low spray volumes (300–800 l/ha).

Glufosinate application in autumn (2.4 kg/ha) allowed a better control of the weed than the spring treatment. This contact herbicide in combination with plant cutting could be an option for situations where a rapid herbicidal action is needed. For example, in wheat fields in LGVFX it is sometimes necessary to sow as early as possible following rainfall. Treatment with glyphosate can be inconvenient due to its slow action.

The costs of a cutting-herbicide combination were not determined. However, considering the long-term control – 2–3 yr – obtained with glyphosate, at least two cutting operations can be spared, which is probably more expensive than herbicide application.

Acknowledgements

We gratefully acknowledge collaboration of 'Associação de Beneficiários da Lezíria Grande de Vila Franca de Xira', in particular, we would like to extend our tanks to its Director António Piçarra who contributed to this project; the technicians Jorge F. Moreira, Eunice Matos and Lídia Alcobia for co-operation and assistance with the field trials; 'Junta Nacional de Investigação Científica' on the Programme STRIDE, 'Integrated control of aquatic weed' Project AGR/STRD/0047 developed by the Consortium – Companhia das Lezírias, Sociedade Agrícola do Faiel and Instituto Superior de Agronomia, for financial support.

References

Arsenovic, M., 1986. Control of aquatic weeds with herbicides in drainage system "DUNAV – TISA-DUNAV" in Yugoslavia. Proc. EWRS/AAB 7th Int. Symp. on Aquatic Weeds: 31–35.

Arsenovic, M. & B. Konstantinovic, 1990. Aquatic weed control with herbicides in drainage system in Yugoslavia. Proc. EWRS 8th Int. Symp. on Aquatic Weeds: 13–17.

Barrett, P. R. F., 1985. Efficacy of glyphosate in the control of aquatic weeds. In Grossbard, E. & D. Atkinson, 1985. The Herbicide Glyphosate. Butterwords, London: 365–374.

Buhler, D. D. & O. C. Burnside, 1983. Effect of water quality, carrier volume and acid on glyphosate phytotoxicity. Weed Sci. 31: 163–169.

Buttler, A., 1992. Permanent plot research in wet meadows and cutting experiment. Vegetatio 103: 113–114.

Caffrey, J. M., 1996. Glyphosate in fisheries management. Hydrobiologia 340: 259–263.

Costa, V. J., 1981. New methods of applying Roundup by wiping. XIII Jornadas de Estudio de la Asociacion Interprofesional para el Desarrollo Agrario: 6 pp.

Evans, D. M., 1978. Aquatic weed control with the isopropylamine salt of N–phosphonomethyl glycine. Proc. EWRS 5th Symp. on Aquatic Weeds: 171–178.

Evans, D. M., 1982. Phragmites control with glyphosate through selective equipment. Proc. EWRS 6th Symp. on Aquatic Weeds: 209–211.

Fernandes, E. & I. Moreira, (1985). Ensaios de luta química para o controlo de infestantes aquáticas. Cited in Moreira, I., T. Ferreira & A. Monteiro, 1989. Aquatic weed bioecology and control in Portugal – A review. In Resultados dos Projectos de Investigação Agrária. Portuguese-German Co-operation in Applied Agricultural Research 1982–1989. Vila Real: 71–106.

Fernandes, J. D., A. R. Guerreiro, T. Vasconcelos & J. M. Moreira, 1978. Essais de lutte contre les plantes aquatiques au Portugal. Proc. 5th EWRS Symp. on Aquatic Weeds: 189–194.

Fernandes, J. D., T. Vasconcelos & J. M. Martins, 1980. Luta química contra o caniço (*Phragmites australis* (Cav.) Steudel). I Congresso Português de Fitiatria e Fitofarmacologia 4: 173–180.

Jambrino, J. I. D. & F. J. Fernández-Anero, 1997. Resultados en la recuperación de la biodiversidade en el Racó de l'Olla (l'Albufera de Valéncia) tras la aplicación selectiva de um herbicida de baja peligrosidad. Bol. San. Veg. Plagas 23: 17–37.

Jordan, T. N., 1981. Effect of diluent volumes and surfactants on the phytotoxicity of glyphosate to bermudagrass (*Cynodon dactylon*). Weed Sci. 29: 79–83.

Kay, S. H., 1995. Efficacy of wipe-on applications of glyphosate and imazapir on common reed in aquatic sites. J. aquat. Pl. Mgmt 33: 25–26.

Kliemand, G., 1974. Proposals for the control of persistent aquatic weeds based on the results of trials. Proc. 4th EWRC Int. Symp. on Aquatic Weeds: 20–27.

Monteiro, A., I. Moreira & E. Sousa, 1999. Effect of prior common reed (*Phragmites australis*) cutting on herbicide efficacy. Hydrobiologia 415 (Dev. Hydrobiol. 147): 305–308.

Nir, A., 1976. Control of perennial weeds in drainage ditches. Phytoparasitica 4: 157.

Sale, P. J. M. & R. G. Wetzel, 1983. Growth and metabolism of Typha in relation to cutting treatments. Aquat. Bot. 15: 321–334.

Sandberg, C. L., W. F. Meggit & D. Penner, 1978. Effect of diluent volume and calcium on glyphosate phytotoxicity. Weed Sci. 26: 476–479.

Seddon, J. C., 1981. The control of aquatic weeds with the isopropylamine salt of N-phosphonomethyl glycine. Proc. Aquatic Weeds and their control: 141–148.

Stahlman, P. W. & W. M. Phillips, 1979. Effect of water quality and spray volume on glyphosate phytotoxicity. Weed Sci. 27: 38–41.

Stankovic, A., M. Arsenovic, M. Dimitrijevic & B. Konstantinovic, 1982. A study of the possibility of the elimination of emerged vegetation in irrigation systems by new herbicides. Proc. 6th EWRS Symp. on Aquatic Weeds: 185–192.

Van Rensburg, E., 1996. Reed eradication under transmission powerlines in South Africa using microlight aircraft. Proc. 2nd Int. Weed Control Cong.: 1095–1101.

Van Thai K., V. V. Vandiver, Jr. & R. D. Connant, 1986. Effect of herbicide rate and carrier volume on glyphosate phytotoxicity. J. aquat. Plant mgmt 24: 66–69.

Wilson, D. B., 1977. Cumbungi control: a field data. Tasmanian J. Agric. 48: 89–33.

Hydrobiologia **415:** 305–308, 1999.
J. M. Caffrey, P. R. F. Barrett, M. T. Ferreira, I.S. Moreira, K. J. Murphy & P. M. Wade (eds),
Biology, Ecology and Management of Aquatic Plants.
© 1999 *Kluwer Academic Publishers. Printed in the Netherlands.*

Effect of prior common reed (*Phragmites australis*) cutting on herbicide efficacy

A. Monteiro, I. Moreira* & E. Sousa
Departamento de Protecção das Plantas e de Fitoecologia, Instituto Superior de Agronomia, P-1349-017 Lisboa, Portugal
E-mail: ilidimor@esoterica.pt

Key words: glyphosate, weed control

Abstract

The study evaluated the effect of prior plant cutting of common reed (*Phragmites australis*), during flowering, on the efficacy of isopropylamine and trimesium salts of glyphosate (2.9 and 2.6 kg a.i./ha, respectively) applied at two spray volumes in the following spring. Efficacy was measured by plant cover (% control), number of plants/m^2 and plant biomass/m^2. A positive effect of the plant cutting was observed when plant cover and number of plants/m^2 were measured. The evaluation of herbicide phytotoxicity on plant biomass did not show significant differences. Common reed susceptibility to isopropylamine and trimesium salts of glyphosate was similar. Significant differences between the spray volumes of 320 and 400 l/ha were observed only in plant cover estimates. The 400 l/ha spray volume showed higher efficacy.

Introduction

Common reed (*Phragmites australis* (Cav.) Steudel) is a frequent aquatic weed in Portugal. It is found from north to south along the banks of all river basins, but mainly along the Tagus and Sado Rivers (Franco & Afonso, 1980). In drainage channels, it is a troublesome emergent weed (Moreira & Ferreira, 1986; Moreira et al., 1989; Ferreira et al., 1998).

Common reed has recently become a serious problem at Lezíria Grande de Vila Franca de Xira (LGVFX) in the lower Tagus Basin (a fertile alluvial plain near Lisbon, with a total of about 470 km drainage channels). It affects water flow because it grows in the bottom of the channels and causes difficulties for the mechanical removal of other aquatic weeds, particularly water hyacinth (*Eichhornia crassipes* (Mart.) Solms). This is an intensive agricultural area and common reed can be a problem for farmers. Rhizomes more than 50 m long invades fields, causing difficulties for wheat harvest and reducing tomato and melon production. Mechanical control is the main method used at LGVFX to control common reed.

The efficacy of glyphosate on common reed control has been evaluated by many researchers (e.g. Evans, 1978; Fernandes et al., 1978, 1980; Evans, 1982; Murphy & Eaton, 1982; Seddon, 1982; Barrett, 1985; Arsenovic & Konstantinovic, 1990). These have tested efficacy using different rates, different sprayer equipment and different application times. The rates reported to produce an effective control vary and it is usually considered that high rates were required for long-term control.

The increasing need for effective, economic aquatic weed control with limited environmental implications has led to the consideration of the combination of mechanical cutting with glyphosate applications.

This paper presents the results from experiments undertaken in LGVFX during 1993/94. The main objective of this study was: (a) to evaluate the effect of prior plant cutting (flowering) on susceptibility of common reed to isopropylamine and trimesium salts of glyphosate; (b) to compare the efficacy of the isopropylamine and trimesium salts of glyphosate and (c) to investigate the effect of spray volume on the phyto-

* Author for correspondence

toxicity of the two salts of glyphosate on common reed.

Materials and methods

A dense infestation of common reed (54 shoots/m^2 untreated cut plots; 70 shoots/m^2 untreated uncut plots) growing in a drainage channel at LGVFX was used for the study.

Plots 17×5 m were arranged end-to-end on the bank of the drainage channel in a split-split-plot design with three replications. Plant cutting was the main factor, glyphosate salts the subplot factor and spray volumes the sub-subplot factor.

Common reed plants were cut in autumn 1993, during flowering. The study sites were cleared of vegetation to within 15 cm of the surface using a hand-held weed trimmer equipped with a saw blade. Isopropylamine and trimesium salts of glyphosate, both at 8 l/ha commercial formulation (2.9 and 2.6 kg a.i./ha, respectively), were applied at 400 and 320 l/ha spray volume (i.e. 2% and 2.5% commercial formulation). Applications were made with the mistblower sprayer on 28 April 1994. The air temperature was 29 °C and air humidity 46%.

Stand reduction was measured as percentage cover by two independent observers, number of plants/m^2 and by plant biomass/m^2 (aerial part) at the end of July, of the year following treatment (i.e. 3 months after treatment). Number of plants/m^2 and plant biomass/m^2 were assessed in each plot by counting and weighted the stems on four randomised square meters. Percent common reed control was rated visually, based on a scale of 0–100 (0 = no reduction in weed biomass and 100 = no living weeds present).

Estimates were averaged before the data were subjected to analysis of variance. Cover percentages were subject to angular transformation before analysis of variance. Means of percentage cover (transformed), number of plants and plant biomass were separated using Fisher's Protected LSD Test at the 5% level of significance.

Results and discussion

Cutting improved control of the common reed (Table 1). There were significant differences between cut and uncut plots using both percentage efficacy, estimated by two independent observers (Figure 1), and

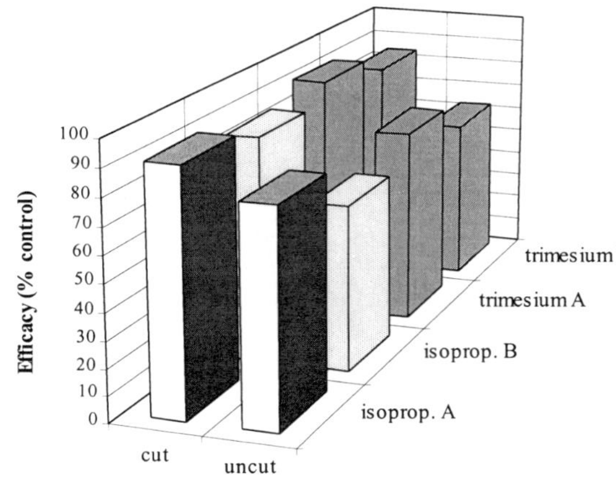

Figure 1. Efficacy (% of the control) on common reed of glyphosate salts, applied in early spring, at spray volumes of 400 l/ha (A) and 320 l/ha (B).

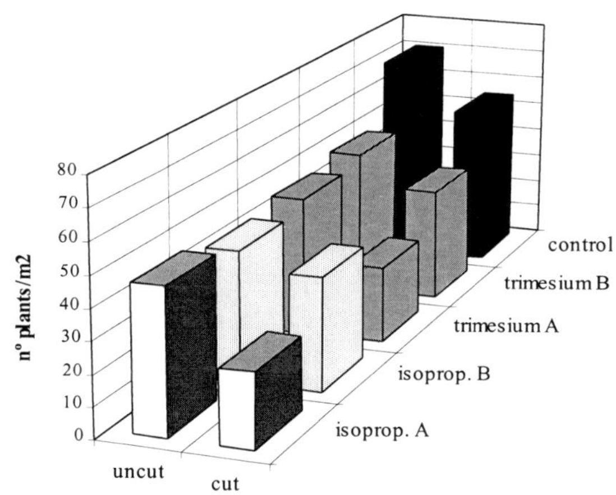

Figure 2. Number of plants/m^2 of common reed 3 months after the early spring application of glyphosate salts at spray volumes of 400 l/ha (A) and 320 l/ha (B).

by counting the number of plants/m^2 (Figure 2). The use of plant fresh weight/m^2 (aerial part) did not show significant differences (Figure 3). Herbicide trials in North Carolina also indicated that common reed cutting provided a long-term control of this weed (Kay, 1995).

The effect of common reed cutting on density and vitality of plants seems to be influenced by ecological conditions and time of cutting. In the U.K. reeds were shorter and occurred at higher densities following

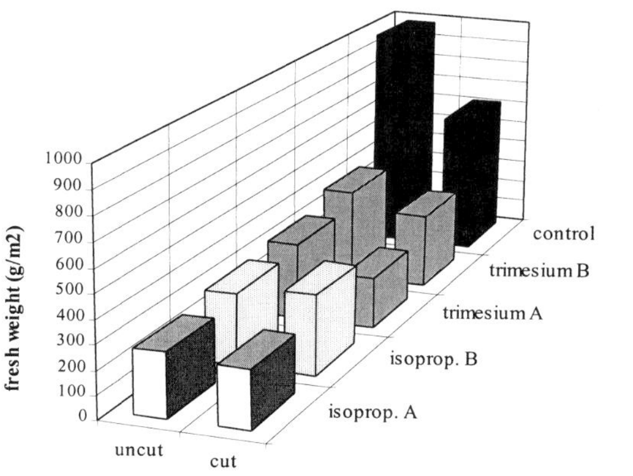

Figure 3. Fresh weight/m^2 of common reed, 3 months after the early spring application of glyphosate salts at spray volumes of 400 l/ha (A) and 320 l/ha (B).

Table 1. Analysis of variance output for the effect of plant cutting, salt of glyphosate and spray volume on the control of common reed (*Phragmites australis*)

Factor	Efficacy (% of the control)	Fresh weight g/m^2 (aerial part)	no of plants/ m^2
Cutting (Cut)			
cut	89.6 a	390	38.8 b
uncut	70.1 b	544	54.4 a
significance	*	**	*
Glyphosate salt (Herb)			
isopropylamine	80.6	458	46.0
trimesium	79.7	475	47.2
significance	ns	ns	ns
Spray volume (l/ha) (Vol)			
400	85.2 a	264	35.4 b
320	75.0 b	339	42.1 b
Control	–	797	62.3 a
significance	**	***	***
Cut.herb	ns	ns	ns
Cut.vol	ns	***	ns
Herb.vol	ns	ns	ns
Cut.herb.vol	ns	ns	ns

ns, *, **, ***, Non significant or significant at the 5%, 1% or 0.1% level, respectively. Mean separation within columns using the Fisher's Protected LSD test at the 5% level of significance. Brackets indicate that the factor is involved in significant interactions; therefore no mean separation was performed for the main effect.

summer cuts (Cowie et al., 1992) and in Switzerland cuts induced a better vitality especially in winter cuts (Buttler, 1992). The autumn cut at LGVFX induced lower numbers of individuals than were present in uncut plots. The lower density of plants and the removal of the litter in the cut plots could have a positive effect on herbicide efficacy.

No differential response of common reed was observed with the isopropylamine and trimesium salts of glyphosate for any of the response variables measured (Table 1).

Spray volumes of 320 and 400 l/ha affected the phytotoxicity of glyphosate salts. The high volume provided better results, >90% control, resulting in less plant biomass and less plant density. Many workers have reported that the phytotoxicity of glyphosate is reduced as the carrier volume is increased (Sanderg et al., 1978; Stahlman & Phillips, 1979; Jordan, 1981; Buhler & Burnside, 1983). This relationship is especially evident with low or marginal rates of glyphosate, as is the case in this study (2.8 kg/ha). Such reduced phytotoxicity has been attributed to the quality of water used in the spray solution (Sanderg et al., 1978; Stahlman & Phillips, 1979; Buhler & Burnside, 1983) and to differences in the dispersion pattern of the solution on leaf surfaces (Van Thai et al., 1986).

Often, the addition of a surfactant to glyphosate formulations increases the activity and masks the effect of spray volume, especially at low rates of glyphosate application (Reimer, 1976; Van Thai et al., 1986; Mueller & Womac, 1997). In our studies, the high spray volume (400 l/ha) provided a better control than the low one (320 l/ha). The efficacy was significantly lower in the uncut plots. Perhaps, the low spray volume was not sufficient to allow a uniform dispersion pattern inside the high plant biomass (uncut plots: 70 shoots/m^2 and 983 g/m^2 fresh weight).

Conclusions

The results show that isopropylamine and trimesium salts of glyphosate at 2.9 and 2.6 kg/ha, respectively, effectively control common reed. Cutting (during flowering), in the year before herbicide treatment, improved the efficacy of the herbicides probably because it may affect accumulation of reserves into the rhizomes. This effect may be caused, externally, the lower plant infestation on the cut plots observed at spring regrowth.

Spray volumes under 400 l/ha should be chosen with care, because with high plant stands the dispersion pattern could be affected.

The use of plant fresh weight/m^2 (aerial part) is probably not a reliable measurement of the effect of glyphosate on the control of common reed because it is difficult to separate green from dry parts of the plants. Plant cover (% control) and number of plants/m^2 are preferable.

Acknowledgements

Experiment carried out on the project "Integrated control of aquatic weeds", developed by the Consortium – Companhia das Lezírias, Sociedade Agrícola do Faiel and Instituto Superior de Agronomia, sponsored by Junta Nacional de Investigação Científica on the Programme STRIDE, with the co-operation in field trials of the technicians Jorge F. Moreira, Eunice Matos and Lídia Alcobia.

References

Arsenovic, M. & B. Konstantinovic, 1990. Aquatic weed control with herbicides in drainage system in Yugoslavia. Proc. EWRS 8th Int. Symp. Aquat. Weeds: 13–17.

Barrett, P. R. F., 1985. Efficacy of glyphosate in the control of aquatic weeds. In Grossbard, E. & D. Atkinson (eds), The Herbicide Glyphosate. Butterwords. London: 365–374.

Buhler, D. D & O. C. Burnside, 1983. Effect of water quality, carrier volume and acid on glyphosate phytotoxicity. Weed Sci. 31: 163–169.

Buttler, A., 1992. Permanent plot research in wet meadows and cutting experiment. Vegetatio 103: 113–124.

Cowie, N. R., W. J. Sutherland, M. K. M Ditlhogo & R. James, 1992. The effects of conservation management of reed beds. II. The flora and litter disappearance. J. appl. Ecol. 29: 277–284.

Evans, D. M., 1978. Aquatic weed control with the isopropylamine salt of N-phosphonomethyl glycine. Proc. EWRS 5th Symp. Aquat. Weeds: 171–178.

Evans, D. M., 1982. *Phragmites* control with glyphosate through selective equipment. Proc. EWRS 6th Symp. Aquat. Weeds: 209–211.

Fernandes, J. D., A. R. Guerreiro, T. Vasconcelos & I. Moreira, 1978. Essais de lutte contre les plantes aquatiques au Portugal. Proc. EWRS 5th Symp. Aquat. Weeds: 189–194.

Fernandes, J. D., T. Vasconcelos & J. M. Martins, 1980. Luta química contra o caniço (*Phragmites australis* (Cav.) Steudel). I Cong. Portugês de Fitiatria e Fitofarmacologia 4: 173–180.

Ferreira, M. T., L. Catarino & I. Moreira, 1998. Aquatic weed assemblages in an Iberian drainage channel system and related environmental factors. Weed Res. 38: 291–300.

Franco, J. A. & M. L. A. Afonso, 1980. Distribuição em Portugal das principais infestantes. I Cong. Português de Fitiatria e Fitofarmacologia 3: 1–22.

Jordan, T. N., 1981. Effect of diluent volumes and surfactants on the phytotoxicity of glyphosate to bermudagrass (*Cynodon dactylon*). Weed Sci. 29: 79–83.

Kay, S. H., 1995. Efficacy of wipe-on applications of glyphosate and imazapyr on common reed in aquatic sites. J. Aquat. Plant Mgmt 33: 25–26.

Moreira, I. & T. Ferreira, 1986. Aquatic weeds in Portugal. Aquaphyte 6: 6.

Moreira, I., T. Ferreira & A. Monteiro, 1989. Aquatic weed bioecology and control in Portugal – a review. In Resultados dos Projectos de Investigação Agrária. Portuguese-German Co-operation in Applied Agricultural Research 1982–1989. Vila Real: 71–106.

Mueller, T. C. & A. R. Womac, 1997. Effect of formulation and nozzle type on droplet size with isopropylamine and trimesium salts of glyphosate. Weed Techn. 11: 639–643.

Murphy, K. J. & J. W. Eaton, 1982. Ecological effects of four herbicides and two mechanical clearance methods used for aquatic weed control in canals. Proc. EWRS 6th Symp. Aquat. Weeds: 201–217.

Reimer, D. N., 1976. Long-term effects of glyphosate applications to Phragmites. J. Aquat. Plant Mgmt 14: 39–43.

Sandberg, C. L., W. F. Meggit & D. Penner, 1978. Effect of diluent volume and calcium on glyphosate phytotoxicity. Weed Sci. 26: 476–479.

Seddon, J. C., 1982. The control of aquatic weeds with the isopropylamine salt of N-phosphonomethyl glycine. Proc. EWRS 6th Symp. Aquat. Weeds: 141–148.

Stahlman, P. W. & W. M. Phillips, 1979. Effect of water quality and spray volume on glyphosate phytotoxicity. Weed Sci. 27: 38–41.

Van Thai K., V. V. Vandiver, Jr. & R. D. Connant, 1986. Effect of herbicide rate and carrier volume on glyphosate phytotoxicity. J. Aquat. Plant Mgmt 24: 66–69.

Hydrobiologia **415**: 309–313, 1999.
J. M. Caffrey, P. R. F. Barrett, M. T. Ferreira, I.S. Moreira, K. J. Murphy & P. M. Wade (eds),
Biology, Ecology and Management of Aquatic Plants.
© 1999 Kluwer Academic Publishers. Printed in the Netherlands.

Long-term algal control in a reservoir using barley straw

P. R. F. Barrett[1,*], J. W. Littlejohn[2] & J. Curnow[3]
[1]*Consultant in Aquatic Plant Management, 8 Sunderland Avenue, Oxford OX2 8DX, U.K.*
[2]*Water Quality and Regulations Manager, North of Scotland Water Authority, Turriff Laboratory, Shandscross, Turriff, Aberdeenshire AB53 7PL, U.K.*
[3]*Consultant in Environmental Medicine, Grampian Health Board, Department of Public Health Medicine, Foresterhill, Westburn Road, Aberdeen AB25 2XH, U.K.*

Key words: Algal control, barley straw, cyanobacteria, diatoms

Abstract

Populations of cyanobacteria, diatoms and unicellular green algae in a potable supply reservoir have been suppressed continuously since 1993 by repeated treatments of barley straw. Algal cell numbers dropped soon after the straw was introduced and have remained at approximately one quarter of those recorded prior to treatment. All types of algae, including diatoms and cyanobacteria, have been uniformly affected and no new or resistant species have developed. Taint and odour problems in the potable water have been reduced and filters at the water treatment works require less frequent cleaning.

Introduction

Algal blooms create problems in potable supply reservoirs by reducing water quality and affecting the efficiency of filtration and cleaning processes. In addition, cyanobacteria have the potential to produce toxins that are dangerous to humans and livestock that have access to the water. Such blooms may necessitate the temporary closure of the reservoir. There are various methods of controlling algae, not all of which are acceptable in potable supply reservoirs, but one method which appears to have few, if any, adverse effects is to introduce barley straw (Barrett, 1994). Barley straw releases substances during aerobic decay in water which inhibit algal growth but which appear to have no adverse effects on vascular plants, invertebrate animals or fish. Numerous trials with barley straw indicate that it could offer a cheap and effective form of treatment in a wide range of aquatic habitats including potable supply reservoirs.

Barley straw was first applied to a potable supply reservoir near Aberdeen, Scotland in 1993, which previously had a long history of diatom and cyanobacterial blooms. The effects of that treatment and subsequent additions of straw in December 1993 and June 1994 on algal growth and some aspects of water chemistry were described by Barrett et al. (1996). Those treatments reduced the mean algal cell counts in the first year of treatment to about one quarter of their previous levels without any serious adverse effects on water quality or treatment. More importantly, blooms of cyanobacteria were almost completely eliminated. Similar effects resulting from the use of barley straw have been reported elsewhere (Everall & Lees, 1996; Harriman et al., 1997). However, these studies monitored algal numbers for only 1–2 years after the straw was added and did not show what, if any, long-term effect on the algal populations might result from continuous exposure to the anti-algal substances released from straw.

The results of the early treatments of the reservoir near Aberdeen in 1993 and 1994 were sufficiently encouraging for the treatments to be continued. The main objective of this study was to assess the long-term effects on algal populations of continuous exposure to the products released by decaying barley straw. A secondary objective was to derive an effective treatment regime in which the efficacy of the straw could be maximised while keeping the handling costs of the straw treatments to a minimum. This paper describes

* Author for correspondence

the ongoing results of this long-term study on the algal counts and other observations on the effects of straw.

Materials and methods

The site of this experiment was a potable supply reservoir with a surface area of 25 000 m^2 and a nominal volume of 250 000 m^3. The reservoir was fed by two streams which drain surrounding land used for mixed farming, and the reservoir supplied water to a treatment works which provided 3.5 megalitres per day to a nearby town.

Barley straw was first applied in March 1993. In this treatment, a total of 7 tonnes of straw was applied at three locations in the reservoir and retained in marine trawler nets, each containing 2.3 tonnes. Since then, both the quantity of straw and method of application have been modified on a number of occasions to improve efficacy and to facilitate handling of the straw. A second treatment, applied in December 1993, was made up of 1.6 tonnes of straw which was retained in 50 nets each holding about 32 kg and supported by floats. This method of application increased the number of individual units of straw and so improved the distribution of the straw throughout the reservoir. However, in both the first and second treatments, the straw rotted only slowly and it was eventually necessary to remove and dispose of the old nets still containing large quantities of waterlogged straw.

The third treatment of 1.9 tonnes of straw was applied in tubular, high density polyethylene netting (Tildenet Ltd, Bristol, U.K.) which had a diameter of approximately 500 mm and a mesh size of 10–12 mm. It was filled with straw using a tree wrapping device. The netting was cut into 10 m lengths and each 10 m segment was filled with about 20 kg of loosely packed straw. The nets were attached to floats. When first filled, these loosely packed nets of straw had the appearance of giant sausages and were buoyant so that they could be easily towed into position before being anchored at either end. This method of treatment gave a good distribution of straw throughout the reservoir and allowed water to move easily through the straw mass, keeping it well oxygenated and encouraging speedy and complete decomposition of the straw. Subsequently, it was only necessary to remove the empty nets when the straw had rotted, which greatly facilitated handling and replacement of the straw nets.

Table 1. Dates and quantities of barley straw applied to the reservoir

Date of Treatments	Quantity of barley straw (tonnes)	Equivalent dose (g/m^3)	Method of application
March 1993	7.0	28.0	Marine trawler nets
December 1993	1.6	6.4	Netting sacks
June 1994	1.9	7.6	Tubular netting
November 1994	1.5	6.0	∼ ∼
May 1995	1.5	6.0	∼ ∼
July 1995	1.5	6.0	∼ ∼
November 1995	1.5	6.0	∼ ∼
March 1996	1.5	6.0	∼ ∼
August 1996	1.5	6.0	∼ ∼
November 1996	1.5	6.0	∼ ∼
May 1997	1.5	6.0	∼ ∼
November 1997	2.3	9.2	∼ ∼
February 1998	3.0	12.0	∼ ∼

In November 1994, this treatment technique was repeated but the dose of straw was reduced to 1.5 tonnes of straw and this rate was maintained until November 1997 when the dose was increased.

The doses of barley straw and timing of treatments over the six years of this trial are shown in Table 1.

Water samples were collected and analysed throughout the year by the North of Scotland Water Authority to establish algal type and cell numbers. The frequency of filter cleaning before and after straw treatments was noted.

Results

Prior to the first treatment in 1993, water samples contained mixed populations of species but, in general, the samples taken each year between January and July were dominated by diatoms, particularly *Asterionella* spp., while those taken from August to December contained cyanobacteria, particularly *Anabaena* spp. The spring blooms of diatoms frequently reached levels in excess of 40 000 cells per ml and, on a number of occasions, the reservoir had been taken out of service because of summer and autumn blooms of cyanobacteria. Other algal species which were regularly found in samples included *Melosira, Golenkenia, Stephanodiscus, Synedra, Tabellaria, Pediastrum, Staurastrum* and *Chlorella* spp.

Table 2. Mean Monthly Algal Counts (cells per ml)

Year	1991	1992	1993	1994	1995	1996	1997	1998
January	N/C	10 000	10 000	400	6900	N/C	2900	N/C
February	13 000	18 000	17 500	6200	5600	4800	8800	8700
March	21 500	28 000	22 800	7800	6700	12 000	6100	2900
April	57 400	38 000	29 000	8700	13 900	9400	5100	1100
May	67 500	25 200	14 500	3400	10 600	6400	5000	1300
June	45 000	16 500	6800	106	2000	3200	8800	100
July	N/C	17 700	3000	57	60	150	2600	100
August	2000	10 500	4500	440	500	40	3800	200
September	N/C	OS	1500	900	600	1700	11 400	
October	4000	OS	1000	270	300	10 200	2900	
November	10 000	4000	1000	160	6700	7200	N/C	
December	2000	7000	1000	2300	N/C	400	N/C	
Mean annual cell numbers								
	24 700	17 500	9400	2600	4900	5000	5700	2100

Notes 1. Underlined figures (e.g. 22 800) indicate the months in which straw treatments were applied. 2. N/C - no algal counts were made. OS indicates when dense blooms of *Anabaena* spp. caused the reservoir to be taken temporarily out of service and cell counts were suspended.

The mean monthly algal cell counts (including those for 1991–1994 previously published by Barrett et al., 1996) are shown in Table 2. After the straw treatment was initiated in 1993, the cell numbers were reduced to approximately one half to one quarter of comparable monthly figures prior to treatment. However, the range of algal species found throughout the year after the straw was introduced remained similar to that found before treatment. In addition, the population dynamics remained broadly similar with spring blooms of diatoms and summer blooms of green algae and cyanobacteria, although at a lower density than before. The main benefit from the treatment was that cyanobacterial numbers were reduced with the consequent reduction in risk to humans, livestock and wildlife.

Other benefits resulting from the straw treatment were observed. At the water treatment works, the reduction in algal numbers reduced the frequency with which filters needed to be cleaned. Before straw treatments started, it had been essential to clean filters at least three times per day, whereas by May 1993 when algal numbers started to fall, this was reduced to the minimum level which management protocols would allow, that is twice daily, although less frequent cleaning would have been adequate. The reduction in diatom numbers also produced a noticeable reduction in odour levels at the water treatment works. This was noted soon after the first straw treatment (Barrett et al., 1996) and has remained as a consistent improvement since then. Circumstantial evidence also suggests that there was a reduction in the number of complaints about odour and water quality from the public. Another effect of the straw treatment was that the clarity of the water in the reservoir increased.

The second objective of the treatment was to develop a satisfactory method of applying straw to large water bodies. It had already been established that bales of barley straw were too tightly compressed and that straw controlled algae most effectively if applied in a loose form with a large surface to volume ratio (Barrett & Newman, 1992). The first treatment consisted of three very large nets of loose straw. Although algal numbers started to drop soon after the first treatment had been applied, it became clear that much of the straw in the trawler nets was rotting very slowly, if at all, so that only a small proportion of the straw was contributing to the algal control. After one year, these nets were removed along with a large mass of straw. In the second treatment, the number of nets of straw was increased, with smaller quantities of straw in each net, to improve the distribution of the straw throughout the reservoir and in the expectation that this would allow the straw to rot more quickly. After this treatment, algal numbers remained low but, again, the straw did not rot quickly and much of the straw had to be removed. In the third and subsequent treatments, the straw was packed loosely into tubular netting with a diameter of about 0.5 m. The straw in these nets rotted completely over a period ranging from approximately

4–6 months after which the empty nets could be easily removed. Thus, after an effective treatment had been established, the dose of straw had been reduced from an initial level of 28 g/m^3–6 g/m^{3}without any apparent increase in algal numbers.

However, a dose of 6 g/m^3 was known to be at the lower end of the effective range for algal control and so, in November 1997, the dose was increased to an equivalent of 9.2 g/m^3 and further increasing in February 1998 to 12 g/m^3 in order to determine if the algal counts would be further reduced by the increased dose of straw. The algal counts up to August 1998 (the latest data available) gave an indication that cell numbers had been reduced although the higher treatment level will be maintained for at least another year in order to confirm this effect.

Discussion

In a trial such as the one described in this paper, the effect of the straw treatment can only be assessed by comparing algal numbers before and after the straw was introduced. Although the data for only two years prior to treatment has been shown, severe algal blooms had been a regular feature of this reservoir for many years. After straw was first introduced and over the subsequent six years of continuous treatment, these blooms have been greatly reduced. During this period, the algal populations have followed a pattern similar to that seen prior to treatment, although at a lower density. This suggests that, while the low dose of barley straw suppressed algal growth, it did not totally prevent the natural expression of growth cycles throughout the season.

This trial is the longest continuous treatment of a water body dominated by diatoms and cyanobacteria known to the authors, although some ditches in East Anglia have been treated with barley straw to control the filamentous alga *Vaucheria dichotoma* for a similar period (T.G.Cave, pers comm). The long-term and continuous exposure of algae to the effects of barley straw in this reservoir has shown two important aspects of the properties of barley straw. First, there has been no noticeable increase in algal numbers since the treatments began. Repeated use of other algicides has sometimes resulted in the development of resistant strains of algae (Gibson & Barrett, 1989), but this does not appear to have occurred to date in this reservoir. Second, there has been no noticeable change in the species of algae found in the reservoir, which might have been expected if the anti-algal products released by the straw were selective to only a limited range of algae. Similar observations of continuous control of *V. dichotoma* without recolonisation by other algal species have been made in the fenland ditches. These two criteria are important if barley straw is to be developed as a long term treatment of reservoirs and other water bodies where algal problems occur.

The method of applying the straw is also important in ensuring efficacy and in minimising handling and labour costs. When the first treatment was applied, it seemed convenient to apply a very large mass of straw which would last a long time and so reduce the need for frequent retreatment. However, the three, very large, nets of straw resulted in a poor distribution of the straw throughout the reservoir and in a very slow rate of decomposition. In the second treatment, the distribution of the straw was improved, but the individual units of straw were again too big and too tightly compressed to allow rapid decomposition and, eventually, had to be removed. Removing the waterlogged, semi-rotten and sometimes anaerobic straw was necessary but was difficult and time consuming. Since the primary objective of applying the straw was to allow it to rot in order to release the anti-algal substances, it was clearly inefficient to pack the straw in a way that impedes this process.

So far, in this paper, the term ‘dose of straw’ has been used to indicate the amount of straw applied on each treatment date. However, algae are not controlled directly by the straw but by substances released as the straw decomposes so that a more accurate measure of dose would be to calculate the amount of straw which had rotted over a measured time. Unfortunately, this cannot be done for the first two applications because an unknown quantity of old straw was taken out when the nets were removed. This old straw was waterlogged and it was not possible to dry it sufficiently to compare the weight of material removed with the original weight in order to determine how much had decomposed. For subsequent treatments, however, the nets were empty when removed so that all the straw in them had decomposed sufficiently to wash out of the nets. This escaping straw had probably not decomposed completely but it is likely that further decomposition would have taken place as the fragments were suspended in the water or sank onto the sediment. It is, therefore, possible to make an approximate and retrospective calculation of the rate of decomposition of the straw as a Mean Daily Dose (MDD) by dividing the quantity of straw applied by the number of days it took to rot. For example, over the period between November 1994 and May 1997, a total of eight straw treatments were applied to the reservoir, each equivalent to 6 g/m^3. By November 1997, all the straw, including that applied in the May 1997 treatment, had rotted sufficiently at least to escape from the nets so that, over the entire period, 48 g/m^3 had rotted in 3 years (approximately 1100 days). Therefore, the MDD over this period was 0.04 g/d/m^3. This may even be a slight over-estimate because there might have been some undecomposed straw remaining on or in the sediment. Nevertheless, it is an indication of the level of activity of the anti-algal substances released during straw decomposition. It also emphasises the importance of applying the straw in a manner which allows decomposition to occur. The disadvantage of applying straw in a form which rots quickly is that it is necessary to repeat the treatments at more frequent intervals, thus adding to the work load.

The amounts of straw applied between November 1994 and May 1997 were known to be at the low end of the effective range for barley straw and there were occasions when sudden, but very temporary, blooms of algae started to appear (e.g. September, 1997). One factor which could affect decomposition rates of barley straw in water is the weather. In particular, high water temperature and turbulence induced by wind and wave action may help the decomposition process. Although no correlation has been made between weather conditions and the appearance of these blooms, it is possible that they occurred during periods when straw decomposition was slow and this may have allowed a temporary bloom to develop until conditions for straw decomposition improved and algal growth was again suppressed.

The objective in increasing the dose of straw in November 1997 and again in February 1998, while retaining the same net size and method of application, was to increase the MDD and to test if this would further suppress algal growth and prevent these temporary blooms. Assuming that decompostion rates remain similar to previous treatments, the treatment applied in February 1998 should give a MDD of 0.08g/d/m^3. Algal counts in 1998 appear to support this hypothesis as, despite record high temperatures in February, diatom numbers did not exceed those found in the spring of several previous years and, subsequently, algal numbers have been at, or below, any values previously recorded.

Barley straw is being used to control algal blooms in other reservoirs by the North of Scotland Water Authority. It is also being used increasingly by other water managers to control a range of algal problems.

Acknowledgement

Our thanks to Mr J. Cockburn, Managing Director Water Services, North of Scotland Water Authority, for permission to publish this work.

References

Barrett, P. R. F., 1994. Field and laboratory experiments on the effects of barley straw on algae. BCPC Monograph No. 59: Comparing glasshouse and field pesticide performance ii: 191–200.

Barrett, P. R. F. & J. R. Newman, 1992. Control of algae. Progress Report, Aquatic Weeds Research Unit, Broadmoor Lane, Sonning, Reading, RG4 0TH. U.K.

Barrett, P. R. F., J. C. Curnow & J. W. Littlejohn, 1996. The control of diatom and cyanobacterial blooms in reservoirs using barley straw. Hydrobiologia 340: 307–311.

Everall, N. C. & D. R. Lees, 1996. The use of barley straw to control general and blue-green algal growth in a Derbyshire reservoir. Wat. Res. 30: 269–276.

Gibson, M. T. & P. R. F. Barrett, 1989. Terbutryn resistance in *Vaucheria dichotoma*. J. Appl. Phycol. 1: 167 –170.

Harriman, R., E. A. Adamson, R. G. J. Shelton & G. Moffett, 1997. An assessment of the effectiveness of straw as an algal inhibitor in an upland Scottish loch. Biocontrol Sci. Technol. 7: 287–296.

Newman, J. R. & P. R. F. Barrett, 1993. Control of *Microcystis aeruginosa* by decomposing barley straw. J. Aquat. Plant Mgmt. 31: 203–206.

Hydrobiologia **415:** 315–318, 1999.
J. M. Caffrey, P. R. F. Barrett, M. T. Ferreira, I.S. Moreira, K. J. Murphy & P. M. Wade (eds), Biology, Ecology and Management of Aquatic Plants.
© 1999 *Kluwer Academic Publishers. Printed in the Netherlands.*

Filamentous algal control using barley straw

J. M. Caffrey* & C. Monahan
Central Fisheries Board, Mobhi Boreen, Glasnevin, Dublin, 9, Ireland

Key words: canals, coarse fisheries, angling, boating

Abstract

Year-round problems with dense growths of filamentous algae reduce the amenity and conservation value of Irish canals. Because algal control operations were relatively ineffective, trials using barley straw were undertaken. These commenced in October 1990 on a section of the Royal Canal where filamentous algae continuously interfered with amenity exploitation and water management. Bales of barley straw were anchored along the canal banks at roughly 50 m intervals. Further straw applications were made in 1991 and 1992. Algal growth in the control section broadly followed a cyclical pattern, with peak biomass between July and September and low production in February and March. In the treated section, however, algal biomass decreased from the time the straw was first introduced. Thereafter, as long as rotted straw was present, no filamentous algae were recorded. The absence of algae in this section between August 1991 and spring/summer 1993 permitted the recolonisation of higher plants, which are commonly less troublesome and more ecologically useful than algae.

Introduction

Dense growths of filamentous algae pose problems for amenity and recreational exploitation of long sections of canals in Ireland. The canals are managed as amenity waterways, where angling and boating are the main recreational pursuits. The canal corridors are also extensively used for walking, picnicing and nature study. The presence of unsightly and often foul-smelling algae floating on the water surface detracts from the aesthetic appeal of the canals and seriously impacts any water-based activity. The canals are meso- to eutrophic and, where competing phanerogams are reduced or absent, algal populations have the potential to develop to nuisance proportions. Thus, where weed control operations denude or seriously reduce the rooted vegetation, dense filamentous algal infestations commonly occur. A survey conducted on the Royal and Grand Canals in 1994 estimated that 7% of the channel (c. 23 km) supported nuisance levels of filamentous algae (Caffrey & Allison, 1998).

Excessive algal growth impeded boat traffic, obstructs angling, clogs sluices and lock chambers, in addition to creating unsightly and malodorous masses. The growth of large algal populations can result in serious diurnal fluctuations in dissolved oxygen levels and in fish kills (Caffrey, 1992). Likewise, the death and decay of a large algal biomass can deoxygenate the water, killing fish and other aquatic fauna (Barrett & Newman, 1993; Barrett & Banks, 1995). Floating mats of algae can reduce the level of incident light that reaches submerged plants, thereby restricting growth.

Control of filamentous algae in the canals using mechanical methods (cutting, raking, harvesting) has been largely unsuccessful. This is because large numbers of fragments remain and regrowth is rapid (Caffrey, 1990). The use of terbutryn and diquat in certain sections of canal provided moderate to good results against algae but, because these herbicides are not selective, they also killed useful submerged plant species (Caffrey & Monahan, 1995).

There is a considerable body of research that demonstrates the antialgal properties of rotted barley straw, properties bestowed without having any discernible adverse impact on higher plants, fish or

* Author for correspondence

invertebrate fauna (e.g. Welch et al., 1990; Newman & Barrett, 1993; Barrett et al., 1999). A primary requirement for the successful use of barley straw is the maintenance of aerobic conditions (Ridge et al., 1995). Unstable, short-lived algal inhibitors are released during the aerobic decomposition of the straw. These are highly selective against planktonic and filamentous algae (Newman et al., 1994; Barrett et al., 1996) and are algistatic rather than algicidal. There is strong evidence that these algal inhibitors are derived from oxidised polyphenolics released from solubilsed lignin (Ridge & Pillinger, 1996), although the precise nature of the inhibitors or their mode of action remains unknown.

The effects of more than 100 barley straw treatments in the U.K. and Ireland were documented by Barrett & Newman (1993) and results revealed that algal control was achieved to at least some extent, in all types of water body, but was better in smaller waterbodies (< 5 ha).

In order to quantitatively assess the efficacy of barley straw in controlling nuisance filamentous algae in amenity canals, a trial was established in a section where algal growth annually impaired all beneficial use of the waterway. The effects on the algae, on the plant community structure in the area and implications for canal management are discussed.

Methods and materials

Site description

The trial was conducted on the 16th level of the Royal Canal between October 1990 and October 1994. The physical and physico-chemical characteristics of the canal are described in Caffrey et al. (1998). Essentially, the canal is 13 m wide, 1.5 m deep, supports a negligible flow and a deep, mud substrate. The banks are open and a boat traffic density does not exceed 100 movements ha^{-1} yr^{-1} (*sensu* Murphy & Eaton, 1983). Filamentous algae are present throughout the level all year round and generally reach nuisance proportions between April and October.

For trial purposes, the 2 km level was divided in two. The downstream portion was treated with barley straw while the upstream section functioned as an untreated control. This permitted comparisons to be made between treated and untreated waters, a facility not available in most other documented studies.

Straw treatment

In October 1990, straw was first applied to the canal. Bales of barley straw, retained in garden netting, were anchored along the margins of the canal. The straw was loosened to give a good surface-to-volume ratio and to allow the straw to rot quickly and aerobically (Barrett et al., 1996). Polypropylene bottles provided buoyancy which kept the straw mass off the bottom, thus ensuring maximum liberation of antialgal factor into the water. The straw mattresses were spaced at 50 m intervals and alternated from bank to bank. A quantity of straw to provide a dose of 10 g m^{-3} was applied (Barrett & Banks, 1995; Ridge et al., 1995). Subsequent treatments were applied as the submerged straw rotted away. Applications were made at intervals of between 4 and 7 months, up to February 1993. Between this and late November no straw was applied (by omission rather than intent). A final straw application was made in November 1993.

Biomass sampling

The biomass of filamentous algae in the control and treated sections was quantitatively assessed on 21 occasions over the four year study. On each site visit 10 replicate stovepipe samples (Caffrey, 1990) were taken from each of the control and treated sections. The stovepipe, made from heavy duty plastic material, measured 80 cm in height and 23 cm in internal diameter. It was thrust forcefully into the vegetation and pushed deep into the substratum. All the vegetation contained within the stovepipe was removed by hand and placed into labelled bags. The plant biomass was recorded as g dry weight m^{-2}.

Results

In the control section of canal, the filamentous algae, mainly *Cladophora glomerata*, broadly followed the normal annual growth cycle, with peak standing crops between July and September and biomass troughs in February and March (Figure 1). This section remained virtually unnavigable and unfishable, and posed considerable maintenance problems to the canal managers. In the straw-treated section, by contrast, there was no significant resurgence of growth in spring/summer 1991 and by August of that year no algae were present. On this same sampling date, a standing crop of 113 g m^{-2} was recorded in the contiguous, untreated section.

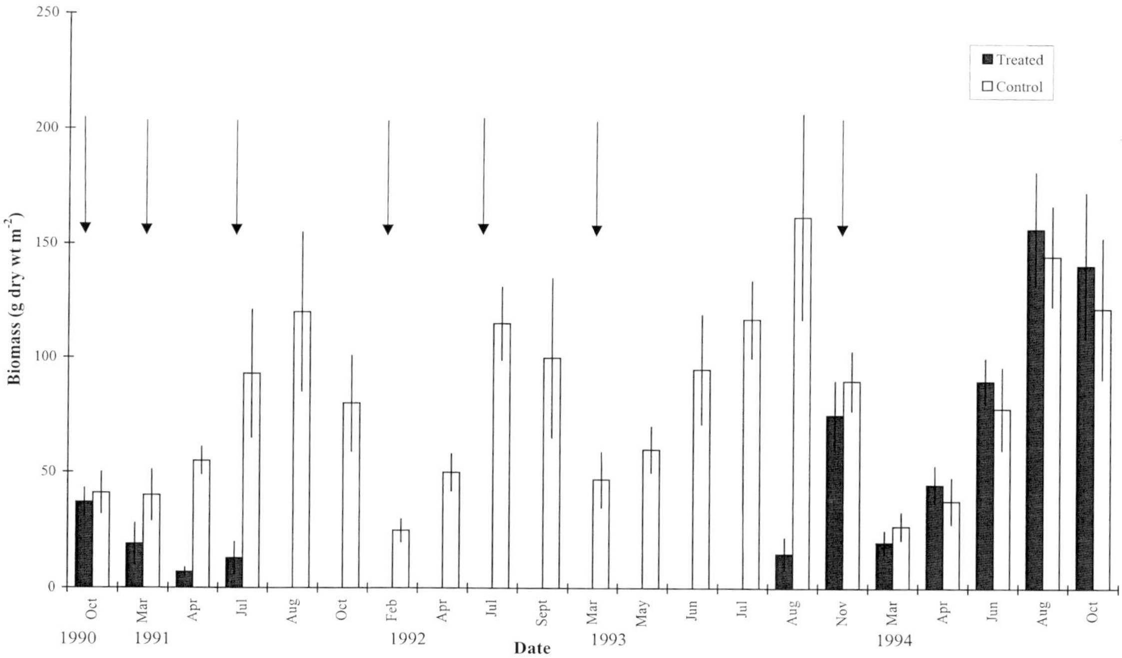

Figure 1. Effect of successive barley straw treatments on the biomass of filamentous algae in the Royal Canal between 1990 and 1994. S.E. bars are included. ↓ Straw replacement.

Regular biomass sampling between August 1991 and August 1993 failed to produce any filamentous algae in the treated section. In November of that year an algal crop of 62 g m^{-2} was taken in this section. This was unexpected as the straw had, to this date, successfully suppressed algal growth. An enquiry revealed that the replacement straw due in July had not been added and, consequently, the antialgal factor in the water decreased to a level below which it suppressed algae. A single straw treatment in November 1993 elicited no significant response among the resident algae (Figure 1).

Vegetation sampling in the treated section in April 1992 revealed localised stands of *Myriophyllum verticillatum*, presumably established from turions washed in from canal levels upstream. These turions had not previously established, probably because of competitive exclusion by the carpet-forming algae. No *M. verticillatum* was present in the control section. By July 1993, *Myriophyllum* had established a healthy population in the treated section and occupied *circa* 25% cover. A maximum standing crop of 85 g m^{-2} for this plant was recorded on this date.

Discussion

Following the introduction of barley straw into the canal, there was a significant reduction in the standing crop of filamentous algae. Continuous exposure to barley straw and the unstable antialgal factor over a two-and-a-half year period eliminated the nuisance algae and allowed for desirable changes in plant community structure. No such changes were recorded in the contiguous untreated section. Failure to replace straw that had rotted away and no longer produced antialgal substances resulted in the regrowth of filamentous algae and a reversal to a monospecific plant community.

The implications of these findings for the management of the canals are significant. Mechanical control of algal infested channels is costly, disruptive and ineffective, while chemical control commonly results in the elimination of algae and higher plants, leaving them available for subsequent recolonisation by opportunistic algae. The use of barley straw, on the other hand, is effective, cost-efficient, selective and environmentally-friendly. As such, it offers a valuable tool to the water manager for the control of nuisance algae.

In addition to providing effective control against algae, barley straw produces conditions that favour the establishment of diverse plant and faunal communities (Welch et al., 1990; Barrett & Newman, 1993; Caffrey, 1993a; Everall & Lees, 1996). This effect was clearly demonstrated in a trial conducted on a 12 ha eutrophic lake in the south of Ireland (Champ & Caffrey, 1994). Dense algal blooms annually restricted trout angling on this watercourse and, in May 1990, 120 loosened bales of straw were placed in corrals throughout the lake. By September 1990, water transparency had improved and stands of *Myriophyllum spicatum* and *Elodea canadensis* were evident. In 1991, this rooted vegetation gained a strong foothold and restricted the establishment of algae.

Different growth form categories of aquatic plants support different mean standing crops of fish (Caffrey, 1993b). The highest figures were recorded in canal sections where submerged plants with broad or complex leaf types (e.g. *Myriophyllum* spp.) predominated. Sections where filamentous algae were the principal species supported less than one third of the mean fish crop present in the above areas. Likewise, the number and diversity of macroinvertebrate species is increased in channels where diverse phanerogam vegetation is present and reduced where filamentous algae prevail (Monahan & Caffrey, 1996). The changes brought about by barley straw treatment therefore increase ecological diversity and conservation value of the channel.

The addition of straw also provides a substrate for invertebrate colonisation and, in the present trial, large numbers of *Asellus aquaticus* were observed in and beneath the straw mattresses. Similar observations were made by Welch et al. (1990), Barrett & Newman (1993), Everall & Lees (1995) and Harriman et al. (1997).

Results from the present trial and other studies offer conclusive proof that, in most instances, algal problems in static or semi-static watercourses can be resolved using barley straw. It must be borne in mind, however, that this effect may only be cosmetic and that the underlying causes of the algal infestation, whether pollution or ecological disruption, must be addressed before long-term control will be achieved.

References

Barrett, P. R. F. & J. R. Newman, 1993. The control of algae with barley straw. In Straw – A Valuable New Material, PIRA: Paper 41.

Barrett, P. R. F. & N. Banks, 1995. Aquatic weed control operation. Best practice guidelines. National Rivers Authority, R&D Note 395.

Barrett, P. R. F., J. C. Curnow & J. W. Littlejohn, 1996. The control of diatom and cyanobacterial blooms in reservoirs using barley straw. Hydrobiologia 340: 307–311.

Barrett, P. R. F., J. W. Littlejohn & J. C. Curnow, 1999. Long-term algal control in a reservoir using barley straw. Hydrobiologia 415 (Dev. Hydrobiol. 147): 309–313.

Caffrey, J. M., 1990. The classification, ecology and dynamics of aquatic plant communities in some Irish rivers. Ph.D. Thesis, University College, Dublin: 254 pp.

Caffrey, J. M., 1992. Management of aquatic and invasive riparian vegetation in Irish amenity watercourses. In Mollan, C. (ed.), Water of Life. Royal Dublin Society, Dublin: 130–141.

Caffrey, J. M., 1993a. Plant management as an integral part of Ireland's aquatic resources. Hydroecol. Appliq.: 77–96.

Caffrey, J. M., 1993b. Aquatic plant management in relation to Irish recreational fisheries development. J. Aquat. Plant Mgmt. 31: 162–168.

Caffrey, J. M & J. P. Allison, 1998. Eutrophication in canals. In Mollan, C. (ed.), Eutrophication in Irish Waters. Royal Irish Academy, Dublin: 1–10.

Caffrey, J. M. & C. Monahan, 1995. Aquatic plant management in Irish canals, 1990–1995. Office of Public Works commissioned report. Centr. Fish. Bd, Dublin: 102 pp.

Caffrey J. M., J. P. Allison, R. E. Donnelly, D. M. Tierney & C. Monahan, 1998. Development and management of recreational angling in Irish canals. In Hickey, P. (ed.), Recreational Fisheries: Social, Economic and Management Aspects. EIFAC/FAO: 214–226.

Champ, W. S. T. & J. Caffrey, 1994. Preliminary results from barley straw trials in Irish watercourses. 9th EWRS Symp. on Aquat. Weeds. Abstracts: 12 pp.

Everall, N. C. & D. R. Lees, 1996. The use of barley straw to control general and blue-green algal growth in a Derbyshire reservoir. Wat. Res. 30: 269–276.

Harriman, R., E. A. Adamson, R. G. Shelton & G. Moffett, 1997. An assessment of the effectiveness of straw as an algal inhibitor in an upland Scottish loch. Biocontr. Sci. Technol. 7: 287–296.

Monahan, C. & J. M. Caffrey, 1996. The effect of weed control practices on macroinvertebrate communities in Irish canals. Hydrobiologia 340: 205–211.

Murphy, K. J. & J. W. Eaton, 1983. Effects of pleasure boat traffic on macrophyte growth in canals. J. appl. Ecol. 20: 713–729.

Newman, J. R. & P. R. F. Barrett, 1993. Control of *Microcystis aeruginosa* by decomposing barley straw. J. Aquat. Plant Mgmt. 31: 203–206.

Newman, J. R., P. R. F. Barrett & F. G. Cave, 1994. The use of straw to control algae in drainage ditches – an ecological survey. Proc. Nature Conservation in Drainage Habitats, Nottingham University.

Ridge, I. & J. M. Pillinger, 1996. Towards understanding the nature of algal inhibitors from barley straw. Hydrobiologia 340: 301–305.

Ridge, I., J. M. Pillinger & J. Walters, 1995. Alleviating the problems of excessive algal growth. In Harper, D. M. & A. J. D. Ferguson (eds), The Ecological Basis for River Management. John Wiley & Sons Ltd: 211–218.

Welch, I. M., P. R. F. Barrett, M. T. Gibson & I. Ridge, 1990. Barley straw as an inhibitor of algal growth I: studies in the Chesterfield Canal. J. appl. Phycol. 2: 231–239.

Hydrobiologia **415:** 319–322, 1999.
J. M. Caffrey, P. R. F. Barrett, M. T. Ferreira, I.S. Moreira, K. J. Murphy & P. M. Wade (eds), Biology, Ecology and Management of Aquatic Plants.
© 1999 *Kluwer Academic Publishers. Printed in the Netherlands.*

Preliminary observations on the control of algal growth by magnetic treatment of water

Jonathan R. Newman* & Rachael C. Watson
IACR-Centre for Aquatic Plant Management, Broadmoor Lane, Sonning, Reading, Berkshire RG4 6TH, U.K.

Key words: calcium, plasmalemma

Abstract

Cladophora glomerata (L.) Kz. was grown in Jaworski's algal culture medium and exposed to a fluctuating electromagnetic field which induced development of dark colouration, increased density and brittleness. After 4 weeks exposure, algae were unable to grow when transferred to unexposed culture medium. In addition, algal growth in culture medium exposed to a fluctuating magnetic field before the algae were introduced, was inhibited significantly compared with an appropriate control after 4 days, indicating a transfer of effect from the brief exposure to a magnetic field of less than 1 second to the 4-day period when growth was active.

Introduction

The control of filamentous algae is one of the most expensive and time-consuming aquatic weed control operations in temperate regions. With increased public awareness of pesticide residues in water and pressure from Water Utility Companies to prevent pesticide pollution of water, the development of non-chemical methods of algal control has become increasingly desirable and feasible.

Reduction of algal growth by exposure of water to magnetic fields has been known for 20 years. An American patent (Rigby, 1977) relates to the placement of magnets on pipes to reduce the growth of algae in swimming pools, . But this provides no data on how the control effect is produced in algae. In other situations, algae have been removed from water by magnetic filtration (Bitton et al., 1975). Magnetic treatment of water has also been shown to reduce the growth of zebra mussels (Smythe et al., 1997).

In contrast to many of the experiments on the effects of magnetic fields on biological systems (Jones & McLeod, 1995), the system described in this paper is one in which the magnetic field acts on water which subsequently comes into contact with cells and not directly on the cells themselves. There is a spatial separation of the exposure site from the growth site. Any effect on the algae will therefore imply that algae can sense the induction of a change in the physical or chemical properties of ions in water exposed to magnetic fields.

Although there are data on how magnetic field affect precipitation of ionic compounds from water, nothing is known about how dissolved ions are affected, or how these ions react with plant cell membranes. Plant membranes rely on the development of a specific trans-membrane potential to recognise particular dissolved ions (Maathuis et al., 1997). If magnetic treatment of water interferes with this recognition of ions, then ion transport difficulties may arise, with consequences such as super-accumulation of ions, reduction or inhibition of ion uptake or uncontrollable leakage of ions out of the cell. These mechanisms would imbalance ionic concentrations within the cell, altering intracellular pH, affecting intracellular signalling and transport of ions and disrupting cell function. The result would be reduced growth, possibly cell death and eventual control of nuisance algal species exposed to such fields.

The present work demonstrates that growth of the filamentous alga, *C. glomerata,* and associated growth of mixed unicellular algae is reduced in water exposed to low strength magnetic fields.

* Author for correspondence

Materials and methods

An electronic water treatment system (Scalewatcher II, Fast Systems, Henley, UK), which treats water with variable magnetic fields, was used in the experiments. The system passes a fluctuating DC current through a wire coil wrapped 11 times around the water supply pipe producing a magnetic field of fluctuating strength. In a second system, two ferrous magnets (100 × 50 × 25.4 mm) were used to create a constant magnetic flux density of 1900 gauss through a supply pipe, by positioning them 10 mm apart on either side of the pipe with the long broad axes parallel to the pipe.

Sterile Jaworski's algal culture medium (JM1) (Thompkins et al., 1995) was pumped at 6.5 litres min^{-1} through a clear plastic tube (12 mm internal diameter) surrounded by the coils of the Scalewatcher II system. The culture medium was passed through the pipe once in Experiment 1 and pumped past the coils again in Experiment 2, thereby doubling the exposure time in the latter case. Exposure time was calculated as the time 50 cm^3 took to pass by the coils of the system, which occupied 6 cm of supply pipe. Treated culture medium was then sterilized by membrane filtration (0.2 μm) of 50 cm^3 aliquots into autoclaved 250 cm^3 flasks and inoculated with 0.5 g fresh weight (fwt) of *Cladophora glomerata* (L.) Kz. filaments previously grown under controlled conditions in the laboratory. Growth (fwt) was measured after 96 h. In a third experiment, JM 1 (70l) in a tank was inoculated with 50 g of *C. glomerata* and circulated continuously through a tube of 12 mm internal diameter to which the Scale-Watcher II apparatus was fitted, at a flow rate of 3.0 l min^{-1}. All experiments were done at 23 °C and 150 μmol photons m^{-2} s^{-1} PAR. . A control was established with identical flow regime, light, temperature and algal inoculum. Both treatment and control systems were open to the atmosphere and subject to airborne contamination. A fourth experiment was set up to compare the effects of ferrous magnets (constant magnetic field strength) with those produced by the electromagnetic system. Three sets of three tanks (50 l) were used with water recirculating through 12 mm internal diameter plastic pipe. The tanks were arranged on three levels, so that the water flowed at 1.4 l min^{-1} from the bottom tank to the top and down through the system. Tanks were inoculated with 2 g fwt of *C. glomerata* and left undisturbed for 10 weeks, apart from topping up on a weekly basis, at 23 °C under natural irradiance. Growth of phytoplankton was measured by absorbance of water samples at 678 nm in a Cecil 3000 series spectrophotometer on a weekly basis.

Regrowth of filamentous algae was assessed by inoculation of 0.2 g of alga from treated and control experiments into new JM 1 medium and measurement of fwt after 96 h.

Ca^{2+} concentrations in water were measured according to the methods in Mackereth et al. (1989). pH was measured using a gel combination electrode attached to a Jenway 3020 pH meter, automatically adjusted for temperature.

Results

C. glomerata inoculated into JM1 medium that had received one exposure to the fluctuating magnetic field (Experiment 1) appeared to grow normally during the 4 d incubation period. The trichomes were clearly elongated and separate from each other. In contrast, when grown for 4 d in JM1 medium previously exposed twice to the magnetic field (Experiment 2), there was a significant decrease in biomass (Table 1) and the filaments were clumped into small spherical aggregates. In both treatments the algae regrew normally when transferred to untreated JM 1 medium (Table 1).

In Experiment 3, continuous cycling of the medium through the fluctuating electromagnetic field for 28 d not only prevented growth of *C. glomerata* but also significantly reduced final fresh weight values below that of the initial inoculum (Table 2). The biomass reduction (67%) was accompanied by discolouration and necrosis of the algal filaments. The algal mat sank to the bottom of the experimental chamber after 4 d and did not re-float throughout the experiment. This sunken mat became black, brittle and fragmented easily. The algae in untreated JM 1 formed a healthy floating mass at the surface with a biomass increase of 348%. The algae from treated medium did not regrow when transferred to fresh medium for 1 week at the end of the 28 d experiment. Growth of contaminating suspended algae in the treated system was less than that in the control.

Growth of unicellular algae was also measured in Experiment 4 (Table 3). While growth was not reduced in water exposed to the fixed magnetic field, the fluctuating electromagnetic field system reduced suspended (mainly unicellular) algal growth to 17% of that in the control.

Table 1. Experiments 1 & 2. Growth of *C. glomerata* (g fwt) after 96 h in JMI algal culture medium previously exposed to afluctuating magnetic field. Data are the means of three replicate samples ± standard deviation (n=3)

Treatment Regime	Experiment 1 (1 pass)	Experiment 2 (2 passes)
Exposure time of water (seconds)	0.063	0.125
Exposure time of alga to water (days)	4	4
Control	1.27 ± 0.12	1.09 ± 0.09
Treatment	1.26 ± 0.07	0.78 ± 0.13
Regrowth after 1 week	2.98 ± 0.13	2.68 ± 0.36

Table 2. Experiment 3. Biomass of *C. glomerata* after 28 d growth in medium continuously circulated through an electromagnetic field. Data from single control and treatment

Treatment	Weight of *C. glomerata* (g)
Control	174.0
Electromagnetic	16.4

Table 3. Experiment 4. Suspended algae measured by chlorophyll content, and Ca^{2+} concentrations after 2 weeks. Data are means ± standard deviation (n=5).

Parameter	Control	Fixed Magnet	Electromagnet
Chl *a* (μg l^{-1})	19.40 ± 1.70	22.855 ± 1.35	7.77 ± 2.07
(Ca^{2+}) (mmol l^{-1})	34.87 ± 1.84	30.13 ± 1.30	32.00 ± 1.17

Discussion

Although these data are from limited preliminary experiments, the treatment of flowing JM1 algal culture medium in a fluctuating magnetic field clearly affected its ability to support the growth and survival of *C. glomerata*. Increased exposure time and continuous exposure to magnetic fields exacerbated the effects. The reasons for the observed treatment effects of biomass reduction, colony morphology changes and cell necrosis have not been established, but the effects are sufficient to be exploited for the control of algae in any situation, such as ornamental ponds or fountains, where recirculating or piped water supplies are used.

The primary mode of action of magnetic fields on flowing water is believed to be linked to an increased rate of precipitation of mineral ions around suitable nuclei in water (Wang et al., 1997). These nuclei are mainly Ca^{2+} ions associated with $CO_3{}^{2-}$ ions. However, dissolved [Ca^{2+}] did not alter significantly as a result of the treatment (Table 3). A consequence of exposure to magnetic fields increased above background levels is that $CaCO_3$ changes its crystal structure becoming more easily precipitated; this results in larger crystal sizes and more rapid precipitation due to increased individual crystal mass (Parsons et al., 1997a). What happens to the calcium which is left in solution is, at present, unclear. As an effect can be demonstrated on crystalline calcium carbonate, then there may also be an effect on dissolved calcium carbonate which is, in effect, very small crystals of a few molecules in size. Barrett & Parsons (1998) assert that the magnetic effect is maintained for at least 60 h after exposure. This could account for the observed growth suppression in Experiment 2. They also state that magnetic exposure exerts a greater effect on Na_2CO_3 than on $CaCl_2$. Thus, there may be effects on other metal ion complexes in water which could interact with algal cell membranes.

The clumping of filaments observed in Experiment 2 may be associated with a change in the cell surface charge resulting in attraction between filaments. A similar observation has been made for yeast cells. In magnetically treated beer lines, yeast cells clump and remain in suspension longer, rather than adhering to the walls of polythene tubing, as occurs in untreated systems (Bland, pers. comm., 1997). Effects on cell surface charges are theoretically possible (Kovacs et al., 1997).

A measurable change in pH of treated water has been reported by Parsons et al. (1997b). A similar effect (reduction of 1 pH unit) has been noted in our work. This would increase the solubility of $CaCO_3$ and reduce the build up of carbonate scale. In buffered systems, no reduction in scale is evident (Parsons et al., 1997b). It would also increase the solubility of existing scale deposits and account for reductions in these deposits, which is observed when magnetic water treatment is used. A reduction of pH from 9 to 8 would increase both photosynthesis and growth rates of submerged plants, due to increased availability of 'free' CO_2, the carbon source for the photosynthetic enzyme, ribulose bisphosphate carboxylase oxygenase (Rubisco). However, photosynthetic rates were not measured and future experiments will provide these data.

Filamentous algae growing in continuously treated water first show reduced growth rate and morphological change as described in this paper but then, after about 6 months, they are reported to grow back to their original state. If the magnetic field is then removed, the new crop of algae undergoes exactly the same process of growth alteration and death as when the magnetic field was first applied (Fast Systems, pers. comm., 1997). This implies that there are at least two states which can exist in the same algal species with respect to recognition of extracellular ions. This may be related to a switch in extracellular charge.

This area of research promises to offer several insights into how plant cell membranes recognise ions and how membrane function is related to internal processes. Adaptation to altered magnetic field strength or direction may be an inherited trait. The polarity of the Earth's magnetic field has changed several times over geological time and algae have been in existence during these changes. Therefore, they may have an inherent ability to adapt to changing magnetic field strength and direction, which may be expressed by altered mineral uptake processes or changes in membrane function.

Acknowledgements

IACR-Centre for Aquatic Plant Management is funded by a consortium of UK subscribers. IACR receives grant-aided support from the Biotechnology and Biological Sciences Research Council of the UK. The authors thank Fast Systems of Henley for information, Martin Cummins of 'Cermag Ltd' and Hugh Bland of 'Magnaflow' for providing fixed magnet equipment and experimental reports.

References

Barrett, R. A & S. A. Parsons, 1998. The influence of magnetic fields on calcium carbonate precipitiation. Wat. Res. 32: 609–612.

Bitton, G., J. L. Fox & H. G. Strickland, 1975. Removal of algae from lakes by magnetic filtration. Microbiology 30: 905–908.

Jones, D. & B. McLeod, 1995. Electromagnetic cell stimulation. In Lynch, P. T. & M. R. Davey (eds), Electrical Manipulation of Cells. Chapman & Hall, New York: 223–247.

Kovacs, P. E., R. L. Valentine & P. J Alvarez, 1997. The effect of static magnetic fields on biological systems: Implications for enhanced biodegradation. Critical Rev. Envir. Sci. Techn. 27: 319–382.

Maathuis, F. J. M., D. Sanders & D. Gradmann, 1997. Kinetics of high affinity K^+ uptake in plants, derived from K^+-induced changes in current-voltage relationships. A modelling approach to the analysis of carrier-mediated transport. Planta 203: 229–236.

Mackereth, F. J. H., J. Heron & J. F. Talling, 1989. Water analysis. Some revised methods for limnologists. Freshwat. Biol. Assoc. Scient. Publ. 36.

Parsons, S. A., S. J. Judd, T. Stephenson, S. Udol & B.-L. Wang, 1997a. Magnetically augmented water treatment. Transac. Inst. Chem. Engin. 75: 98–104.

Parsons, S. A., B.-L. Wang, S. J. Judd & T. Stephenson, 1997b. Magnetic treatment of calcium carbonate scale – effect of pH control. Wat. Res. 31: 339–342.

Rigby, R. A., 1977. Algae growth control. Official Gazette of the United States Patent Office 965: 1487.

Smythe, A. G., C. L. Lange, T. M. Short, E. A., Dardeau & M. Brown-Hester, 1997. Evaluation of an in-line magnetic system as a control for zebra mussels. In Ditri, F. M. (ed.), Zebra Mussels and Aquatic Nuisance Species, Ann Arbor Press Inc.: 439–465.

Thompkins, J., M. M. De Ville, J. G. Day & M. F. Turner, 1995. Culture Collection of Algae and Protozoa. Catalogue of Strains. Natural Environment Research Council.

Wang, Y., A. J. Babchin, L. T. Chernyi, R. S. Chow & R. P. Sawatzky, 1997. Rapid onset of calcium carbonate crystallization under the influence of a magnetic field. Wat. Res. 31: 346–350.

Hydrobiologia **415:** 323–327, 1999.
J. M. Caffrey, P. R. F. Barrett, M. T. Ferreira, I.S. Moreira, K. J. Murphy & P. M. Wade (eds),
Biology, Ecology and Management of Aquatic Plants.
© 1999 *Kluwer Academic Publishers. Printed in the Netherlands.*

Biomass characterization of *Azolla filiculoides* grown in natural ecosystems and wastewater

M. Lourdes Costa[1], M. Conceição Santos[2] & Francisco Carrapiço[3]
[1]*Escola Superior Agrária de Coimbra, I.P.C., Bencanta 3000 Coimbra, Portugal*
E-mail: mlcosta@mail.esac.pt
[2]*Departamento de Ciências e Engenharia do Ambiente, Faculdade de Ciências e Tecnologia, Universidade Nova de Lisboa, 2825 Campus de Caparica, Portugal*
E-mail: mcrs@mail.fct.unl.pt
[3]*Centro de Biologia Ambiental, Departamento de Biologia Vegetal, Faculdade de Ciências da Universidade de Lisboa, Ed. C2, Campo Grande, 1749-016 Lisboa, Portugal*
E-mail: F.Carrapico@fc.ul.pt

Key words: biofilter, productivity, phosphorus, water quality, *Azolla*

Abstract

The aquatic pteridophyte *Azolla*, a small-leaf floating plant, which lives in symbiosis with a nitrogen fixing cyanobacteria, *Anabaena azollae*, was widespread throughout water channels and hydrographic basins of Portugal. *Azolla* is also the aim of a study for its utilization as a biofilter for wastewater purification, namely for phosphorus removal (± 36%). The goal of this work is to compare the growth characteristics and biomass composition of this water fern in natural ecosystems with those obtained in some wastewaters. Plant growth rate (0.107± 0.037 d^{-1}) and productivity (5.8 g dw m^{-2} d^{-1}) suggest that *Azolla* can grow well in partially treated domestic wastewater, but not in diluted pig wastes. This fact, associated to its biomass composition, namely in phosphorus content (1.38 ±0.20%), increase the possibility of this plant being used to improve wastewater discharge quality. It may also be possible to use the biomass as a biofertiliser or as a feed supplement for aquatic and terrestrial animals due to its protein, crude fiber and mineral content.

Introduction

In Portugal, the aquatic pteridophyte *Azolla*, a small-leaf floating fern that lives in symbiosis with a nitrogen fixing cyanobacteria, is considered as a weed and can be found in many hydrographic basins, namely those of the Tagus, Sado, Mondego, Vouga, Coa and Guadiana rivers. Climatic conditions and eutrophication of many aquatic ecosystem aid its development.

Due to the existence of the symbiosis, *Azolla* can grow in aquatic environments devoid of combined nitrogen, if there is enough phosphorus. In fact, phosphorus is for *Azolla*, as for many other photoautrotrophic aquatic organisms, the limiting nutrient for growth. On the other hand, many eutrophication problems in our country are due to the discharge of domestic or industrial wastewaters only submitted to secondary biological treatment. These effluents contains high levels of nutrients and tertiary chemical treatments for nutrient removal are very expensive.

Macrophyte utilization for wastewater nutrient removal had been already reported and discussed frequently (Fisher, 1988; Kumar & Garde, 1989; Bishop & Eighmy, 1989; DeBusk et al., 1989; Oron, 1990). In fact, the alternative of using macrophytes to treat and polish effluents may present some advantages such as low costs and easy maintenance. The success of this depends on adequate plant growth all over the year and on the rate of nutrient uptake.

Azolla sp. has been intensively studied during the last few years, due to its potential use as a green manure in rice fields of Asia and Africa and as a feed supplement for aquatic and terrestrial animals (Lumpkin & Plucknett, 1980). The plant utilization in wastewater treatment to remove contaminants, particularly phosphorus, was also suggested by some authors (Reddy & DeBusk, 1985; Fisher, 1988). *Azolla* sp. is unique among floating macrophytes, because it can grow in waters devoid of combined nitrogen, due to the symbiosis with a N_2 fixing cyanobacterium, *Anabaena azollae*, that lives in the dorsal lobe cavity of its leaf. So, this plant can grow even after the exaustion of combined nitrogen in secondary effluents, improving an adequate phosphorus removal (Kitoh et al., 1993).

Azolla has not yet been studied for phosphorus removal from wastewater, and as it seemed to be a good phosphorus assimilator, even in the absence of nitrogen, we studied its performance in this field. Besides, the biomass of *Azolla* produced in wastewater treatment plants could be easily harvested and utilized as a fertilizer for agriculture, or as feedstock for animals.

It is known that biomass composition limits its utilization possibilities. So, the goal of this work is to assess and compare the growth rate, productivity, nitrogen and phosphorus uptake and the chemical composition of *Azolla* grown in natural aquatic environments, like Samora Correia irrigation channels (about 3 m width and 1 m depth), Guadiana river (flow rate less than 10 $m^3 s^{-1}$) and Alverca lagoon–Golegã (about 20 m width and 6–7 m depth), with that produced in a maturation pond effluent from a sewage treatment plant, and in a solution with 5% pig wastes.

Material and methods

The growth rate and productivity of *Azolla* grown in Samora Correia irrigation channels had been evaluated under confined conditions with 0.16 m^2 circular floating PVC net frames, during seven weeks, from February to April, under the climatic conditions prevailing in the rural area. The natural photoperiod during this period averaged 12 h d^{-1}, with light intensities from less than 25 Wm^{-2} to about 51 Wm^{-2}. The air temperature ranged from 11.5 °C to 20 °C, the corresponding water temperature ranged from 10.5 °C to 15 °C, and the mean relative humidity was about 85%.

Biomass and water samples were collected in Samora Correia irrigation channels during two plant growth cycles, from 1991 to 1993, in the Guadiana river during the 1993 bloom and in Alverca lagoon (Golegã) at 1995. *Azolla*'s productivity in Guadiana river and Alverca lagoon have been estimated from Samora's *Azolla* growth rate, because the growth period and climatic conditions in those sites are similar. Water samples were analysed according to *Standard Methods for Examination of Water and Wastewater Analysis* (APHA, 1992).

The growth of *Azolla* in a solution of 5% pig waste and in a maturation pond effluent, were, carried out in PVC reactors (detention time, tr=10 d), with 18.3l capacity, within confining PVC frames (A=0.11 m^2), inside a Fitoclima 750 E culture chamber. Controlled conditions were: 25±0.5 °C, during 14 h d light (31 Wm^{-2}) and 18±0.5 °C during 10/h night period; humidity was between 70 and 75%.The methodology for growth rate, density and productivity evaluation has been described earlier (Costa et al., 1994, 1996).

The phosphorus and nitrogen uptake (mgP m^{-2} d^{-1} and mgN m^{-2} d^{-1}) were calculated based on growth parameters and on biomass phosphorus or nitrogen content. In order to evaluate phosphorus bioaccumulation, phosphorus yield [mgP (g $dw)^{-1}$] had been determined based on phosphorus uptake and productivity.

Analysis of dry and organic matter, ash content, crude fiber, cellulose, hemicellulose, lignin, calcium and magnesium were performed according to AOAC (1990), regarding the potential uses of *Azolla*. Heavy metals (Cu, Zn, Pb, Ni, Cd and Cr) were also determined in order to evaluate possible toxicity for animals when it is used for animal feedstock. The results of biomass composition are expressed on a dry weight basis.

Multiple correlations between different *Azolla*'s growth parameters, biomass composition and medium nutrients concentration, in natural ecosystems and wastewaters, had been calculated.

Results and discussion

Under field conditions, *Azolla filiculoides* grows between November–December and May–June, and its sporulation began in March–April. In these ecosystems, water salinity changed from 0.8 to 1.2 mg L^{-1}. The highest values observed did not affect significantly the fern growth.

Azolla growth rates (Table 1) in maturation pond effluent and in Samora Correia channels were similar, but lower than those observed in Hoagland medium H-40 under the same controlled conditions (0.146±0.044 d^{-1}) (Costa, unpublished data), for densities ranging between 6.82 g dw m^{-2} and 68.2 g dw m^{-2}. Never-

Table 1. Growth parameters, phosphorus and nitrogen uptake of *Azolla*

Variables	Samora	Correia	Guadiana river	Alverca lagoon	Pig waste	Maturation pond effluent
	Confined	Field				
Growth rate (d^{-1})	0.106±0.020 (n=2)	–	–	–	0.054±0.012 (n=14)	0.107±0.037 (n=7)
Density (g dw m^{-2})	66.4	243±10 (n=3)	175	258	46.4±11.3 (n=14)	54.2±19.7 (n=7)
Productivity (g dw $m^{-2}d^{-1}$)	7.04	25.8	18.6	27.4	2.55	5.80
Phosphorus uptake (mgP $m^{-2}d^{-1}$)	23.9	87.7	46.9	119	12.2	80.0
Phosphorus Yield (mgP (g dw$)^{-1}$)	3.40	3.40	2.51	4.34	4.80	13.8
Nitrogen uptake (mgN $m^{-2}d^{-1}$)	227	831	584	718	71.7	242

theless, the maximum value observed in wastewater medium for low densities (13.6 g dw m^{-2}) had been quite high 0.162±0.013 d^{-1}. In diluted pig waste, the growth rate had been much lower, perhaps because in this medium algae grew much more than in domestic effluent.

The productivity values obtained in confined systems (Samora channels and wastewaters) were similar to those reported by Vicenzini et al. (1985) and De-Busk &Ryther (1987). Higher values have been observed under field conditions (Samora Correia field, Guadiana river and Alverca lagoon) due to the higher densities found in these aquatic ecosystems, where wind and hydraulic patterns could promote local biomass accumulation. Nevertheless, these values are similar to those repported by Moorhead & Reddy (1987) for other macrophytes. Phosphorus uptake in the Guadiana river and in Samora Correia confined assays were similar to that reported by Reddy et al. (1985), but other field values (Samora Correia and Alverca lagoon) were much higher because of high levels of productivity found there.

The nitrogen uptake values under field conditions were higher than those reported by Reddy et al. (1985) and Kumarasinghe & Eskew (1993), but more consistent with those obtained by Vicenzini et al. (1985), what could be a result of the high productivities calculated under these conditions. The value calculated for Samora Correia assays performed under confined conditions, must be more realistic due to the densities observed which led to lower productivity. Even considering the lowest values of nitrogen uptake obtained with *Azolla*, these were much greater than those reported for nitrogen fixation by leguminous plants (15mg m^{-2} d^{-1} to 138 mg m^{-2} d^{-1}) (Barber, 1995; MADRP, 1997). Indeed, the nitrogen uptake was found to be positively correlated with density (r = 0.972; p = 0.01) and productivity (r = 0.985; p = 0.01). Nevertheless, the high nitrogen and phosphorus uptake observed under field conditions were in the same order of magnitude to those repported by DeBusk (1989), in wastewater tertiary treatment using water hyacinth.

The analysis of growth characteristics indicates that *Azolla* can grow quite well in treated urban wastewater, having growth rates and productivities great enough to encourage its utilisation as a biofilter for nutrient removal and biomass production. Indeed, its productivity is even greater than those of some conventional cultures in Portugal, like rice and alfalfa (1.3 g m^{-2} d^{-1} to 8.2 g m^{-2} d^{-1}, SAPEC, 1989).

The macronutrient content of *Azolla* (Table 2) is high enough to allow it to be used as a fertiliser. When *Azolla* grew in domestic effluent, its phosphorus content was about five times higher than when it grew in natural waters. This value is also greater than that reported by Lumpkin & Plucknett (1980). This has happened because *Azolla's* phosphorus content depends on growth medium phosphate concentration (Reddy & DeBusk, 1985; Mian &Azmal, 1989; Costa et al., 1994). In fact, the phosphate level in natural waters were less (0.93±0.72 mg P–PO_4^{3-} L^{-1}) than the observed in maturation pond effluent (8.26

Table 2. Composition of *Azolla* biomass

Variables	Samora Correia channels (n=8)	Guadiana river (n= 3)	Alverca lagoon	Diluted pig waste (n=10)	Maturation pond effluent (n=7)
Dry matter (%)	6.83±1.50	7.41±1.31	4.69	6.80±1.50	5.43±0.24
Organic matter(%)	90.01±1.5	87.97±0.51	91.29	84.15	85.65
Ash (%)	9.99±1.67	12.21±0.51	8.71	16.85±0.07	14.35
Phosphorus(%)	0.34±0.09	0.25±0.01	0.434	0.48±0.03	1.38±0.20
Nitrogen (%)	3.22±0.55	3.14	2.62	2.81±0.02	4.31±0.65
K (%)	1.97±2.97	1.38	1.25	1.91±0.53	3.17±0.79
Protein (%)	20.49±3.45	19.36±1.31	16.38	17.56±2.00	26.96±4.04
Crude fat (%)	3.60±1.64	5.34±0.82	3.25	5.82±0.10	4.04±0.81
Crude fiber (%)	13.99±1.40	–	16.55	13.35±4.30	12.53±0.83
Cellulose (%)	19.87±3.92	22.48±1.79	22.72	–	11.03±2.05
Hemicellulose (%)	15.05±4.66	10.75±0.65	14.82	–	18.05±7.90
Lignin (%)	35.51±4.39	39.12±1.73	40.99	–	19.52±1.74

– Not determined.

± 1.07 mgP–PO_4^{3-} L^{-1}). Indeed, phosphorus biomass content was highly significantly correlated with nutrients present in the growth medium (N–NH_4^+- r=0.916, p=0.01; P–PO_4^{3-} - r=0.979, p=0.01). We had observed earlier that *Azolla* can accumulate phosphorus by luxury uptake (Costa, 1994). Results from Phosphorus Yield [mgP (g dw$_)$$^{-1}$] confirm that observation.

Biomass content of protein and fiber weren't very different regardless of where the plants were grown (Table 2). Crude protein values were even greater than those observed in some other fodder plants (CEIP, 1980) like maize (Martins, 1996). Nevertheless, it will be important to determine lysine, methionine and histamine content. Fiber was high, but less than in some animal feed (Costa, unpublished data). Cellulose and lignin from *Azolla* developed in maturation pond effluent were lower than that observed in natural conditions, which could favor its digestability.

In order to use *Azolla* as feedstock, it is important to analyse its mineral content. Heavy metals content of *Azolla* harvested from natural environments (Table 3) were in the same order of magnitude of those reported by Lumpkin & Plucknett (1980). The copper and zinc levels were similar to those of oligoelements concentrations in some cereals and leguminous plants (CEIP, 1980). Zinc and lead contents of biomass grown in treated sewage and pig wastes were quite high, especially zinc. Other heavy metals, like Ni, Cd and Cr, were not detected, but special attention should always be paid when this plant is grown in wastewater, because of its potential bioaccumulation of some metals.

Table 3. Metals in *Azolla*

Metal content (mg (kg dw$_)$$^{-1}$)	Samora Correia channels (n = 5)	Guadiana river	Alverca lagoon	Diluted pig waste	Maturation pond effluent (n=6)
Ca	17708±2664	–	–	–	5355±1084
Mg	28161±10812	20000	12500	–	3383±491
Cu	29.00±11.54	34	21	26	15.34±11.01
Zn	37.00±23.00	20	81	115	52.31±31.30
Pb		n.d.	n.d.	–	6.07±3.07

– Not determined.
n.d. – Not detected.

The mineral content of all biomass samples (K, Ca and Mg – Tables 2 and 3) was normal when compared with what has been found in reported earlier studies (Lumpkin & Plucknett, 1980), but comparatively higher than other traditionally used forage plants, especially magnesium in *Azolla* grown under field conditions.

Conclusions

Azolla can grow quite well in treated urban wastewater, having growth rates and productivities great enough to encourage its utilisation. Indeed, its productivity is even greater than those of some conventional crops like rice and alfalfa (SAPEC, 1989). The maximum plant growth rate in secondary treated wastewater was 0.161 d^{-1} and the phosphorus removal efficiency was about 36% (Costa et al., 1996). So, the *Azolla* utilisation as a biofilter, namely for phosphorus removal, can be improved for wastewater purification.

The biomass analysis shows that the great nitrogen, phosphorus, potassium and organic content could favor the use of *Azolla* as a biofertiliser, especially when it is growing in domestic wastewaters, but also when it is harvested from natural environments. The nitrogen content was even greater than that found in some other nitrogen fixing plants, and because it can bioaccumulate phosphorus, the concentration of this element in the plant was as higher as the growth medium was rich on phosphate.

Azolla's biomass can be used in feedstock, but further research is needed to determine protein and fat quality, and also digestability and toxicological aspects. With *Azolla* grown in wastewaters, it would be necessary to pay special attention to heavy metals content, especially copper, zinc and lead.

As *Azolla* presents some interest for its utilization as biofertilizer or feed supplement for animals, harvesting could have less economical constraints what could decrease this plant bloom impacts on aquatic ecosystems. In the other hand, *Azolla*'s growth in wastewaters, namely due to its phosphorus removal capacity, could improve discharged water quality helping to control natural waters eutrophication.

Acknowledgements

This research was supported by JNICT/FCT Projects (PEAM/P/APR/121/91 and PEAM/C/GRH/581/95) and INAG.

References

APHA, AWWA, APCF, 1992. Standard methods for the examination of water and wastewater. 20th American Public Health Association, Washington.

AOAC, 1990. Official methods of analysis, 15th edn., Vol II.

Barber, S. A., 1995. Soil nutrient bioavailability, 2nd edn., John Wiley & Sons, Inc., New York: 157–179.

Bishop, P. L. & T. T. Eighmy, 1989. Aquatic wastewater treatment using *Elodea nuttallii*. Journal WPCF 61: 641–648.

C.E.I.P., 1980. Tables de composition des matiers premiers destinées à alimentation animale. 1 – valeurs chimiques.

DeBusk, T. A. & J. H. Ryther, 1987. Biomass production and yields of aquatic plants. In Reddy, K. R. & W. H. Smith (eds), Aquatic Plants for Water Treatment and Research Recovery. Florida: 579–598.

DeBusk, T. A., K. R. Reddy, T. D. Hayes & B. R. Schwegler, 1989. Performance of a pilot-scale water hyacinth-based secondary treatment system. Journal WPCF 61: 1217–1224.

Costa, M. L., F. Carrapiço & M. C. R. Santos, 1994. Biomass and growth characterization of *Azolla filiculoides* in natural and artificial environments. In Hegazi, N. A., M. Fayez & M. Monib (eds), Nitrogen Fixation with Non-Legumes. The American University in Cairo Press: 455–461.

Costa, M. L., F. Carrapiço & M. C. R. Santos, 1996. Contribuição para o estudo da utilização de *Azolla* no tratamento de águas residuais domésticas. In Borrego, C., C. Coelho, L. Arroja, C. Bóia & E. Figueiredo (eds), Proceedings 5[a] Conferência Nacional Sobre a Qualidade do Ambiente 2. Aveiro, Portugal: 1945–1954.

Fisher, J. P., 1988. Wastewater treatment using aquatic plants. In (eds), Alternative Waste Treatment Systems, Rao Bhamidimarri, Elsevier Applied Science, Palmerston North, New Zealand: 34–44.

Kitoh, S., N. Shiomi & E. Uheda, 1993. The growth and nitrogen fixation of *Azolla filliculoides* Lam. in polluted water. Aquat. Bot. 46:129–139.

Kumar, P. & R. J. Garde, 1989. Potentials of water hyacinth for sewage treatment. Journal WPCF 61: 1702–1706.

Kumarasinghe, K. S. & D. L. Eskew, 1993. Isotopic studies of *Azolla* and nitrogen fertilization of rice, Kluwer Academic Publishers, London: 42–54.

Lumpkin, T. A. & D. L. Plucknett, 1980. *Azolla*: botany, physiology and use as green manure. Econom. Bot. 34: 111–153.

M.A.D.R.P., 1997. Código de boas práticas agrícolas para a protecção da água contra a poluição com nitratos de origem agrícola, Ministério da Agricultura, do Desenvolvimento Rural e das Pescas, Lisboa.

Martins, M. I. F. M., 1996. Maize nitrogen fertilisation in fluvisoles of the lower Mondego valley – Field trials, In Van Cleemput, O. G. Hofman & A. Vermossen (eds), Progress in Nitrogen Cycling Studies, Kluwer Academic Publishers, London: 223–229.

Mian, M. H. & A. K. M. Azmal, 1989. The response of *Azolla pinnata* R. Brown to the split application of phosphorus and the transfer of assimilated phosphorus to flooded rice plants, Plant Soil 119: 211–216.

Moorhead, K. K. & K. R. Reddy, 1987. Nitrogen cycling in an integrated 'biomass for energy' system. In Reddy, K. R. & W. H. Smith (eds), Aquatic Plants for Water Treatment and Research Recovery, Florida: 535–542.

Oron, G., 1990. Economic considerations in wastewater treatment with duckweed for effluent and nitrogen renovation. Journal WPCF 62: 692–696.

Reddy, K. R. & W. F. DeBusk, 1985. Growth characteristics of aquatic macrophytes cultured in nutrient-enriched water:II. *Azolla*, duckweed and *Salvinia*, Econom. Bot. 39: 200–208.

S.A.P.E.C., 1989. Manual de adubação, 2nd. edn., SAPEC – Adubos Agro, S.A.

Vicenzini, M., M. C. Margueri &C. Sili, 1985. Outdoor mass culture of *Azolla* spp.: yields and efficiency of nitrogen fixation. Plant Soil 86: 57–67.

Watanabe, I. & C. M. Ramirez, 1984. Relationship between soil phosphorus availability and *Azolla* growth. Soil Sci. Plant Nutr. 30: 595–598.

Hydrobiologia **415:** 329–333, 1999.
J. M. Caffrey, P. R. F. Barrett, M. T. Ferreira, I.S. Moreira, K. J. Murphy & P. M. Wade (eds), Biology, Ecology and Management of Aquatic Plants.
© 1999 *Kluwer Academic Publishers. Printed in the Netherlands.*

Residual effects of herbicide-treated *Eichhornia crassipes* used as a soil amendment

Randall K. Stocker & William T. Haller
IFAS Center for Aquatic and Invasive Plants and Department of Agronomy, University of Florida, 7922 NW 71st Street, Gainesville, FL 32653-3071, U.S.A.

Key words: water hyacinth, utilization, herbicide, mulch, *Lycopersicon esculentum*

Abstract

The use of *Eichhornia crassipes* (Mart.) Solms (water hyacinth) as a soil amendment in fruit and vegetable production has been proposed in several areas of the world. To assess possible residual effects, *E. crassipes* was treated with maximum label rates of diquat, glyphosate, and 2,4-dichlorophenoxyacetic acid (2,4-D) herbicides. *Placement Experiment: E. crassipes* was shredded into a fine mulch $\frac{1}{2}$ hour, and 1, 2, and 7 days after herbicide or water (control) treatment (DAT) and added to soil in pots as a top mulch, a layered mulch, or a mixture, with seedling *Lycopersicon esculentum* (Mill; tomato) as assay species. Negative effects of glyphosate and 2,4-D were noted for *L. esculentum* growth only for the mixed treatment, and only at 2 DAT. *Duration Experiment*: coarsely chopped herbicide-treated *E. crassipes* was thoroughly mixed with soil. Reductions in *L. esculentum* plant biomass, including mortality, persisted for 1 DAT with diquat, 7 DAT with 2,4-D, and 14 DAT with glyphosate. *Ratio/Soil Type Experiment*: one day after herbicide application to *E. crassipes*, *L. esculentum* seedlings were planted into ratios (4:0, 3:1, 2:2, 1:3, 0:4) of herbicide-treated to untreated *E. crassipes*, and mixed with clay, loam, or sand. Glyphosate and 2,4-D reduced *L. esculentum* growth in proportion to the ratio of treated to untreated *E. crassipes*. Soil type did not influence herbicide effects on *L. esculentum* growth.

Introduction

Eichhornia crassipes (Mart.) Solms (water hyacinth) is a weedy floating aquatic herb found in tropical and warm temperate areas. Holm et al. (1977) list *E. crassipes* as the eighth most troublesome pest plant in the world, and the distribution list continues to expand (for example, Mansor 1996). As governments face the negative impacts of *E. crassipes* on native plant communities, water quality, fisheries, transportation, human health and recreation, efforts frequently turn to identifying ways of utilizing this species to partly offset high control costs, including use as an agricultural soil amendment. *E. crassipes* mulch has been credited with reducing soil moisture loss (Abdalla & Abdel Hafeez 1969; National Council for Research et al., 1975), reducing weeds (Abdalla & Abdel Hafeez, 1969), increasing plant growth rates by reducing night time soil heat loss (NCR et al., 1975), and has been identified as a promising compost component (Kamal & Little, 1970) mixing partially-dried plants and cow dung, with a covering of earth.

The benefits of mulching agricultural soils have been well documented (for example, Black, 1968). Dry weight plant nutrient content of *E. crassipes* is similar to that of many agricultural crops (Parra & Hortenstine, 1974), and has "fairly high concentrations" of plant nutrients compared with other plant species (Parra & Hortenstine, 1976). Addition of *E. crassipes* mulch positively affected pearl millet growth in sandy soils (Parra & Hortenstine, 1974; Parra & Hortenstine, 1976). *E. crassipes* mulch applied at 22 metric tons/ha (air dry weight) reduced evaporation by 15% from a 1:2 sand-soil mixture, increased soil water penetrability into clay soils, and enhanced salt leaching (Rahman et al., 1996).

The major disadvantage to using *E. crassipes* as a soil amendment is its 0.05 ratio of dry weight to

* Published as Journal Series No. R-06613 of the Florida Agricultural Experiment Station.

Table 1. Summary (F-values) of analysis of variance showing the significance of time after treatment (DAT), mulching method (PLACEMENT), herbicide or water treatment (CHEMICAL), and interactions, on total dry weight biomass of *Lycopersicon esculentum*

Source	DF[a]	Mean Square	F Value	Pr > F^b
Model	47	0.108	2.68	0.0001
Error	335	0.040		
Corrected Total	382			
DAT	3	0.140	3.47	0.016
PLACEMENT	2	0.361	8.94	0.0002
DAT*PLACEMENT	6	0.080	1.98	0.068
CHEMICAL	3	0.104	2.56	0.055
DAT*CHEMICAL	9	0.086	2.14	0.026
PLACEMENT*CHEMICAL	6	0.148	3.67	0.002
DAT*PLACEMENT*CHEMICAL	18	0.084	2.08	0.007

[a] Degrees of freedom
[b] Probability of a larger value of F by chance. Values less than 0.05 are statistically significant at $\alpha = 0.05$; values less than 0.01 are statistically significant at $\alpha = 0.01$

wet weight (Parra & Hortenstine, 1976), and transportation requirements to move biomass from water bodies to agricultural sites. Alternative sources of organic matter are frequently available in agricultural areas that do not require the expenses associated with use of *E. crassipes*.

Herbicide application is a common method of controlling the growth of this noxious weed, and the potential damage to agricultural crops that could result from soil incorporation of herbicide-treated *E. crassipes* mulch has not been investigated. Soil-residual herbicidal effects after application of glyphosate and 2,4-dichlorophenoxyacetic acid (2,4-D) have been documented in several studies (Anderson & Baker, 1950; Carlyle & Thorpe, 1947; Devlin et al., 1986), but none were associated with *E. crassipes*.

The objective of this study was to assess the impact of herbicide-treated *E. crassipes* used as a soil amendment on the subsequent growth of *L. esculentum* seedlings. It was selected as the assay species because it is sensitive to many herbicides, and has been shown to respond positively to the addition of many types of mulch (Monks et al., 1997).

Materials and methods

Placement Experiment

E. crassipes plants were collected from ponds near the University of Florida, Gainesville, FL, sufficient to cover nearly 100% of the surface area of four concrete tanks (77 × 219 × 65 cm, filled with water to a depth of 50 cm).

Test materials were maximum label rates of diquat [4.5 kg/ha (cation) with surfactant at 0.25 per cent by volume], glyphosate [4.2 kg/ha (acid) with surfactant at 0.5 per cent by volume], and 2,4-D amine [at 4.3 kg/ha (acid)], mixed with water, and applied at a rate equivalent to 701 l/ha. The surfactant used was a nonionic organosilicone wetting agent (polyalkyleneoxide modified heptamethyltrisiloxane and nonionic surfactants; Silkin®; Riverside/Terra Corporation). Individual 946 ml plastic spray bottles were used to apply herbicide and water control treatments. One herbicide was applied to all the *E. crassipes* plants contained within one tank; the fourth tank was treated with water. Plants were dry approximately 20 minutes after herbicide application.

Ten *E. crassipes* plants were removed from each tank $\frac{1}{2}$ hour and 1, 2, and 7 days after herbicide application. After draining excess water, plant pieces were placed in a food processor until most pieces were less than 1 mm.

Three combinations of shredded *E. crassipes* (mulch) and commercial topsoil were made for each herbicide or control treatment: (1) a mixture of mulch and soil (4 parts soil to 1 part mulch by wet weight); (2) a layer of mulch (20 g, approximately 1 cm deep) added to the potting soil surface after a 17-day old *L. esculentum* seedling was transplanted into the pot;

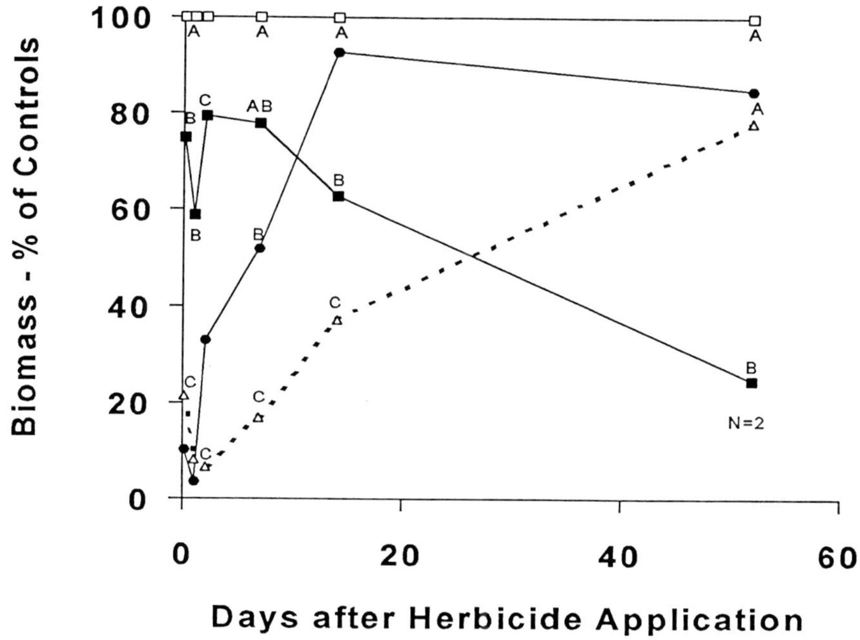

Figure 1. *L. esculentum* dry weight biomass as a per cent of controls, at $^1/_2$ hour, and 1, 2, 7, 14, and 52 days after treatment [water (□); diquat (■); glyphosate (△); 2,4-D (●)]. N = 8 for all treatments except N = 2 for diquat at 52 days. Within each planting date (days after treatment application), values labeled with the same letter are not significantly different at the 5% level, according to Duncan's New Multiple Range Test.

and (3) a layer of mulch (20 g, approximately 1 cm deep) placed at a level that would be just below the roots of transplanted *L. esculentum* seedlings. Each combination was replicated in eight pots (9.5 × 9.5 × 11-depth cm, plastic). After transplanting, pots were placed in a greenhouse and watered as needed.

Root and shoot biomass and length were measured for each *L. esculentum* plant 16 days after transplanting into treatment/soil combinations. An analysis of variance (ANOVA) was performed on four treatments (water, diquat, glyphosate, 2,4-D), three placements (surface mulch, soil layer, mixed), and time since herbicide application [$^1/_2$ hour, and 1, 2, and 7 days after treatment (DAT)]. Duncan's Multiple Range Test was used to compare means within the ANOVA.

Duration experiment

The procedure was similar to the placement experiment, with only one method of incorporation used – coarsely chopped *E. crassipes* mixed 2:1 (by wet weight) with soil. The objective was to determine how long, if at all, herbicides remained active in the *E. crassipes* mulch.

Diquat, glyphosate, 2,4-D, and a water control were applied to *E. crassipes* plants on 11 November 1996 at rates described previously. Fifteen *E. crassipes* plants were removed from each tank $\frac{1}{2}$ hour after herbicide application and were chopped by hand into pieces approximately 1–2 cm in length. Wet weight of the 15 plants was between 1.7 and 2.8 kg.

Chopped *E. crassipes* was added to commercial topsoil in a ratio of two parts plants to one part soil, by wet weight. Into this mixture one seedling was planted in each of eight pots per treatment. This process was repeated 1, 2, 7, 14 and 52 DAT. Transplant seedlings were fertilized with 20-20-20 (NPK) every other week. Root and shoot dry weight were measured 35 days after transplanting. Data were analyzed using ANOVA and Duncan's Multiple Range Test to compare means.

Ratio experiment

The final experiment was conducted to determine if any effects measured in previous experiments were due to herbicides in, or on, the *E. crassipes*, and not due to organic decay products within the *E. crassipes*. To ensure that *E. crassipes* would be decaying at approximately the same rate while the *L. esculentum* seedlings were growing, all *E. crassipes* was collected and coarsely chopped 24 to 48 hours after herbicide application (8 July 1997). Herbicide-treated

Table 2. Regression statistics of average *Lycopersicon esculentum* dry weight biomass on percent herbicide treated *Eichhornia crassipes* soil amendment for three soil types. Data for Y intercept and slope are plus/minus standard error

Soil	Diquat			Glyphosate			2,4-D		
	R^2	Intercept	Slope	R^2	Intercept	Slope	R^2	Intercept	Slope
Clay	0.08	0.65 ± 0.05	0.04 ± 0.02	0.40	0.61 ± 0.07	−0.15 ± 0.03	0.64	0.47 ± 0.04	−0.12 ± 0.01
Loam	0.02	0.96 ± 0.05	−0.02 ± 0.02	0.77	1.04 ± 0.07	−0.27 ± 0.03	0.53	0.82 ± 0.07	−0.18 ± 0.03
Sand	0.19	0.33 ± 0.06	0.07 ± 0.02	0.27	0.23 ± 0.05	−0.07 ± 0.02	0.22	0.13 ± 0.02	−0.27 ± 0.01

(same products and rates as in previous experiments) mulch was added to soils in ratios of 4:0, 3:1, 2:2, 1:3 and 0:4 treated to untreated mulch. Treatments were conducted with clay, loam, and commercial sand at rates of 2:1 wet weight mulch to soil. Seven-day-old *L. esculentum* seedlings were planted into the ratio mixes, and grown for 30 days before final dry weight biomass was determined.

For overall treatment effects an ANOVA was conducted on three variables (herbicide/water application, treated:untreated *E. crassipes* ratio, and substrate type), with eight replications of each. Regression analysis of total *L. esculentum* dry weight on fraction of herbicide treated *E. crassipes* was conducted for each herbicide and soil type.

Results and discussion

Placement experiment

No differences were found for total (root plus shoot) biomass or for total length of *L. esculentum* plants in the treatment media. There were significant differences for shoot biomass by time, for root biomass overall, and by time, placement, and time*placement. Increases in one tissue (shoot or root) tended to be offset by decreases in the other tissue. A somewhat similar pattern emerged with shoot and root length data. On the basis of this analysis, it was decided to report total dry weight biomass for all experiments.

For the ANOVA of total biomass (Table 1), the overall F value was significant ($Pr > F = 0.0001$), as were all main effects except chemical, and interactions except time*placement. Total biomass was significantly lower 2 DAT, with no significant differences among the other DAT. At 2 DAT, the herbicide may be bound more strongly (than earlier DAT) to the plant material, and yet remain available for uptake by *L. esculentum* seedling roots that come into contact with the herbicide treated *E. crassipes*. The effect does not appear to be related to breakdown of the *E. crassipes* as a separate action from the herbicide action, because diquat causes very rapid *E. crassipes* breakdown, and the main effect for chemical was not significant.

The application of finely chopped herbicide-treated *E. crassipes* to the soil surface at rates equivalent to 8000 kg/ha (dry weight) did not affect the growth of *L. esculentum* seedlings. Incorporation or layering of this amount of treated *E. crassipes* into the soil, however, did reduce seedling growth with *E. crassipes* mulched up to 2 DAT. Because mixed (soil incorporated) treatments were easier and more typical of what farmers might do, for the following experiments we increased the mulch application rate to an equivalent 16 000 kg/ha dry weight, increased the size of *E. crassipes* mulch pieces, and mixed the mulch into the soil.

Duration experiment

Chopped (1–2 cm) herbicide-treated (or control) *E. crassipes* pieces were used to reduce possible inactivation of herbicide by soil particles. An ANOVA performed on the water control plants showed significantly reduced growth of *L. esculentum* at 52 DAT, probably due to cooler greenhouse temperatures. To offset differences in greenhouse conditions over this time period, results for herbicide treatments are presented as percent of controls for each starting date (Figure 1). Total biomass for *L. esculentum* seedlings was significantly reduced for $\frac{1}{2}$ hour, 1, 2, and at least 7 DAT for 2,4-D, and for $\frac{1}{2}$ hour, 1, 2, and at least 14 days after application for glyphosate.

The case for diquat was very different. The effects for treatments differed significantly from control for the *E. crassipes* collected $\frac{1}{2}$ hour, and 1, 14, and 52 days after treatment. Because diquat causes *E. crassipes* to break down very quickly after application, very little *E. crassipes* was available to use as mulch by day 52, and only two replicates are included in the assessment for that day. It is possible that some breakdown products of *E. crassipes* itself inhibit *L. esculentum* seedling growth, since it is unlikely that any residual herbicide effect would increase that steadily over time.

The results of the placement experiment were very different from the results of the duration experiment. Doubling the mulch application rate (8000 to 16 000 hg/ha dry weight), chopping into coarser pieces, and soil incorporation, singly or in combination resulted in negative effects to *L. esculentum* seedling growth for mulch applied up to 7 DAT for 2,4-D, 14 DAT for glyphosate and 52 DAT for diquat (two replications).

Ratio Experiment

Per cent organic matter for the three substrate types was 0.3, 3.1, and 26.7 for the sand, clay, and loam respectively. *L. esculentum* growth on the sand substrate was the lowest of the three soil types for each herbicide tested, even though fertilizer was provided weekly. Higher biomass was produced in the loam substrate than the clay for each herbicide (Table 2). The coefficient of determination (r^2) was above 0.5 for only glyphosate-treated mulch in loam soil, and 2,4-D-treated mulch in clay and loam. The increased surface area of the clay soil was expected to reduce the negative effects of increasing herbicide treated *E. crassipes* ratio in the mulch mixes, but this did not occur.

The lowest ratio of diquat:untreated *E. crassipes* resulted in reduction of *L. esculentum* growth. Higher ratios of treated *E. crassipes* resulted in increased *L. esculentum* seedling growth in clay and sand. Rapid breakdown of *E. crassipes* by diquat may free more organic material and nutrients for *L. esculentum* seedling uptake. Higher ratios of glyphosate and 2,4-D-treated *E. crassipes* resulted in negative impacts to *L. esculentum* seedling growth, with fatal effects on many replicates.

The ratio experiments indicated that 2,4-D and glyphosate had negative effects when applied as mulch 1–2 days after herbicide treatment. Further, decaying *E. crassipes* and soil type did not affect seeding growth or reduce damage caused by herbicides, respectively.

In summary, the application rate of treated *E. crassipes* mulch, length of time after herbicide treatment, type of herbicide applied, and possibly even the degree of chopping all influenced subsequent growth of *L. esculentum* seedlings. Pending results of further studies, these data support not using herbicide-treated *E. crassipes* for mulch because of possible damage to crops.

Acknowledgments

The authors wish to thank the following for their assistance: Brian Smith, Mike Netherland, Neal Hill, Joe Zolczynski, Gidget Zolczynski, Shelley Stocker, Shevaun Stocker, Margaret Glenn, and Alison Fox. We also appreciate review and recommendations provided by Donn Shilling.

References

Abdalla, A. A. & A. T. Abdel Hafeez, 1969. Some aspects of utilization of water hyacinth (Eichhornia crassipes). Pest Articles and News Summaries 15: 204–7.

Anderson, G. R. & G. O. Baker, 1950. Some effects of 2,4-D in representative Idaho soils. Agr. J. 42: 456–458.

Black, C. A., 1968. Soil-plant relationships. 2nd edn. John Wiley and Sons, Inc. New York, NY. 792 pp.

Carlyle, R. E. & J. D. Thorpe, 1947. Some effects of ammonium and sodium 2,4-dichlorophenoxyacetates on legumes and the Rhizobium bacteria. J. am. Soc. Agr. 39: 929–936.

Devlin, R. M., S. J. Karczmarczyk, I. I Zbiec & Z. K. Koszanski, 1986. Initial and residual activity of glyphosate and SC-0224 in a sandy soil. Crop Protect. 5: 293–296.

Holm, L. G., D. L. Plucknett, J. V. Pancho & J. P. Herberger, 1977. The World's Worst Weeds: Distribution and Biology. University Press of Hawaii. 609 pp.

Kamal, I. A. & E. C. S. Little, 1970. The potential utilization of water hyacinth for horticulture in the Sudan. Pest Articles and News Summaries 16: 488–96.

Mansor, M., 1996. Noxious floating weeds of Malaysia. Hydrobiologia 340: 121–125.

Monks, C., D. Monks, T. Basden, A. Selders, S. Poland & E. Rayburn, 1997. Soil temperature, soil moisture, weed control, and tomato (*Lycopersicon esculentum*) response to mulching. Weed Technol. 11: 561–566.

National Council for Research, Agricultural Research Council of the Sudan. 1975. Aquatic weed management: some perspectives for the Sudan and the Nile Basin. NTIS No. PB 259–990: 65 pp.

Parra, J. V. & C. C. Hortenstine, 1974. Plant nutrient content of some Florida waterhyacinths and response by pearl millet to incorporation of waterhyacinths in three soil types. J. aquat. Plant Mgmt 12: 85–90.

Parra, J. V. & C. C. Hortenstine, 1976. Response by pearl millet to soil incorporation of waterhyacinths. J. aquat. Plant Mgmt. 14: 75–79.

Rahman, H. A. A., M. H. Dahab & M. A. Mustafa, 1996. Impact of soil amendments on intermittent evaporation, moisture distribution and salt redistribution in saline-sodic clay soil columns. Soil Sci. 161: 793–802.

Hydrobiologia **415:** 335–339, 1999.
J. M. Caffrey, P. R. F. Barrett, M. T. Ferreira, I.S. Moreira, K. J. Murphy & P. M. Wade (eds), Biology, Ecology and Management of Aquatic Plants.
© 1999 *Kluwer Academic Publishers. Printed in the Netherlands.*

A simple model for evaluating the costs and benefits of aquatic macrophytes

Egbert H. Van Nes[1,*], Marcel. S. van den Berg[2], John S. Clayton[3], Hugo Coops[2], Marten Scheffer[1] & Ekko van Ierland[4]
[1]*Wageningen Agricultural University, Aquatic Ecology and Water Quality Management Group, P.O. Box 8080, 6700 DD Wageningen, The Netherlands*
E-mail: egbert.vannes@aqec.wkao.wau.nl
[2]*RIZA, P.O. Box 17, 8200 AA Lelystad, The Netherlands*
[3]*NIWA, Gate 10 Silverdale Road, Box 11-115, Hamilton, New Zealand*
[4]*Wageningen Agricultural University, Environmental Economics Group, P.O. Box 8130, 6700 EW Wageningen, The Netherlands*

Key words: aquatic weeds; welfare; recreation

Abstract

Dense beds of aquatic macrophytes often cause nuisance to boaters and swimmers, and may obstruct water flow. Management of aquatic vegetation is, therefore, often aimed at reducing the biomass of the plants. If the nuisance is caused by exotic invasive species, there usually is no controversy with nature conservation aims. In shallow lakes, however, the interests of recreational users may conflict with nature conservation because the promotion of indigenous submerged vegetation is considered an important tool for lake restoration. Aiming at intermediate vegetation biomass seems a good solution for this controversy at first sight. However, we argue that such a compromise is often not the best policy from a welfare economic point of view. We present preliminary results of a graphical model, showing that the overall benefit for all ecosystem users may be minimal at the intermediate vegetation biomass. Furthermore, even if there is an optimal benefit at an intermediate macrophyte biomass, we argue that it is may not always be feasible to force the vegetation biomass to the desired level. Due to ecological feedback mechanisms, the system can have two alternative stable states: one with high vegetation biomass and one with little or no vegetation. It is concluded that it will often be better to realise a management strategy aimed at keeping some lakes (or parts of lakes) free of aquatic plants, whereas allowing others to be densely vegetated.

Introduction

Dense beds of macrophytes can be a nuisance for boating, fishing, swimming in lakes and may obstruct the water flow in irrigation channels. Therefore, aquatic weed control is the topic of much research (e.g. Caffrey & Wade, 1996) and management of aquatic vegetation is often aimed at reducing the biomass of vegetation. If the nuisance is caused by exotic invasive species, there usually will be no controversy with nature conservation aims, as these species are a threat for native species (De Winton & Clayton, 1996).

In many shallow lakes, however, lake managers are trying to restore the vegetation that vanished as a result of eutrophication, which has led to turbid water and cyanobacterial blooms. Establishing a stable aquatic vegetation is considered a very important tool to restore eutrophic lakes (Crawford, 1979; Perrow et al., 1997a; Scheffer, 1998). The main benefit of an abundant vegetation is that vegetation reduces turbidity in shallow lakes (e.g. Hasler & Jones, 1949; Timms & Moss, 1984; Van den Berg et al., 1998). Also, bird abundance and biodiversity of many groups of animals is usually higher in vegetated lakes (Scheffer, 1998).

Aiming at an intermediate biomass of macrophytes seems the obvious solution to restore lakes without

* Author for correspondence

causing nuisance, as shown by the graphical model of Clayton (1999). In this paper, we present an extension of this model, which shows that such compromise is not always the optimal policy (i.e. the policy that maximises the mean benefit for all ecosystem users).

A graphical model of the costs and benefits of aquatic vegetation

Aquatic macrophytes are of importance to several different groups of ecosystem users, for example people interested in nature conservation, recreational users, farmers and boaters. The interests of these groups often conflict, but it is hard to compare the interests of these different groups. We use the approach of environmental economists to account for all interests in a common unit, namely 'welfare' or 'benefit' (Hanley & Spash, 1993; Varian, 1996; Perman et al., 1996). Environmental economists have ways to quantify the welfare of different groups including the value of environmental services (e.g. Carson & Mitchell, 1993; Dixon et al., 1994). Determining detailed quantitative welfare functions is beyond the scope of this paper; we will only discuss qualitatively different possibilities.

In our simple model we included the welfare functions of two groups of users:

1. Nature conservationists aiming at the restoration of eutrophic waters. In general, this group promotes indigenous macrophytes, possibly up to an optimum level, above which dense macrophyte beds could have a negative impact on biodiversity, water quality and conservation values. This group includes recreationists that are interested in nature: 'eco-recreation'.
2. Recreational users that are hindered by dense vegetation. This group includes boaters, wind surfers and swimmers.

Of course there may be more groups of stake holders involved in many lakes. For simplicity, we ignore these groups in the current analysis, but our approach may be extended to include any group whose welfare depends in a known way on vegetation abundance in the lake. For simplicity, we also ignore in the welfare function of biomass harvesting costs and other costs to maintain biomass at a desired level.

In our model, the welfare functions of both groups link the biomass of vegetation to the benefit or cost of these two ecosystem users. The welfare functions are defined as functions ranging from 0 (minimal benefit or maximal nuisance) to 1 (maximal benefit or minimal nuisance). The overall welfare of the society is composed of the contribution from welfare by users in each group, weighted by the number of individuals in that group.

The optimal strategy from a 'rational social planner's' point of view is to aim at the biomass where the total welfare function is optimal.

For restoration of eutrophic lakes, the benefit of aquatic vegetation is strongly connected to the effect of macrophytes on turbidity and vice versa. Although the exact shape of especially the clearing effect of vegetation is yet unclear, the positive feedback of macrophytes on their own growth is known to cause alternative stable states (Scheffer, 1998). If a lake has alternative stable states, the benefit of vegetation is typically a threshold function. Below a certain biomass, the vegetation is not able to cause a shift from the turbid state to the clear water phase. The exact value of the critical biomass is not only dependent on the trophic state of the ecosystem, but also on the growth form of the macrophytes. Most beneficial are probably low growing species like charophytes (Crawford, 1979; Clayton & Tanner, 1988; Coops & Doef, 1996). Besides the clearing effect of macrophytes, there are also other benefits associated to macrophytes, such as an increase of habitats for invertebrates (e.g. Hargeby et al., 1994) and food sources for birds (e.g. Perrow et al., 1997b) and fish (e.g. Van Donk et al., 1994). These effects are probably more gradual. The total nature benefit of most types of indigenous macrophytes is probably most appropriately characterised by an increasing sigmoidal curve (Figure 1b, solid line).

The nuisance caused by aquatic plants for recreational users increases with macrophyte biomass. To our knowledge, the shape of this nuisance function has not yet been assessed. It seems reasonable, however, to assume a sigmoidal curve for this function also. Below a certain biomass level, boating and swimming is not hindered. If the vegetation stands exceed a certain density, and especially if the plants cover the water surface, vegetation becomes a nuisance (Figure 1 b, dotted line). The growth form of the macrophytes is important for this group too. The tolerance of recreational users for low growing species will be higher than for floating or canopy forming species.

Hill functions offer a convenient way to model such welfare functions. The Hill function has two parameters: the half saturation H which defines, in our case, the biomass (B) of plants where the welfare (w) is 0.5 and the exponent p which defines the steepness

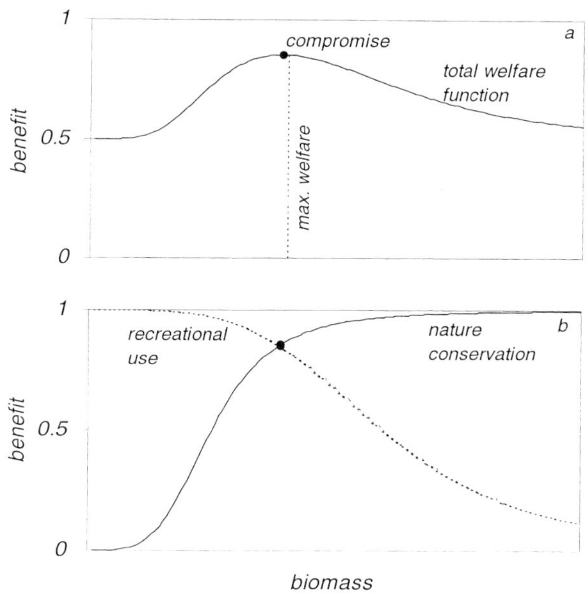

Figure 1. Vegetation with a low growth form. (a) the total welfare function (mean of functions in lower panel). The point with a compromise between both groups (closed circle) coincides with the maximum total welfare. (b) functions of benefit for recreational use and nature conservation.

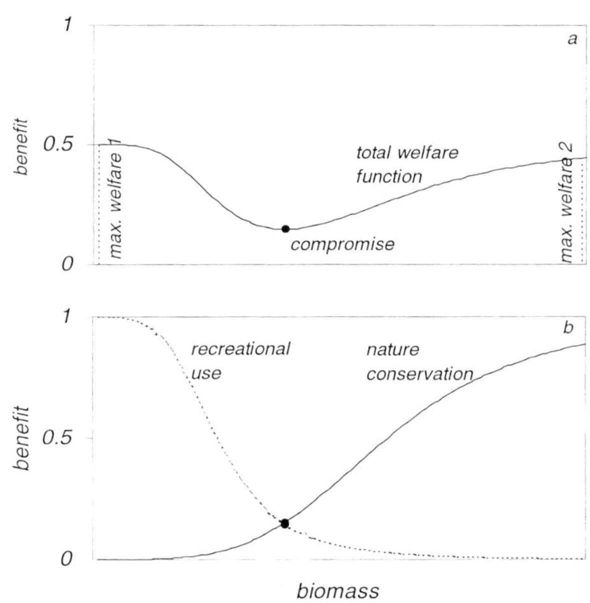

Figure 2. Canopy forming or floating vegetation. (a) the total welfare function (mean of functions in lower panel). The point with a compromise between both groups (closed circle) coincides with the maximum total welfare. (b) functions of benefit for recreational use and nature conservation.

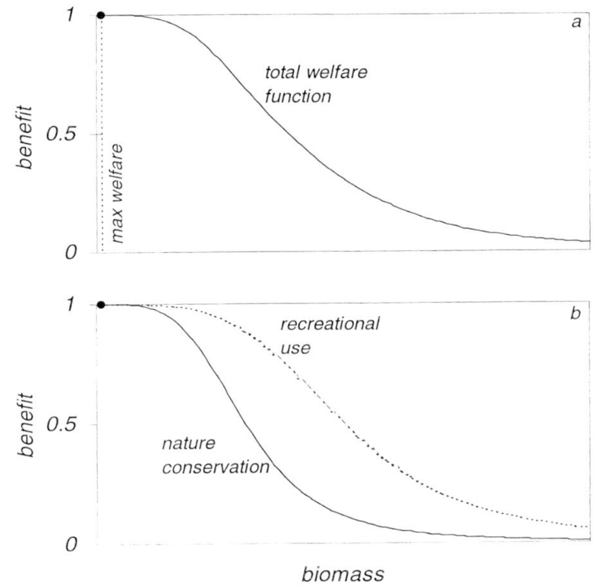

Figure 3. Exotic invasive species. (a) the total welfare function (mean of functions in lower panel). The point with a compromise between both groups (closed circle) coincides with the maximum total welfare. (b) functions of benefit for recreational use and nature conservation. Both groups have the same interests, the target of the management should be control (or eradication) of the plants.

of the function:

$$w = \frac{B^p}{H^p + B^p}.$$

This increasing function can be mirrored to become a decreasing function by changing it the following way:

$$w = \frac{H^p}{H^p + B^p}.$$

We explored the effect of different combinations of welfare functions on the resulting total welfare. For simplicity, we give both user groups an equal weight. Several different situations may arise, of which we highlight three examples in our analysis: 1. Vegetation with a low growth form (Figure 1). The benefit of low growing plants for water quality is relatively large, whereas the nuisance is relatively low. Therefore, the half saturation of nature benefit is assumed to be lower than the half saturation of recreational nuisance. This large overlap in welfare functions causes an optimum in total welfare at intermediate vegetation densities. Thus, the compromise between both types of ecosystem users coincides with an optimum overall welfare and aiming at intermediate biomass seems the optimal policy in such a case. 2. Canopy forming or floating vegetation (Figure 2). This type of vegetation causes more nuisance for boating and swimming and is less beneficial for lake restoration. As a result, a moderate amount of such plants may be too much from a recreational point of view and insufficient from a nature conservation point of view. In the model, we represent this by setting the half saturation of nature benefit higher than the half saturation of recreational nuisance. This causes the total welfare to be at minimum at an intermediate biomass. Compromising between both groups is obviously a bad policy in this case, provided that the same strategy is applied to the entire water body. Designing separate areas for recreation and nature conservation can a better policy in such situations, if free-floating species are not dominant. 3. Vegetation dominated by exotic invasive species (Figure 3). As mentioned before, usually the interests of nature conservationists and recreationists do not conflict in this case. We have mirrored the Hill function of nature conservation to represent this. The exact shape of the curves and the value of the half saturation coefficients is not essential in such situation. The optimum policy is always to eradicate the species.

In conclusion, aiming at an intermediate vegetation biomass is only a good strategy for maximising overall welfare in a limited class of cases (Figure 1).

Discussion

Whether or not aiming at an intermediate vegetation biomass is a good policy from an environmental economist's point of view, depends on the way in which welfare of different groups of lake users varies with vegetation biomass and the cost of harvesting biomass or other management costs. For a thorough analysis, one would need to assess the actual shape of these curves to analyse any field situation in practice and the management characteristics. However, our analysis shows that, at least, the qualitative point can be made that a compromise with intermediate vegetation biomass will only lead to a high average welfare for all users of the lake in a limited subset of cases.

Obviously, the benefit curves will not always have the shape as presented in our model. The Hill curves that we used in our model are not appropriate to describe all complex relations that exist in reality. We disregarded the fact that very high biomass levels of canopy forming and floating vegetation can have a detrimentally impact on water quality, biodiversity and conservation values. The benefit curve of nature conservation will have an optimum then. Furthermore, the nature value of a water body that is dominated with exotic species can still have a substantial value for nature conservation. Again, there will be situations where an optimum curve would be a better description. For clarity, and because we lack detailed information on the real shape of the benefit curves, we presented these simplified relationships.

Another aspect that we did not consider in the current benefit analysis is that loss of vegetation usually invokes algal blooms in shallow eutrophic lakes. For swimmers and other recreational users that often complain about high vegetation biomass, such algal blooms would also cause a nuisance (Herath, 1997; Boggess et al., 1997). This implies that recreational users may wish for a situation (a clear lake with little vegetation) that is ecologically impossible under the high nutrient conditions. With our model, it would mean that recreational users really have a more complex welfare function: aiming at the lowest biomass that is sufficient for preventing algal blooms that would yield the highest welfare for recreational users in this case. Also, in reality, recreational users are a diverse group with often conflicting interests. For instance, power boating is incompatible with many other recreational activities (Nuttall & Richardson, 1989).

An interesting feedback could result from damage to aquatic vegetation by recreational users, which is observed frequently (Murphy & Eaton, 1983; Best, 1987; Mumma et al., 1996). Murphy & Eaton (1983) showed that heavy pleasure-boat traffic can even be a principal factor in suppressing macrophyte abundance in British canals. In such cases, it is likely that recreational users have a positive effect on the suitability of the water for recreation, implying a potential positive feedback in recreational use, which could lead to alternative stable states. Moreover, Murphy & Eaton (1983) found a critical level of boat traffic above which the macrophyte community was affected. This is a further indication that there could be alternative stable states (Scheffer, 1998). This phenomenon would be a much like the maintenance of 'grazing lawns' in which invasion by large unpalatable (hindering) plants is prevented by intense use by terrestrial and aquatic grazers (McNaughton, 1984).

From the environmental economic analyses, we concluded that, in certain situations (with low growing macrophyte species), the optimal management strategy could be to aim at intermediate biomass. However, if the system has alternative stable states (Scheffer, 1990; Scheffer, 1998), it will be impossible to set vegetation to any desired level. The vegetation biomass could suddenly drop to zero after a certain critical biomass is exceeded. One should, therefore, keep in mind that the best policy economically may not be feasible due to ecological constraints.

In conclusion, environmental economic aspects suggest that aiming at intermediate vegetation biomass is only a good policy for maximising overall welfare of different types of lake users in a limited number of situations. An example of such a situation is for vegetation with a low growth form, as illustrated in Figure 1. In situations where recreation and nature conservation are competing, a compromise can be better achieved by assigning lakes or parts of lakes to recreational use and other areas for nature conservation. Alternatively, a selective control of submerged macrophytes aimed at a shift in the species composition towards low growing species could be a good solution for the controversy.

Acknowledgements

We want to thank the organisation of the 10th EWRS International Symposium on Aquatic Weeds for bringing together researchers working on the control of aquatic weeds with people studying the functional ecology of macrophytes. Discussions during the symposium led to the model described in this paper. An anonymous referee made useful suggestions to improve the manuscript.

References

Best, E. P. H., 1987. The submerged macrophytes in Lake Maarsseveen I, the Netherlands: changes in species composition and biomass over a six-year period. Hydrobiol. Bull. 21: 55–60.

Boggess, W. G., G. Johns & C. Meline, 1997. Economic impacts of water quality programs in the Lake Okeechobee Watershed of Florida. J.Diary Sci. 80: 2682–2691.

Caffrey, J. M. & P. M. Wade, 1996. Preface: The European Weed Research Society and the management and ecology of freshwater plants. Hydrobiologia 340: IX–XIII.

Carson, R. G. & R. J. Mitchell, 1993. The value of clean water: the public's willingness to pay for boatable, fishable and swimmable quality water. Wat. Resour. Res. 29: 2445–2454.

Clayton, J. S., 1998. Aquatic plant management in New Zealand: achievements and directions. In Monteiro, A., T. Vasconcelos & L. Catarino (eds), Proceedings of the 10th EWRS International Symposium on Aquatic Weeds: Towards an Integrated Aquatic Plant Management. Lisbon 21–25 September 1999. European Weed Research Society: 335–340.

Clayton, J. S. & C. C. Tanner, 1988. Selective control of submerged aquatic plants to enhance recreational uses for water bodies. Verh. int. Ver. Theor. Angew. Limnol. 23: 1518–1521.

Coops, H. & R. W. Doef, 1996. Submerged vegetation development in two shallow, eutrophic lakes. Hydrobiologia 340: 115–120.

Crawford, S. A., 1979. Farm pond restoration using *Chara vulgaris* vegetation. Hydrobiologia. 62: 17–32.

De Winton, M. D. & J. S. Clayton, 1996. The impact of invasive submerged weed species on seed banks in lake sediments. Aquat. Bot. 53: 31–45.

Dixon, J. A., L. Fallon, R. A. Carpenter & J. A. Sherratt, 1994. Economic Analysis of Environmental Impacts. Earthscan Publications, London: 210 pp.

Hanley, N. & C. Spash, 1993. Cost-benefit analysis and the environment. Edward Elgar Publishing Limited, Aldershot: 278 pp.

Hargeby, A., G. Andersson, I. Blindow & S. Johansson, 1994. Trophic web structure in a shallow eutrophic lake during a dominance shift from phytoplankton to submerged macrophytes. Hydrobiologia 280: 83–90.

Hasler, A. D. & E. Jones, 1949. Demonstration of the antagonistic action of large aquatic plants on algae and rotifers. Ecology 30: 346–359.

Herath, G., 1997. Freshwater algal blooms and their control: Comparison of the European and Australian experience. J. envir. Mgmt 51: 217–227.

McNaughton, S. J., 1984. Grazing lawns animals in herds plant form and coevolution. Am. Nat. 124: 863–886.

Mumma, M. T., C. E. Cichra & J. T. Sowards, 1996. Effects of recreation on the submersed aquatic plant community of Rainbow River, Florida. J.Aquat.Plant Mgmt. 34: 53–56.

Murphy, K. J. & J. W. Eaton, 1983. Effects of pleasure-boat traffic on macrophyte growth in canals. J. appl. Ecol. 20: 713–729.

Nuttall, P. M. & J. L. Richardson, 1989. Incompatible recreation-oriented benefits in a residential tidal canal estate and marina, Victoria (Australia). Wat.Sci.Technol. 21: 273–279.

Perman, R., Y. Ma & J. McGilvray, 1996. Natural Resource and Environmental Economics. Longman, London: 396 pp.

Perrow, M. R., M. L. Meijer, P. Dawidowicz & H. Coops, 1997a. Biomanipulation in the shallow lakes: State of the art. Hydrobiologia 342: 355–365.

Perrow, M. R., J. Schutten, J. R. Howes, T. Holzer, F. J. Madgwick & A. J. D. Jowitt, 1997b. Interactions between coot (*Fulica atra*) and submerged macrophytes: the role of birds in the restoration process. Hydrobiologia 342: 241–255.

Scheffer, M., 1990. Multiplicity of stable states in freshwater systems. Hydrobiologia 200/201: 475–486.

Scheffer, M., 1998. Ecology of Shallow Lakes. Chapman and Hall, London: 357 pp.

Timms, R. M. & B. Moss, 1984. Prevention of growth of potentially dense phytoplankton populations by zooplankton grazing in the presence of zooplanktivorous fish in a shallow wetland ecosystem. Limnol.Oceanogr. 29: 472–486.

Van den Berg, M. S., H. Coops, M. L. Meijer, M. Scheffer & J. Simons, 1998. Clear water associated with a dense *Chara* vegetation in the shallow and turbid Lake Veluwemeer, the Netherlands. In Jeppesen, E., Ma. Sondergaard, Mo. Sondergaard & K. Kristoffersen. (eds), Structuring Role of Submerged Macrophytes in Lakes. Springer-Verlag, New York: 339–352.

Van Donk, E., E. De Deckere, J. G. P. Klein Breteler & J. T. Meulemans, 1994. Herbivory by waterfowl and fish on macrophytes in a biomanipulated lake: Effects on long-term recovery. Verh. int. Ver. Theor. Angew. Limnol. 21: 2139–2143.

Varian, H. R., 1996. Intermediate Microeconomics, a Modern Approach. Norton & Company, New York: 650 pp.